TRIANGULAR NORMS

TRENDS IN LOGIC
Studia Logica Library

VOLUME 8

SCOPE OF THE SERIES

Trends in Logic is a bookseries covering essentially the same area as the journal *Studia Logica* – that is, contemporary formal logic and its applications and relations to other disciplines. These include artificial intelligence, informatics, cognitive science, philosophy of science, and the philosophy of language. However, this list is not exhaustive, moreover, the range of applications, comparisons and sources of inspiration is open and evolves over time.

The titles published in this series are listed at the end of this volume.

ERICH PETER KLEMENT
Johannes Kepler University,
Linz, Austria

RADKO MESIAR
Slovak University of Technology,
Bratislava, Slovakia and *Czech Academy of Sciences,*
Prague, Czech Republic

and

ENDRE PAP
University of Novi Sad,
Yugoslavia

TRIANGULAR NORMS

KLUWER ACADEMIC PUBLISHERS
DORDRECHT / BOSTON / LONDON

A C.I.P. Catalogue record for this book is available from the Library of Congress.

ISBN 0-7923-6416-3

Published by Kluwer Academic Publishers,
P.O. Box 17, 3300 AA Dordrecht, The Netherlands.

Sold and distributed in North, Central and South America
by Kluwer Academic Publishers,
101 Philip Drive, Norwell, MA 02061, U.S.A.

In all other countries, sold and distributed
by Kluwer Academic Publishers,
P.O. Box 322, 3300 AH Dordrecht, The Netherlands.

Printed on acid-free paper

Printed in the Netherlands.

To

Bernhard, Sissy, Katharina,
Janka, Andrea,
and Danijela

Table of Contents

Appendix

Reference material

Acknowledgments

The authors gratefully acknowledge the generous support of their respective home institutions, namely, the *Johannes Kepler University* (Linz), the *Slovak University of Technology* (Bratislava), the *Czech Academy of Sciences* (Prague), and the *University of Novi Sad*, during the preparation of this work. In particular, the Department of Algebra, Stochastics and Knowledge-Based Mathematical Systems of the Johannes Kepler University, on the one hand, and the Department of Mathematics and Descriptive Geometry of the Faculty of Civil Engineering of the Slovak University of Technology, on the other hand, were the hosts of several extended and fruitful working periods in Linz and Bratislava, respectively. We also appreciate the support of the *Austrian Mathematical Society (ÖMG)*.

We benefitted from European activities, in particular from the *COST Action* 15 *Many-Valued Logics for Computer Science Applications*, from bilateral agreements, such as the *Action Austria-Slovakia* (Projects 18s38 and 26s34) and the *Action Austria-Czech Republic* (Projects 16p12 and 23p16), as well as from several national grants: *FWF* (Austria, Projects P10672-ÖTE and P12900-TEC), *VEGA* (Slovakia, Projects 1/71467/20 and 2/6087/99), *GAČR* (Czech Republic, Project 402/99/0032), and *MNTS* (Yugoslavia, Project 04M05).

The authors are indebted to many friends and colleagues for stimulating discussions, for interesting suggestions, and for their willingness to read earlier versions of the manuscript and to correct a number of mistakes therein. In particular, we sincerely thank in this context *Giuseppina Barbieri, Peter Bauer, Ulrich Bodenhofer, Dan Butnariu, Bernard De Baets, János C. Fodor, Siegfried Gottwald, Petr Hájek, Ulrich Höhle, Sándor Jenei, Gerianne Krause, Milan Mareš, Andrea Mesiarová, Markus Mittendorfer, Mirko Navara, Vilém Novák, Roland Richter, Stephen E. Rodabaugh, Wolfgang Sander, Dana Smutná, Peter Viceník, Hans Weber*, and *Siegfried Weber*.

The whole manuscript was typeset with LaTeX 2_ε, and the powerful computer algebra system *Mathematica*® 4.0 was used to draw 70 out of the 74 figures.

Linz, Bratislava, and Novi Sad

Erich Peter Klement
Radko Mesiar
Endre Pap

Introduction

The history of triangular norms started with the paper "Statistical metrics" [Menger 1942]. The main idea of Karl Menger was to construct metric spaces where probability distributions rather than numbers are used in order to describe the distance between two elements of the space in question. Triangular norms (t-norms for short) naturally came into the picture in the course of the generalization of the classical triangle inequality to this more general setting. The original set of axioms for t-norms was considerably weaker, including among others also the functions which are known today as triangular conorms.

Consequently, the first field where t-norms played a major role was the theory of probabilistic metric spaces (as statistical metric spaces were called after 1964). Berthold Schweizer and Abe Sklar in [Schweizer & Sklar 1958, 1960, 1961] provided the axioms of t-norms, as they are used today, and a redefinition of statistical metric spaces given in [Šerstnev 1962] led to a rapid development of the field. Many results concerning t-norms were obtained in the course of this development, most of which are summarized in the monograph [Schweizer & Sklar 1983].

Mathematically speaking, the theory of (continuous) t-norms has two rather independent roots, namely, the field of (specific) functional equations and the theory of (special topological) semigroups.

Concerning functional equations, t-norms are closely related to the equation of associativity (which is still unsolved in its most general form). The earliest source in this context seems to be [Abel 1826], further results in this direction were obtained in [Brouwer 1909, Cartan 1930, Aczél 1949, Hosszú 1954]. Especially János Aczél's monograph (both the German [Aczél 1961] and the English [Aczél 1966] version) had (and still has) a big impact on the development of t-norms. The main result based on this background was the full characterization of continuous Archimedean t-norms by means of additive generators in [Ling 1965] (for the case of strict t-norms see [Schweizer & Sklar 1961]). Further significant contributions are due to a group of Spanish researchers around Enric Trillas and Claudi Alsina.

Another direction of research was the identification of several parameterized families of t-norms as solutions of some (more or less) natural functional equations. The perhaps most famous result in this context has been proven in [Frank 1979], showing that the family of Frank t-norms and t-conorms (together with ordinal sums thereof) are the only solutions of the so-called Frank functional equation.

The study of a class of compact, irreducibly connected topological semigroups was initiated in [Faucett 1955], including a characterization of such semigroups, where the boundary points (at the same time annihilator and neutral element) are the only idempotent elements and where no nilpotent elements exist. In the language of t-norms, this provides a full representation of strict t-norms. In [Mostert & Shields 1957] all such semigroups, where the boundary points play the role of annihilator and neutral element, were characterized (see also [Paalman-de Miranda 1964]). Again in the language of t-norms, this provides a representation of all continuous t-norms [Ling 1965].

Several construction methods from the theory of semigroups, such as (isomorphic) transformations (which are closely related to generators mentioned above) and ordinal sums [Climescu 1946, Clifford 1954, Schweizer & Sklar 1963], have been successfully applied to construct whole families of t-norms from a few given prototypical examples [Schweizer & Sklar 1963].

Summarizing, starting with only three t-norms, namely, the minimum $T_{\mathbf{M}}$, the product $T_{\mathbf{P}}$ and the Łukasiewicz t-norm $T_{\mathbf{L}}$, it is possible to construct all continuous t-norms by means of isomorphic transformations and ordinal sums [Ling 1965].

Many specific results, such as characterizations of the order or convergence theorems, are based on this general representation for continuous t-norms.

Non-continuous t-norms, such as the drastic product $T_{\mathbf{D}}$, have been considered from the very beginning [Schweizer & Sklar 1960]. In [Ling 1965] even an additive generator for this t-norm was given. However, a general classification of non-continuous t-norms is still not known.

For the construction of not necessarily continuous t-norms, several methods, which are more or less related to those already mentioned, have been proposed recently.

All these topics, together with an investigation of algebraic and analytical properties of t-norms and the relationship between these properties, constitute the content of Part I of this book. The exposition in this part is self-contained in the sense that all necessary concepts are precisely defined, and that the proofs of all stated results are given in full detail (with the only exception of the solution of the Frank functional equation). Also, several important parameterized families of t-norms and a large number of examples for the main properties are given, supported by 66 graphical illustrations.

While the first part is devoted exclusively to the development of the theory of triangular norms, in Part II we present some fields where t-norms play a significant role. To keep this part readable and to avoid lengthy introductions to the notions and notations of these fields, we concentrate on the main concepts and present only the most important results highlighting the usefulness of t-norms. In particular, no proofs of the theorems and propositions are given, but full references are always included, encouraging the interested reader to learn more about the subjects in question.

The first of these fields, dealing with distribution functions, leads us back to the origins of t-norms. We discuss probabilistic metric spaces [Schweizer & Sklar 1983] and related topics.

A very fast developing field is that of general (not necessarily associative) aggregation operators [Zimmermann & Zysno 1980, Yager 1988, Fodor & Roubens 1994, Grabisch 1995]. Among the many examples we pick those which have a close relationship with t-norms, including uninorms [Yager & Rybalov 1996, Klement *et al.* 1996] and nullnorms [Calvo *et al.* 200x].

Based on the seminal work of Jan Łukasiewicz and Kurt Gödel in the twenties and thirties, an extensive theory of many-valued logic has been developed during the past few decades. The crucial role of t-norms in this context is presented in the monographs [Gottwald 1989, Hájek 1998b, Cignoli *et al.* 2000].

Already in his first paper [Zadeh 1965] on fuzzy sets, Lotfi A. Zadeh suggested to use the minimum $T_{\mathbf{M}}$, the product $T_{\mathbf{P}}$ and, in a restricted sense, the Łukasiewicz t-conorm $S_{\mathbf{L}}$. The use of general t-norms and t-conorms for modeling the intersection and the union of fuzzy sets (see, e.g., [Kruse *et al.* 1994a, Nguyen & Walker 1997]) apparently goes back to some seminars held by Enric Trillas and to suggestions given by Ulrich Höhle during some conferences in the late seventies (see also [Dubois & Prade 1980a, Barro *et al.* 1998]). Fuzzy sets recently found many practical applications, in particular in the context of intelligent control (see [Mamdani & Assilian 1975, Takagi & Sugeno 1985] and [Sugeno 1985a, 1985b]).

Finally we discuss a generalized theory of measures and integrals. Here t-norms and t-conorms generalize the standard set operations [Butnariu & Klement 1993], on the one hand, or standard arithmetic operations, on the other hand [Kampé de Fériet & Forte 1967, Sugeno 1974, Weber 1984] (compare also [Sion 1973]).

This last topic, in some sense, was also a starting point of the present book since the original intention of the authors was to write a monograph on measures with some generalized additivities. The usefulness, even indispensability of t-norms and t-conorms in that context, on the one hand, and the lack of a source with complete information of the state of the art of t-norms, on the other hand, were the main reasons for changing our mind and for starting the preparation of the present book. During this process we realized more and more that t-norms form a diversified and challenging topic of its own (including many applications in a variety of other mathematical fields), and we experienced a rapid growth of the volume of this manuscript. Needless to say, we benefitted a lot from numerous stimulating discussions with many colleagues, especially during the annual Linz Seminars on Fuzzy Set Theory.

The reader of this book (especially of Part I) is expected to have some basic knowledge (on the level of a graduate student) in algebra, logic and analysis. Although we sincerely hope that each chapter is of sufficient interest for everybody, here are some suggestions how to maximize the profit for the reader. To extract a basic impression combined with a reasonable amount of information, we suggest to concentrate in a first reading on Chapter 1, Sections 2.1, 3.2 and 3.3, Chapter 4, and Sections 5.1 and 5.3 (without proofs). Readers who want to locate the proper place of t-norms within the field of algebra at large are referred to the complete Chapter 2, those who want to work also with non-continuous t-norms, to Sections 3.1 and 3.4. Section 5.4 lists some of the most prominent functional equations related to t-norms. Finally, Chapters 6–8 present some additional results (partially unpublished so far) on comparison, values, discretization, and convergence of t-norms, and they should be studied when this type of information is required.

The chapters in Part II provide, on the one hand, a feeling of the importance and the role of t-norms in various theoretical and applied fields. On the other hand, they should attract the readers to these particular topics and invite them to obtain a deeper knowledge by studying the main references given there.

Notations used in this book

For the logical operations conjunction, disjunction, negation, and implication, we shall write $\wedge$, $\vee$, $\neg$, and $\rightarrow$, respectively.

The intersection and the union of two sets A and B are denoted by $A \cap B$ and $A \cup B$, respectively, $A \setminus B = \{x \in A \mid x \notin B\}$ stands for the difference of the sets A and B, and $\complement A$ for the complement of A. The symbol $A \subseteq B$ means that A is a subset of B (where $A = B$ is possible), and $A \times B = \{(a, b) \mid a \in A, b \in B\}$ is the Cartesian product of A and B. The empty set is denoted by $\emptyset$, the power set of X, i.e., the set of all subsets of X, by $\mathcal{P}(X)$, and the topological closure of A, i.e., the smallest closed set containing A, by $\mathrm{cl}(A)$. For the set of all functions from a set X into a set Y we shall write Y^X. The cardinality of a set A is denoted by $\mathrm{card}(A)$.

For a subset A of X, its characteristic function $\mathbf{1}_A : X \longrightarrow \{0, 1\}$ is given by

$$\mathbf{1}_A(x) = \begin{cases} 1 & \text{if } x \in A, \\ 0 & \text{otherwise,} \end{cases}$$

and the membership function of a fuzzy subset A of X will be denoted by $\mu_A : X \longrightarrow [0, 1]$ (see Definition 12.1).

The symbol $\mathbb{N}$ is used for the set of positive integers, i.e., $\mathbb{N} = \{1, 2, 3, \dots\}$, $\mathbb{Z}$ for the set of integers, i.e., $\mathbb{Z} = \{0, \pm 1, \pm 2, \dots\}$, $\mathbb{Q}$ for the set of rational numbers, and $\mathbb{R}$ for the set of real numbers. The extended real line, i.e., the set $\mathbb{R} \cup \{-\infty, \infty\}$, is consistently denoted by $[-\infty, \infty]$. For a closed interval we shall write $[a, b]$, for an open interval $]a, b[$, and for half-open intervals $[a, b[$ and $]a, b]$, respectively.

A sequence of elements a_n of a set X is denoted $(a_n)_{n \in \mathbb{N}}$, and a family of elements a_i of X with index set I by $(a_i)_{i \in I}$ (observe that these are elements of $X^{\mathbb{N}}$ and X^I, respectively).

If $f : X \longrightarrow Y$ is a function, then instead of X we sometimes write $\mathrm{Dom}(f)$, and the range of f is given by $\mathrm{Ran}(f) = \{f(x) \mid x \in X\}$. For $A \subseteq X$ and $B \subseteq Y$ the image $f(A)$ and the preimage $f^{-1}(B)$ are given by, respectively,

$$f(A) = \{f(x) \mid x \in A\},$$
$$f^{-1}(B) = \{x \in X \mid f(x) \in B\}.$$

Note that this means $\mathrm{Ran}(f) = f(X)$. The identity function $\mathrm{id}_X : X \longrightarrow X$ is defined by $\mathrm{id}_X(x) = x$. The composition $g \circ f : X \longrightarrow V$ of two functions $f : X \longrightarrow Y$ and $g : U \longrightarrow V$ with $Y \subseteq U$ is given by $g \circ f(x) = g(f(x))$, and the restriction $f|_A : A \longrightarrow Y$ of $f : X \longrightarrow Y$ to a subset A of X by $f|_A = f \circ \mathrm{id}_A$. If a function $f : X \longrightarrow Y$ is bijective then $f^{-1} : Y \longrightarrow X$ denotes its inverse function which satisfies $f^{-1}(y) = x$ if and only if $f(x) = y$ or, equivalently, both $f^{-1} \circ f = \mathrm{id}_X$ and $f \circ f^{-1} = \mathrm{id}_Y$. For a function $f : X \longrightarrow [0, \infty]$ its support is given by $\mathrm{supp}(f) = f^{-1}(]0, \infty])$.

If $f : X \times Y \longrightarrow Z$ is a function in two variables then, for each $x_0 \in X$, the vertical section $f(x_0, \cdot) : Y \longrightarrow Z$ and, for each $y_0 \in Y$, the horizontal section $f(\cdot, y_0) : X \longrightarrow Z$ (which are one-place functions) are given by, respectively,

$$\begin{aligned} f(x_0, \cdot) : Y &\longrightarrow Z \\ y &\longmapsto f(x_0, y), \\[1ex] f(\cdot, y_0) : X &\longrightarrow Z \\ x &\longmapsto f(x, y_0). \end{aligned}$$

For the floor of a real number x, i.e., the greatest integer which is smaller than or equal to x, we shall write $\lfloor x \rfloor$, and for the (natural) logarithm (i.e., the logarithm with respect to e) of $x \in]0, \infty[$ simply $\log x$, whereas, for each number $\lambda \in]0, 1[\cup]1, \infty[$, $\log_\lambda x$ denotes the logarithm of x with respect to the basis λ.

If $f :]a, b[\longrightarrow [-\infty, \infty]$ is a real function then the values $f(a^+)$ and $f(b^-)$ denote the right- and left-hand limit of f in the points a and b, respectively, i.e.,

$$\begin{aligned} f(a^+) &= \lim_{x \searrow a} f(x), \\ f(b^-) &= \lim_{x \nearrow b} f(x). \end{aligned}$$

For a monotone function $f : [a, b] \longrightarrow [c, d]$, $f^{(-1)} : [c, d] \longrightarrow [a, b]$ will denote the pseudo-inverse of f (see Definition 3.2).

In order to obtain well-defined continuous functions on a closed subinterval of the extended real line $[-\infty, \infty]$ (with values in $[-\infty, \infty]$), we shall frequently use the subsequent continuous extension: given a continuous function $f :]a, b[\longrightarrow]c, d[$ (where $]a, b[$ and $]c, d[$ are open subintervals of $[-\infty, \infty]$) such that the two limits $f(a^+)$ and $f(b^-)$ exist in $[c, d]$, then we define the continuous function $\overline{f} : [a, b] \longrightarrow [c, d]$ via continuous extension, i.e.,

$$\overline{f}(x) = \begin{cases} f(a^+) & \text{if } x = a, \\ f(x) & \text{if } x \in]a, b[\,, \\ f(b^-) & \text{if } x = b. \end{cases}$$

To simplify the notation, the function f and its extension $\overline{f}$ are quite often identified.

For example, this means that, extending the function $f :]0,\infty[\longrightarrow]0,\infty[$ given by $f(x) = \frac{1}{x}$, we have

$$\frac{1}{0} = \infty, \qquad \frac{1}{\infty} = 0,$$

but also, extending the addition, for each $x \in \mathbb{R}$

$$x + \infty = \infty, \qquad x - \infty = -\infty,$$

as an extension of the multiplication, for each $x \in \mathbb{R} \setminus \{0\}$

$$x \cdot \infty = \begin{cases} \infty & \text{if } x > 0, \\ -\infty & \text{if } x < 0, \end{cases} \qquad \infty^x = \begin{cases} \infty & \text{if } x > 0, \\ 0 & \text{if } x < 0, \end{cases}$$

and, finally, extending several other important functions,

$$\begin{aligned} e^{-\infty} &= 0, & e^{\infty} &= \infty, \\ \log 0 &= -\infty, & \log(\infty) &= \infty, \\ \arctan(-\infty) &= -\tfrac{\pi}{2}, & \arctan(\infty) &= \tfrac{\pi}{2}. \end{aligned}$$

For expressions which cannot be well-defined via continuous extensions (such as $\frac{0}{0}$ or $\infty - \infty$) we sometimes use individual conventions which are mentioned where they are needed.

In a lattice $(L, \preceq)$, the meet (i.e., the greatest lower bound) and the join (i.e., the least upper bound) of a subset A of L are denoted $\bigwedge A$ (or $\inf A$) and $\bigvee A$ (or $\sup A$), respectively, for the meet and join of two elements $a, b \in L$ we simply write $a \wedge b$ and $a \vee b$, respectively. Observe that each element of L is both an upper and a lower bound of the empty set $\emptyset$, so $\inf \emptyset$ and $\sup \emptyset$ depend on the underlying set L. In particular, in the lattice $([a,b], \leq)$ (with $[a,b] \subseteq [-\infty, \infty]$ and where $\leq$ is the usual order on the (extended) real line) we obtain $\inf \emptyset = b$ and $\sup \emptyset = a$.

Basic references for algebra are, e.g., [Lidl & Pilz 1998] and, concerning ordered and topological semigroups, [Fuchs 1963, Carruth *et al.* 1983], for lattice theory we usually refer to [Birkhoff 1973], and for the necessary background in calculus and real analysis to [Apostol 1965, Hewitt & Stromberg 1965]. In the context of logic our exposition is based on [Rosser 1953], topological concepts and notions are understood in the sense of [Bourbaki 1966, Dugundji 1966], and our basic references in measure and probability theory are [Halmos 1950, Bauer 1981].

PART I

1. Basic definitions and properties

... where $T(\alpha, \beta)$ is a function defined for $0 \leq \alpha \leq 1$ and $0 \leq \beta \leq 1$ such that

(a) $0 \leq T(\alpha, \beta) \leq 1$.
(b) T is non-decreasing in either variable.
(c) $T(\alpha, \beta) = T(\beta, \alpha)$.
(d) $T(1, 1) = 1$.
(e) If $\alpha > 0$, then $T(\alpha, 1) > 0$.

[...] We shall call the function T the triangular norm of the statistical metric ...

KARL MENGER

Giving exactly this definition, Karl Menger in 1942 introduced triangular norms into the mathematical literature. Triangular norms (t-norms for short) were originally used when generalizing the triangle inequality from classical metric spaces to statistical metric spaces (today called probabilistic metric spaces, see Section 9.3) which were the key point of interest in [Menger 1942].

Nowadays, a different set of axioms is used: t-norms are considered to be special semigroup operations on $[0, 1]$ with neutral element 1 (see Section 2.2). This definition is equivalent to that given by Berthold Schweizer and Abe Sklar in [Schweizer & Sklar 1958, 1960, 1961]. Note that the original axioms were much less stringent: no associativity was required (i.e., t-norms were not necessarily semigroup operations), and the boundary condition was significantly weaker (thus including also triangular conorms and other operations on the unit interval).

After giving this definition and presenting the four basic examples (minimum, product, Łukasiewicz t-norm, and drastic product), we mention some rather general properties of t-norms.

Section 1.2 deals with triangular conorms, which can be introduced either axiomatically (as we do), or as operations which are dual to t-conorms, using a kind of De Morgan formula.

Then we shall have a closer look at some analytical properties of triangular norms. In particular, we shall see that the continuity in one component is sufficient for the continuity of t-norms (which are two-place functions, i.e., functions of two variables). Also, some weaker forms of continuity are mentioned, along with some (counter-)examples.

1.1 Triangular norms

Let us start with the definition of t-norms. As already mentioned, the following axioms, as they are used today, go back to [Schweizer & Sklar 1958, 1960, 1961].

1.1 DEFINITION
A *triangular norm* (*t-norm* for short) is a binary operation T on the unit interval $[0,1]$, i.e., a function $T : [0,1]^2 \longrightarrow [0,1]$, such that for all $x, y, z \in [0,1]$ the following four axioms are satisfied:

(T1)	$T(x,y) = T(y,x)$.	(*commutativity*)
(T2)	$T(x, T(y,z)) = T(T(x,y), z)$.	(*associativity*)
(T3)	$T(x,y) \leq T(x,z)$ whenever $y \leq z$.	(*monotonicity*)
(T4)	$T(x,1) = x$.	(*boundary condition*)

Since a t-norm is an algebraic operation on the unit interval $[0,1]$, it is of course also possible to use an infix notation like $x * y$ instead of the prefix notation $T(x,y)$ (which, actually, has been done quite often in the literature, e.g., in [Nguyen & Walker 1997]). In fact, the axioms (T1)–(T4) then would look more familiar to many readers, namely, for all $x, y, z \in [0,1]$

(T1) $x * y = y * x$.
(T2) $x * (y * z) = (x * y) * z$.
(T3) $x * y \leq x * z$ whenever $y \leq z$.
(T4) $x * 1 = x$.

Since many functional aspects of this operation (e.g., continuity) are rather important for us and since we prefer to keep a unique notation throughout this monograph, we shall consistently use the prefix notation for t-norms (and t-conorms).

There exist uncountably many t-norms, and we first list the four basic t-norms. Chapter 4 will be devoted to the presentation of a number of ordered families of t-norms which are interesting from different points of view.

1.2 EXAMPLE
The following are the four basic t-norms $T_{\mathbf{M}}$, $T_{\mathbf{P}}$, $T_{\mathbf{L}}$, and $T_{\mathbf{D}}$ (see Figures 1.1 and 1.2 for 3D and contour plots) given by, respectively:

$$T_{\mathbf{M}}(x,y) = \min(x,y), \qquad (\textit{minimum})$$

$$T_{\mathbf{P}}(x,y) = x \cdot y, \qquad (\textit{product})$$

$$T_{\mathbf{L}}(x,y) = \max(x+y-1, 0), \qquad (\textit{Lukasiewicz t-norm})$$

$$T_{\mathbf{D}}(x,y) = \begin{cases} 0 & \text{if } (x,y) \in [0,1[^2, \\ \min(x,y) & \text{otherwise.} \end{cases} \qquad (\textit{drastic product})$$

Only the associativity of the operations $T_{\mathbf{L}}$ and $T_{\mathbf{D}}$ is not completely trivial. For $T_{\mathbf{L}}$ it is seen as follows:

$$T_{\mathbf{L}}(x, T_{\mathbf{L}}(y,z)) = \max(0, x+y+z-2) = T_{\mathbf{L}}(T_{\mathbf{L}}(x,y),z).$$

For $T_{\mathbf{D}}$ we obtain a value different from 0 on either side only if at least two of the values x, y and z equal 1, in which case we obviously have $\min(x,y,z)$ on both sides.

It should be mentioned that the t-norms $T_{\mathbf{M}}$, $T_{\mathbf{P}}$, $T_{\mathbf{L}}$, and $T_{\mathbf{D}}$ were denoted M, Π, W, and Z, respectively, in [Schweizer & Sklar 1983].

These four basic t-norms are remarkable for several reasons. The drastic product $T_{\mathbf{D}}$ and the minimum $T_{\mathbf{M}}$ are the smallest and the largest t-norm, respectively (with respect to the pointwise order, see Remark 1.5(i)). The minimum $T_{\mathbf{M}}$ is the only t-norm where each $x \in [0,1]$ is an idempotent element (see Example 2.2 and Proposition 1.9), whereas the product $T_{\mathbf{P}}$ and the Łukasiewicz t-norm $T_{\mathbf{L}}$ are prototypical examples of two important subclasses of t-norms, namely, of the strict (see Proposition 5.9) and nilpotent t-norms (see Proposition 5.10), respectively.

We shall frequently visualize t-norms (and functions $F : [0,1]^2 \longrightarrow [0,1]$ in general) in various ways: as 3D plots, i.e., as surfaces in the unit cube, as contour plots showing the curves (or, more generally, the sets) where the function in question has constant (equidistant) values, and, occasionally, as diagonal sections, i.e., as graphs of the function $x \longmapsto F(x,x)$.

The boundary condition (T4) and the monotonicity (T3) were given in their minimal form. In fact, we can say much more:

1.3 REMARK

(i) Directly from Definition 1.1 we can deduce that, for all $x \in [0,1]$, each t-norm T satisfies the following additional boundary conditions:

$$T(0,x) = T(x,0) = 0, \tag{1.1}$$

$$T(1,x) = x. \tag{1.2}$$

Therefore, all t-norms coincide on the boundary of the unit square $[0,1]^2$.

(ii) The monotonicity of a t-norm T in its second component described by (T3) is, together with the commutativity (T1), equivalent to the (joint) monotonicity in both components, i.e., to

$$T(x_1,y_1) \leq T(x_2,y_2) \quad \text{whenever } x_1 \leq x_2 \text{ and } y_1 \leq y_2. \tag{1.3}$$

Indeed, if $x_1 \leq x_2$ and $y_1 \leq y_2$, then we have

$$T(x_1,y_1) \leq T(x_1,y_2) = T(y_2,x_1) \leq T(y_2,x_2) = T(x_2,y_2).$$

Since t-norms are just functions from the unit square into the unit interval, the comparison of t-norms is done in the usual way, i.e., pointwise.

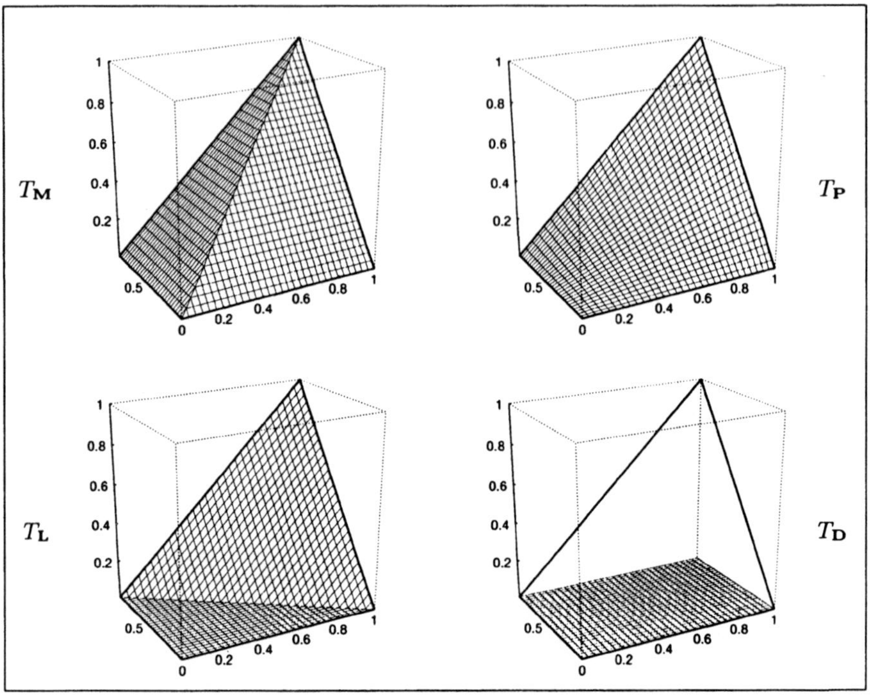

Fig. 1.1. 3D plots of the four basic t-norms

1.4 DEFINITION

(i) If, for two t-norms T_1 and T_2, the inequality $T_1(x,y) \leq T_2(x,y)$ holds for all $(x,y) \in [0,1]^2$, then we say that T_1 is *weaker* than T_2 or, equivalently, that T_2 is *stronger* than T_1, and we write in this case $T_1 \leq T_2$.

(ii) We shall write $T_1 < T_2$ whenever $T_1 \leq T_2$ and $T_1 \neq T_2$, i.e., if $T_1 \leq T_2$ and for some $(x_0, y_0) \in [0,1]^2$ we have $T_1(x_0,y_0) < T_2(x_0,y_0)$.

This, together with the properties of t-norms, yields the following comparisons involving the four basic t-norms. In general, it is not always so easy to determine whether a given t-norm is weaker than another one. In Sections 6.1 and 6.2 this problem will be treated in more detail.

1.5 REMARK

(i) As a consequence of (1.3), for each t-norm T and for each $(x,y) \in [0,1]^2$ we have both $T(x,y) \leq T(x,1) = x$ and $T(x,y) \leq T(1,y) = y$. All t-norms coincide on the boundary of $[0,1]^2$ and for all $(x,y) \in\,]0,1[^2$ we trivially have $T(x,y) \geq 0 = T_{\mathbf{D}}(x,y)$. Therefore, the drastic product $T_{\mathbf{D}}$ is the weakest, and the minimum $T_{\mathbf{M}}$ is the strongest t-norm:

$$T_{\mathbf{D}} \leq T \leq T_{\mathbf{M}}. \tag{1.4}$$

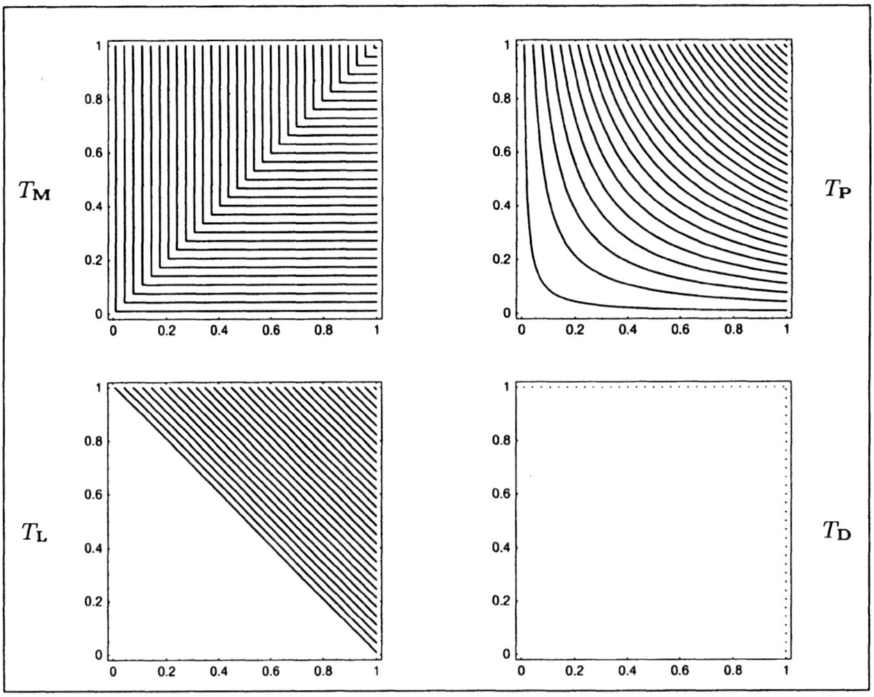

Fig. 1.2. Contour plots of the four basic t-norms (observe that there are no contour lines for $T_{\mathbf{D}}$)

(ii) Since obviously $T_{\mathbf{L}} < T_{\mathbf{P}}$, we get this ordering for the four basic t-norms:

$$T_{\mathbf{D}} < T_{\mathbf{L}} < T_{\mathbf{P}} < T_{\mathbf{M}}. \tag{1.5}$$

It is clear that the most important thing for a t-norm T is that commutativity (T1), associativity (T2) and monotonicity (T3) be satisfied on the open unit square $]0,1[^2$ (of course, together with the boundary conditions (T4), (1.1) and (1.2)). However, this is not sufficient to guarantee the monotonicity on the whole unit square $[0,1]^2$, as the function F_3 in Example 1.12 below shows. Therefore, an additional condition is needed, and we obtain the subsequent general result.

1.6 PROPOSITION
Let A be a set with $]0,1[\, \subseteq A \subseteq [0,1]$, and assume that $*: A^2 \longrightarrow A$ is a binary operation on A such that for all $x, y, z \in A$ the properties (T1)–(T3) and

$$x * y \leq \min(x, y) \tag{1.6}$$

are satisfied. Then the function $T : [0,1]^2 \longrightarrow [0,1]$ defined by

$$T(x,y) = \begin{cases} x * y & \textit{if } (x,y) \in (A \setminus \{1\})^2, \\ \min(x,y) & \textit{otherwise}, \end{cases} \tag{1.7}$$

is a t-norm. Moreover, T is the only t-norm whose restriction to $(A \setminus \{1\})^2$ coincides with the restriction of $*$ to $(A \setminus \{1\})^2$.

Proof: The commutativity (T1) and the boundary condition (T4) are satisfied by definition, and the uniqueness of T follows from Remark 1.3(i). Concerning the associativity (T2), observe that for $x, y, z \in A \setminus \{0,1\}$ we have $T(T(x,y),z) = T(x, T(y,z))$ as a consequence of the associativity of $*$. If $0 \in \{x,y,z\}$ then we clearly get 0 on both sides, and if $1 \in \{x,y,z\}$ then $T(T(x,y),z) = T(x,T(y,z))$ follows from (T4). Concerning the monotonicity (T3), suppose $y \leq z$. In the cases $x, y, z \in A \setminus \{1\}$ or $x \in \{0,1\}$ or $y = 0$, the inequality $T(x,y) \leq T(x,z)$ is inherited from the monotonicity of $*$ and min. The only non-trivial case is when $x, y \in A \setminus \{1\}$ and $z = 1$, in which case $T(x,y) \leq T(x,z)$ follows from (1.6). □

It is interesting to note that, in the cases $A =]0,1[$ or $A =]0,1]$, the product $T_{\mathbf{P}}$ and the minimum $T_{\mathbf{M}}$ can be obtained by the construction in Proposition 1.6, whereas the Łukasiewicz t-norm $T_{\mathbf{L}}$ and the drastic product $T_{\mathbf{D}}$ cannot (simply because the restriction of these two t-norms to the half-open unit square $]0,1]^2$ is not a binary operation on $]0,1]$). If $A = [0,1[$ then every t-norm can be obtained via Proposition 1.6.

In the case $A = [0,1]$ the class of operations considered in Proposition 1.6 was given a special name in [Jenei 1999, Jenei 200x]:

1.7 DEFINITION
A function $F : [0,1]^2 \longrightarrow [0,1]$ which satisfies, for all $x, y, z \in [0,1]$, the properties (T1)–(T3) and (1.6) is called a *t-subnorm*.

Clearly, each t-norm is a t-subnorm, but not vice versa: for example, the zero function, i.e., the function $F : [0,1]^2 \longrightarrow [0,1]$ given by $F(x,y) = 0$, is a t-subnorm but not a t-norm.

An intrinsic part of Proposition 1.6 is that each t-subnorm F can be transformed into a t-norm by redefining (if necessary) its values on the upper right boundary of the unit square.

1.8 COROLLARY
If F is a t-subnorm then the function $T : [0,1]^2 \longrightarrow [0,1]$ defined by

$$T(x,y) = \begin{cases} F(x,y) & \textit{if } (x,y) \in [0,1[^2, \\ \min(x,y) & \textit{otherwise}, \end{cases}$$

is a triangular norm.

An interesting question is whether a t-norm is determined uniquely by its values on the diagonal or some other subset of the unit square. In general, this is not the case, but the extremal t-norms $T_{\mathbf{D}}$ and $T_{\mathbf{M}}$ are completely determined by their their diagonal sections, i.e., by their values on the diagonal of the unit square (for a more extensive treatment of this question see Section 7.2):

1.9 PROPOSITION

(i) *The only t-norm T satisfying $T(x,x) = x$ for all $x \in [0,1]$ is the minimum $T_{\mathbf{M}}$.*

(ii) *The only t-norm T satisfying $T(x,x) = 0$ for all $x \in [0,1[$ is the drastic product $T_{\mathbf{D}}$.*

Proof: If for a t-norm T we have $T(x,x) = x$ for each $x \in [0,1]$, then for all $(x,y) \in [0,1]^2$ with $y \leq x$ the monotonicity (T3) implies

$$y = T(y,y) \leq T(x,y) \leq T_{\mathbf{M}}(x,y) = y,$$

which, together with (T1), means $T = T_{\mathbf{M}}$. To show the validity of (ii), assume $T(x,x) = 0$ for each $x \in [0,1[$. Then for all $(x,y) \in [0,1[^2$ with $y \leq x$ we have $0 \leq T(x,y) \leq T(x,x) = 0$, hence, together with (T1) and (T4), yielding $T = T_{\mathbf{D}}$. □

Since t-norms, introduced in Definition 1.1 as binary operations, are associative, they can be extended to operations with more than two arguments:

1.10 REMARK

(i) The associativity (T2) allows us to extend each t-norm T in a unique way to an n-ary operation in the usual way by induction, defining for each n-tuple $(x_1, x_2, \ldots, x_n) \in [0,1]^n$

$$\operatorname*{\mathsf{T}}_{i=1}^{n} x_i = T\Big(\operatorname*{\mathsf{T}}_{i=1}^{n-1} x_i, x_n\Big) = T(x_1, x_2, \ldots, x_n).$$

If, in particular, we have $x_1 = x_2 = \cdots = x_n = x$, we shall briefly write

$$x_T^{(n)} = T(x, x, \ldots, x).$$

Finally we put, by convention, for each $x \in [0,1]$

$$x_T^{(0)} = 1 \qquad \text{and} \qquad x_T^{(1)} = x.$$

(ii) The fact that each t-norm T is weaker than $T_{\mathbf{M}}$ makes it possible to extend it to a (countably) infinitary operation, putting for each sequence $(x_i)_{i\in\mathbb{N}}$ of elements of $[0,1]$, i.e., for each $(x_i)_{i\in\mathbb{N}} \in [0,1]^{\mathbb{N}}$,

$$\operatorname*{\mathsf{T}}_{i=1}^{\infty} x_i = \lim_{n\to\infty} \operatorname*{\mathsf{T}}_{i=1}^{n} x_i. \tag{1.8}$$

Note that the limit on the right-hand side always exists since the sequence

$$\left(\mathop{\mathsf{T}}_{i=1}^{n} x_i\right)_{n\in\mathbb{N}}$$

is non-increasing and bounded from below. However, in general it may not be possible to group the arguments or to change their order in a completely arbitrary way without influencing the result (this is similar to the properties of conditionally convergent infinite series of real numbers [Mesiar 1997c]).

(iii) We only mention that for an arbitrary (possibly uncountable) index set I and for each $(x_i)_{i\in I} \in [0,1]^I$, i.e., for each family $(x_i)_{i\in I}$ of elements of $[0,1]$, the following is well-defined and a proper generalization of (1.8):

$$\mathop{\mathsf{T}}_{i\in I} x_i = \inf\left\{ \mathop{\mathsf{T}}_{j=1}^{k} x_{i_j} \;\middle|\; (x_{i_1},\ldots,x_{i_k}) \text{ is a finite subfamily of } (x_i)_{i\in I} \right\}.$$

1.11 EXAMPLE

The n-ary extensions of the minimum $T_{\mathbf{M}}$ and the product $T_{\mathbf{P}}$ are obvious. For the Łukasiewicz t-norm $T_{\mathbf{L}}$ and the drastic product $T_{\mathbf{D}}$ we get the following n-ary extensions:

$$T_{\mathbf{L}}(x_1,x_2,\ldots,x_n) = \max\left(\sum_{i=1}^{n} x_i - (n-1), 0\right),$$

$$T_{\mathbf{D}}(x_1,x_2,\ldots,x_n) = \begin{cases} x_i & \text{if } x_j = 1 \text{ for all } j \neq i, \\ 0 & \text{otherwise.} \end{cases}$$

Finally we mention that the axioms (T1)–(T4) in Definition 1.1 are independent from each other, as can be seen from the following examples of operations on $[0,1]$, where exactly one of the axioms (T1)–(T4) does not hold:

1.12 EXAMPLE

Define for $i = 1,2,3,4$ the functions $F_i : [0,1]^2 \longrightarrow [0,1]$ by, respectively,

$$F_1(x,y) = \begin{cases} 0 & \text{if } (x,y) \in [0,0.5] \times [0,1[\,, \\ \min(x,y) & \text{otherwise,} \end{cases}$$

$$F_2(x,y) = x \cdot y \cdot \max(x,y),$$

$$F_3(x,y) = \begin{cases} 0.5 & \text{if } (x,y) \in \,]0,1[^2\,, \\ \min(x,y) & \text{otherwise,} \end{cases}$$

$$F_4(x,y) = 0.$$

Obviously, each F_i satisfies the axioms (T1)–(T4) with the exception of (Ti).

1.2 Triangular conorms

In [Schweizer & Sklar 1961] triangular conorms were introduced as dual operations of t-norms. We give here an independent axiomatic definition.

1.13 DEFINITION
A *triangular conorm* (*t-conorm* for short) is a binary operation S on the unit interval $[0,1]$, i.e., a function $S : [0,1]^2 \longrightarrow [0,1]$, which, for all $x, y, z \in [0,1]$, satisfies (T1)–(T3) and

$$\text{(S4)} \qquad S(x,0) = x. \qquad \textit{(boundary condition)}$$

From an axiomatical point of view, t-norms and t-conorms differ only with respect to their respective boundary conditions. In fact, the concepts t-norms and t-conorms are dual in some sense, as we shall see very soon.

1.14 EXAMPLE
The following are the basic t-conorms $S_{\mathbf{M}}$, $S_{\mathbf{P}}$, $S_{\mathbf{L}}$, and $S_{\mathbf{D}}$ (see Figures 1.3 and 1.4 for 3D and contour plots) given by, respectively:

$$S_{\mathbf{M}}(x,y) = \max(x,y), \qquad \textit{(maximum)}$$

$$S_{\mathbf{P}}(x,y) = x + y - x \cdot y, \qquad \textit{(probabilistic sum)}$$

$$S_{\mathbf{L}}(x,y) = \min(x+y,1), \qquad \textit{(Łukasiewicz t-conorm, bounded sum)}$$

$$S_{\mathbf{D}}(x,y) = \begin{cases} 1 & \text{if } (x,y) \in \,]0,1]^2, \\ \max(x,y) & \text{otherwise.} \end{cases} \qquad \textit{(drastic sum)}$$

The t-conorms $S_{\mathbf{M}}$, $S_{\mathbf{P}}$, $S_{\mathbf{L}}$, and $S_{\mathbf{D}}$ were denoted M^*, Π^*, W^* and Z^*, respectively, in [Schweizer & Sklar 1983].

The original definition of t-conorms given in [Schweizer & Sklar 1961] is completely equivalent to the axiomatic definition given above:

1.15 PROPOSITION
A function $S : [0,1]^2 \longrightarrow [0,1]$ is a t-conorm if and only if there exists a t-norm T such that for all $(x,y) \in [0,1]^2$

$$S(x,y) = 1 - T(1-x, 1-y). \tag{1.9}$$

Proof: If T is a t-norm then obviously the operation S defined by (1.9) satisfies (T1)–(T3) and (S4) and is, therefore, a t-conorm. On the other hand, if S is a t-conorm, then define the function $T : [0,1]^2 \longrightarrow [0,1]$ by

$$T(x,y) = 1 - S(1-x, 1-y). \tag{1.10}$$

Again, it is trivial to check that T is a t-norm and that (1.9) holds. □

The t-conorm given by (1.9) is called the *dual t-conorm* of T and, analogously, the t-norm given by (1.10) is said to be the *dual t-norm* of S.

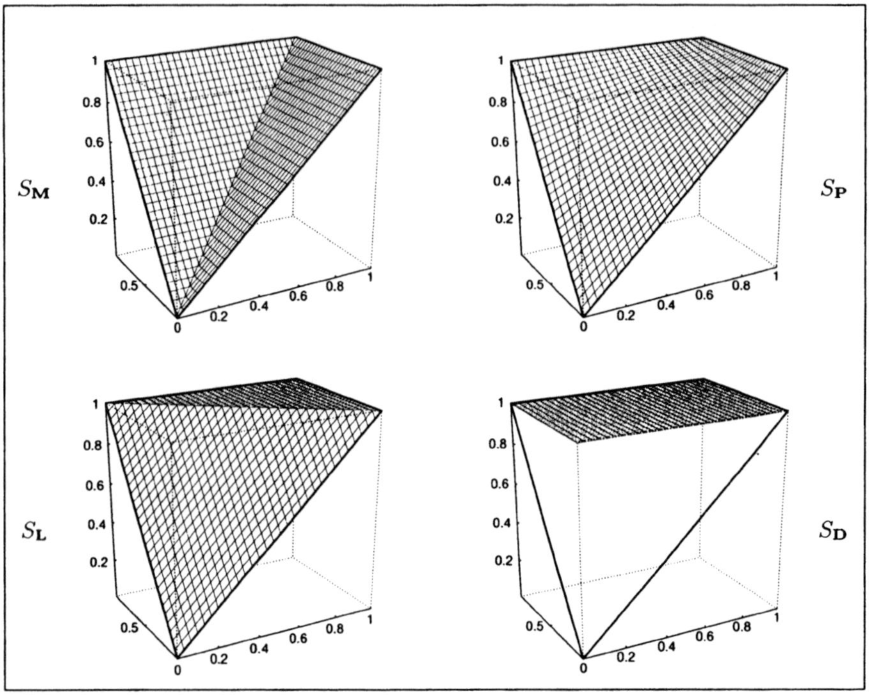

Fig. 1.3. 3D plots of the four basic t-conorms

Obviously, the name t-conorm goes back to the fact that, in the unit interval, $1 - x$ acts in a similar way as a complement of x. We shall keep this original notion and avoid the term *s-norm* which sometimes is used synonymously in the literature.

The duality expressed in (1.9) allows us to translate many properties of t-norms into the corresponding properties of t-conorms.

1.16 REMARK

(i) The proof of Proposition 1.15 makes it clear that also each t-norm is the dual operation of some t-conorm. Note that $(T_\mathbf{M}, S_\mathbf{M})$, $(T_\mathbf{P}, S_\mathbf{P})$, $(T_\mathbf{L}, S_\mathbf{L})$, and $(T_\mathbf{D}, S_\mathbf{D})$ are pairs of t-norms and t-conorms which are mutually dual to each other.

(ii) All t-conorms coincide on the boundary of $[0,1]^2$, as a consequence of these additional boundary conditions which hold for all $x \in [0,1]$:

$$S(1,x) = S(x,1) = 1,$$
$$S(0,x) = x.$$

As in the case of t-norms (see Proposition 1.6), it would be sufficient to define a t-conorm S on the open or half-open unit square only and,

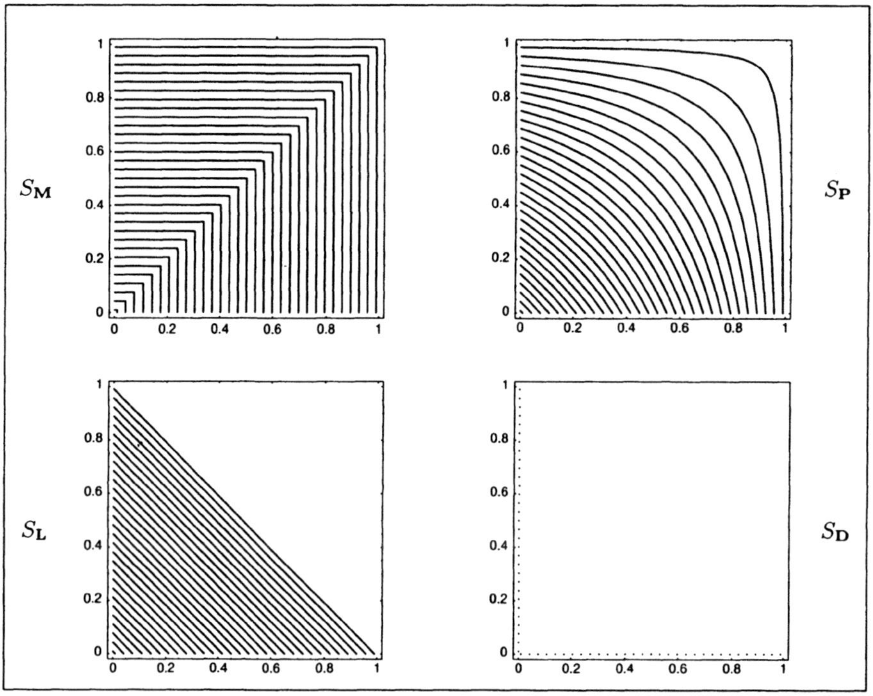

Fig. 1.4. Contour plots of the four basic t-conorms

provided that (S1)–(S3) and $S(x,y) \geq \max(x,y)$ hold for all (x,y) in the open or half-open unit square, respectively, to define (where necessary) the values at the boundary of the unit square according to the boundary conditions. However, in this book we always shall define t-conorms on the closed unit square.

(iii) The duality changes the order: if, for some t-norms T_1 and T_2 we have $T_1 \leq T_2$, and if S_1 and S_2 are the dual t-conorms of T_1 and T_2, respectively, then we get $S_1 \geq S_2$. Consequently, for each t-conorm S we have

$$S_{\mathbf{M}} \leq S \leq S_{\mathbf{D}},$$

i.e., the maximum $S_{\mathbf{M}}$ is the weakest and the drastic sum $S_{\mathbf{D}}$ is the strongest t-conorm. For the t-conorms in Example 1.14 we get this ordering:

$$S_{\mathbf{M}} < S_{\mathbf{P}} < S_{\mathbf{L}} < S_{\mathbf{D}}.$$

(iv) Given a t-conorm S, we can, in complete analogy to Remark 1.10, extend this operation to n-tuples $(x_1, x_2, \ldots, x_n) \in [0,1]^n$, sequences $(x_i)_{i \in \mathbb{N}} \in [0,1]^{\mathbb{N}}$ and families $(x_i)_{i \in I} \in [0,1]^I$ with an arbitrary (even uncountable) index set I as follows (note that everything is well-defined):

$$\mathop{\mathsf{S}}_{i=1}^{n} x_i = S\Big(\mathop{\mathsf{S}}_{i=1}^{n-1} x_i, x_n\Big) = S(x_1, x_2, \ldots, x_n),$$

$$\mathop{\mathsf{S}}_{i=1}^{\infty} x_i = \lim_{n\to\infty} \mathop{\mathsf{S}}_{i=1}^{n} x_i,$$

$$\mathop{\mathsf{S}}_{i\in I} x_i = \sup\left\{ \mathop{\mathsf{S}}_{j=1}^{k} x_{i_j} \;\Big|\; (x_{i_1}, \ldots, x_{i_k}) \text{ is a finite subfamily of } (x_i)_{i\in I}\right\}.$$

In the special case $x_1 = x_2 = \cdots = x_n = x$, we shall briefly write $x_S^{(n)}$ instead of $S(x, x, \ldots, x)$, and $x_S^{(0)} = 0$ and $x_S^{(1)} = x$ for each $x \in [0,1]$.

1.17 EXAMPLE

Again, the n-ary extension of the maximum $S_{\mathbf{M}}$ is obvious. For the probabilistic sum $S_{\mathbf{P}}$, the Łukasiewicz t-conorm $S_{\mathbf{L}}$ and the drastic sum $S_{\mathbf{D}}$ we have the following n-ary extensions:

$$S_{\mathbf{P}}(x_1, x_2, \ldots, x_n) = 1 - \prod_{i=1}^{n}(1 - x_i),$$

$$S_{\mathbf{L}}(x_1, x_2, \ldots, x_n) = \min\Big(\sum_{i=1}^{n} x_i, 1\Big),$$

$$S_{\mathbf{D}}(x_1, x_2, \ldots, x_n) = \begin{cases} x_i & \text{if } x_j = 0 \text{ for all } j \neq i, \\ 1 & \text{otherwise.} \end{cases}$$

1.18 REMARK

Note that, if (T, S) is a pair of mutually dual t-norms and t-conorms, then the dualities (1.9) and (1.10) can be generalized as follows (here K can an arbitrary finite, countable or uncountable index set):

$$\mathop{\mathsf{S}}_{k\in K} x_k = 1 - \mathop{\mathsf{T}}_{k\in K} (1 - x_k),$$

$$\mathop{\mathsf{T}}_{k\in K} x_k = 1 - \mathop{\mathsf{S}}_{k\in K} (1 - x_k).$$

1.3 Continuity

This section is devoted to analytic properties a t-norm or a t-conorm may have in addition to the four respective axioms. As can be seen from the drastic product $T_{\mathbf{D}}$ and its dual $S_{\mathbf{D}}$, t-norms and t-conorms (viewed as functions in two variables) need not be continuous. In Example 3.75 we shall see that they need not even be Borel measurable functions. Nevertheless, for a number of reasons continuous t-norms and t-conorms play an important role. Therefore, we shall discuss here continuity as well as lower and upper semicontinuity.

Recall that a function $F : [0,1]^2 \longrightarrow [0,1]$ is continuous if for all convergent sequences $(x_n)_{n\in\mathbb{N}}, (y_n)_{n\in\mathbb{N}} \in [0,1]^{\mathbb{N}}$ we have

$$F\Big(\lim_{n\to\infty} x_n, \lim_{n\to\infty} y_n\Big) = \lim_{n\to\infty} F(x_n, y_n).$$

Since the unit square $[0,1]^2$ is a compact subset of the real plane $\mathbb{R}^2$, the continuity of a function $F : [0,1]^2 \longrightarrow [0,1]$ is equivalent to its uniform continuity.

Obviously, the basic t-norms $T_{\mathbf{M}}, T_{\mathbf{P}}$ and $T_{\mathbf{L}}$ as well as their dual t-conorms $S_{\mathbf{M}}, S_{\mathbf{P}}$ and $S_{\mathbf{L}}$ are continuous, and the drastic product $T_{\mathbf{D}}$ and the drastic sum $S_{\mathbf{D}}$ are not continuous.

Since it is trivial that a t-conorm is continuous if and only if its dual t-norm is continuous, we shall state all results concerning continuity for t-norms only (for weaker forms of continuity, such as lower semicontinuity or left-continuity, some modifications are necessary when switching from t-norms to t-conorms, see Remark 1.21(ii) below).

In general, a real function of two variables, e.g, with domain $[0,1]^2$, may be continuous in each variable without being continuous on $[0,1]^2$. Triangular norms (and conorms, of course) are exceptions from this, a fact which follows from the subsequent more general result for monotone real functions (stated here for non-decreasing functions on the unit square).

1.19 PROPOSITION
A function $F : [0,1]^2 \longrightarrow [0,1]$ which is non-decreasing, i.e., which satisfies (1.3), is continuous if and only if it is continuous in each component, i.e., if for all $x_0, y_0 \in [0,1]$ both the vertical section $F(x_0, .) : [0,1] \longrightarrow [0,1]$ and the horizontal section $F(., y_0) : [0,1] \longrightarrow [0,1]$ are continuous functions in one variable.

Proof: If a function $F : [0,1]^2 \longrightarrow [0,1]$ is continuous, then it is obviously continuous in each component.

Conversely, if F is continuous in each component, fix $(x_0, y_0) \in [0,1]^2$, $\varepsilon > 0$ and let $(x_n)_{n\in\mathbb{N}}$ and $(y_n)_{n\in\mathbb{N}}$ be sequences in $[0,1]$ converging to x_0 and y_0, respectively. From this we can construct four monotone sequences $(a_n)_{n\in\mathbb{N}}, (b_n)_{n\in\mathbb{N}}, (c_n)_{n\in\mathbb{N}}$ and $(d_n)_{n\in\mathbb{N}}$ such that for all $n \in \mathbb{N}$ we have

$$\begin{aligned} a_n \le x_n \le b_n \quad &\text{and} \quad (a_n)_{n\in\mathbb{N}} \nearrow x_0, \quad (b_n)_{n\in\mathbb{N}} \searrow x_0,\\ c_n \le y_n \le d_n \quad &\text{and} \quad (c_n)_{n\in\mathbb{N}} \nearrow y_0, \quad (d_n)_{n\in\mathbb{N}} \searrow y_0. \end{aligned}$$

The continuity of F in the second component implies the continuity of the vertical section $F(x_0, .)$, which means that there exists an $N \in \mathbb{N}$ such that, as a consequence of the monotonicity of F, for all $n \ge N$

$$F(x_0, y_0) - \varepsilon < F(x_0, c_N) \le F(x_0, y_n) \le F(x_0, d_N) < F(x_0, y_0) + \varepsilon.$$

Since F is continuous in the first component, also the two functions $F(., c_N)$ and $F(., d_N)$ are continuous and, consequently, there is a number $M \in \mathbb{N}$ such that for all $m \geq M$ and $n \geq N$ (again taking into account the monotonicity of F) we get

$$F(x_0, c_N) - \varepsilon < F(a_M, c_N) \leq F(x_m, y_n) \leq F(b_M, d_N) < F(x_0, d_N) + \varepsilon.$$

Putting $K = \max(M, N)$, then for all $k \geq K$ we have

$$F(x_0, y_0) - 2\varepsilon < F(x_k, y_k) < F(x_0, y_0) + 2\varepsilon,$$

proving that $(F(x_k, y_k))_{k \in \mathbb{N}}$ converges to $F(x_0, y_0)$, i.e., that F is continuous in (x_0, y_0). □

Obviously, because of the commutativity (T1), for a t-norm or a t-conorm its continuity is equivalent to its continuity in the first component.

For applications, e.g., in probabilistic metric spaces, many-valued logics or decomposable measures, quite often weaker forms of continuity are sufficient.

1.20 Definition

A function $F : [0,1]^2 \longrightarrow [0,1]$ is called *lower (upper) semicontinuous* if for each point $(x_0, y_0) \in [0,1]^2$ and for each $\varepsilon > 0$ there is a $\delta > 0$ such that, respectively,

$$\begin{aligned} F(x,y) &> F(x_0,y_0) - \varepsilon \qquad \text{whenever } (x,y) \in \left]x_0 - \delta, x_0\right] \times \left]y_0 - \delta, y_0\right], \\ F(x,y) &< F(x_0,y_0) + \varepsilon \qquad \text{whenever } (x,y) \in \left[x_0, x_0 + \delta\right[\times \left[y_0, y_0 + \delta\right[. \end{aligned}$$

1.21 Remark

(i) The *nilpotent minimum* $T^{\mathbf{nM}}$ (mentioned in [Perny 1992, Perny & Roy 1992, Fodor 1995], for a visualization see Figure 1.5) defined by

$$T^{\mathbf{nM}}(x,y) = \begin{cases} 0 & \text{if } x + y \leq 1, \\ \min(x,y) & \text{otherwise,} \end{cases} \tag{1.11}$$

is a t-norm (this is easy to check, but can also be seen from Proposition 3.63 and Example 3.65(i)). It is lower semicontinuous but not upper semicontinuous (and, therefore, not continuous). The drastic product $T_{\mathbf{D}}$ is upper semicontinuous but not lower semicontinuous.

(ii) A t-norm T is lower (upper) semicontinuous if and only if its dual t-conorm given by (1.9) is upper (lower) semicontinuous, respectively.

Similar to the situation we had in the case of continuity, the lower semicontinuity of monotone real functions can be described by their left-continuity in each component (again stated here for non-decreasing functions on the unit square).

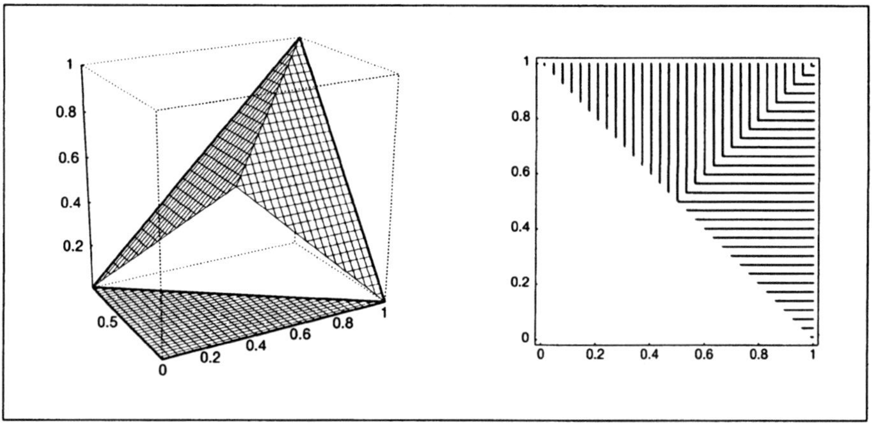

Fig. 1.5. 3D plot (left) and contour plot of the nilpotent minimum $T^{\mathbf{nM}}$ defined by (1.11)

1.22 PROPOSITION

A non-decreasing function $F : [0,1]^2 \longrightarrow [0,1]$ *is lower semicontinuous if and only if it is left-continuous in each component, i.e., if for all* $x_0, y_0 \in [0,1]$ *and for all sequences* $(x_n)_{n\in\mathbb{N}}, (y_n)_{n\in\mathbb{N}} \in [0,1]^{\mathbb{N}}$ *we have*

$$\begin{aligned}\sup\{F(x_n, y_0) \mid n \in \mathbb{N}\} &= F\big(\sup\{x_n \mid n \in \mathbb{N}\}, y_0\big),\\ \sup\{F(x_0, y_n) \mid n \in \mathbb{N}\} &= F\big(x_0, \sup\{y_n \mid n \in \mathbb{N}\}\big).\end{aligned}$$

Proof: This is shown in complete analogy to the proof of Proposition 1.19. □

By the same token, the upper semicontinuity of a non-decreasing function $F : [0,1]^2 \longrightarrow [0,1]$ is equivalent to its right-continuity in each component. Note also that the left- and right-continuity mean exactly the interchangeability of the supremum and infimum, respectively, with the monotone function.

In particular, a t-norm or a t-conorm is lower (upper) semicontinuous if and only if it is left-continuous (right-continuous) in its first component. Proposition 1.22 allows us to speak about *left-continuous* and *right-continuous* t-norms or t-conorms instead of lower and upper semicontinuous t-norms or t-conorms, respectively.

Sometimes, even weaker forms of continuity of t-norms and t-conorms are sufficient (compare [Pap 1991]).

1.23 DEFINITION

A t-norm T (a t-conorm S) is said to be *border continuous* if it is continuous on the boundary of the unit square $[0,1]^2$, i.e., on the set $[0,1]^2 \setminus]0,1[^2$.

Observe that for the border continuity of a t-norm T, it is sufficient to require the continuity on the upper right boundary (and, for a t-conorm S,

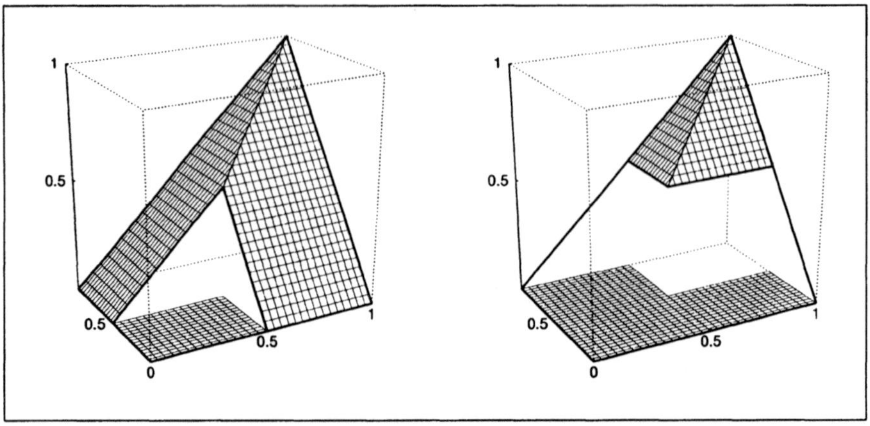

Fig. 1.6. The border continuous t-norm defined in Example 1.24(i) (left) and the t-norm which is continuous in $(1, 1)$ (but not border continuous) given in Example 1.24(ii)

on the lower left boundary) of the unit square $[0, 1]^2$. Indeed, as an immediate consequence of Remarks 1.5(i) and 1.16, for each $y \in [0, 1]$ we have

$$T(0^+, y) = T(0, y),$$
$$S(1^-, y) = S(1, y).$$

The following t-norms (see Figure 1.6) are special examples of the t-norms of type T_A which will be introduced in Proposition 3.63.

1.24 EXAMPLE

(i) The following function $T : [0, 1]^2 \longrightarrow [0, 1]$ is a t-norm which is border continuous but not left-continuous:

$$T(x, y) = \begin{cases} 0 & \text{if } (x, y) \in]0, 0.5[^2, \\ \min(x, y) & \text{otherwise.} \end{cases}$$

(ii) The following function $T : [0, 1]^2 \longrightarrow [0, 1]$ is a t-norm which is continuous in the point $(1, 1)$ but not border continuous:

$$T(x, y) = \begin{cases} 0 & \text{if } (x, y) \in]0, 1[^2 \setminus [0.5, 1[^2, \\ \min(x, y) & \text{otherwise.} \end{cases}$$

Another analytical property closely related to continuity is the following (again we restrict ourselves to functions on the unit square).

1.25 DEFINITION

A function $F : [0, 1]^2 \longrightarrow [0, 1]$ satisfies the *Lipschitz property* if there is a constant $c \in]0, \infty[$ such that for all pairs $(x_1, y_1), (x_2, y_2) \in [0, 1]^2$

$$|F(x_1, y_1) - F(x_2, y_2)| \leq c \cdot (|x_1 - x_2| + |y_1 - y_2|).$$

In general, the Lipschitz property of a real function F implies the uniform continuity and, therefore, the continuity of F. The converse of this assertion is not true, as can be seen from the following t-norm.

1.26 EXAMPLE
Consider the t-norm T given by $T(x,y) = \sqrt{\max(x^2+y^2-1,0)}$ (in Example 4.3 we shall see that T is just the Schweizer-Sklar t-norm $T_2^{\mathbf{SS}}$) which is obviously (uniformly) continuous. However, the assumption that T satisfies the Lipschitz property for some constant c, for $d = \max(c,2)$ leads to

$$|T(\tfrac{\sqrt{2}}{2}+\tfrac{1}{d^2},\tfrac{\sqrt{2}}{2}) - T(\tfrac{\sqrt{2}}{2},\tfrac{\sqrt{2}}{2})| \leq \frac{c}{d^2} \leq \frac{1}{d}$$

and, because of $T(\frac{\sqrt{2}}{2},\frac{\sqrt{2}}{2}) = 0$, to the contradiction

$$T(\tfrac{\sqrt{2}}{2}+\tfrac{1}{d^2},\tfrac{\sqrt{2}}{2}) = \frac{1}{d}\sqrt{\sqrt{2}+\tfrac{1}{d^2}} > \frac{1}{d}.$$

An immediate consequence of the boundary condition (T4) is that, whenever a t-norm T has the Lipschitz property, the corresponding constant c must belong to the interval $[1,\infty[$.

2. Algebraic aspects

> Algebra is essentially concerned with *calculating*, that is, performing, on elements of a set, "algebraic operations", the most well-known example of which is provided by the "four rules" of arithmetic. [...] It is no doubt the possibility of these successive extensions, in which the *form* of the calculations remained the same, whereas the *nature* of the mathematical entities subjected to these calculations varied considerably, which was responsible for the gradual isolation of the guiding principle of modern mathematics, namely that mathematical entities in themselves are of little importance; what matters are their *relations* ...
>
> NICOLAS BOURBAKI

We start with the introduction and investigation of some elementary algebraic properties of t-norms: idempotent and nilpotent elements and zero divisors, then strict monotonicity, the cancellation law and, most importantly, the Archimedean property as well as strict and nilpotent t-norms. The relationship between these properties is described in full detail, including many (counter-)examples. An important result is that each continuous Archimedean t-norm is either strict or nilpotent.

In the rest of this chapter, we shall place t-norms and t-conorms into more general algebraic frameworks. First, it is clear that they are special semigroup operations on $[0,1]$. In the sequel, the role of semigroup isomorphisms and involutions (the latter allow us to switch between t-norms and t-conorms), is explained.

In Section 2.3 we discuss topological semigroups, i.e., semigroups with continuous operations. In particular we show that, for so-called I-semigroups, we can drop the commutativity axiom (T1) and weaken the boundary condition (T4). Moreover, each I-semigroup operation turns out to be isomorphic either to a continuous t-norm or to a continuous t-conorm.

Finally, we consider special l-monoids, a concept on the junction of lattice and semigroup theory. A key feature is the existence of a residuum (which plays an important role in many-valued logics, see Chapter 11), and the main result of this section is the equivalence between commutative, residuated integral l-monoids (on $[0,1]$) and left-continuous t-norms.

2.1 Elementary algebraic properties

Now we turn our attention to some algebraic aspects of t-norms, most of which are well-known from the general theories of semigroups and lattices (see Sections 2.2–2.4). Recall, for a t-norm T, $x \in [0,1]$ and $n \in \mathbb{N}_0$, the notion $x_T^{(n)}$ of the n-th power of x (with respect to T) introduced in Remark 1.10(i). Our first focus are idempotent and nilpotent elements, and zero divisors. Since for each $n \in \mathbb{N}$ we trivially have $0_T^{(n)} = 0$ and $1_T^{(n)} = 1$, only elements of $]0,1[$ will be considered as candidates for nilpotent elements and zero divisors in the following definition.

2.1 DEFINITION
Let T be a t-norm.

(i) An element $a \in [0,1]$ is called an *idempotent element* of T if $T(a,a) = a$. The numbers 0 and 1 (which are idempotent elements for each t-norm T) are called *trivial* idempotent elements of T, each idempotent element in $]0,1[$ will be called a *non-trivial* idempotent element of T.
(ii) An element $a \in\ ]0,1[$ is called a *nilpotent element* of T if there exists some $n \in \mathbb{N}$ such that $a_T^{(n)} = 0$.
(iii) An element $a \in\ ]0,1[$ is called a *zero divisor* of T if there exists some $b \in\ ]0,1[$ such that $T(a,b) = 0$.

2.2 EXAMPLE
(i) Each $a \in [0,1]$ is an idempotent element of the minimum $T_{\mathbf{M}}$ (actually, $T_{\mathbf{M}}$ is the only t-norm whose set of idempotent elements equals $[0,1]$ (as a consequence of Proposition 1.9), and each $a \in\ ]0,1[$ is both a nilpotent element and a zero divisor of the Łukasiewicz t-norm $T_{\mathbf{L}}$ as well as of the drastic product $T_{\mathbf{D}}$. The minimum $T_{\mathbf{M}}$ has neither nilpotent elements nor zero divisors, and $T_{\mathbf{L}}$ and $T_{\mathbf{D}}$ both possess only trivial idempotent elements. The product $T_{\mathbf{P}}$ has neither non-trivial idempotent nor nilpotent elements nor zero divisors.
(ii) For the nilpotent minimum $T^{\mathbf{nM}}$ defined in (1.11), a number a is an idempotent element if and only if $a \in \{0\} \cup\ ]0.5,1]$, a is a nilpotent element if and only if $a \in\ ]0,0.5]$, and a is a zero divisor if and only if $a \in\ ]0,1[$.
(iii) If $*$ is a binary operation on $]0,1[$ or on $]0,1]$ which satisfies for all $x,y,z \in A$ the properties (T1)–(T3) and (1.6) then the t-norm constructed by means of Proposition 1.6 has neither zero divisors nor nilpotent elements.

The idempotent elements of t-norms can be characterized in the following way, which involves the minimum operation.

2.3 PROPOSITION
(i) *An element $a \in [0,1]$ is an idempotent element of a t-norm T if and only if for all $x \in [a,1]$ we have $T(a,x) = \min(a,x)$.*

(ii) If T is a continuous t-norm, then $a \in [0,1]$ is an idempotent element of T if and only if for all $x \in [0,1]$ we have $T(a,x) = \min(a,x)$.

Proof: If for all $x \in [a,1]$ we have $T(a,x) = \min(a,x)$ then also $T(a,a) = a$. The converse in (i) is an immediate consequence of the monotonicity (T3) and the boundary condition (T4) of T.

When proving (ii), because of (i) it suffices to show that for an idempotent element $a \in [0,1]$ we have $T(a,x) = x$ for all $x \in [0,a]$. Since the function $T(a,.) : [0,a] \longrightarrow [0,a]$ satisfies $T(a,0) = 0$ and $T(a,a) = a$, the continuity of T implies that there is some $z \in [0,a]$ with $T(a,z) = x$, leading to

$$T(a,x) = T(a,T(a,z)) = T(T(a,a),z) = T(a,z) = x = \min(a,x),$$

which completes the proof. □

For arbitrary t-norms some general observations concerning idempotent and nilpotent elements and zero divisors can be formulated.

2.4 REMARK

(i) If $a \in [0,1]$ is an idempotent element of a t-norm T then, by induction, we also have $a_T^{(n)} = a$ for all $n \in \mathbb{N}$. In particular, this means that no element of $]0,1[$ can be both idempotent and nilpotent.

(ii) Each nilpotent element a of a t-norm T is also a zero divisor of T (if $n > 1$ is the smallest integer such that $a_T^{(n)} = 0$ then $T\big(a, a_T^{(n-1)}\big) = 0$ with $a_T^{(n-1)} > 0$), but not conversely (e.g., $T^{\mathbf{nM}}$ is a counterexample).

(iii) If a t-norm T has a nilpotent element a then there is always an element $b \in]0,1[$ such that $b_T^{(2)} = 0$. Indeed, if $n > 1$ is the smallest integer such that $a_T^{(n)} = 0$ then $b = a_T^{(n-1)}$ satisfies $b_T^{(2)} = 0$.

(iv) If $a \in]0,1[$ is a nilpotent element (a zero divisor) of a t-norm T then each number $b \in]0,a[$ is also a nilpotent element (a zero divisor) of T.

(v) Therefore, both the set of nilpotent elements and the set of zero divisors of a t-norm T can either be the empty set (as for $T_{\mathbf{M}}$ or $T_{\mathbf{P}}$) or an interval of the form $]0,c[$ or $]0,c]$. For example, for the nilpotent minimum $T^{\mathbf{nM}}$ the set of nilpotent elements is $]0,0.5]$, and the set of zero divisors equals $]0,1[$. For the t-norm T (see [Schweizer & Sklar 1983, Example 5.3.13] and Figure 2.1) given by

$$T(x,y) = \begin{cases} 0 & \text{if } (x,y) \in [0,0.5]^2\,, \\ 2(x-0.5)(y-0.5)+0.5 & \text{if } (x,y) \in]0.5,1]^2\,, \\ \min(x,y) & \text{otherwise,} \end{cases} \tag{2.1}$$

its set of nilpotent elements and its set of zero divisors both equal $]0,0.5]$, and for each element of the family $(T_c)_{c\in]0,1]}$ of t-norms (observe that T_c is just the ordinal sum $(\langle 0,c,T_{\mathbf{L}}\rangle)$, see Definition 3.44) defined by

$$T_c(x,y) = \begin{cases} \max(0, x+y-c) & \text{if } (x,y) \in [0,c]^2, \\ \min(x,y) & \text{otherwise,} \end{cases}$$

the set of nilpotent elements and the set of zero divisors of T_c equal $]0, c[$.

Although the set of nilpotent elements is in general a subset of the set of zero divisors, for each t-norm the existence of zero divisors is equivalent to the existence of nilpotent elements.

2.5 PROPOSITION
For each t-norm T the following are equivalent:

(i) T has zero divisors.
(ii) T has nilpotent elements.

Proof: If T has a zero divisor, i.e., if $T(a,b) = 0$ for some $a > 0$ and $b > 0$, then for $c = \min(a,b) > 0$ we obtain $T(c,c) = 0$, showing that c is a nilpotent element of T. The converse follows directly from Remark 2.4(ii). □

For right-continuous t-norms it is possible to obtain each idempotent element as the limit of the powers of a suitable $x \in [0,1]$ (in fact, the right-continuity of T on the diagonal of the unit square is sufficient).

2.6 PROPOSITION
Let T be a t-norm which is right-continuous on the diagonal $\{(x,x) \mid x \in [0,1]\}$ of the unit square $[0,1]^2$, and let $a \in [0,1]$. The following are equivalent:

(i) a is an idempotent element of T.
(ii) There exists an $x \in [0,1]$ such that $a = \lim\limits_{n\to\infty} x_T^{(n)}$.

Proof: Obviously, for each idempotent element a of T and for each $n \in \mathbb{N}$ we have $a_T^{(n)} = a$, so $a = \lim_{n\to\infty} a_T^{(n)}$.

Conversely, because of the monotonicity (T3), for each $x \in [0,1]$ the sequence $\left(x_T^{(n)}\right)_{n\in\mathbb{N}}$ is non-increasing and, consequently, has some limit $a \in [0,1]$. The right-continuity of T on $\{(x,x) \mid x \in [0,1]\}$ implies

$$T(a,a) = \lim_{n\to\infty} T\left(x_T^{(n)}, x_T^{(n)}\right) = \lim_{n\to\infty} x_T^{(2n)} = a,$$

showing that a is an idempotent element of T. □

A full characterization of the sets of idempotent elements of t-norms can be given as follows:

2.7 THEOREM
For each set I with $\{0,1\} \subseteq I \subseteq [0,1]$ the following are equivalent:

(i) There exists a t-norm T such that I coincides with the set of idempotent elements of T.

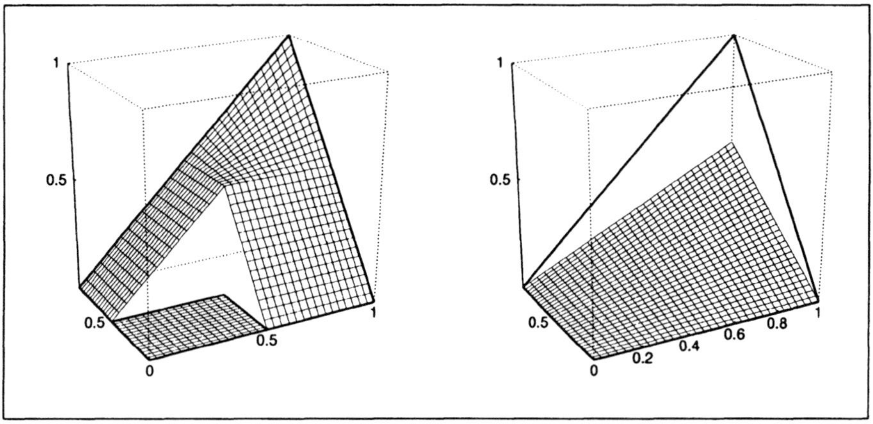

Fig. 2.1. The non-continuous t-norms defined by (2.1) (left) and (2.3) (right)

(ii) There exists a finite or countably infinite index set A and a family $(]a_\alpha, b_\alpha[)_{\alpha\in A}$ of pairwise disjoint open subintervals of $[0,1]$ such that

$$\bigcup_{\alpha\in A}]a_\alpha, b_\alpha[\subseteq [0,1]\setminus I \subseteq \bigcup_{\alpha\in A}]a_\alpha, b_\alpha] . \tag{2.2}$$

Proof: Let T be a t-norm, fix $x_0 \in [0,1]\setminus I$, i.e., $T(x_0, x_0) < x_0$, and put

$$\begin{aligned} a_{x_0} &= \sup\{x \in [0, x_0[\mid T(x,x) = x\}, \\ b_{x_0} &= \inf\{x \in]x_0, 1] \mid T(x,x) = x\}. \end{aligned}$$

Then a_{x_0} is an idempotent element of T because of

$$\begin{aligned} T(a_{x_0}, a_{x_0}) &\geq \sup\{T(x,x) \mid x \in [0, x_0[, T(x,x) = x\} \\ &= \sup\{x \in [0, x_0[\mid T(x,x) = x\} \\ &= a_{x_0}, \end{aligned}$$

and we have $\{x_0\} \subseteq]a_{x_0}, x_0] \subseteq]a_{x_0}, b_{x_0}[\cup \{x_0\} \subseteq [0,1]\setminus I$, showing that

$$\bigcup_{x_0\in[0,1]\setminus I}]a_{x_0}, b_{x_0}[\subseteq [0,1]\setminus I \subseteq \bigcup_{x_0\in[0,1]\setminus I}]a_{x_0}, x_0] \subseteq \bigcup_{x_0\in[0,1]\setminus I}]a_{x_0}, b_{x_0}] .$$

For all $x_0, y_0 \in [0,1]\setminus I$ with $x_0 \neq y_0$, the intervals $]a_{x_0}, b_{x_0}[$ and $]a_{y_0}, b_{y_0}[$ are either disjoint or equal. Each of these intervals is non-empty and, therefore, contains some rational number, which can be used as an index of the corresponding interval. Consequently, the cardinality of the resulting index set A cannot exceed the cardinality of all rational numbers (in $[0,1]$), i.e., A must be a finite or countably infinite subset of $[0,1]\setminus I$ satisfying (2.2).

Conversely, if I is a subset of $[0,1]$ satisfying (2.2), where A is a finite or countably infinite index set and $(]a_\alpha, b_\alpha[)_{\alpha\in A}$ is a family of pairwise disjoint open subintervals of $[0,1]$, this means that $[0,1]\setminus I$ can be represented as a union of a finite or countably infinite family of pairwise disjoint intervals $(J_\alpha)_{\alpha\in A}$, where, for each $\alpha \in A$, either $J_\alpha =]u_\alpha, v_\alpha[$ or $J_\alpha =]u_\alpha, v_\alpha]$ for suitable $u_\alpha, v_\alpha \in [0,1]$, and where $J_\alpha \cup J_\beta$ is not an interval for $\alpha \neq \beta$. Then it is not difficult to check that the function $T : [0,1]^2 \longrightarrow [0,1]$ defined by

$$T(x,y) = \begin{cases} u_\alpha & \text{if } (x,y) \in J_\alpha \times J_\alpha, \\ \min(x,y) & \text{otherwise,} \end{cases}$$

is a t-norm whose set of idempotent elements coincides with I by definition. □

In the case of t-norms which are right-continuous in some specific points of the diagonal of the unit square, the set of idempotent elements of a t-norm can be further specified (obviously, the right-continuity on the diagonal is sufficient for that):

2.8 COROLLARY
Let T be a t-norm such that for each $a \in [0,1[$

$$T(a,a) = a \qquad \textit{whenever} \lim_{x\searrow a} T(x,x) = a.$$

Then the set of idempotent elements of T is a closed subset of $[0,1]$.

Proof: As a consequence of the proof of Theorem 2.7 we get that, for each non-decreasing convergent sequence of idempotent elements of T, also its limit is an idempotent element of T. If $(a_n)_{n\in\mathbb{N}}$ is a non-increasing sequence of idempotent elements of T which converges to $a \in [0,1[$, then

$$a = \lim_{n\to\infty} a_n = \lim_{n\to\infty} T(a_n, a_n)$$

implies $a = T(a,a)$, showing that a is an idempotent element of T too. □

The t-norm T given in (2.1) shows that the converse implication does not necessarily hold in Corollary 2.8 (just consider the case $a = 0.5$).

Some t-norms have additional algebraic properties. The first group of such properties centers around the notions of strict monotonicity and the Archimedean property, which play an important role in many algebraic concepts (e.g., in semigroups, see Section 2.1).

2.9 DEFINITION
For an arbitrary t-norm T we consider the following properties:

(i) The t-norm T is said to be *strictly monotone* if

(SM) $\quad T(x,y) < T(x,z) \quad$ whenever $x > 0$ and $y < z$.

(ii) The t-norm T satisfies the *cancellation law* if

$$\text{(CL)} \qquad T(x,y) = T(x,z) \qquad \text{implies} \quad x = 0 \text{ or } y = z.$$

(iii) The t-norm T satisfies the *conditional cancellation law* if

$$\text{(CCL)} \qquad T(x,y) = T(x,z) > 0 \qquad \text{implies} \quad y = z.$$

(iv) The t-norm T is called *Archimedean* if

$$\text{(AP)} \qquad \text{for each } (x,y) \in]0,1[^2 \text{ there is an } n \in \mathbb{N} \text{ with } x_T^{(n)} < y.$$

(v) The t-norm T has the *limit property* if

$$\text{(LP)} \qquad \text{for all } x \in]0,1[: \quad \lim_{n\to\infty} x_T^{(n)} = 0.$$

2.10 EXAMPLE

(i) The minimum $T_{\mathbf{M}}$ has none of these properties (remember that each $x \in [0,1]$ is an idempotent element of $T_{\mathbf{M}}$), and the product $T_{\mathbf{P}}$ satisfies all of them. The Łukasiewicz t-norm $T_{\mathbf{L}}$ and drastic product $T_{\mathbf{D}}$ are Archimedean and satisfy the conditional cancellation law (CCL) and the limit property (LP), but none of the other properties.
(ii) If a t-norm T satisfies the cancellation law (CL) then it obviously fulfills the conditional cancellation law (CCL), but not conversely (see, e.g., $T_{\mathbf{L}}$).
(iii) The algebraic properties introduced in Definition 2.9 are independent of the continuity: the continuous t-norm $T_{\mathbf{M}}$ shows that continuity implies none of these properties. Conversely, $T_{\mathbf{D}}$ and the non-continuous t-norm T given by (see Figure 2.1)

$$T(x,y) = \begin{cases} \frac{xy}{2} & \text{if } (x,y) \in [0,1[^2, \\ \min(x,y) & \text{otherwise,} \end{cases} \tag{2.3}$$

which is strictly monotone and satisfies the cancellation law (CL), are examples demonstrating that none of the algebraic properties implies the continuity of the t-norm under consideration.

The strict monotonicity of a t-norm is related to the other properties as follows:

2.11 PROPOSITION
Let T be a t-norm. Then we have:

(i) T is strictly monotone if and only if it satisfies the cancellation law (CL).
(ii) If T is strictly monotone then it has only trivial idempotent elements.
(iii) If T is strictly monotone then it has no zero divisors.

Proof: Obviously, the strict monotonicity (SM) of T implies the validity of the cancellation law (CL). Conversely, the strict monotonicity follows from the cancellation law together with the monotonicity (T3), thus proving (i). Assertion (ii) follows immediately from $T(x,x) < T(x,1) = x$ for all $x \in]0,1[$. Concerning (iii), the assumption that $a \in]0,1[$ is a zero divisor, i.e., $T(a,b) = 0$

for some $b \in\,]0,1[$, leads to $T(a, \frac{b}{2}) = T(a,b) = 0$ and, therefore, violates the strict monotonicity of T. □

The Archimedean property (AP) of a t-norm can be characterized in the following way.

2.12 THEOREM
For a t-norm T the following are equivalent:

(i) T is Archimedean.
(ii) T satisfies the limit property (LP).
(iii) T has only trivial idempotent elements and, whenever

$$\lim_{x \searrow x_0} T(x,x) = x_0$$

for some $x_0 \in\,]0,1[$, there exists a $y_0 \in\,]x_0,1[$ such that $T(y_0,y_0) = x_0$.

Proof: In order to show that (i) implies (iii) let T be Archimedean. The assumption that some $a \in\,]0,1[$ is an idempotent element of T implies $a_T^{(n)} = a$ for all $n \in \mathbb{N}$, so T can have only trivial idempotent elements. If $\lim_{x \searrow x_0} T(x,x) = x_0$ for some $x_0 \in\,]0,1[$ and for all $y \in\,]x_0,1[$ we have $T(y,y) > x_0$ then, by induction, we also have $y_T^{(n)} > x_0$ for all $y \in\,]x_0,1[$ and for all $n \in \mathbb{N}$, again violating (AP).

If T satisfies (iii), fix an arbitrary $x \in\,]0,1[$ and put $x_0 = \lim_{n\to\infty} x_T^{(n)}$. Then, because of the monotonicity (T3), we also have $\lim_{y \searrow x_0} T(y,y) = x_0$. If $x_0 > 0$ then there is some $y_0 \in\,]x_0,1[$ such that $T(y_0,y_0) = x_0$ and also $x_T^{(n)} < y_0$ for some $n \in \mathbb{N}$, implying that we must have $x_T^{(2n)} = x_0$ for all sufficiently large $n \in \mathbb{N}$, leading to the contradiction

$$T(x_0,x_0) = x_T^{(4n)} = x_0.$$

Consequently, the only possibility is $x_0 = 0$. Since $x \in\,]0,1[$ was chosen arbitrarily, T satisfies (LP).

Finally, assume that T satisfies (ii) and choose $x, y \in\,]0,1[$. Because of $\lim_{n\to\infty} x_T^{(n)} = 0$, there exists an $n \in \mathbb{N}$ such that $x_T^{(n)} < y$, i.e., T is Archimedean, showing that (ii) implies (i). □

Combining the continuity with some algebraic properties, we obtain two extremely important classes of t-norms.

2.13 DEFINITION
(i) A t-norm T is called *strict* if it is continuous and strictly monotone.
(ii) A t-norm T is called *nilpotent* if it is continuous and if each $a \in\,]0,1[$ is a nilpotent element of T.

2.14 EXAMPLE

(i) The product $T_{\mathbf{P}}$ is a strict t-norm, and the Łukasiewicz t-norm $T_{\mathbf{L}}$ is a nilpotent t-norm. In Corollary 5.7 or, equivalently, in Propositions 5.9 and 5.10 we shall see that each strict t-norm is isomorphic to $T_{\mathbf{P}}$ and that each nilpotent t-norm is isomorphic to $T_{\mathbf{L}}$.

(ii) Because of Proposition 2.11(i), a t-norm T is strict if and only if it is continuous and satisfies the cancellation law (CL).

(iii) As a consequence of (ii), each strict t-norm fulfills the conditional cancellation law (CCL).

(iv) Also, each nilpotent t-norm T satisfies the conditional cancellation law (CCL). To verify this assume $T(x,y) = T(x,z)$ and $y < z$. Then, because of the continuity of T, there must be a $u \in [0,1[$ such that $y = T(z,u)$. Because of the associativity (T2) we get

$$T(x,z) = T(x,y) = T(x,T(z,u)) = T(T(x,z),u)$$

and, by induction, $T(x,z) = T\big(T(x,z), u_T^{(n)}\big)$ for each $n \in \mathbb{N}$. Since T is nilpotent, the only possibility is then $T(x,z) = 0$, i.e., T satisfies (CCL).

The following result gives a number of sufficient conditions for a t-norm to be Archimedean.

2.15 PROPOSITION

For an arbitrary t-norm T we have:

(i) If T is right-continuous and has only trivial idempotent elements then it is Archimedean.

(ii) If T is right-continuous and satisfies the conditional cancellation law (CCL) then it is Archimedean.

(iii) If $\lim_{x \searrow x_0} T(x,x) < x_0$ for each $x_0 \in]0,1[$ then T is Archimedean.

(iv) If T is strict then it is Archimedean.

(v) If each $x \in]0,1[$ is a nilpotent element of T then T is Archimedean.

Proof: If a right-continuous t-norm T has only trivial idempotent elements then Proposition 2.6 implies that T satisfies (LP) and, because of Theorem 2.12, T is Archimedean, thus proving (i).

Concerning (ii), for an arbitrary $x \in]0,1[$ the sequence $\big(x_T^{(n)}\big)_{n\in\mathbb{N}}$ converges to some $a \in [0,1[$ which is an idempotent element of T because of Proposition 2.6. Consequently, we have $T(a,a) = T(a,1)$ and, because of (CCL), $a = 0$, i.e., T has the limit property and, because of Theorem 2.12, is Archimedean.

To show (iii), assume $\lim_{x \searrow x_0} T(x,x) < x_0$ for each $x_0 \in]0,1[$. Then, because of the monotonicity (T3), we have $T(x_0,x_0) \leq \lim_{x \searrow x_0} T(x,x) < x_0$ for all $x_0 \in]0,1[$, and T is Archimedean as a consequence of Theorem 2.12.

Since a strict t-norm is continuous and can have only trivial idempotent elements, (iv) is a special case of (i).

Finally, if each $x \in]0,1[$ is a nilpotent element of T then T satisfies (LP) and is Archimedean. □

In the case of an Archimedean t-norm, left-continuity implies continuity [Kolesárová 1999]:

2.16 PROPOSITION
For each Archimedean t-norm T the following are equivalent:

(i) T is left-continuous.
(ii) T is continuous.

Proof: Trivially, (ii) implies (i). In order to show the converse, assume that T is left-continuous and Archimedean, but not right-continuous in some point $(x_0, y_0) \in]0,1[^2$ (recall that each t-norm is right-continuous on the lower left boundary of $[0,1]^2$). Then fix an arbitrary strictly increasing sequence $(z_n)_{n\in\mathbb{N}}$ in $[0,1]$ with $\lim_{n\to\infty} z_n = 1$. Since T is Archimedean, for each $n \in \mathbb{N}$ there exist numbers $k_n, l_n \in \mathbb{N}$ such that

$$(z_n)_T^{(k_n)} \leq x_0 < (z_n)_T^{(k_n-1)},$$
$$(z_n)_T^{(l_n)} \leq y_0 < (z_n)_T^{(l_n-1)},$$

implying that for all $n \in \mathbb{N}$

$$(z_n)_T^{(k_n+l_n)} \leq T(x_0, y_0) < T(x_0^+, y_0^+) \leq (z_n)_T^{(k_n+l_n-2)}.$$

The left-continuity of T yields $\lim\limits_{n\to\infty} (z_n)_T^{(2)} = 1$ and, consequently,

$$\lim_{n\to\infty} T\big(T(x_0^+, y_0^+), (z_n)_T^{(2)}\big) = T(x_0^+, y_0^+).$$

But then there is some $n \in \mathbb{N}$ such that

$$\begin{aligned}(z_n)_T^{(k_n+l_n)} &\leq T(x_0, y_0)\\ &< T\big(T(x_0^+, y_0^+), (z_n)_T^{(2)}\big)\\ &\leq T\big((z_n)_T^{(k_n+l_n-2)}, (z_n)_T^{(2)}\big)\\ &= (z_n)_T^{(k_n+l_n)},\end{aligned}$$

which is a contradiction. □

All the implications between the algebraic properties of t-norms considered so far are summarized and visualized in Figure 2.2. The following are counterexamples showing that there are no other logical relations between these algebraic properties.

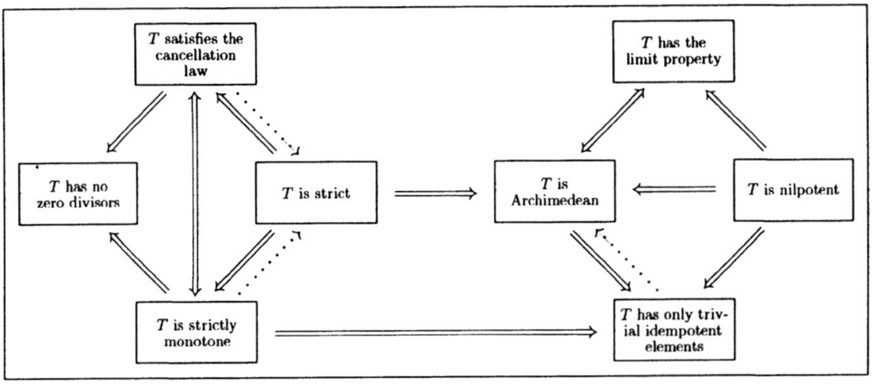

Fig. 2.2. The logical relationship between various algebraic properties of t-norms: a double arrow indicates an implication, a dotted arrow means that the corresponding implication holds for *continuous* t-norms

2.17 EXAMPLE

(i) The Łukasiewicz t-norm $T_{\mathbf{L}}$ shows that an Archimedean t-norm need not be strictly monotone, and that the limit property (LP) does not imply the cancellation law (CL). The product $T_{\mathbf{P}}$ is an example of a continuous Archimedean t-norm without nilpotent elements. The drastic product $T_{\mathbf{D}}$ is an example of a non-continuous Archimedean t-norm for which each $a \in]0,1[$ is a nilpotent element.

(ii) The t-norm T given in (2.3) shows that a strictly monotone t-norm need not be continuous and, subsequently, not necessarily strict.

(iii) In order to see that the non-continuous t-norm T given in (2.1) has only trivial idempotent elements, observe that for $x \in]0, 0.5]$ we have $T(x,x) = 0 < x$, whereas for $x \in]0.5, 1[$ we get $T(x,x) = 2x^2 - 2x + 1$ which is strictly smaller than x on $]0.5, 1[$. Also, T is not strictly monotone since it is constant on $[0, 0.5]^2$. Moreover, it has neither the limit property (LP), since for each $x \in]0.5, 1[$ we have $\lim_{n\to\infty} x_T^{(n)} = 0.5$, nor is it Archimedean (as a consequence of Theorem 2.12). It therefore shows that a t-norm with only trivial idempotent elements is not necessarily strictly monotone or Archimedean.

(iv) A t-norm may satisfy both the strict monotonicity (SM) and the Archimedean property (AP) without being continuous and, subsequently, without being strict. One example for this is the t-norm introduced in (2.3), another t-norm with these features is the following [Budinčević & Kurilić 1998]: recall that each $(x, y) \in]0,1]^2$ is in a one-to-one correspondence with a pair $\big((x_n)_{n\in\mathbb{N}}, (y_n)_{n\in\mathbb{N}}\big)$ of strictly increasing sequences of natural numbers given by the unique infinite dyadic representations

$$x = \sum_{n=1}^{\infty} \frac{1}{2^{x_n}} \qquad \text{and} \qquad y = \sum_{n=1}^{\infty} \frac{1}{2^{y_n}} \tag{2.4}$$

of the numbers x and y, respectively. Using this notion, then the function $T_{+} : [0,1]^2 \longrightarrow [0,1]$ given by

$$T_{+}(x,y) = \begin{cases} \sum\limits_{n=1}^{\infty} \dfrac{1}{2^{x_n+y_n}} & \text{if } (x,y) \in \,]0,1[^2, \\ \min(x,y) & \text{otherwise,} \end{cases}$$

is a t-norm (this will be proved in Example 2.29) which is strictly monotone and Archimedean. Next, let $x, y \in \,]0,1[$ be given as in (2.4). Define the sequence $(z_n)_{n\in\mathbb{N}} \in \,]0,1[^{\mathbb{N}}$ by

$$z_n = \sum_{i=1}^{n-1} \frac{1}{2^{x_i}} + \sum_{i=n}^{\infty} \frac{1}{2^{x_n-n+1+i}}.$$

Obviously, for each $n \in \mathbb{N}$, $z_n < z_{n+1} < x$ and $\sup\{z_n \mid n \in \mathbb{N}\} = x$. A simple computation yields $\sup\{T_{+}(z_n, y) \mid n \in \mathbb{N}\} = T_{+}(x,y)$, implying that, since $x, y \in \,]0,1[$ were chosen arbitrarily, T_{+} is left-continuous on $]0,1[^2$. However, T_{+} is discontinuous in each point $(x,y) \in \,]0,1]^2$ where at least one coordinate is a dyadic rational number (i.e., of the form $\frac{m}{2^n}$ for some $m, n \in \mathbb{N}$ with $m \leq 2^n$; observe that the set of discontinuity points of T_{+} is dense in $[0,1]^2$). Consequently, T_{+} is not strict.

(v) A modification of the t-norm in (iv) yields a t-norm which is strictly monotone but neither Archimedean nor continuous (compare [Smutná 1998b]): keeping the notation of (iv), the function $T_{\diamond} : [0,1]^2 \longrightarrow [0,1]$, which is defined by

$$T_{\diamond}(x,y) = \begin{cases} \sum\limits_{n=1}^{\infty} \dfrac{1}{2^{x_n+y_n-n}} & \text{if } (x,y) \in \,]0,1]^2, \\ 0 & \text{otherwise,} \end{cases}$$

is a t-norm (this will also be proved in Example 2.29) which is, using similar arguments as in (iv), strictly monotone, left-continuous on $[0,1]^2$, but discontinuous in each point $(x,y) \in \,]0,1[^2$ where at least one coordinate is a dyadic rational number. However, $T_{\diamond}$ is not Archimedean, a fact which follows from Theorem 2.12, since for each $x \in \,]0.5,1[$ we have $\lim_{n\to\infty} x_T^{(n)} \geq 0.5$.

(vi) The function $T : [0,1]^2 \longrightarrow [0,1]$ defined by

$$T(x,y) = \begin{cases} xy & \text{if } (x,y) \in [0,0.5]^2, \\ 2(x-0.5)(y-0.5)+0.5 & \text{if } (x,y) \in \,]0.5,1]^2, \\ \min(x,y) & \text{otherwise,} \end{cases}$$

is a t-norm which has only trivial idempotent elements, no zero divisors, is not Archimedean and not strictly monotone.

It turns out that among the continuous Archimedean t-norms there are only two classes: the nilpotent and the strict t-norms. They can be completely characterized by checking whether they have nilpotent or zero elements:

2.18 THEOREM
Let T be a continuous Archimedean t-norm. Then the following are equivalent:

(i) T is nilpotent.
(ii) There exists some nilpotent element of T.
(iii) There exists some zero divisor of T.
(iv) T is not strict.

Proof: It is trivial that (i) implies (ii) and that (ii) implies (iv). From Proposition 2.5 we know that (ii) and (iii) are equivalent.

In order to show that (iv) implies (iii), assume that T is continuous and Archimedean but not strict. Then there are numbers $u, v, w \in [0, 1]$ with $u > 0$ and $v < w$ such that $T(u, v) = T(u, w)$. Since also $T(v, w) \le v < w = T(1, w)$, the continuity of T implies that there exists a number $z \in [v, 1[$ such that $v = T(z, w) = T(w, z)$. Then $T(u, w) = T(u, v) = T(u, T(w, z)) = T(T(u, w), z)$ and, by induction, for each $n \in \mathbb{N}$ we get $T(u, w) = T\big(T(u, w), z_T^{(n)}\big)$. Using the continuity and Theorem 2.12, u and w are zero divisors of T because of

$$T(u, w) = \lim_{n\to\infty} T\big(T(u, w), z_T^{(n)}\big) = T\big(T(u, w), \lim_{n\to\infty} z_T^{(n)}\big) = T(T(u, w), 0) = 0.$$

Finally, to show that (ii) implies (i), assume that $a \in]0, 1[$ is a nilpotent element of T, i.e., $a_T^{(m)} = 0$ for some $m \in \mathbb{N}$. Then for each $x \in]0, 1[$, because T is Archimedean, we have $x_T^{(n)} < a$ for some $n \in \mathbb{N}$ and, as a consequence, $x_T^{(mn)} \le a_T^{(m)} = 0$, showing that T is nilpotent. □

2.19 REMARK

(i) An immediate consequence of Theorem 2.18 and of Example 2.14(iii) and (iv) is that a continuous t-norm is Archimedean if and only if it satisfies the conditional cancellation law (CCL).

(ii) A consequence of Theorem 2.12 is that a t-norm T is Archimedean if and only if it fulfills the limit property (LP), i.e., if for each $x \in]0, 1[$ the sequence $\big(x_T^{(n)}\big)_{n\in\mathbb{N}}$ converges to 0. Note that, e.g., for topological semigroups (see Section 2.2), this is a usual way to define the Archimedean property (see [Mostert & Shields 1957, Carruth *et al.* 1983]).

(iii) From Theorem 2.18 it follows that a continuous t-norm T is strict if and only if for each $x \in]0, 1[$ the sequence $\big(x_T^{(n)}\big)_{n\in\mathbb{N}}$ is strictly decreasing and converges to 0. Again, this is the usual way to define the strictness of topological semigroups.

2.20 REMARK
The strict monotonicity of t-conorms as well as strict, Archimedean and nilpotent t-conorms can be introduced using the dualities (1.9) and (1.10). Without presenting all the technical details, we only mention that it suffices to interchange the words t-norm and t-conorm and the roles of 0 and 1, respectively, in order to obtain the proper definitions and results for t-conorms. For instance, a t-conorm S is strictly monotone if

(SM*) $\quad S(x,y) < S(x,z) \quad$ whenever $x < 1$ and $y < z$.

For the Archimedean property, however, it is necessary to reverse the inequalities involved, so a t-conorm S is Archimedean if

(AP*) $\quad$ for each $(x,y) \in \,]0,1[^2$ there is an $n \in \mathbb{N}$ such that $x_S^{(n)} > y$.

Of course, a t-conorm fulfills any of these properties if and only if the dual t-norm fulfills it.

Finally let us have a brief look at the distributivity of t-norms and t-conorms.

2.21 DEFINITION
Let T be a t-norm and S be a t-conorm. Then we say that T *is distributive over* S if for all $x, y, z \in [0,1]$

$$T(x, S(y,z)) = S(T(x,y), T(x,z)), \tag{2.5}$$

and that S *is distributive over* T if for all $x, y, z \in [0,1]$

$$S(x, T(y,z)) = T(S(x,y), S(x,z)). \tag{2.6}$$

If T is distributive over S and S is distributive over T, then (T, S) is called a *distributive pair* (of t-norms and t-conorms).

In the context of distributivity the minimum $T_{\mathbf{M}}$ and the maximum $S_{\mathbf{M}}$ play a distinguished role (compare also [Bellman & Giertz 1973]).

2.22 PROPOSITION
Let T be a t-norm and S a t-conorm. Then we have:

(i) S is distributive over T if and only if $T = T_{\mathbf{M}}$.
(ii) T is distributive over S if and only if $S = S_{\mathbf{M}}$.
(iii) (T, S) is a distributive pair if and only if $T = T_{\mathbf{M}}$ and $S = S_{\mathbf{M}}$.

Proof: Obviously, each t-conorm is distributive over $T_{\mathbf{M}}$ because of the monotonicity (S3) of the t-conorm. Conversely, if S is distributive over T then for all $x \in [0,1]$ we have $x = S(x, T(0,0)) = T(S(x,0), S(x,0)) = T(x,x)$, and from Proposition 1.9 we obtain $T = T_{\mathbf{M}}$. An analogous argument proves (ii), and (iii) is just the combination of (i) and (ii). □

An immediate consequence of Proposition 2.22 is that the distributive pair $(T_{\mathbf{M}}, S_{\mathbf{M}})$ can also be characterized as follows:

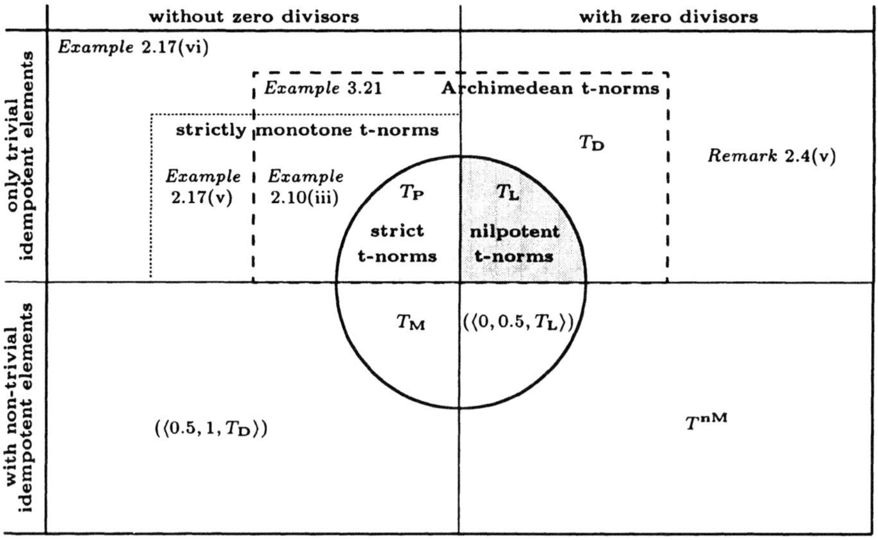

Fig. 2.3. Different classes of t-norms, each of them with a typical representative: within the central circle one finds the continuous t-norms, and the classes of strict and nilpotent t-norms are marked in grey (for the definition of the ordinal sums $(\langle 0, 0.5, T_{\mathbf{L}}\rangle)$ and $(\langle 0.5, 1, T_{\mathbf{D}}\rangle)$ see Definition 3.44)

2.23 REMARK
If T is a t-norm, S the dual t-conorm and if T is distributive over S (or S is distributive over T) then we necessarily have $T = T_{\mathbf{M}}$ and $S = S_{\mathbf{M}}$.

2.2 Semigroups and t-norms

It is clear from the axioms (T1)–(T4) and (S4), respectively, that both triangular norms and conorms turn the unit interval $[0,1]$ into a special semigroup (even a special monoid). In order to study this relationship carefully we first recall the most relevant notions from the theory of abstract semigroups (for more extensive descriptions of semigroups see, e.g., [Clifford & Preston 1961, Fuchs 1963]).

2.24 DEFINITION

(i) Let X be a non-empty set and $*$ a binary operation on X, i.e., a function $* : X^2 \longrightarrow X$. The pair $(X, *)$ is called a *semigroup* if the operation $*$ is associative, i.e., for all $x, y, z \in X$ we have $(x * y) * z = x * (y * z)$.

(ii) A semigroup $(X, *)$ is called *commutative* if $*$ is commutative, i.e., if for all $x, y \in X$ we have $x * y = y * x$.

(iii) An element $e \in X$ is called a *neutral* or *unit element* of the semigroup $(X, *)$ if for each $x \in X$ we have $x * e = e * x = x$. A semigroup which has a neutral element is also called a *monoid.*

(iv) An element $a \in X$ is called an *annihilator* or *zero element* of the semigroup $(X, *)$ if for each $x \in X$ we have $x * a = a * x = a$.

(v) An element $x \in X$ is called an *idempotent element* of the semigroup $(X, *)$ if $x * x = x$.

(vi) If the semigroup $(X, *)$ has an annihilator a, then an element $x \in X \setminus \{a\}$ is called a *nilpotent element* of $(X, *)$ if for some $n \in \mathbb{N}$ we have $x^n = a$, where, by convention, $x^1 = x$ and $x^{n+1} = x * x^n$.

From this definition it is not difficult to derive that a semigroup $(X, *)$ can have at most one neutral element and at most one annihilator. The neutral element and the annihilator, if they exist, are always idempotent elements of $(X, *)$, the so-called *trivial* idempotent elements.

The purely algebraic concept of a semigroup may be connected with two other important structural aspects, namely, with the order and with the geometry, i.e., the topology.

2.25 DEFINITION

(i) Let $\preceq$ be a partial or linear order on X. The triple $(X, *, \preceq)$ is called a *partially ordered semigroup* or *fully ordered semigroup*, respectively, if $(X, *)$ is a semigroup such that $x * y \preceq x * z$ and $y * x \preceq z * x$ whenever $y \preceq z$ (i.e., the operation $*$ is *order preserving* in each component).

(ii) Let $\mathcal{T}$ be a topology on X. The triple $(X, *, \mathcal{T})$ is called a *topological semigroup* if $(X, *)$ is a semigroup such that the function $* : X^2 \longrightarrow X$ is continuous with respect to $\mathcal{T}$.

(iii) A partially or fully ordered semigroup $(X, *, \preceq)$ satisfying that $(X, *, \mathcal{T}_{\preceq})$ is a topological semigroup (where $\mathcal{T}_{\preceq}$ denotes the topology induced by the partial (linear) order $\preceq$) is called a *partially ordered topological semigroup* or *fully ordered topological semigroup*, respectively.

We shall use the term fully ordered topological semigroup mostly in the cases where X is a closed subinterval of the extended real line $[-\infty, \infty]$, equipped with the usual order (for which we shall reserve the symbol $\leq$ exclusively) and the usual topology (observe that the usual topology on $[-\infty, \infty]$ is induced by the usual order $\leq$).

Many of the standard operations in mathematics, such as addition and multiplication of integers or of real or complex numbers, composition of functions or multiplication of matrices, are semigroup operations (indeed, they induce much richer algebraic structures such as groups, fields or vector spaces). Here we give some rather unusual examples of semigroups (the properties claimed are easy to check) which shall be useful later on (see Example 2.29 and Proposition 5.10).

2.26 EXAMPLE

(i) If we define the operation $*$ on $[0.5, 1]$ by $x * y = \max(xy, 0.5)$, then $([0.5,1], *, \leq)$ is a commutative, fully ordered topological semigroup with annihilator 0.5 and neutral element 1.

(ii) Let X be the set of all strictly increasing sequences $(a_n)_{n\in\mathbb{N}} \in \mathbb{N}^{\mathbb{N}}$. Define the inverse lexicographic order $\preceq$ on X by $(a_n)_{n\in\mathbb{N}} \preceq (b_n)_{n\in\mathbb{N}}$ if and only if either $(a_n)_{n\in\mathbb{N}} = (b_n)_{n\in\mathbb{N}}$ or there is an $n_0 \in \mathbb{N}$ with $a_{n_0} > b_{n_0}$ and $a_n = b_n$ for all $n < n_0$. Consider the componentwise addition $+$ and the operation $\diamond$ on X defined by

$$(a_n)_{n\in\mathbb{N}} + (b_n)_{n\in\mathbb{N}} = (a_n + b_n)_{n\in\mathbb{N}},$$
$$(a_n)_{n\in\mathbb{N}} \diamond (b_n)_{n\in\mathbb{N}} = (a_n + b_n - n)_{n\in\mathbb{N}}.$$

Then $(X, +, \preceq)$ and $(X, \diamond, \preceq)$ are both commutative, fully ordered semigroups (neither of which has an annihilator). The first semigroup $(X, +, \preceq)$ has no neutral element, and $(n)_{n\in\mathbb{N}}$ is the neutral element of $(X, \diamond, \preceq)$.

An isomorphism between two (possibly partially ordered or fully ordered or topological) semigroups must, of course, preserve the algebraic structure and, if applicable, also the partial (linear) order and the topological structure.

2.27 DEFINITION

(i) Two semigroups $(X, *)$ and $(Y, \diamond)$, are said to be *isomorphic* if there exists a bijection $\varphi : X \longrightarrow Y$ such that for all $x, y \in X$

$$\varphi(x * y) = \varphi(x) \diamond \varphi(y). \tag{2.7}$$

(ii) Two partially or fully ordered semigroups $(X, *, \preceq)$ and $(Y, \diamond, \sqsubseteq)$ are called *isomorphic* if there exists a bijection $\varphi : X \longrightarrow Y$ which satisfies (2.7) and is also *order preserving*, i.e., for all $x, y \in X$

$$\varphi(x) \sqsubseteq \varphi(y) \quad \text{whenever} \quad x \preceq y.$$

(iii) Two topological semigroups $(X, *, \mathcal{T}_1)$ and $(Y, \diamond, \mathcal{T}_2)$ are called *isomorphic* if there exists a homeomorphism $\varphi : X \longrightarrow Y$ (i.e., a bijection such that both φ and φ^{-1} are continuous) which satisfies (2.7).

(iv) Two partially or fully ordered topological semigroups $(X, *, \preceq)$ and $(Y, \diamond, \sqsubseteq)$ are said to be *isomorphic* if there exists a homeomorphism $\varphi : X \longrightarrow Y$ which is order preserving and satisfies (2.7).

Since we shall mostly deal with semigroups defined on closed subintervals of the extended real line $[-\infty, \infty]$, let us mention that in this context order preserving functions are just non-decreasing functions, and that order preserving bijections are necessarily continuous, strictly increasing functions. To emphasize the strict monotonicity in the latter case, we shall usually, in spite of the redundancy carried by this notion, speak about strictly increasing bijections instead of order preserving bijections. Similarly, order reversing bijections between two real intervals will be called strictly decreasing bijections. In the case of two isomorphic semigroups $(X, *)$ and $(X, \diamond)$ we shall sometimes simply say (if no confusion is possible) that the operations $*$ and $\diamond$ themselves are isomorphic.

Obviously, if $\varphi : X \longrightarrow Y$ is an isomorphism between the semigroups $(X, *)$ and $(Y, \diamond)$ and if e and a are the neutral element and the annihilator of $(X, *)$, respectively, then $\varphi(e)$ and $\varphi(a)$ are the neutral element and the annihilator of $(Y, \diamond)$, respectively. Also, if x is an idempotent or nilpotent element of $(X, *)$, then $\varphi(x)$ is an idempotent or nilpotent element of $(Y, \diamond)$, respectively.

It is now immediately clear that t-norms and t-conorms both define special semigroups, even monoids on the unit interval. In particular, if $X = [0,1]$ and $*$ and $\diamond$ are both t-norms or both t-conorms and if $(X, *)$ and $(X, \diamond)$ are isomorphic as semigroups, then we speak of *isomorphic t-norms* or *isomorphic t-conorms*, respectively.

Also the other assertions of the following proposition are straightforward, so the detailed proof thereof is left as an easy exercise for the interested reader. Observe especially that we can use strictly increasing bijections from the unit interval onto itself in order to construct new t-norms from a given one.

2.28 PROPOSITION

(i) *A function $T : [0,1]^2 \longrightarrow [0,1]$ is a t-norm if and only if the triple $([0,1], T, \leq)$ is a fully ordered commutative semigroup with neutral element* 1 *and annihilator* 0.

(ii) *A function $S : [0,1]^2 \longrightarrow [0,1]$ is a t-conorm if and only if the triple $([0,1], S, \leq)$ is a fully ordered commutative semigroup with neutral element* 0 *and annihilator* 1.

(iii) *If $[a,b] \subseteq [-\infty, \infty]$, if $([a,b], *, \leq)$ is a fully ordered commutative semigroup with neutral element b, and if $\varphi : [0,1] \longrightarrow [a,b]$ is a strictly increasing bijection, then the function $T : [0,1]^2 \longrightarrow [0,1]$ given by*

$$T(x,y) = \varphi^{-1}(\varphi(x) * \varphi(y))$$

is a t-norm. Obviously, the two fully ordered semigroups $([a,b], *, \leq)$ and $([0,1], T, \leq)$ are isomorphic.

(iv) *In particular, if T is a t-norm, S a t-conorm and $\varphi : [0,1] \longrightarrow [0,1]$ a strictly increasing bijection, then the operations $T_\varphi, S_\varphi : [0,1]^2 \longrightarrow [0,1]$ given by the respective formulae*

$$T_\varphi(x,y) = \varphi^{-1}(T(\varphi(x), \varphi(y))), \tag{2.8}$$

$$S_\varphi(x,y) = \varphi^{-1}(S(\varphi(x), \varphi(y))), \tag{2.9}$$

are a t-norm and a t-conorm which are isomorphic to T and S, respectively.

2.29 EXAMPLE

(i) Consider the commutative, fully ordered semigroup $([0.5, 1], *, \leq)$ with annihilator 0.5 and neutral element 1 defined in Example 2.26(i). The Łukasiewicz t-norm $T_{\mathbf{L}}$ and the operation $*$ are isomorphic (the function $\varphi : [0,1] \longrightarrow [0.5,1]$ given by $\varphi(x) = 2^{x-1}$ is an appropriate isomorphism).

(ii) Let $(X, +, \preceq)$ and $(X, \diamond, \preceq)$ be the commutative, fully ordered semigroups described in Example 2.26(ii). Then the function $\varphi : X \longrightarrow]0,1]$ given by

$$\varphi((a_n)_{n\in\mathbb{N}}) = \sum_{n=1}^{\infty} \frac{1}{2^{a_n}}$$

is an order preserving bijection, and for all sequences $(a_n)_{n\in\mathbb{N}}, (b_n)_{n\in\mathbb{N}} \in X$

$$\varphi\big((a_n)_{n\in\mathbb{N}} + (b_n)_{n\in\mathbb{N}}\big) \leq \min\big(\varphi((a_n)_{n\in\mathbb{N}}), \varphi((b_n)_{n\in\mathbb{N}})\big),$$
$$\varphi\big((a_n)_{n\in\mathbb{N}} \diamond (b_n)_{n\in\mathbb{N}}\big) \leq \min\big(\varphi((a_n)_{n\in\mathbb{N}}), \varphi((b_n)_{n\in\mathbb{N}})\big).$$

If, for $x, y \in]0,1]$, we write $\varphi^{-1}(x) = (x_n)_{n\in\mathbb{N}}$ and $\varphi^{-1}(y) = (y_n)_{n\in\mathbb{N}}$, then Proposition 1.6 tells us that the functions $T_+, T_\diamond : [0,1]^2 \longrightarrow [0,1]$ which are defined by

$$T_+(x,y) = \begin{cases} \sum_{n=1}^{\infty} \frac{1}{2^{x_n+y_n}} & \text{if } (x,y) \in]0,1[^2, \\ \min(x,y) & \text{otherwise,} \end{cases} \tag{2.10}$$

$$T_\diamond(x,y) = \begin{cases} \sum_{n=1}^{\infty} \frac{1}{2^{x_n+y_n-n}} & \text{if } (x,y) \in]0,1]^2, \\ 0 & \text{otherwise,} \end{cases} \tag{2.11}$$

are t-norms (actually, they were already mentioned in Example 2.17(iv) and (v)), taking into account that for all $x \in]0,1]$ we have

$$T_\diamond(x,1) = \sum_{n=1}^{\infty} \frac{1}{2^{x_n}} = x.$$

2.30 REMARK

(i) For all strictly increasing bijections $\varphi, \psi : [0,1] \longrightarrow [0,1]$ and for each t-norm T we obtain the following identities:

$$(T_\varphi)_\psi = T_{\varphi\circ\psi},$$
$$(T_\varphi)_{\varphi^{-1}} = (T_{\varphi^{-1}})_\varphi = T.$$

(ii) The construction (2.8) in Proposition 2.28 preserves the continuity, the Archimedean property, and the strictness, as well as the existence of idempotent and nilpotent elements, and the existence of zero divisors.

The only invariants under strictly increasing bijections are the two extremal t-norms $T_{\mathbf{M}}$ and $T_{\mathbf{D}}$:

2.31 PROPOSITION
Let T be a t-norm. The following are equivalent:

(i) *For each strictly increasing bijection $\varphi : [0,1] \longrightarrow [0,1]$ we have $T_\varphi = T$.*
(ii) *$T = T_{\mathbf{M}}$ or $T = T_{\mathbf{D}}$.*

Proof: It is not difficult to check that for each strictly increasing bijection $\varphi : [0,1] \longrightarrow [0,1]$ we have $(T_{\mathbf{M}})_\varphi = T_{\mathbf{M}}$ and $(T_{\mathbf{D}})_\varphi = T_{\mathbf{D}}$. Conversely, let T be a t-norm with $T_\varphi = T$ for each strictly increasing bijection $\varphi : [0,1] \longrightarrow [0,1]$. If $T(x,x) = x$ for all $x \in [0,1]$ then from Proposition 1.9(i) we know that $T = T_{\mathbf{M}}$. Assume $T(x_0,x_0) < x_0$ for some $x_0 \in \,]0,1[$. Now fix an arbitrary element $a \in \,]0,1[$ and let $\varphi_a, \psi_a : [0,1] \longrightarrow [0,1]$ be the two strictly increasing bijections defined by

$$\varphi_a(x) = \begin{cases} \frac{a}{x_0} x & \text{if } x \in [0,x_0], \\ a + \frac{1-a}{1-x_0}(x - x_0) & \text{otherwise,} \end{cases}$$

$$\psi_a(x) = \begin{cases} \frac{a}{\sqrt{x_0}} \sqrt{x} & \text{if } x \in [0,x_0], \\ a + \frac{1-a}{1-x_0}(x - x_0) & \text{otherwise.} \end{cases}$$

Then from $T_{\varphi_a} = T = T_{\psi_a}$ we obtain

$$\tfrac{T(a,a)}{a} \cdot x_0 = T_{\varphi_a}(x_0,x_0) = T_{\psi_a}(x_0,x_0) = \left(\tfrac{T(a,a)}{a}\right)^2 \cdot x_0 = T(x_0,x_0) < x_0,$$

which is only possible if $T(a,a) = 0$. Since a can be chosen arbitrarily in $]0,1[$, Proposition 1.9(ii) implies $T = T_{\mathbf{D}}$. □

2.32 PROPOSITION
Let T be a t-norm.

(i) *If $a \in [0,1[$ is an idempotent element of T, if $b \in \,]a,1]$ and if the function $\varphi : [a,b[\longrightarrow [0,1[$ is a strictly increasing bijection then the following function $T^* : [0,1]^2 \longrightarrow [0,1]$ is a t-norm:*

$$T^*(x,y) = \begin{cases} \varphi(T(\varphi^{-1}(x), \varphi^{-1}(y))) & \text{if } (x,y) \in [0,1[^2, \\ \min(x,y) & \text{otherwise.} \end{cases}$$

(ii) *If $a,b \in [0,1]$ are idempotent elements of T with $a < b$ and such that for all $x \in [a,b]$ we have $T(x,b) = x$, and if $\varphi : [a,b] \longrightarrow [0,1]$ is a strictly increasing bijection then the function $T_{ab} : [0,1]^2 \longrightarrow [0,1]$ is a t-norm:*

$$T_{ab}(x,y) = \varphi(T(\varphi^{-1}(x), \varphi^{-1}(y))).$$

Proof: If $a \in [0,1[$ is an idempotent element of T and $b \in]a,1]$, then, because for all $(x,y) \in [a,b[^2$ we have $a = T(a,a) \leq T(x,y) \leq x < b$, the restriction $T|_{[a,b[^2}$ is a commutative, associative and monotone operation on $[a,b[$. Consequently, for each strictly increasing bijection $\varphi : [a,b[\longrightarrow [0,1[$ the operation $* : [0,1[^2 \longrightarrow [0,1[$ defined by $x * y = \varphi(T(\varphi^{-1}(x), \varphi^{-1}(y)))$ is a commutative, associative and monotone operation on $[0,1[$ satisfying $x * y \leq \min(x,y)$, and (i) follows immediately from Proposition 1.6.

If the hypotheses of (ii) are satisfied, then for each $x \in [0,1]$ we obtain $T_{ab}(x,1) = \varphi(T(\varphi^{-1}(x), b)) = \varphi(\varphi^{-1}(x)) = x$, so (ii) follows from (i). □

When talking about fully ordered semigroups, we have only considered strictly increasing bijections so far. Strictly decreasing bijections, on the other hand, lead to a concept of duality:

2.33 DEFINITION

Let $(X, *, \preceq)$ and $(Y, \diamond, \sqsubseteq)$ be fully ordered semigroups and $\varphi : X \longrightarrow Y$ a strictly decreasing bijection. Then $(X, *, \preceq)$ is said to be φ*-dual* to $(Y, \diamond, \sqsubseteq)$ if for all $x, y \in X$ we have $\varphi(x * y) = \varphi(x) \diamond \varphi(y)$.

It is immediately clear that $(X, *, \preceq)$ is φ-dual to $(Y, \diamond, \sqsubseteq)$ if and only if $(Y, \diamond, \sqsubseteq)$ is φ^{-1}-dual to $(X, *, \preceq)$. Observe also that if $(X, *, \preceq)$ is φ*-dual* to $(Y, \diamond, \sqsubseteq)$ and if we ignore the orders $\preceq$ and $\sqsubseteq$, then $(X, *)$ and $(Y, \diamond)$ are isomorphic.

The concept of duality allows us to construct t-conorms from t-norms and vice versa, as is summarized in the following statement, whose proof is obvious:

2.34 PROPOSITION

Let $F : [0,1]^2 \longrightarrow [0,1]$ be a function such that the triple $([0,1], F, \leq)$ is a fully ordered semigroup, let $\varphi : [0,1] \longrightarrow [0,1]$ be a strictly decreasing bijection, and define the operation F_φ on $[0,1]$ given by

$$F_\varphi(x,y) = \varphi^{-1}(F(\varphi(x), \varphi(y))). \tag{2.12}$$

Then $([0,1], F_\varphi, \leq)$ is a fully ordered semigroup. Moreover we have:

(i) If F is a t-norm then F_φ is a t-conorm which is φ-dual to F.
(ii) If F is a t-conorm then F_φ is a t-norm which is φ-dual to F.

2.35 REMARK

(i) For an arbitrary strictly decreasing bijection $\varphi : [0,1] \longrightarrow [0,1]$ we have $(T_\mathbf{M})_\varphi = S_\mathbf{M}$, $(S_\mathbf{M})_\varphi = T_\mathbf{M}$, $(T_\mathbf{D})_\varphi = S_\mathbf{D}$ and $(S_\mathbf{D})_\varphi = T_\mathbf{D}$.
(ii) For all strictly decreasing bijections $\varphi, \psi : [0,1] \longrightarrow [0,1]$ and for an arbitrary t-norm T we have again the following identities:

$$(T_\varphi)_\psi = T_{\varphi \circ \psi},$$
$$(T_\varphi)_{\varphi^{-1}} = (T_{\varphi^{-1}})_\varphi = T.$$

(iii) The construction (2.12) in Proposition 2.34 preserves the continuity, the Archimedean property as well as the strictness of the original semigroup $([0,1], F, \leq)$.

Starting from a t-norm T and a strictly decreasing bijection $\varphi : [0,1] \longrightarrow [0,1]$, $(T_\varphi)_\varphi$ is always a t-norm which is isomorphic to T ($\varphi \circ \varphi$ is a strictly increasing bijection). In general, however, the construction in (2.12) is not involutive, i.e., we do not always have $(T_\varphi)_\varphi = T$:

2.36 EXAMPLE

(i) Consider the strictly decreasing bijection $\varphi : [0,1] \longrightarrow [0,1]$ defined by $\varphi(x) = 1 - x^2$, then we obtain

$$(T_{\mathbf{P}})_\varphi(x,y) = \sqrt{x^2 + y^2 - x^2y^2},$$

$$((T_{\mathbf{P}})_\varphi)_\varphi(x,y) = \sqrt{1 - \sqrt{1 - 4x^2y^2 + 2x^4y^2 + 2x^2y^4 - x^4y^4}},$$

i.e., $T_{\mathbf{P}} \neq ((T_{\mathbf{P}})_\varphi)_\varphi$ (however, these two t-norms are isomorphic).

(ii) In general, if $\varphi : [0,1] \longrightarrow [0,1]$ is a strictly decreasing bijection, if T is a t-norm, and if we write $T_1 = T_\varphi$, $T_{n+1} = (T_n)_\varphi$ and $T_{n-1} = (T_n)_{\varphi^{-1}}$, then we get for each $n \in \mathbb{Z}$ and for all $(x,y) \in [0,1]^2$

$$T_n(x,y) = \varphi^{-n}\big(T(\varphi^n(x), \varphi^n(y))\big),$$

where $\varphi^0 = \mathrm{id}_{[0,1]}$, $\varphi^{n+1} = \varphi \circ \varphi^n$ and $\varphi^{-n} = (\varphi^n)^{-1}$. Clearly, T_n is a t-norm if n is an even integer, and a t-conorm if n is odd.

2.37 DEFINITION

Let $[a,b]$ be a closed subinterval of $[-\infty, \infty]$. An *involution* (on $[a,b]$) is a non-increasing function $\varphi : [a,b] \longrightarrow [a,b]$ such that $\varphi \circ \varphi = \mathrm{id}_{[a,b]}$.

It is evident that each involution is a bijection and, consequently, strictly decreasing. A full characterization of involutions on $[0,1]$ was given in [Trillas 1979] (for a short proof see [Nguyen & Walker 1997]), a result which can be immediately generalized to arbitrary closed subintervals of $[-\infty, \infty]$.

2.38 PROPOSITION

Let $[a,b]$ be a closed subinterval of $[-\infty,\infty]$ and $\varphi : [a,b] \longrightarrow [a,b]$ a function. The following are equivalent:

(i) φ is an involution.

(ii) There is a monotone bijection $g : [a,b] \longrightarrow [0,1]$ such that for all $x \in [a,b]$

$$\varphi(x) = g^{-1}(1 - g(x)). \tag{2.13}$$

Proof: Obviously, for each monotone bijection $g : [a,b] \longrightarrow [0,1]$ construction (2.13) yields an involution. Conversely, given a bounded closed interval $[a,b] \subseteq \mathbb{R}$ and an involution $\varphi : [a,b] \longrightarrow [a,b]$, define the strictly increasing bijection $g : [a,b] \longrightarrow [0,1]$ by

$$g(x) = \frac{1}{2}\Big(1 - \frac{\varphi(x) - x}{b - a}\Big).$$

Then it is easily checked that equality (2.13) holds for all $x \in [a,b]$. For an unbounded interval $[a,\infty]$ with $-\infty < a$ and an involution $\varphi : [a,\infty] \longrightarrow [a,\infty]$ define the strictly increasing bijection $g : [a,\infty] \longrightarrow [0,1]$ by

$$g(x) = \frac{x - a}{x + \varphi(x) - 2a}.$$

Again it is easy to see that we have (2.13) for all $x \in [a,\infty]$. The other cases of unbounded intervals are completely similar. □

2.39 REMARK
It is immediately clear that, for an involution $\varphi : [0,1] \to [0,1]$ and for the monotone bijection $g : [0,1] \to [0,1]$ (constructed as in the proof of Proposition 2.38), condition (2.13) can be rewritten as $g \circ \varphi = 1 - g$, which is equivalent to $g + g \circ \varphi = 1$. This means that each strictly increasing bijection $g^* : [0,x_0] \longrightarrow [0,0.5]$, where $x_0 \in \,]0,1[$ is the unique fixed point of φ, can be extended to a monotone bijection $g : [0,1] \longrightarrow [0,1]$ satisfying (2.13) via

$$g(x) = \begin{cases} g^*(x) & \text{if } x \in [0,x_0]\,, \\ 1 - g^*(\varphi(x)) & \text{if } x \in \,]x_0,1]\,. \end{cases}$$

2.3 Topological semigroups and continuous t-norms

Obviously, continuous t-norms and t-conorms are always operations in topological semigroups (for a basic reference see, e.g., [Carruth *et al.* 1983]). Since we shall always work with closed subintervals of the extended real line $[-\infty,\infty]$, equipped with the usual topology, we do not need to mention the topology explicitly.

We shall see that the operation in the following special topological semigroup (first studied in [Faucett 1955]), based on a closed interval and equipped with an additional property (but without requiring commutativity and monotonicity) is isomorphic to either a continuous t-norm or t-conorm.

2.40 DEFINITION
A topological semigroup $([a,b],*)$, where $[a,b]$ is a closed subinterval of the extended real line $[-\infty,\infty]$ with $a < b$, such that one of the boundary points of $[a,b]$ acts as a neutral element and the other one as an annihilator, is called an *I-semigroup*.

A non-trivial example of an I-semigroup is given by $([0.5, 1], *)$ as defined in Example 2.26(i). Of course, also each continuous t-norm T and each continuous t-conorm S induce I-semigroups. It will be more surprising that each I-semigroup operation is isomorphic to a t-norm or a t-conorm (see Theorem 2.43 below).

Clearly, if $([a, b], \diamond)$ is an I-semigroup, and if we consider the order preserving homeomorphism $\varphi : [a, b] \longrightarrow [0, 1]$ given by

$$\varphi(x) = \frac{\arctan(x) - \arctan(a)}{\arctan(b) - \arctan(a)}$$

and the operation $*$ on $[0, 1]$ defined by

$$x * y = \varphi(\varphi^{-1}(x) \diamond \varphi^{-1}(y))$$

then the I-semigroups $([a, b], \diamond)$ and $([0, 1], *)$ are obviously isomorphic. Therefore, we can restrict ourselves to the case $[a, b] = [0, 1]$.

First we show that even a set of slightly weaker axioms leads to an I-semigroup.

2.41 PROPOSITION

Let $([0, 1], *)$ *be a topological semigroup with annihilator* 0 *such that* $1 * 1 = 1$. *Then* $([0, 1], *)$ *is an* I*-semigroup with annihilator* 0 *and neutral element* 1.

Proof: In order to see that 1 is a neutral element, fix an $x \in [0, 1]$ and observe $0 * 1 = 0$ and $1 * 1 = 1$. Since $*$ is continuous there is some $x_0 \in [0, 1]$ with $x_0 * 1 = x$, and we get

$$x * 1 = (x_0 * 1) * 1 = x_0 * (1 * 1) = x_0 * 1 = x.$$

Analogously we have $1 * x = x$, and $([0, 1], *)$ is an I-semigroup. □

Now we are able to show that each I-semigroup is necessarily a fully ordered semigroup.

2.42 PROPOSITION

If $([0, 1], *)$ *is an* I*-semigroup then* $([0, 1], *, \leq)$ *is a fully ordered semigroup.*

Proof: Without loss of generality, we assume that 0 is the annihilator of the I-semigroup $([0, 1], *)$. First we show that $x * y \leq \min(x, y)$ for all $(x, y) \in [0, 1]^2$. Fix an element $x \in]0, 1[$ and consider the set

$$M_x = \{z \in [0, 1] \mid \text{for all } (u, v) \in [0, z] \times [0, 1] : u * v \leq x\}.$$

Because of $0 \in M_x$ this set is non-empty, so we have $z_0 = \sup M_x \in M_x$ as a consequence of the continuity of $*$. In particular, $z_0 * v \leq x$ for all $v \in [0, 1]$.

In order to show that there exists a number $v_0 \in [0, 1]$ with $z_0 * v_0 = x$, assume $z_0 * v < x$ for all $v \in [0, 1]$. Then the uniform continuity of $*$ implies that there is some $z > z_0$ such that $u * v < x$ for all $(u, v) \in [0, z] \times [0, 1]$, i.e., $z \in M_x$, contradicting the fact that z_0 is the least upper bound of M_x.

As x was chosen arbitrarily in $]0, 1[$, for all $x \in]0, 1[$ and for all $y \in [0, 1]$ we get $x * y = (z_0 * v_0) * y = z_0 * (v_0 * y) \leq x$. Since $x * y \leq x$ is obviously satisfied for $x \in \{0, 1\}$ and since $x * y \leq y$ can be derived in complete analogy, this proves $x * y \leq \min(x, y)$ for all $(x, y) \in [0, 1]^2$.

To show the monotonicity of $*$, choose $(x, y) \in [0, 1]^2$ with $x \leq y$. Then, because of $0 * y = 0$ and $1 * y = y$ the continuity of $*$ implies that there is some $x_0 \in [0, 1]$ such that $x_0 * y = x$. Then, since $*$ is bounded from above by min, for all $z \in [0, 1]$ we obtain $x * z = (x_0 * y) * z = x_0 * (y * z) \leq (y * z)$. The monotonicity in the second component is shown in complete analogy. □

This result tells us that an I-semigroup $([0, 1], *)$ already satisfies three of the four properties of a continuous triangular norm or conorm: associativity (T2), monotonicity (T3), and the boundary conditions (T4) or (S4). In the following we prove that $*$ is necessarily commutative (a result which essentially goes back to [Hölder 1901]), thus showing that each I-semigroup operation $*$ on the unit interval $[0, 1]$ is indeed either a continuous t-norm or t-conorm.

2.43 THEOREM
Let $*$ be a binary operation on $[0, 1]$. The following are equivalent:

(i) $([0, 1], *)$ is an I-semigroup.
(ii) The operation $*$ is either a continuous t-norm or a continuous t-conorm.

Proof: Obviously (ii) implies (i). For the converse, without loss of generality, let $([0, 1], *)$ be an I-semigroup with annihilator 0. From Proposition 2.42 we know that $*$ is a continuous operation satisfying (T2)–(T4), so the only missing part is the proof of the commutativity of $*$ which will be carried out in two steps.
Assume first that $([0, 1], *)$ is a non-commutative I-semigroup, i.e., there are elements $x_0, y_0 \in]0, 1[$ with, say, $x_0 < y_0$ such that $x_0 * y_0 < y_0 * x_0$ (the case $x_0 * y_0 > y_0 * x_0$ is completely analogous). Further, assume that $*$ is Archimedean, i.e., for all $x, y \in]0, 1[$ there exists an $n \in \mathbb{N}$ such that $x^n < y$.

Case 1.1: $x_0 * y_0 > 0$.
Since $*$ is continuous, there are an $x \in]0, 1[$ with $x_0 * y_0 = y_0 * x_0 * x$ and a $z \in]0, 1[$ such that $z * z \geq x$ and $z \geq y_0$ (for an arbitrary $u \in]x, 1[$ the continuity of $*$ implies the existence of a $v \in]0, 1[$ such that $x = u * v$; then we may choose $z = \max(u, v, y_0)$). Since $*$ is Archimedean, there exist integers $m, n \in \mathbb{N}$ such that

$$z^m \geq x_0 > z^{m+1} \qquad \text{and} \qquad z^n \geq y_0 > z^{n+1}. \tag{2.14}$$

Next we show that in an Archimedean I-semigroup a restricted cancellation law always holds in the sense that if we have both $0 < a*b = a*c$ and $0 < b*a = c*a$ then necessarily $b = c$. Assume the contrary, i.e., that there exist numbers

$a, b, c \in]0,1[$ such that, say, $b < c$ and $0 < a * b = a * c$, and put $v = a * b > 0$. Then (again by the continuity of $*$) there exists a $u \in]0,1[$ such that $b = c * u$, from which we get $v = a * b = a * c * u = v * u$ and, by induction, $v = v * u^n$ for all $n \in \mathbb{N}$. Because of $v \neq 0$, for any $w \in]0, v[$ there exists an integer $k \in \mathbb{N}$ such that $u^k \leq w$, and we obtain the contradiction $v = v * u^k \leq v * w \leq w < v$, showing that the restricted cancellation law holds. This, together with (2.14) yields the contradiction $x_0 * y_0 = y_0 * x_0 * x \leq z^{m+n+2} < x_0 * y_0$, showing that the assumption $x_0 * y_0 < y_0 * x_0$ is wrong in this case.

Case 1.2: $x_0 * y_0 = 0$.
Since $*$ is Archimedean, there is an integer $k \in \mathbb{N}$ such that ${y_0}^k \geq x_0 > {y_0}^{k+1}$. Then, because of the continuity of $*$, there exists a number $z \in [x_0, 1]$ such that $x_0 = {y_0}^k * z$, thus leading to

$$y_0 * z \geq {y_0}^k * z = x_0 > 0.$$

Because of $y_0 * z > 0$, we have $z * y_0 = y_0 * z$ as a consequence of Case 1.1. Subsequently we obtain

$$y_0 * x_0 = y_0 * {y_0}^k * z = {y_0}^k * y_0 * z = {y_0}^k * z * y_0 = x_0 * y_0 = 0,$$

showing $x_0 * y_0 = y_0 * x_0$ also in this case, and completing the proof of the commutativity in the case of an Archimedean I-semigroup.

In the second step, let $([0,1], *)$ be an arbitrary, i.e., not necessarily Archimedean, I-semigroup with annihilator 0 and neutral element 1, and fix two elements $x_0, y_0 \in]0,1[$ with, say, $x_0 < y_0$. This time we have to distinguish four cases.

Case 2.1: x_0 is an idempotent element of $([0,1], *)$.
In this case we get $\min(x_0, y_0) = x_0 = x_0 * x_0 \leq x_0 * y_0 \leq \min(x_0, y_0)$ and, analogously, $y_0 * x_0 = \min(y_0, x_0)$.

Case 2.2: y_0 is an idempotent element of $([0,1], *)$.
Because of the continuity of the operation $*$ there exist elements $u, v \in]0,1[$ such that $x_0 = u * y_0 = y_0 * v$. Then we obtain

$$x_0 * y_0 = u * y_0 * y_0 = u * y_0 = x_0 = \min(x_0, y_0),$$
$$y_0 * x_0 = y_0 * y_0 * v = y_0 * v = x_0 = \min(x_0, y_0).$$

Case 2.3: There is an idempotent element $z_0 \in]x_0, y_0[$ of $([0,1], *)$.
We get $x_0 = \min(x_0, z_0) = x_0 * z_0 \leq x_0 * y_0 \leq \min(x_0, y_0) = x_0$ because of cases 2.1 and 2.2 and, analogously, $y_0 * x_0 = \min(x_0, y_0)$.

Case 2.4: There is no idempotent element of $([0,1], *)$ in $[x_0, y_0]$.
For the sake of brevity, put

$$a = \inf\{z \in]y_0, 1] \mid z * z = z\},$$
$$b = \sup\{z \in [0, x_0[\mid z * z = z\}.$$

Then $0 \leq a < x_0 < y_0 < b \leq 1$, and for all $u, v \in [a, b]$ we get $u * v \in [a, b]$ because of $a = a * a \leq u * v \leq b * b = b$. Then $([a, b], *|_{[a,b]^2})$ is an Archimedean I-semigroup which is isomorphic to some Archimedean I-semigroup on the unit interval which was shown to be commutative in the first step of this proof. Subsequently, we obtain also in this case $x_0 * y_0 = y_0 * x_0$, thus completing the proof. □

This means that, for arbitrary I-semigroups, we get the following immediate consequence of Theorem 2.43:

2.44 COROLLARY
The following are equivalent:

(i) $([a, b], *)$ *is an* I*-semigroup.*
(ii) The operation $*$ *is isomorphic either to a continuous t-norm or to a continuous t-conorm.*

2.4 Lattice-ordered monoids and left-continuous t-norms

Throughout this section, let L be a non-empty set. Firstly, we always assume that $(L, \preceq)$ is a lattice, i.e., a partially ordered set which, for any two elements $x, y \in L$, also contains their join $x \vee y$ (i.e., the least upper bound of the set $\{x, y\}$) and their meet $x \wedge y$ (i.e., the greatest lower bound of $\{x, y\}$). Secondly, we always assume that $(L, *)$ is a semigroup with neutral element. Following [Birkhoff 1973, Höhle 1995] we introduce the subsequent notions:

2.45 DEFINITION
Let $(L, \preceq)$ be a lattice and $(L, *)$ a semigroup with neutral element.

(i) The triple $(L, *, \preceq)$ is called a *lattice-ordered monoid* (or an *l-monoid*) if for all $x, y, z \in L$ we have
(LM1) $x * (y \vee z) = (x * y) \vee (x * z)$,
(LM2) $(x \vee y) * z = (x * z) \vee (y * z)$.
(ii) An l-monoid $(L, *, \preceq)$ is said to be *commutative* if the semigroup $(L, *)$ is commutative.
(iii) A commutative l-monoid $(L, *, \preceq)$ is called a *commutative, residuated l-monoid* if there exists a further binary operation $\to_*$ on L, i.e., a function $\to_*: L^2 \longrightarrow L$ (the *$*$-residuum*), such that for all $x, y, z \in L$ we have
(R) $x * y \preceq z$ if and only if $x \preceq y \to_* z$.
(iv) An l-monoid $(L, *, \preceq)$ is called *integral* if there is a greatest element in the lattice $(L, \preceq)$ (often called the universal upper bound) which coincides with the neutral element of the semigroup $(L, *)$.

Obviously, each l-monoid $(L, *, \preceq)$ is a partially ordered semigroup, and in the case of commutativity the axioms (LM1) and (LM2) are equivalent.

Since we shall concentrate on the lattice $([0,1], \leq)$, we will usually work with a complete lattice, i.e., for each subset A of L its join $\bigvee A$ and its meet $\bigwedge A$ exist and are contained in L. In this case, L always has a greatest element (also called the universal upper bound).

Let us give some examples of l-monoids, some of which will play a role also in Sections 7.3 and 10.2.

2.46 Example

(i) If, for some $n \in \mathbb{N}$, $\Gamma_{(n)} = \{0, \frac{1}{n}, \frac{2}{n}, \ldots, \frac{n-1}{n}, 1\}$ then $(\Gamma_{(n)}, *, \leq)$, where $* : \Gamma_{(n)}^2 \longrightarrow \Gamma_{(n)}$ is given by

$$\frac{i}{n} * \frac{j}{n} = \frac{1}{n} \max(i + j - n, 0),$$

is a commutative, residuated integral l-monoid, and a simple computation verifies that the $*$-residuum $\to_*$ is given by

$$\frac{i}{n} \to_* \frac{j}{n} = \frac{1}{n} \min(n - i + j, n).$$

In Example 7.23(ii) we shall see that $*$ is a special discrete t-norm, namely, $D_{\mathbf{L}}^{(n)}$. Note that, if L is a finite set, then each lattice $(L, \preceq)$ is complete and each l-monoid $(L, *, \preceq)$ is necessarily residuated.

(ii) If we define $* : [0,1]^2 \longrightarrow [0,1]$ by

$$x * y = \begin{cases} \min(x, y) & \text{if } x + y \leq 1, \\ \max(x, y) & \text{otherwise,} \end{cases}$$

then $([0,1], *, \leq)$ is a commutative, residuated l-monoid (which is not integral since the neutral element is 0.5), and the $*$-residuum $\to_*$ is given by

$$x \to_* y = \begin{cases} \max(1 - x, y) & \text{if } x \leq y, \\ \min(1 - x, y) & \text{otherwise.} \end{cases}$$

The operation $*$ turns out to be a so-called uninorm (see Section 10.2 and [De Baets & Fodor 1999]).

(iii) The triple $(]-\infty, 0]^2, *, \preceq)$, where the binary operation $*$ on $]-\infty, 0]^2$ is the pointwise addition, i.e., $(x_1, x_2) * (y_1, y_2) = (x_1 + y_1, x_2 + y_2)$, and where the partial order $\preceq$ on $]-\infty, 0]^2$ is the pointwise extension of the usual order $\leq$ on $]-\infty, 0]$, i.e.,

$$(x_1, x_2) \preceq (y_1, y_2) \quad \text{if and only if} \quad x_1 \leq y_1 \text{ and } x_2 \leq y_2,$$

is a commutative, residuated integral l-monoid (observe that the lattice $(]-\infty, 0]^2, \preceq)$ is not complete), and the $*$-residuum $\to_*$ is given by

$$(x_1, x_2) \to_* (y_1, y_2) = (\min(y_1 - x_1, 0), \min(y_2 - x_2, 0)).$$

(iv) The triple $([0,1], T_{\mathbf{D}}, \leq)$ is a commutative integral l-monoid which is not residuated. Indeed, the validity of (R) implies that for all $y \in [0,1[$ and for all $z \in [0,1]$ we would have $y \rightarrow_{T_{\mathbf{D}}} z = 1$ because of $T_{\mathbf{D}}(x,y) = 0$ for all $(x,y) \in [0,1[^2$. However, if $z < y$, this contradicts $T_{\mathbf{D}}(1,y) = y > z$.

In each commutative, residuated l-monoid $(L, *, \preceq)$, the $*$-residuum $\rightarrow_*$ is uniquely determined by the formula

$$x \rightarrow_* y = \bigvee \{z \in L \mid x * z \preceq y\}, \tag{2.15}$$

where the join exists in L and is the greatest element of the set in question. Indeed, from (R) it follows that the $*$-residuum $x \rightarrow_* y$ is an upper bound of the set $\{z \in L \mid x * z \preceq y\}$ which is also an element of this set, therefore coinciding with its join.

Moreover, in a commutative, residuated l-monoid $(L, *, \preceq)$, the semigroup operation $*$ can be reconstructed from the $*$-residuum $\rightarrow_*$ by

$$x * y = \bigwedge \{z \in L \mid x \preceq y \rightarrow_* z\}. \tag{2.16}$$

We only mention that, for a commutative l-monoid $(L, *, \preceq)$ equipped with an additional binary operation $\rightarrow_*$, the validity of (R) for all $x, y, z \in L$ is equivalent to the simultaneous validity of (2.15) and (2.16) for all $x, y \in L$ (this follows from the fact that the relationship between $*$ and $\rightarrow_*$ is based on the Galois connection [Gierz *et al.* 1980]).

Turning back to t-norms, it is evident that $([0,1], T, \leq)$ is a commutative integral l-monoid if and only if the function $T : [0,1]^2 \longrightarrow [0,1]$ is a t-norm. It turns out that the the left-continuity of a t-norm can be characterized by the fact that the corresponding l-monoid is residuated.

2.47 PROPOSITION
For each function $T : [0,1]^2 \longrightarrow [0,1]$ the following are equivalent:

(i) $([0,1], T, \leq)$ is a commutative, residuated integral l-monoid.
(ii) T is a left-continuous t-norm.

In this case, the T-residuum $\rightarrow_T$ is given by

$$x \rightarrow_T y = \sup\{z \in [0,1] \mid T(x,z) \leq y\}. \tag{2.17}$$

Proof: We already mentioned that $([0,1], T, \leq)$ is a commutative integral l-monoid if and only if T is a t-norm. Therefore, in order to prove that (i) implies (ii), assume that $([0,1], T, \leq)$ is residuated, fix an arbitrary sequence $(x_n)_{n\in\mathbb{N}}$ in $[0,1]$ and an element $y_0 \in [0,1]$, and put $x_0 = \sup\{x_n \mid n \in \mathbb{N}\}$ and $z_0 = \sup\{T(x_n, y_0) \mid n \in \mathbb{N}\}$. Obviously, $z_0 \leq T(x_0, y_0)$, and (R) implies $y_0 \rightarrow_T z_0 \geq x_n$ for all $n \in \mathbb{N}$ and, subsequently, $y_0 \rightarrow_T z_0 \geq x_0$. Applying again (R), we obtain $T(x_0, y_0) \leq z_0$ and, because of the monotonicity (T3), $T(x_0, y_0) = z_0$, i.e.,

$$T(\sup\{x_n \mid n \in \mathbb{N}\}, y_0) = \sup\{T(x_n, y_0) \mid n \in \mathbb{N}\}.$$

Now the left-continuity of T follows from Proposition 1.22.

Conversely, if the t-norm T is left-continuous, define the operation $\rightarrow_T$ by (2.17). Then clearly, for all $x, y, z \in [0,1]$, $x \leq y \rightarrow_T$ whenever $T(x,y) \leq z$. The left-continuity of T then implies $T(y \rightarrow_T z, y) \leq z$ which, together with the monotonicity (T3), ensures that $\rightarrow_T$ is indeed the T-residuum. □

Because of its interpretation in $[0,1]$-valued logics (it can be considered as an extension of the Boolean implication, see Section 11.1), the T-residuum $\rightarrow_T$ is also called a *residual implication* (or briefly an *R-implication*, compare also [Pedrycz 1982, 1983, 1989], [Dubois & Prade 1985], [Gottwald 1989, 200x], [Fodor 1991a, 1993a], and [Kruse *et al.* 1994a]). Observe that for each left-continuous t-norm T we have $x \rightarrow_T y = 1$ if and only if $x \leq y$.

2.48 Example

For the left-continuous basic t-norms $T_{\mathbf{M}}$, $T_{\mathbf{P}}$ and $T_{\mathbf{L}}$ the corresponding residuum (we shall write briefly $\rightarrow_{\mathbf{M}}$, $\rightarrow_{\mathbf{P}}$ and $\rightarrow_{\mathbf{L}}$ instead of $\rightarrow_{T_{\mathbf{M}}}$, $\rightarrow_{T_{\mathbf{P}}}$ and $\rightarrow_{T_{\mathbf{L}}}$, respectively) is given by:

$$x \rightarrow_{\mathbf{M}} y = \begin{cases} 1 & \text{if } x \leq y, \\ y & \text{otherwise,} \end{cases} \qquad \text{(\textit{Gödel implication})}$$

$$x \rightarrow_{\mathbf{P}} y = \begin{cases} 1 & \text{if } x \leq y, \\ \frac{y}{x} & \text{otherwise,} \end{cases} \qquad \text{(\textit{Goguen implication})}$$

$$x \rightarrow_{\mathbf{L}} y = \min(1 - x + y, 1). \qquad \text{(\textit{Łukasiewicz implication})}$$

The fact that the formula of $\rightarrow_{\mathbf{L}}$ coincides with that of the implication introduced in [Łukasiewicz 1930] within the framework of a three-valued logic, justifies calling $T_{\mathbf{L}}$ and $S_{\mathbf{L}}$ the Łukasiewicz t-norm and t-conorm, respectively, although these operations do not appear explicitly in the work of Jan Łukasiewicz.

We would like to emphasize that formula (2.17) can be computed for arbitrary t-norms T but that, as pointed out in Proposition 2.47, the resulting operation $\rightarrow_T$ equals the T-residuum, i.e., satisfies axiom (R), only if the t-norm T is left-continuous (compare also [Fodor 1991], [Gottwald 1989, 200x], and [Kruse *et al.* 1994a]).

Finally, let us have a look at a notion which, on the one hand, provides a link to the lattice operations $\wedge$ and $\vee$ and, on the other hand, will turn out to have a close relationship to the continuity of the underlying semigroup operation $*$.

2.49 Definition

A commutative integral l-monoid $(L, *, \preceq)$ is called *divisible* if

(DLM) For all $x, y \in L$ with $x \preceq y$ there is a $z \in L$ such that $y * z = x$.

We only mention that, as was shown in [Höhle 1995, Lemma 2.5], a commutative, residuated integral l-monoid $(L, *, \preceq)$ is divisible if and only if the meet $\wedge$ in the lattice $(L, \preceq)$ can be expressed in terms of $*$ and $\rightarrow_*$ as follows:

$$x \wedge y = x * (x \rightarrow_* y), \tag{2.18}$$

As an immediate consequence of Proposition 2.47 and the Intermediate Value Theorem we see that $([0,1], T, \leq)$ is a commutative, residuated, divisible integral l-monoid if and only if T is a continuous t-norm, in which case we also have for all $(x, y) \in [0,1]^2$

$$x \vee y = ((x \rightarrow_T y) \rightarrow_T y) \wedge ((y \rightarrow_T x) \rightarrow_T x). \tag{2.19}$$

3. Construction of t-norms

> But the fullest constructional beauty is the *introspective beauty of mathematics,* where instead of elements of playful causal acting, the basic intuition of mathematics is left to free unfolding. This unfolding is not bound to the exterior world, and thereby to finiteness and responsibility; consequently its introspective harmonies can attain any degree of richness and clearness.
>
> L. E. J. BROUWER

Generalizing the inverse of a bijective function, we consider the pseudo-inverse of monotone functions from $[0, 1]$ to $[0, 1]$. Introducing a construction similar to the one given in [Schweizer & Sklar 1983, Theorem 5.2.1] with the help of the so-called quasi-inverses, we state a rather general method to construct new t-norms from known t-norms using the pseudo-inverses of appropriate functions (compare also [Jenei 1999, Klement *et al.* 1999]). A number of examples (together with visualizations of prototypical cases) for this construction is provided.

An old problem going back to [Abel 1826] (compare also [Schweizer & Sklar 1961, 1963]) is whether there are constructions involving only a one-place real function and the usual addition (or multiplication), which induce two-place real functions (i.e., operations on the real line) having nice algebraic properties, in particular, the associativity (T2). In Section 3.2 we give the definition of additive generators of t-norms [Schweizer & Sklar 1961, 1963], thus providing a most powerful tool for the construction of new t-norms, and we discuss the relationship between the continuity of the additive generators and the continuity of the corresponding t-norms. Most importantly, we show that continuous additive generators always lead to continuous Archimedean t-norms, and that the value of the generator in the point 0 determines whether the t-norm is strict or nilpotent. In Sections 5.1 and 5.2 we shall even see that a t-norm is continuous Archimedean if and only if it has a continuous additive generator. We also mention that there exist additive generators for the drastic product $T_{\mathbf{D}}$ (and for other non-continuous t-norms), but none for the minimum $T_{\mathbf{M}}$ [Arnold 1957, Ling 1965] (or for any other t-norm with a non-trivial idempotent element [Krause 1983]). Finally, we demonstrate the complete duality between additive and multiplicative generators, on the one hand, and between generators of t-norms and t-conorms, on the other hand.

Another construction of new t-norms from a family of given t-norms is based on the so-called ordinal sum of a family of semigroups [Climescu 1946, Clifford 1954, Mostert & Shields 1957, Clifford & Preston 1961, Fuchs 1963, Schweizer & Sklar 1963, Ling 1965, Frank 1979]. In the t-norm case, this construction means that we can glue together arbitrary (appropriately transformed) t-norms along the diagonal of the unit square (and take the minimum otherwise), and that this ordinal sum always is a t-norm with (except for trivial cases) non-trivial idempotent elements, i.e., a non-Archimedean t-norm. Again, several examples and illustrations of this general construction are given. In Section 5.3 we shall see that each continuous t-norm can be represented in a unique way as an ordinal sum of continuous Archimedean t-norms.

Finally, in Section 3.4 we consider some methods which, in general, will produce non-continuous t-norms, most of which are used in a more theoretical context when it comes to build counterexamples.

As a curiosity, we end this chapter with a t-norm which, as a function from the unit square into the unit interval, is not even Borel measurable.

3.1 Pseudo-inverses of monotone functions

The construction (2.8) as described in Proposition 2.28(iv) uses the inverse of the function $\varphi : [0,1] \longrightarrow [0,1]$ and, therefore, requires φ to be bijective. In order to relax this strong requirement of bijectivity and to replace it by the weaker monotonicity, we first recall some general properties of monotone functions.

If $f : [a,b] \longrightarrow [c,d]$, where $[a,b]$ and $[c,d]$ are closed subintervals of the extended real line $[-\infty,\infty]$, is a monotone (i.e., non-decreasing or non-increasing) function, then it is well-known that the set of discontinuity points of f is a countable (i.e., a finite or countably infinite) subset of $[a,b]$.

The following result will be crucial for the proper definition of pseudo-inverse of monotone functions. The range $\mathrm{Ran}(f)$ of a function $f : X \longrightarrow Y$ is, as usual, the set $\{f(x) \mid x \in X\}$. Note also that, whenever a subset of $[a,b]$ happens to be empty, its supremum equals the lower bound a.

3.1 Proposition

Let $f : [a,b] \longrightarrow [c,d]$ with $[a,b],[c,d] \subseteq [-\infty,\infty]$ be a non-constant monotone function. Then for each $y \in [c,d] \setminus \mathrm{Ran}(f)$ we have

$$\sup\{x \in [a,b] \mid (f(x)-y)\cdot(f(b)-f(a)) < 0\} = \inf\{x \in [a,b] \mid (f(x)-y)\cdot(f(b)-f(a)) > 0\}.$$

Proof: Assume, without loss of generality, that $f : [a,b] \longrightarrow [c,d]$ is non-decreasing. Then the assertion $(f(x) - y)(f(b) - f(a)) < 0$ is equivalent to $f(x) < y$. Choose $y \in [f(a), f(b)] \setminus \text{Ran}(f)$ and denote

$$\begin{aligned} u &= \sup\{x \in [a,b] \mid f(x) < y\}, \\ v &= \inf\{x \in [a,b] \mid f(x) > y\}. \end{aligned}$$

Clearly we have $u \leq v$, and the assumption $u < v$ would imply $f(x) = y$ for all $x \in]u,v[$, thus contradicting the fact $y \notin \text{Ran}(f)$. □

This allows us to introduce the following generalization of an inverse of monotone functions (including constant functions).

3.2 DEFINITION

Let $[a,b]$ and $[c,d]$ be two closed subintervals of the extended real line $[-\infty,\infty]$, and let $f : [a,b] \longrightarrow [c,d]$ be a monotone function. Then the *pseudo-inverse* $f^{(-1)} : [c,d] \longrightarrow [a,b]$ is defined by

$$f^{(-1)}(y) = \sup\left\{x \in [a,b] \mid (f(x) - y)(f(b) - f(a)) < 0\right\}.$$

From Definition 3.2 we immediately get the following formulae for the pseudo-inverse of monotone functions.

3.3 COROLLARY

Let $f : [a,b] \longrightarrow [c,d]$ be a monotone function, where $[a,b]$ and $[c,d]$ are closed subintervals of the extended real line $[-\infty,\infty]$.

(i) If $f(a) < f(b)$, i.e., if f is non-decreasing and non-constant, then for all $y \in [c,d]$ we obtain the simpler formula

$$f^{(-1)}(y) = \sup\{x \in [a,b] \mid f(x) < y\}. \tag{3.1}$$

Moreover, the function $f^{(-1)}$ is non-decreasing and left-continuous, and for each $y \in [c, f(a)]$ we have $f^{(-1)}(y) = a$, and for each $y \in]f(b), d]$ we get $f^{(-1)}(y) = b$. Also, $(f^{(-1)})^{(-1)} = f$ if and only if f is left-continuous and if $f(a) = c$.

(ii) If $f(a) > f(b)$, i.e., if f is non-increasing and non-constant, then for all $y \in [c,d]$ we obtain the simpler formula

$$f^{(-1)}(y) = \sup\{x \in [a,b] \mid f(x) > y\}. \tag{3.2}$$

Moreover, the function $f^{(-1)}$ is non-increasing and right-continuous, and for each $y \in [c, f(b)[$ we have $f^{(-1)}(y) = b$, and for each $y \in [f(a), d]$ we get $f^{(-1)}(y) = a$. Also, $(f^{(-1)})^{(-1)} = f$ if and only if f is right-continuous and if $f(b) = c$.

(iii) If $f(a) = f(b)$, i.e., if f is a constant function, then for all $y \in [c,d]$ we have $f^{(-1)}(y) = a$.

A simple illustration of the pseudo-inverse is given in Figure 3.1 which also indicates how to construct the graph of the pseudo-inverse $f^{(-1)}$ of a non-constant monotone function $f: [a,b] \longrightarrow [c,d]$:

- Draw vertical line segments at discontinuities of f.
- Reflect the graph of f at the first median, i.e., at the graph of the identity function $\mathrm{id}_{[-\infty,\infty]}$.
- Remove any vertical line segments from the reflected graph except for their lowest points.

In the following we list some properties of pseudo-inverses of monotone functions (see [Sherwood & Taylor 1974, Klement *et al.* 1999]).

3.4 Remark

Let $f: [a,b] \longrightarrow [c,d]$ be a monotone function, where $[a,b]$ and $[c,d]$ are closed subintervals of the extended real line $[-\infty,\infty]$, and let $f^{(-1)}$ be its pseudo-inverse.

(i) Obviously, the pseudo-inverse $f^{(-1)}$ of f coincides with the inverse function f^{-1} of f if and only if f is a bijection.

(ii) The pseudo-inverse $f^{(-1)}$ of f is continuous if and only if f is strictly monotone on the set $f^{(-1)}([c,d[)$.

(iii) We always have the inequality

$$f^{(-1)} \circ f \leq \mathrm{id}_{[a,b]}. \tag{3.3}$$

Indeed, for all $u \in [a,b]$ we get

$$\begin{aligned} f^{(-1)} \circ f(u) &= \sup\{x \in [a,b] \mid (f(x) - f(u)) \cdot (f(b) - f(a)) < 0\} \\ &\leq \sup\{x \in [a,b] \mid x < u\} \\ &= u. \end{aligned}$$

(iv) The equality $f \circ f^{(-1)} \circ f = f$ holds if and only if $f(x_0) = \lim_{x \searrow x_0} f(x)$ whenever there are numbers $x_0 \in [a,b[$ and $\varepsilon > 0$ such that the restriction $f|_{]x_0,x_0+\varepsilon]}$ is a constant function ([Viceník 1999c]).

(v) In particular, if f is either right-continuous or strictly monotone then $f \circ f^{(-1)} \circ f = f$ or, equivalently, $f \circ f^{(-1)}|_{\mathrm{Ran}(f)} = \mathrm{id}_{\mathrm{Ran}(f)}$.

(vi) If f is strictly monotone then the restriction of $f^{(-1)}$ to $\mathrm{Ran}(f)$, i.e., the function $f^{(-1)}|_{\mathrm{Ran}(f)} : \mathrm{Ran}(f) \longrightarrow [a,b]$, is also strictly monotone. Moreover, in this case we have the following identities:

$$\begin{aligned} f \circ f^{(-1)}|_{\mathrm{Ran}(f)} &= \mathrm{id}_{\mathrm{Ran}(f)}, \\ f^{(-1)} \circ f &= \mathrm{id}_{[a,b]}. \end{aligned}$$

(vii) If f is surjective then we even have $f \circ f^{(-1)} = \mathrm{id}_{[c,d]}$.

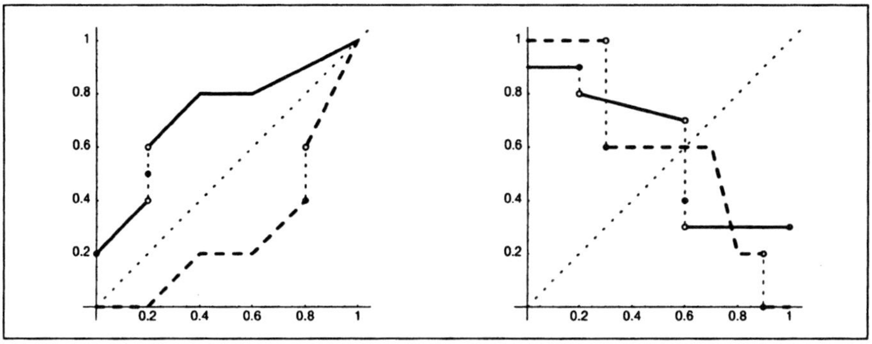

Fig. 3.1. A non-decreasing function (left) and a non-increasing function (right) together with their respective pseudo-inverses (dashed graphs)

(viii) If $\varphi : [a,b] \longrightarrow [a,b]$ and $\psi : [c,d] \longrightarrow [c,d]$ are monotone bijections then

$$(f \circ \varphi)^{(-1)} = \varphi^{-1} \circ f^{(-1)},$$

$$(\psi \circ f)^{(-1)} = f^{(-1)} \circ \psi^{-1}.$$

An interesting application of the pseudo-inverse is the n-th root with respect to a continuous t-norm and, subsequently, powers with rational exponents:

3.5 Remark

For a continuous t-norm T and for each $n \in \mathbb{N}$ the function $f : [0,1] \longrightarrow [0,1]$ defined by $f(x) = x_T^{(n)}$ is continuous, non-decreasing and surjective. Its pseudo-inverse $f^{(-1)}$ can be seen as the n-th root with respect to T, leading to the following notation and explicit formula:

$$x_T^{(\frac{1}{n})} = \sup\{z \in [0,1] \mid z_T^{(n)} < x\}. \tag{3.4}$$

As an immediate consequence of Remark 3.4(vii) we get $(x_T^{(\frac{1}{n})})_T^{(n)} = x$ for all $x \in [0,1]$. If T is a continuous Archimedean t-norm then

$$\lim_{n \to \infty} x_T^{(\frac{1}{n})} = \begin{cases} 0 & \text{if } x = 0, \\ 1 & \text{if } x \in \,]0,1]. \end{cases} \tag{3.5}$$

By duality, for a continuous t-conorm S, the n-th root of x is given by

$$x_S^{(\frac{1}{n})} = \inf\{z \in [0,1] \mid z_S^{(n)} > x\}. \tag{3.6}$$

Putting for $x \in [0,1]$ and $m, n \in \mathbb{N}$

$$x_T^{(\frac{m}{n})} = \bigl(x_T^{(\frac{1}{n})}\bigr)_T^{(m)}, \tag{3.7}$$

it is not hard to see that for all $x \in [0,1]$ and $k, m, n \in \mathbb{N}$ we have $x_T^{(\frac{km}{kn})} = x_T^{(\frac{m}{n})}$, i.e., the function $g : [0,1] \times \mathbb{Q} \longrightarrow [0,1]$ given by $g(x,r) = x_T^{(r)}$ is indeed well-defined. Moreover, for all $x \in [0,1]$ and for all $r, s \in \mathbb{Q}$ we have

$$x_T^{(r+s)} = T\bigl(x_T^{(r)}, x_T^{(s)}\bigr). \tag{3.8}$$

The pseudo-inverse of a non-decreasing function $f : [0,1] \longrightarrow [0,1]$ can be used for a fairly general construction of t-norms. The following construction is similar to the construction in [Schweizer & Sklar 1983, Theorem 5.2.1] with the help of so-called *quasi-inverses*. The construction in [Schweizer & Sklar 1983], when starting from a t-norm and restricting ourselves to the interval $[0,1]$, yields a commutative, fully ordered semigroup on $[0,1]$ which is not necessarily a t-norm (it may happen that the semigroup operation is not bounded from above by the minimum $T_{\mathbf{M}}$; for an extensive treatment of this problem see [Jenei 1999, Klement *et al.* 1999]).

The main idea is to impose additional conditions which ensure that the semigroup operation satisfies (1.6) on the half-open square $[0,1[^2$ and to use Proposition 1.6 to produce a t-norm.

3.6 THEOREM

Let $f : [0,1] \longrightarrow [0,1]$ be a non-decreasing function and T a t-norm such that

$$T(f(x), f(y)) \in \mathrm{Ran}(f) \cup [0, f(0^+)], \tag{3.9}$$

for all $x, y \in [0,1[$ and, for all $(x,y) \in [0,1]^2$ with $T(f(x), f(y)) \in Ran(f)$,

$$f \circ f^{(-1)}(T(f(x), f(y))) = T(f(x), f(y)). \tag{3.10}$$

Then the following function $T_{[f]} : [0,1]^2 \longrightarrow [0,1]$ is a t-norm:

$$T_{[f]}(x,y) = \begin{cases} f^{(-1)}(T(f(x), f(y))) & \text{if } (x,y) \in [0,1[^2, \\ \min(x,y) & \text{otherwise.} \end{cases} \tag{3.11}$$

Proof: The commutativity of $T_{[f]}$ and the existence of the neutral element 1 are obvious. The monotonicity of $T_{[f]}$ on the half-open square $[0,1[^2$ follows from the monotonicity of T, f and $f^{(-1)}$. Further, (3.3) yields for each $(x,y) \in [0,1[^2$

$$\begin{aligned} T_{[f]}(x,y) &= f^{(-1)}(T(f(x), f(y))) \\ &\leq f^{(-1)}(\min(f(x), f(y)) \\ &= \min(f^{(-1)} \circ f(x), f^{(-1)} \circ f(y)) \\ &\leq \min(x,y). \end{aligned}$$

Concerning the associativity of $T_{[f]}$, it suffices to show it on the open unit square $]0,1[^2$ because of Proposition 1.6. Choose $x, y, z \in]0,1[$ and assume that both $T(f(x), f(y))$ and $T(f(y), f(z))$ belong to $\mathrm{Ran}(f)$. Then due to (3.10) we get

$$\begin{aligned} T_{[f]}(T_{[f]}(x,y), z) &= f^{(-1)}(T(f \circ f^{(-1)}(T(f(x), f(y))), f(z))) \\ &= f^{(-1)}(T(T(f(x), f(y)), f(z))) \\ &= f^{(-1)}(T(f(x), T(f(y), f(z)))) \\ &= f^{(-1)}(T(f(x), f \circ f^{(-1)}(T(f(y), f(z))))) \\ &= T_{[f]}(x, T_{[f]}(y,z)). \end{aligned}$$

If $T(f(x), f(y)) \leq f(0^+)$, then $T_{[f]}(x, y) = 0$ and, consequently,

$$0 \leq T_{[f]}(x, T_{[f]}(y, z)) \leq T_{[f]}(x, y) = 0 = T_{[f]}(T_{[f]}(x, y), z).$$

The case $T(f(y), f(z)) \leq f(0^+)$ is analogous, hence $T_{[f]}$ is also associative. □

From the proof of Theorem 3.6 we can immediately see that construction (3.11) leads to a t-subnorm (as introduced in Definition 1.7).

3.7 COROLLARY
Let $f : [0,1] \longrightarrow [0,1]$ be a non-decreasing function and T a t-norm such that (3.9) holds for all $x, y \in [0,1[$ and (3.10) holds for all $(x,y) \in [0,1]^2$ with $T(f(x), f(y)) \in Ran(f)$. Then the function $F : [0,1]^2 \longrightarrow [0,1]$ given by

$$F(x, y) = f^{(-1)}(T(f(x), f(y)))$$

is a t-subnorm.

3.8 REMARK

(i) If we have $f(0) < f(0^+)$ for the function $f : [0,1] \longrightarrow [0,1]$ in Theorem 3.6, then we may replace the function f by the non-decreasing function $h : [0,1] \longrightarrow [0,1]$ given by

$$h(x) = \begin{cases} f(0^+) & \text{if } x = 0, \\ f(x) & \text{if } x \in \,]0,1], \end{cases}$$

and we get $T_{[f]} = T_{[h]}$. This means that the actual value $f(0)$ plays no actual role: we may always assume that f is right-continuous in 0, in which case the condition (3.9) is simplified to

$$T(f(x), f(y)) \in \mathrm{Ran}(f) \cup [0, f(0)[\,. \tag{3.12}$$

(ii) In particular, if for the function $f : [0,1] \longrightarrow [0,1]$ in Theorem 3.6 the range of its restriction $f|_{]0,1[}$ is an interval, then this theorem can be applied to an arbitrary t-norm.

(iii) If $f, g : [0,1] \longrightarrow [0,1]$ are two non-decreasing functions and if T is a t-norm, then (provided all the expressions exist) we have

$$(T_{[f]})_{[g]} = T_{[f \circ g]}.$$

This means in particular that if f and $T_{[f]}$ are known then the original T can be reconstructed whenever $f \circ g = \mathrm{id}_{[0,1]}$, e.g., if $\mathrm{Ran}(f) = [0,1]$ and $g = f^{(-1)}$, in which case we get

$$T = (T_{[f]})_{[f^{(-1)}]}.$$

(iv) The minimum $T_{\mathbf{M}}$ and the drastic product $T_{\mathbf{D}}$ are no longer invariant under the construction of Theorem 3.6 (see Figure 3.2).

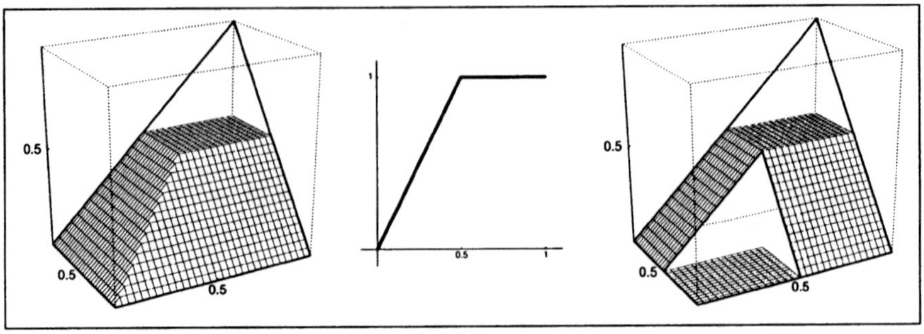

Fig. 3.2. The t-norms $(T_{\mathbf{M}})_{[f_{0.5}]}$ (left) and $(T_{\mathbf{D}})_{[f_{0.5}]}$ (right) induced by the function $f_{0.5}$ (center), illustrating Example 3.11 and the claim of Remark 3.8(iv) (observe also that these are examples for a kind of *dilatation* [Demant 1994, Mesiar 1995b])

(v) If $f : [0,1] \longrightarrow [0,1]$ is a constant function then, for an arbitrary t-norm T, formula (3.11) always yields the drastic product $T_{\mathbf{D}}$.

The following result is a direct consequence of Remark 3.4(v) and Theorem 3.6.

3.9 COROLLARY
Let $f : [0,1] \longrightarrow [0,1]$ be a non-decreasing function and T a t-norm such that for all $x, y \in [0,1[$ we have (3.9), i.e., $T(f(x), f(y)) \in \operatorname{Ran}(f) \cup [0, f(0^+)]$. If f is either a right-continuous or a strictly monotone function then the function $T_{[f]} : [0,1]^2 \longrightarrow [0,1]$ given by (3.11) is a triangular norm.

Obviously, the conditions in Corollary 3.9 are automatically satisfied if the non-decreasing function f is continuous.

3.10 COROLLARY
Let $f : [0,1] \longrightarrow [0,1]$ be a non-decreasing continuous function. Then, for each t-norm T, the function $T_{[f]} : [0,1]^2 \longrightarrow [0,1]$ given by (3.11) is a t-norm.

The general result of Theorem 3.6 can be applied in many special cases which cover a wide range of constructions of t-norms mentioned in the literature. In the following series of examples we shall restrict ourselves to the case of piecewise linear functions, the obvious modifications for general non-decreasing functions are left to the reader.

3.11 EXAMPLE
Let T be an arbitrary t-norm, fix an element $c \in]0,1[$, and consider the function $f_c : [0,1] \longrightarrow [0,1]$ which is given by

$$f_c(x) = \min(\tfrac{x}{c}, 1)$$

(i.e., $f_c|_{[0,c]}$ is a strictly increasing linear bijection from $[0,c]$ to $[0,1]$, so the construction of Theorem 3.6 is applicable). Then $T_{[f_c]}$ is given by

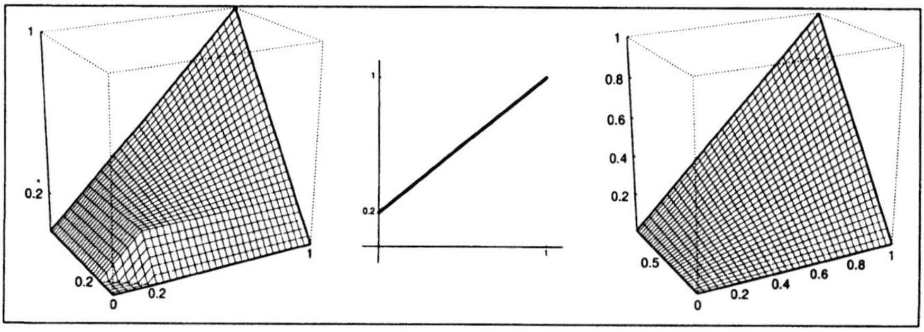

Fig. 3.3. The right-hand picture shows the linear truncation $T_{f_{0.2}} = T_{\mathbf{P}}$ (which is strict) (see Example 3.12) induced by the ordinal sum (see Definition 3.44) $T = (\langle 0, 0.2, T_{\mathbf{L}}\rangle, \langle 0.2, 1, T_{\mathbf{P}}\rangle)$ (left, which has nilpotent elements) and the function $f_{0.2}$ (center)

$$T_{[f_c]}(x,y) = \begin{cases} c \cdot T\left(\frac{x}{c}, \frac{y}{c}\right) & \text{if } (x,y) \in [0,c[^2, \\ c & \text{if } (x,y) \in [c,1[^2, \\ \min(x,y) & \text{otherwise.} \end{cases}$$

It turns out that $T_{[f_c]}$ is always a non-trivial ordinal sum (see Definition 3.44). In particular, the t-norm $T_{[f_c]}$ is neither strictly monotone nor Archimedean (obviously, c is always a non-trivial idempotent element) nor continuous (see Figure 3.2). The t-norm $T_{[f_c]}$ has zero divisors if and only if the original t-norm T has zero divisors.

3.12 EXAMPLE
Let T be an arbitrary t-norm, fix an element $c \in \,]0,1[$, and consider the function $f_c : [0,1] \longrightarrow [0,1]$ which is given by

$$f_c(x) = c + (1-c)x$$

(i.e., f_c is a strictly increasing linear bijection from $[0,1]$ to $[c,1]$, so the construction of Theorem 3.6 is applicable). Then a simple computation shows that the resulting t-norm $T_{[f_c]}$ is given by

$$T_{[f_c]}(x,y) = \max\left(\tfrac{T(c+(1-c)x,\, c+(1-c)y) - c}{1-c}, 0\right).$$

From a geometric point of view, $T_{[f_c]}$ can be viewed as a *linear truncation* of the original t-norm T. If T is continuous, Archimedean or nilpotent, so is $T_{[f_c]}$. However, if the original t-norm T is Archimedean then $T_{[f_c]}$ always has zero divisors and, subsequently, is not strict. Moreover, it is interesting to note that each t-norm can be viewed as a linear truncation: starting, e.g., with an ordinal sum T (see Definition 3.44) which contains the summand $\langle c, 1, T^*\rangle$ then we obtain $T_{[f_c]} = T^*$ (this fact is visualized in Figure 3.3).

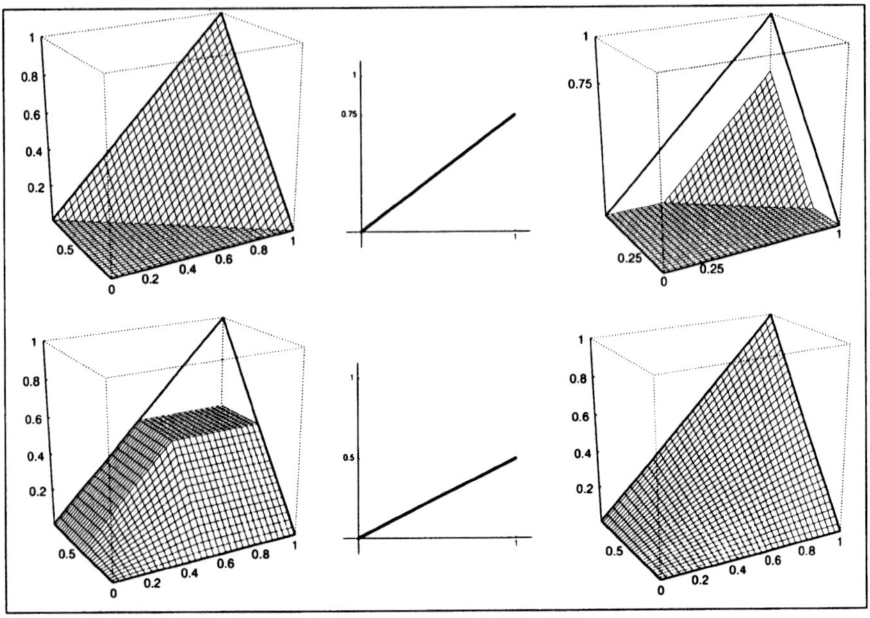

Fig. 3.4. The non-continuous linear contraction (top right, see Example 3.13) induced by the continuous t-norm $T_{\mathbf{L}}$ (top left) and the function $f_{0.75}$ (top center), and the continuous linear contraction $T_{\mathbf{P}}$ (bottom right) induced by the non-continuous ordinal sum $(\langle 0, 0.5, T_{\mathbf{P}}\rangle, \langle 0.5, 1, T_{\mathbf{D}}\rangle)$ (bottom left, see Definition 3.44) and the function $f_{0.5}$

3.13 EXAMPLE

Let T be an arbitrary t-norm, fix an element $c \in]0,1[$, and consider the function $f_c : [0,1] \longrightarrow [0,1]$ which is defined by

$$f_c(x) = cx$$

(i.e., f_c is a strictly increasing linear bijection from $[0,1]$ to $[0,c]$, so the construction of Theorem 3.6 is applicable). Then $T_{[f_c]}$, obviously given by

$$T_{[f_c]}(x,y) = \begin{cases} \dfrac{1}{c} \cdot T(cx, cy) & \text{if } (x,y) \in [0,1[^2, \\ \min(x,y) & \text{otherwise,} \end{cases}$$

can be viewed as a *linear contraction* (see also [Demant 1994]) in the following sense: it takes the values of T on the half-open square $[0,c[^2$ (they are always bounded by c) and enlarges them proportionally to the half-open square $[0,1[^2$ (see Figure 3.4). Hence the values of T of arguments outside of $[0,c[^2$ have no influence on the resulting contraction $T_{[f_c]}$.

If the original t-norm T is Archimedean, strictly monotone, or continuous on the half-open square $[0,1[^2$, then $T_{[f_c]}$ has the same properties. Moreover, $T_{[f_c]}$

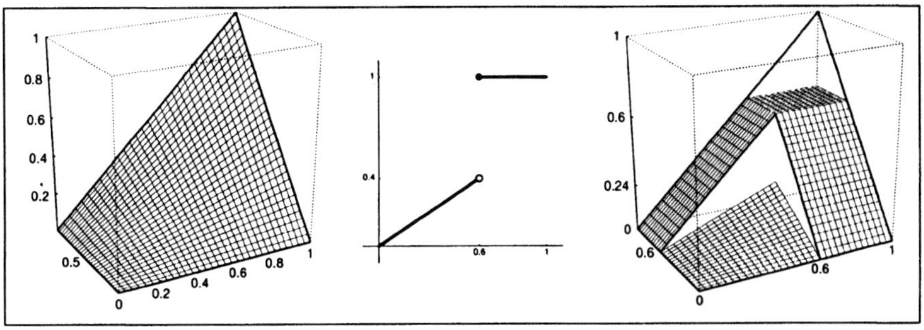

Fig. 3.5. The right-hand picture shows the result of the construction of Example 3.14 applied to the product $T_{\mathbf{P}}$ (left), using the function $f_{0.6,0.4}$ (center)

has zero divisors if and only if T has zero divisors.
However, the continuity on the right and upper boundary points of the form $(x, 1)$ and $(1, x)$ may be destroyed by a linear contraction. As an example, $(T_{\mathbf{L}})_{[f_{0.75}]}$ equals the t-norm defined in (2.3), and it is obviously non-continuous on the right and upper boundary of $[0,1]^2$ (see Figure 3.4).
There are also examples of non-continuous t-norms with continuous contractions: the ordinal sum (see Definition 3.44) $T = (\langle 0, 0.5, T_{\mathbf{P}}\rangle, \langle 0.5, 1, T_{\mathbf{D}}\rangle)$, which is non-continuous, has the continuous contraction $T_{[f_{0.5}]} = T_{\mathbf{P}}$ (see Figure 3.4).

3.14 EXAMPLE
Le T be an arbitrary t-norm, fix elements $a, c \in \,]0,1[$, and consider the function $f_{a,c} : [0,1] \longrightarrow [0,1]$ which is given by

$$f_{a,c}(x) = \begin{cases} \frac{cx}{a} & \text{if } x < a, \\ 1 & \text{if } x \geq a, \end{cases}$$

(i.e., $f_{a,c}|_{[0,a[}$ is a strictly increasing linear bijection from $[0, a[$ to $[0, c[$ and for $x, y \in [a, 1]$ we always have $T(f_{a,c}(x), f_{a,c}(y)) = 1 \in \text{Ran}(f_{a,c})$, so the construction of Theorem 3.6 is applicable). Then the resulting t-norm $T_{[f_{a,c}]}$ is given by

$$T_{[f_{a,c}]}(x,y) = \begin{cases} \frac{a}{c} \cdot T\left(\frac{cx}{a}, \frac{cy}{a}\right) & \text{if } (x,y) \in [0,a[^2, \\ a & \text{if } (x,y) \in [a,1[^2, \\ \min(x,y) & \text{otherwise.} \end{cases}$$

Again, $T_{[f_{a,c}]}$ is always a non-trivial ordinal sum (see Definition 3.44). In particular, the t-norm $T_{[f_{a,c}]}$ is neither strictly monotone nor Archimedean (clearly, a is a non-trivial idempotent element) nor continuous (this is illustrated by Figure 3.5). The t-norm $T_{[f_{a,c}]}$ has zero divisors if and only if the original t-norm T has zero divisors.

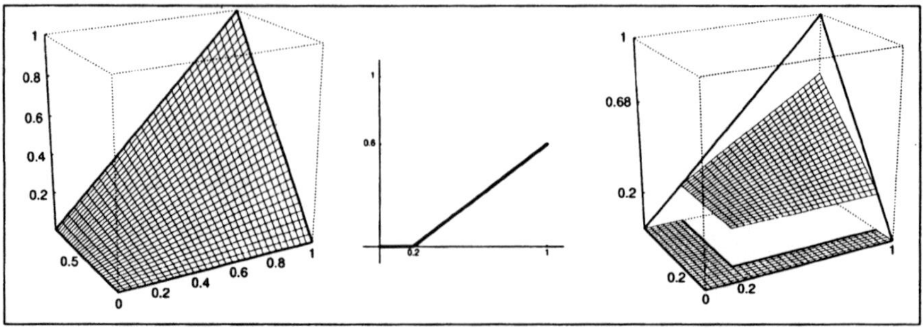

Fig. 3.6. The right-hand picture shows the result of the construction of Example 3.15 applied to the product $T_{\mathbf{P}}$ (left), using the function $f_{0.2,0.6}$ (center)

3.15 EXAMPLE
Let T be an arbitrary t-norm, fix elements $a \in \,]0,1[\,, c \in \,]0,1[$, and consider the function $f_{a,c} : [0,1] \longrightarrow [0,1]$ given by

$$f_{a,c}(x) = \frac{c}{1-a} \cdot \max(x-a, 0)$$

(i.e., $f_{a,c}|_{[a,1]}$ is a strictly increasing linear bijection from $[a,1]$ to $[0,c]$, so the construction of Theorem 3.6 is applicable). Then $T_{[f_{a,c}]}$ is given by

$$T_{[f_{a,c}]}(x,y) = \begin{cases} a + \frac{1-a}{c} \cdot T(c \cdot \frac{x-a}{1-a}, c \cdot \frac{y-a}{1-a}) & \text{if } (x,y) \in \,]a,1[^2, \\ \min(x,y) & \text{if } \max(x,y) = 1, \\ 0 & \text{otherwise.} \end{cases}$$

The t-norm $T_{[f_{a,c}]}$ always has zero divisors, it is neither strictly monotone nor continuous, and $T_{[f_{a,c}]}$ is Archimedean if and only if the set of nilpotent elements of T equals $]0,1[$ (see Figure 3.6).

3.16 EXAMPLE
Let T be an arbitrary t-norm, fix elements $a \in \,]0,1[\,, c \in \,]0,1[$, and consider the function $f_{a,c} : [0,1] \longrightarrow [0,1]$ given by

$$f_{a,c}(x) = c + \frac{1-c}{a} \cdot \min(x, a)$$

(i.e., $f_{a,c}|_{[0,a]}$ is a strictly increasing linear bijection from $[0,a]$ to $[c,1]$, so the construction of Theorem 3.6 is applicable). Then $T_{[f_{a,c}]}$ is given by

$$T_{[f_{a,c}]}(x,y) = \begin{cases} \frac{a}{1-c} \cdot \max\left(T\left(c + \frac{1-c}{a} \cdot x, c + \frac{1-c}{a} \cdot y\right) - c, 0\right) & \\ & \text{if } (x,y) \in [0,a]^2, \\ a & \text{if } (x,y) \in \,]a,1[^2, \\ \min(x,y) & \text{otherwise,} \end{cases}$$

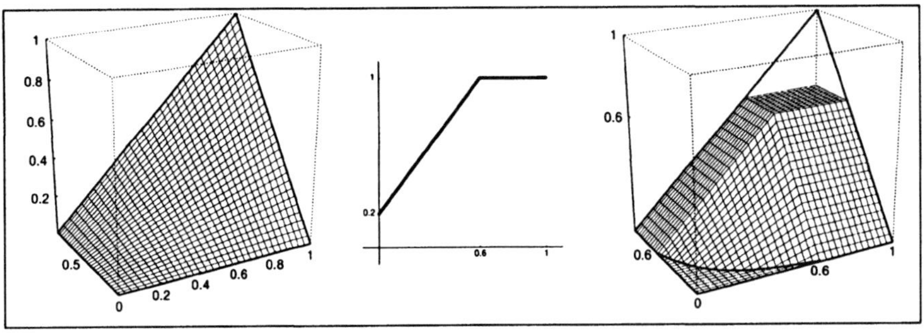

Fig. 3.7. The right-hand picture shows the result of the construction of Example 3.16 applied to the product $T_{\mathbf{P}}$ (left), using the function $f_{0.6,0.2}$ (center)

which is always a non-trivial ordinal sum (see Definition 3.44). In particular, $T_{[f_{a,c}]}$ is neither strictly monotone nor Archimedean (a is a non-trivial idempotent element) nor continuous (see Figure 3.7).

3.17 Example

Let T be an arbitrary t-norm, fix elements $a \in]0,1[$, $c \in]0,1[$, and consider the function $f_{a,c} : [0,1] \longrightarrow [0,1]$ given by

$$f_{a,c}(x) = \max\left(c + \tfrac{1-c}{1-a} \cdot (x-a), c\right),$$

(i.e., $f_{a,c}|_{[a,1]}$ is a strictly increasing linear bijection from $[a,1]$ to $[c,1]$, so the construction of Theorem 3.6 is applicable). Then $T_{[f_{a,c}]}$ is given by

$$T_{[f_{a,c}]}(x,y) = \begin{cases} 0 & \text{if } (x,y) \in ([0,a] \times [0,1[) \cup ([0,1[\times [0,a]), \\ a + \frac{1-a}{1-c} \cdot \max(T(c + \frac{1-c}{1-a} \cdot (x-a), c + \frac{1-c}{1-a} \cdot (y-a)) - c, 0) & \\ & \text{if } (x,y) \in]a,1]^2, \\ \min(x,y) & \text{otherwise.} \end{cases}$$

The t-norm $T_{[f_{a,c}]}$ always has zero divisors. It is neither strictly monotone nor continuous (see Figure 3.8).

If the two parameters are equal then for each t-norm T the t-norm $T_{[f_{a,a}]}$ is obtained by cutting T at level a, i.e.,

$$T_{[f_{a,a}]}(x,y) = \begin{cases} 0 & \text{if } (x,y) \in [0,1[^2 \text{ and } T(x,y) \leq a, \\ T(x,y) & \text{otherwise.} \end{cases}$$

3.18 Example

Let T be an arbitrary t-norm, fix elements $a, b, c \in]0,1]$ with $a < b$, and consider the continuous function $f_{a,b,c} : [0,1] \longrightarrow [0,1]$ given by

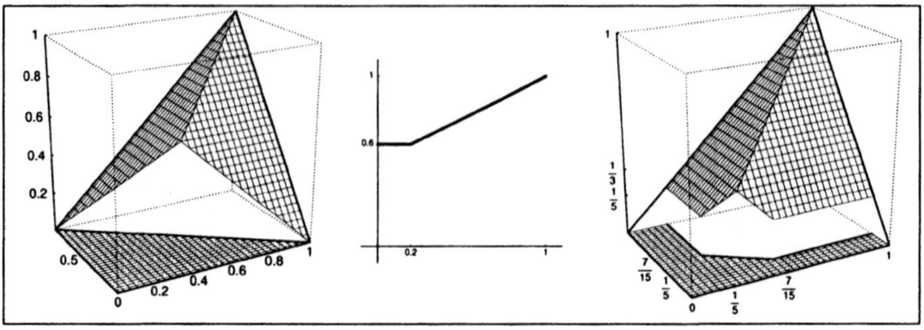

Fig. 3.8. The right-hand picture shows the result of the construction of Example 3.17 applied to the nilpotent minimum $T^{\mathbf{nM}}$ (left), using the function $f_{0.2,0.6}$ (center)

$$f_{a,b,c}(x) = \begin{cases} \frac{c}{a}x & \text{if } x \in [0,a], \\ c & \text{if } x \in \,]a,b]\,, \\ c + \frac{1-c}{1-b} \cdot (x-b) & \text{if } x \in \,]b,1]\,, \end{cases}$$

(i.e., $f_{a,b,c}$ is a piecewise linear surjection, so the construction of Theorem 3.6 is applicable). Then $T_{[f_{a,b,c}]}$ is given by

$$T_{[f_{a,b,c}]}(x,y) = \begin{cases} b + \frac{1-b}{1-c} \cdot \left(T\left(f_{a,b,c}(x), f_{a,b,c}(y)\right) - c\right) \\ \qquad \text{if } (x,y) \in \,]b,1]^2 \text{ and } T\left(f_{a,b,c}(x), f_{a,b,c}(y)\right) > c, \\ \min(x,y) & \text{if } \max(x,y) = 1, \\ \frac{a}{c} \cdot T\left(f_{a,b,c}(x), f_{a,b,c}(y)\right) & \text{otherwise.} \end{cases}$$

Observe that (see Figure 3.9) for all $(x,y) \in [a,b]^2$ we obtain the constant value

$$T_{[f_{a,b,c}]}(x,y) = \frac{a}{c} \cdot T(c,c).$$

3.19 Example

Let T be a t-norm with a non-trivial idempotent element $c \in \,]0,1[$, fix an element $a \in \,]0,1[$, and consider the function $f_{a,c} : [0,1] \longrightarrow [0,1]$ given by

$$f_{a,c}(x) = \begin{cases} 0 & \text{if } x \in [0,a[\,, \\ c + \frac{1-c}{1-a} \cdot (x-a) & \text{otherwise,} \end{cases}$$

(i.e., $f_{a,c}|_{[a,1]}$ is a strictly increasing linear bijection from $[a,1]$ to $[c,1]$; since c is an idempotent element of T we have for all $(x,y) \in [0,1]^2$ that $T(f_{a,c}(x), f_{a,c}(y)) \in \operatorname{Ran}(f_{a,c})$, so the construction of Theorem 3.6 is applicable). Then $T_{[f_{a,c}]}$ is given by

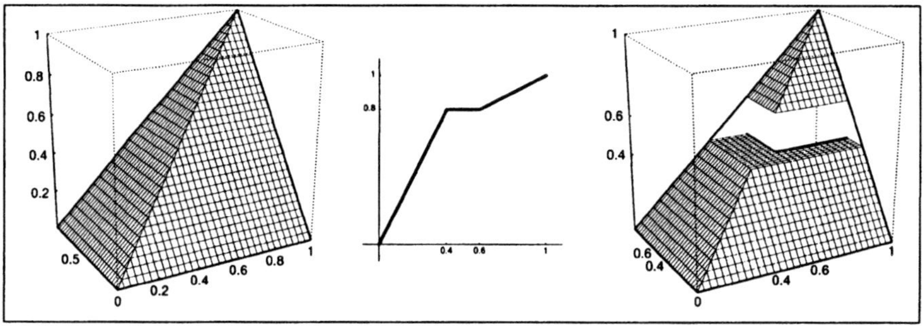

Fig. 3.9. The right-hand picture shows the result of the construction of Example 3.18 applied to the minimum $T_{\mathbf{M}}$ (left), using the function $f_{0.4,0.6,0.8}$ (center)

$$T_{[f_{a,c}]}(x,y) = \begin{cases} 0 & \text{if } (x,y) \in ([0,a[\times [0,1[) \cup ([0,1[\times [0,a[), \\ a + \frac{1-a}{1-c} \cdot (T(c + \frac{1-c}{1-a} \cdot (x-a), c + \frac{1-c}{1-a} \cdot (y-a)) - c) & \\ & \text{if } (x,y) \in [a,1]^2, \\ \min(x,y) & \text{otherwise.} \end{cases}$$

The t-norm $T_{[f_{a,c}]}$ always has zero divisors. It is neither strictly monotone nor Archimedean (a is a non-trivial idempotent element) nor continuous (see Figure 3.10).

3.20 REMARK

If $f : [0,1] \longrightarrow [0,1]$ is a strictly increasing function with $f(1) = 1$ such that

$$f(x) \cdot f(y) \in \operatorname{Ran}(f) \cup [0, f(0^+)]$$

for all $(x,y) \in [0,1]^2$, then the construction of Theorem 3.6 is applicable to the product $T_{\mathbf{P}}$, and we obtain for all $(x,y) \in [0,1]^2$ the simpler expression

$$(T_{\mathbf{P}})_{[f]}(x,y) = f^{(-1)}(f(x) \cdot f(y)).$$

This is easily seen since, for all $x \in [0,1]$, we get in this case

$$f^{(-1)}(f(x) \cdot f(1)) = f^{(-1)}(f(x)) = x = \min(x,1) = (T_{\mathbf{P}})_{[f]}(x,1).$$

This topic will be treated in more detail in the subsequent section.

We conclude this section with a counterexample (based on Theorem 3.6 and presented in [Smutná 1998a]) of an Archimedean t-norm which has no zero divisors and which is not strictly monotone. A more complicated example of this type is the Krause t-norm (see Appendix B.1).

3.21 EXAMPLE

Recall that each number $x \in \,]0,1]$ has a unique infinite dyadic representation $x = \sum_{n=1}^{\infty} \frac{1}{2^{x_n}}$, where $(x_n)_{n\in\mathbb{N}}$ is a strictly increasing sequence of natural

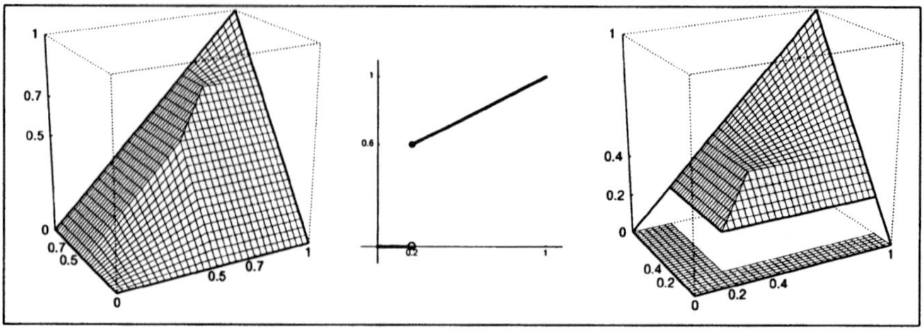

Fig. 3.10. The right-hand picture shows the result of the construction of Example 3.19 applied to the ordinal sum $T = (\langle 0, 0.5, T_{\mathbf{P}}\rangle, \langle 0.7, 1, T_{\mathbf{P}}\rangle)$ (left, see Definition 3.44), using the function $f_{0.2,0.6}$ (center; observe that 0.6 is a non-trivial idempotent element of T as required)

numbers (compare Examples 2.26(ii) and 2.29(ii)), and consider the function $f : [0,1] \longrightarrow [0,1]$ defined by

$$f(x) = \begin{cases} \sum\limits_{n=1}^{\infty} \frac{2}{3^{x_n}} & \text{if } x = \sum\limits_{n=1}^{\infty} \frac{1}{2^{x_n}}, \\ 0 & \text{if } x = 0. \end{cases}$$

Since f is strictly monotone, left-continuous and satisfies both $f(0) = 0$ and $f(1) = 1$, its pseudo-inverse $f^{(-1)} : [0,1] \longrightarrow [0,1]$ is continuous and we have $\left(f^{(-1)}\right)^{(-1)} = f$ and $f^{(-1)} \circ f = \mathrm{id}_{[0,1]}$. Then $(T_{\mathbf{P}})_{[f^{(-1)}]} : [0,1]^2 \longrightarrow [0,1]$ which is given by

$$(T_{\mathbf{P}})_{[f^{(-1)}]}(x,y) = \begin{cases} f\left(f^{(-1)}(x) \cdot f^{(-1)}(y)\right) & \text{if } (x,y) \in [0,1[^2, \\ \min(x,y) & \text{otherwise,} \end{cases}$$

is a t-norm (because of Remark 3.8(ii) or Corollary 3.10) which has no zero divisors. As a consequence of $f^{(-1)}(\frac{1}{3}) = f^{(-1)}(\frac{2}{3}) = 0.5$, $(T_{\mathbf{P}})_{[f^{(-1)}]}$ cannot be strictly monotone. However, for all $x \in \,]0,1[$ we have $f^{(-1)}(x) < 1$ and, subsequently, $(T_{\mathbf{P}})_{[f^{(-1)}]}$ is Archimedean because of

$$\lim_{n\to\infty} x_T^{(n)} = \lim_{n\to\infty} f\left((f^{(-1)})^n(x)\right) = 0.$$

As a consequence of $f(1) = 1$, the left-continuity of f and the continuity of $f^{(-1)}$, the t-norm $(T_{\mathbf{P}})_{[f^{(-1)}]}$ is continuous in the point $(1,1)$.

3.2 Additive and multiplicative generators

In Section 3.1 real functions in one variable were used to construct new t-norms from given t-norms. In this section, we start with the best-known operations, the usual addition and multiplication of real numbers, and construct t-norms with the help of functions in one variable.

The basic idea is already contained in the following classical result [Abel 1826] giving a sufficient condition for operations on the real line to be associative (stated here without proof in a modern mathematical language):

3.22 THEOREM
Let $f : \mathrm{Dom}(f) \longrightarrow \mathbb{R}$ *with* $\mathrm{Dom}(f) \subseteq \mathbb{R}$ *be a continuous strictly monotone function whose range is closed under addition. Then the two-place function* $F : (\mathrm{Dom}(f))^2 \longrightarrow \mathrm{Dom}(f)$ *defined by*

$$F(x, y) = g(f(x) + f(y)),$$

where $g : \mathrm{Ran}(f) \longrightarrow \mathrm{Dom}(f)$ *is the inverse function of* f*, is associative.*

It is easy to see that a multiplication of f by a positive constant has no influence on the resulting associative function F.

If we want to obtain a t-norm with a construction similar to Theorem 3.22, it is obvious that some additional requirements for f are necessary (after all, a t-norm is a rather special associative function). However, some generalizations are also possible, leading to the following result (recall that $f^{(-1)}$ denotes the pseudo-inverse of f as given in Definition 3.2). So, on the one hand, this result is more general than the one in [Abel 1826] in the sense that we do not need the continuity of the one-place function f , that we may relax the requirement that the range of f be closed under addition, and that we may replace the inverse of f by its pseudo-inverse. On the other hand, it is slightly more special since we want to construct an operation on the unit interval with neutral element 1.

3.23 THEOREM
Let $f : [0, 1] \longrightarrow [0, \infty]$ *be a strictly decreasing function with* $f(1) = 0$ *such that*

$$f(x) + f(y) \in \mathrm{Ran}(f) \cup [f(0^+), \infty] \tag{3.13}$$

for all $(x, y) \in [0, 1]^2$*. The following function* $T : [0, 1]^2 \longrightarrow [0, 1]$ *is a t-norm:*

$$T(x, y) = f^{(-1)}(f(x) + f(y)). \tag{3.14}$$

Proof: The commutativity (T1) and the monotonicity (T3) of T are obvious. Also, the boundary condition (T4) holds since for all $x \in [0, 1]$

$$T(x, 1) = f^{(-1)}(f(x) + f(1)) = f^{(-1)}(f(x) + 0) = f^{(-1)}(f(x)) = x.$$

For the associativity (T2), choose $x, y, z \in [0, 1]$. If both values $f(x) + f(y)$ and $f(y) + f(z)$ belong to $\mathrm{Ran}(f)$, then

$$\begin{aligned} T(T(x, y), z) &= f^{(-1)}(f \circ f^{(-1)}(f(x) + f(y)) + f(z)) \\ &= f^{(-1)}(f(x) + f(y) + f(z)) \\ &= f^{(-1)}(f(x) + f \circ f^{(-1)}(f(y) + f(z))) \\ &= T(x, T(y, z)). \end{aligned}$$

If $f(x) + f(y) \geq f(0^+)$ then $T(x,y) = 0$ and, consequently,

$$0 \leq T(x, T(y,z)) \leq T(x,y) = 0 = T(T(x,y), z).$$

The case $f(y) + f(z) \geq f(0^+)$ is analogous and, hence, T is associative. □

3.24 REMARK

(i) In Theorem 3.23, the pseudo-inverse $f^{(-1)}$ may be replaced by any monotone function $g : [0,\infty] \longrightarrow [0,1]$ with $g|_{\mathrm{Ran}(f)} = f^{(-1)}|_{\mathrm{Ran}(f)}$. In some very abstract settings (see, e.g., [Schweizer & Sklar 1983]), such a function g (which may be non-monotone) is called a quasi-inverse of f.

(ii) If $f : [0,1] \longrightarrow [0,\infty]$ satisfies the hypotheses of Theorem 3.23 and if a strictly decreasing function $g : [0,1] \longrightarrow [0,\infty]$ coincides with f on the half-open interval $]0,1]$ then it is clear that we have $f^{(-1)} = g^{(-1)}$ and $\mathrm{Ran}(f) \cup [f(0^+),\infty] = \mathrm{Ran}(g) \cup [g(0^+),\infty]$. Hence, for all $(x,y) \in [0,1]^2$

$$f^{(-1)}(f(x) + f(y)) = g^{(-1)}(g(x) + g(y)).$$

This means that the value $f(0)$ plays no role in this construction, so we shall restrict ourselves to functions which are right-continuous in 0, i.e., which satisfy $f(0) = f(0^+)$ (see Remark 3.8(i)).

(iii) Again it is obvious that a multiplication of f in Theorem 3.23 by a positive constant does not change the resulting t-norm T.

The difference between Theorem 3.23 and Theorem 3.6 is twofold. Firstly, in Theorem 3.6 we start from a (more or less arbitrary) t-norm and a rather general function to derive a new t-norm, whereas in Theorem 3.23 the starting point is always the addition of real numbers. Secondly, in the construction (3.11) of Theorem 3.6 it is in general necessary to define the resulting t-norm by distinguishing two cases ($(x,y) \in [0,1[^2$ or (x,y) on the upper right boundary of $[0,1]^2$), whereas in Theorem 3.23 a single formula suffices.

3.25 DEFINITION

An *additive generator* $t : [0,1] \longrightarrow [0,\infty]$ of a t-norm T is a strictly decreasing function which is also right-continuous in 0 and satisfies $t(1) = 0$, such that for all $(x,y) \in [0,1]^2$ we have

$$t(x) + t(y) \in \mathrm{Ran}(t) \cup [t(0),\infty], \tag{3.15}$$

$$T(x,y) = t^{(-1)}(t(x) + t(y)). \tag{3.16}$$

Now let us see what can be said about the continuity of additive generators and of the t-norm constructed by (3.16).

3.26 PROPOSITION

Let T be a t-norm which has some additive generator $t : [0,1] \longrightarrow [0,\infty]$. Then the following are equivalent:

(i) T is continuous.
(ii) T is left-continuous in the point $(1,1)$.
(iii) t is continuous.
(iv) t is left-continuous in 1.

Proof: It is trivial that (iii) implies (i) and that (i) implies (ii). In order to prove that (ii) implies (iv) assume that the additive generator t is not left-continuous in 1, i.e., $t(1^-) > 0$. Then $T(1^-, 1^-) = t^{(-1)}(2t(1^-)) < 1$, contradicting the left-continuity of T in the point $(1,1)$.

To show that (iv) implies (iii), assume that t is not continuous on $]0,1]$, i.e., that there is some interval $]a,b[\subseteq]0,t(0)[$ with $]a,b[\cap \mathrm{Ran}(t) = \emptyset$. Then the properties of $\mathrm{Ran}(t)$ imply that, for each $n \in \mathbb{N}$, also $]\frac{a}{n}, \frac{b}{n}[\cap \mathrm{Ran}(t) = \emptyset$. Denote by n_0 the smallest integer which is greater than $\frac{a}{b-a}$, then for all $n \geq n_0$ we have $\frac{a}{n} < \frac{b}{n+1}$, i.e., $]\frac{a}{n}, \frac{b}{n}[$ and $]\frac{a}{n+1}, \frac{b}{n+1}[$ are overlapping. Therefore,

$$\bigcup_{n=n_0}^{\infty}]\tfrac{a}{n}, \tfrac{b}{n}[=]0, \tfrac{b}{n_0}[,$$

which means that $t(1^-) \geq \frac{b}{n_0}$, i.e., t is not left-continuous in 1. □

An obvious and useful consequence of Proposition 3.26 is that a left-continuous t-norm which has an additive generator is automatically continuous.

3.27 PROPOSITION
Let T be a t-norm which has some additive generator $t : [0,1] \longrightarrow [0,\infty]$. The following are equivalent:

(i) T is right-continuous.
(ii) T has some right-continuous additive generator.

Proof: If there exists a right-continuous additive generator $t : [0,1] \longrightarrow [0,\infty]$ of T then its pseudo-inverse $t^{(-1)} : [0,\infty] \longrightarrow [0,1]$ is continuous, and T is right-continuous.

Conversely, if $t : [0,1] \longrightarrow [0,\infty]$ is an arbitrary additive generator of the right-continuous t-norm T then define the function $\bar{t} : [0,1] \longrightarrow [0,\infty]$ by

$$\bar{t}(x) = \begin{cases} t(x^+) & \text{if } x \in [0,1[, \\ 0 & \text{if } x = 1. \end{cases}$$

By construction, $\bar{t}$ is a right-continuous additive generator of some t-norm. Both $\mathrm{Ran}(t)$ and $\mathrm{Ran}(\bar{t})$ can be written as union of intervals with the same endpoints and that, in the case of $\mathrm{Ran}(\bar{t})$, these intervals (with the possible exception of the first one) are of the form $]a,b]$, so (3.15) also holds for $\bar{t}$. Moreover, we have $\bar{t}^{(-1)} = t^{(-1)}$. Then, for all $(x,y) \in]0,1[^2$ and for all $\varepsilon > 0$ with $(x+\varepsilon, y+\varepsilon) \in]0,1[^2$ we get

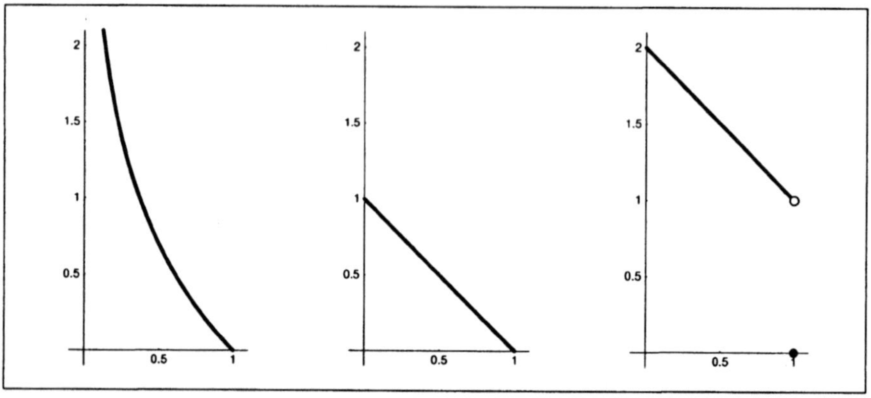

Fig. 3.11. Additive generators of $T_{\mathbf{P}}$ (left), $T_{\mathbf{L}}$ (center) and $T_{\mathbf{D}}$ (right)

$$\begin{aligned} T(x+\varepsilon, y+\varepsilon) &= t^{(-1)}(t(x+\varepsilon)+t(y+\varepsilon)) \\ &\geq \bar{t}^{(-1)}(\bar{t}(x)+\bar{t}(y)) \\ &\geq t^{(-1)}(t(x)+t(y)) \\ &= T(x,y), \end{aligned}$$

which, taking into account the right-continuity of T, shows that $\bar{t}$ is indeed an additive generator of T. □

3.28 EXAMPLE

We give now some strictly decreasing functions $t : [0,1] \longrightarrow [0,\infty]$ such that $t(1) = 0$ and $t(x)+t(y) \in \operatorname{Ran}(t) \cup [t(0),\infty]$ for all $(x,y) \in [0,1]^2$ (so they can be used as additive generators), and list the t-norms induced by (3.16).

(i) For $t(x) = 1-x$ we get the Łukasiewicz t-norm $T_{\mathbf{L}}$ (see Figure 3.11).

(ii) The drastic product $T_{\mathbf{D}}$ (which is right-continuous but not continuous) is obtained putting (see Figure 3.11)

$$t(x) = \begin{cases} 2-x & \text{if } x \in [0,1[\,, \\ 0 & \text{if } x = 1. \end{cases}$$

(iii) If we put for each $\lambda \in \,]0,1]$

$$t_\lambda(x) = \begin{cases} -\log(\lambda x) & \text{if } x \in [0,1[\,, \\ 0 & \text{if } x = 1, \end{cases}$$

then (3.16) induces

$$T_\lambda(x,y) = \begin{cases} \lambda xy & \text{if } (x,y) \in [0,1[^2\,, \\ \min(x,y) & \text{otherwise.} \end{cases}$$

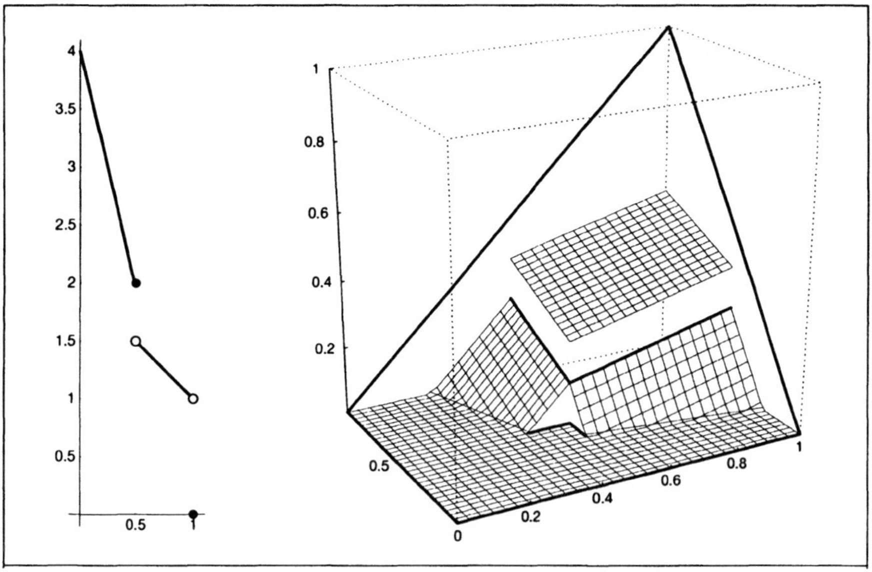

Fig. 3.12. A non-continuous t-norm together with its additive generator (see Example 3.28(iv))

For $\lambda = 1$ we get the continuous t-norm $T_1 = T_{\mathbf{P}}$ (see Figure 3.11), in all the other cases the t-norm T_λ is right-continuous but not continuous (see (2.3) and the t-norm generated by $t_{0.5}$ in Figure 2.1).

(iv) The function t (which is neither left- nor right-continuous) given by

$$t(x) = \begin{cases} 4 - 4x & \text{if } x \in [0, 0.5], \\ 2 - x & \text{if } x \in \,]0.5, 1[\,, \\ 0 & \text{if } x = 1, \end{cases}$$

leads to the non-continuous t-norm (see Figure 3.12)

$$T(x, y) = \begin{cases} \min(x, y) & \text{if } \max(x, y) = 1, \\ \frac{1}{4}(x + y) & \text{if } (x, y) \in \,]0.5, 1[^2\,, \\ \frac{1}{4}(x + 4y - 2) & \text{if } x \in \,]0.5, 1[\text{ and } y \in \left[\frac{2-x}{4}, 0.5\right], \\ \frac{1}{4}(4x + y - 2) & \text{if } y \in \,]0.5, 1[\text{ and } x \in \left[\frac{2-y}{4}, 0.5\right], \\ 0\,. & \text{otherwise.} \end{cases}$$

(v) If T is a t-norm, if $t : [0, 1] \longrightarrow [0, \infty]$ is an additive generator of T, if $c \in \,]0, 1[$ and if the function $f_c : [0, 1] \longrightarrow [0, 1]$ is given by $f_c(x) = cx$, then $t_c : [0, 1] \longrightarrow [0, \infty]$ is an additive generator of the linear contraction $T_{[f_c]}$ (see Example 3.13):

$$t_c(x) = \begin{cases} t(cx) & \text{if } x \in [0,1[, \\ 0 & \text{otherwise.} \end{cases}$$

(vi) If T is a continuous Archimedean t-norm and $t : [0,1] \longrightarrow [0,\infty]$ an additive generator of T then the T-residuum $\rightarrow_T$ (as defined by (2.17)) can be obtained by the simpler formula

$$x \rightarrow_T y = t^{(-1)}(\max(t(y) - t(x), 0)).$$

Triangular norms constructed by means of additive generators are always Archimedean. The converse, however, is not true: the t-norm introduced in Example 3.21 is Archimedean, continuous in $(1,1)$ but not continuous whence, because of Proposition 3.26, it cannot have an additive generator.

3.29 PROPOSITION
If a t-norm T has an additive generator $t : [0,1] \longrightarrow [0,\infty]$, then T is necessarily Archimedean. Moreover, we have:

(i) The t-norm T is strictly monotone if and only if $t(0) = \infty$.
(ii) Each element of $]0,1[$ is a nilpotent element of T if and only if $t(0) < \infty$.
(iii) If t is not continuous and satisfies $t(0) < \infty$ then there exists an $n_0 \in \mathbb{N}$ such that for all $x \in [0,1[$ we have $x_T^{(n_0)} = 0$.

Proof: For arbitrary numbers $x, y \in]0,1[$ we have $t(x) > 0$ and $t(\frac{y}{2}) < \infty$. Consequently, there is some $n \in \mathbb{N}$ such that $n \cdot t(x) > t(\frac{y}{2})$ (observe that this is just the Archimedean property of the semigroup $([0,\infty[, +))$. This implies

$$x_T^{(n)} = t^{(-1)}(n \cdot t(x)) < t^{(-1)}(t(\tfrac{y}{2})) = \tfrac{y}{2} < y,$$

showing that T is Archimedean.

If $t(0) = \infty$ then $\text{Ran}(f) \cup [f(0), \infty] = \text{Ran}(f)$, and for all $x > 0$ and for all $y, z \in [0,1]$ with $y < z$ we get $t(x) + t(y) \in \text{Ran}(f)$, $t(x) + t(z) \in \text{Ran}(f)$, and $t(x) + t(y) > t(x) + t(z)$. Since $f^{(-1)}|_{\text{Ran}(f)}$ is strictly decreasing as a consequence of Remark 3.4(vi), this implies

$$T(x,y) = t^{(-1)}(t(x) + t(y)) < t^{(-1)}(t(x) + t(z)) = T(x,z),$$

i.e., T is strictly monotone.

If $t(0) < \infty$ then, for each $x \in]0,1[$, we have $t(x) > 0$ and, subsequently, $n \cdot t(x) \geq t(0)$ for some $n \in \mathbb{N}$. This means $x_T^{(n)} = 0$, i.e., x is a nilpotent element of T, and T cannot be strictly monotone, thus proving (i) and (ii).

If t is not continuous, i.e., $t(1^-) > 0$ (compare Proposition 3.26(iv)), then put $n_0 \geq \frac{t(0)}{t(1^-)}$, and for all $x \in [0,1[$ we get $x_T^{(n_0)} = 0$. □

As an immediate consequence of Proposition 3.29(i) and (ii), we obtain in the particular case of continuous additive generators of a (necessarily continuous Archimedean) t-norm (see Proposition 3.26) the following.

3.30 Corollary
If $t : [0,1] \longrightarrow [0,\infty]$ *is a continuous additive generator of a continuous Archimedean t-norm* T*, then we have:*

(i) T *is strict if and only if* $t(0) = \infty$.
(ii) T *is nilpotent if and only if* $t(0) < \infty$.

For an arbitrary Archimedean t-norm T and a strictly increasing bijection $\varphi : [0,1] \longrightarrow [0,1]$, Remark 2.30(ii) implies that $T_\varphi : [0,1]^2 \longrightarrow [0,1]$ defined by (2.8) is also an Archimedean t-norm which is isomorphic to T. If the function $t : [0,1] \longrightarrow [0,\infty]$ is an additive generator of T then it is easy to determine an additive generator of T_φ.

3.31 Proposition
Let T *be an Archimedean t-norm with additive generator* $t : [0,1] \longrightarrow [0,\infty]$.

(i) If $\varphi : [0,1] \longrightarrow [0,1]$ *is a strictly increasing bijection then the function* $t\circ\varphi : [0,1] \longrightarrow [0,\infty]$ *is an additive generator of the Archimedean t-norm* T_φ *given in (2.8).*
(ii) If the additive generator t *is continuous and if* $\psi : [0,\infty] \longrightarrow [0,\infty]$ *is a strictly increasing bijection then the function* $\psi \circ t : [0,1] \longrightarrow [0,\infty]$ *is an additive generator of some continuous Archimedean t-norm which is isomorphic to* T.

Proof: Obviously, $t \circ \varphi$ is strictly decreasing and satisfies $(t \circ \varphi)(1) = 0$. Also, for all $(x,y) \in [0,1]^2$ we have $t \circ \varphi(x) + t \circ \varphi(y) \in \text{Ran}(t \circ \varphi) \cup [t \circ \varphi(0), \infty]$. Taking into account $t \circ \varphi(0) = t(0)$, we obtain for all $(x,y) \in [0,1]^2$

$$T_\varphi(x,y) = \varphi^{-1} \circ t^{(-1)}(t \circ \varphi(x) + t \circ \varphi(y)),$$

i.e., $t \circ \varphi$ is an additive generator of T_φ. The proof of (ii) is obvious. □

The result of Proposition 3.31(ii) makes it possible to construct, starting from a continuous Archimedean t-norm, interesting families of t-norms.

3.32 Example
Let T be a continuous Archimedean t-norm and $t : [0,1] \longrightarrow [0,\infty]$ an additive generator of T.

(i) For each index $\lambda \in \,]0,\infty[$ the function $t^\lambda : [0,1] \longrightarrow [0,\infty]$ defined by $t^\lambda(x) = (t(x))^\lambda$ is an additive generator of a continuous Archimedean t-norm we shall denote by $T^{(\lambda)}$, and the family $(T^{(\lambda)})_{\lambda\in]0,\infty[}$ is strictly increasing with respect to the parameter λ (this will be shown in Example 6.4). Adding the limit cases $T^{(0)} = T_{\mathbf{D}}$ and $T^{(\infty)} = T_{\mathbf{M}}$ (see Proposition 8.5), we obtain some well-known families of t-norms in this way, depending on the t-norm T we start with. For instance, $((T_{\mathbf{L}})^{(\lambda)})_{\lambda\in[0,\infty]}$ is the family of Yager t-norms (see Section 4.5), $((T_{\mathbf{P}})^{(\lambda)})_{\lambda\in[0,\infty]}$ is the

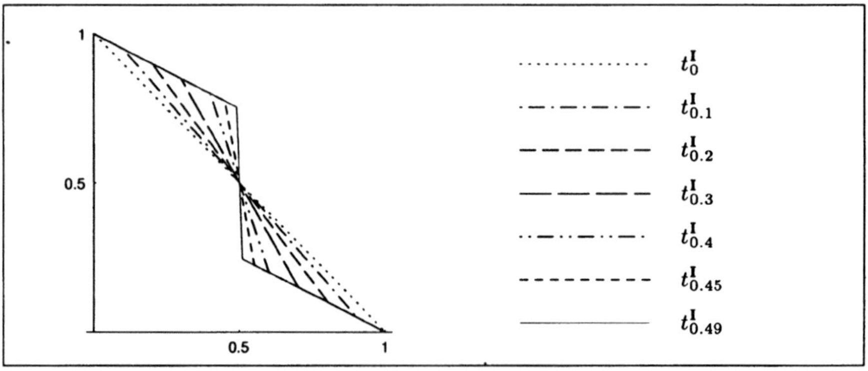

Fig. 3.13. Additive generators of some incomparable t-norms (Example 3.33)

family of Aczél-Alsina t-norms (see Section 4.8), and $((T_0^{\mathbf{H}})^{(\lambda)})_{\lambda\in[0,\infty]}$, where $T_0^{\mathbf{H}}$ is the Hamacher t-norm with parameter 0 (see Section 4.3), is just the family of Dombi t-norms (see Section 4.6).

(ii) Let T^* be a strict t-norm with additive generator $t^* : [0,1] \longrightarrow [0,\infty]$. Then, for each $\lambda \in\,]0,\infty[$, the function $t_{(t^*,\lambda)} : [0,1] \longrightarrow [0,\infty]$ defined by

$$t_{(t^*,\lambda)}(x) = t\big((t^*)^{-1}(\lambda t^*(x))\big)$$

is an additive generator of a continuous Archimedean t-norm we shall denote by $T_{(T^*,\lambda)}$. As a concrete example, for each $\lambda \in\,]0,\infty[$ we obtain $T_{\mathbf{P}(T_0^{\mathbf{H}},\lambda)} = T_\lambda^{\mathbf{H}}$ (see Section 4.3).

(iii) If in (ii) we take $T^* = T_{\mathbf{P}}$ and, subsequently, $t^*(x) = -\log x$ for all $x \in [0,1]$, then we obtain $t_{(t^*,\lambda)}(x) = t(x^\lambda)$, and the t-norm $T_{(T_{\mathbf{P}},\lambda)}$ will be denoted $T_{(\lambda)}$ for simplicity, and its additive generator by $t_{(\lambda)}$. As a concrete example, the subfamily $(T_\lambda^{\mathbf{SS}})_{\lambda\in]-\infty,\infty]}$ of the family of Schweizer-Sklar t-norms (see Section 4.2) is obtained as follows: for each $\lambda \in\,]0,\infty[$ we have $T_\lambda^{\mathbf{SS}} = (T_{\mathbf{L}})_{(\lambda)}$, and for $\lambda \in\,]-\infty,0[$ we have ($T_0^{\mathbf{H}}$ is again the Hamacher t-norm with parameter 0) $T_\lambda^{\mathbf{SS}} = (T_0^{\mathbf{H}})_{(-\lambda)}$. As a consequence of Proposition 8.5(iii) and (iv) we get

$$\lim_{\lambda\searrow 0} T_\lambda^{\mathbf{SS}} = \lim_{\lambda\searrow 0} (T_{\mathbf{L}})_{(\lambda)} = T_{\mathbf{P}} = T_0^{\mathbf{SS}},$$

$$\lim_{\lambda\nearrow 0} T_\lambda^{\mathbf{SS}} = \lim_{-\lambda\searrow 0} (T_0^{\mathbf{H}})_{(-\lambda)} = T_{\mathbf{P}} = T_0^{\mathbf{SS}},$$

$$\lim_{\lambda\to\infty} T_\lambda^{\mathbf{SS}} = \lim_{\lambda\to\infty} (T_{\mathbf{L}})_{(\lambda)} = T_{\mathbf{D}} = T_\infty^{\mathbf{SS}}.$$

Most of the families in Example 3.32 will turn out to be monotone with respect to the parameter λ (see Chapter 6). However, also families of t-norms, which are obviously non-monotone, can be constructed easily by means of additive generators.

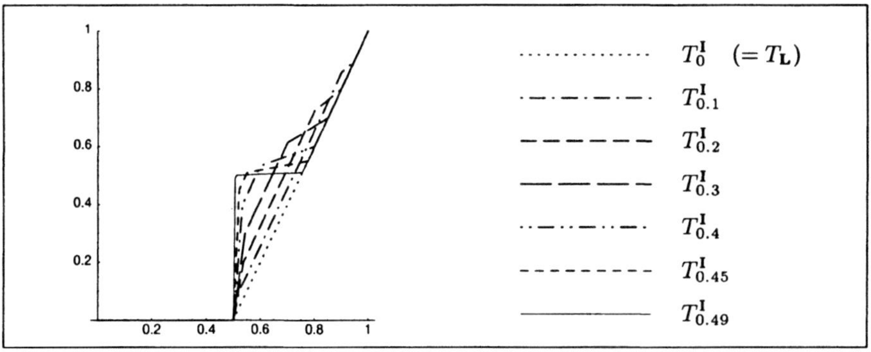

Fig. 3.14. Diagonal sections of some incomparable t-norms (Example 3.33)

3.33 EXAMPLE
Starting with the family $(t_\lambda^{\mathbf{I}} : [0,1] \longrightarrow [0,\infty])_{\lambda\in[0,0.5[}$ of continuous strictly decreasing functions defined by

$$t_\lambda^{\mathbf{I}}(x) = \min\left(1 - \tfrac{x}{2}, \max\left(\tfrac{1-x}{2}, \tfrac{2-3\lambda-(2-2\lambda)x}{2-4\lambda}\right)\right),$$

one sees immediately that they can be considered as additive generators (see Figure 3.13) of a family of nilpotent t-norms we shall denote $(T_\lambda^{\mathbf{I}})_{\lambda\in[0,0.5[}$. For $\lambda,\mu \in\ ]0,0.5[$ with $\lambda \neq \mu$ the corresponding t-norms $T_\lambda^{\mathbf{I}}$ and $T_\mu^{\mathbf{I}}$ are incomparable (see Figure 3.14). More details concerning this family of t-norms (together with some 3D and contour plots) are given in Appendix B.2 (compare also [Viceník 1998a, 1998b]).

An immediate consequence of Proposition 3.29 is that a t-norm with a non-trivial idempotent element (e.g., a non-trivial ordinal sum, see Definition 3.44) cannot have an additive generator. In particular, the minimum $T_{\mathbf{M}}$ has no additive generator, a fact which was mentioned first in [Ling 1965]. In this context also the classical result in [Arnold 1957] (for an extension see [Krause 1983]) is of interest where it was shown that there are no continuous real functions $f : \mathbb{R} \longrightarrow \mathbb{R}$ and $g, h : [0,1] \longrightarrow \mathbb{R}$ such that for all $(x,y) \in [0,1]^2$

$$\min(x,y) = f(g(x) + h(y)).$$

There is a concept completely dual to additive generators, the so-called multiplicative generators of t-norms. Of course, the background of this duality is the well-known fact that exponential function and logarithm provide natural isomorphisms between the additive semigroup $([0,\infty],+)$ and the multiplicative semigroup $([0,1],\cdot)$.

3.34 REMARK
If $t : [0,1] \longrightarrow [0,\infty]$ is an additive generator of the t-norm T and if we define the strictly increasing function $\theta : [0,1] \longrightarrow [0,1]$ by

$$\theta(x) = e^{-t(x)},$$

then it is obvious that for all $(x,y) \in [0,1]^2$

$$T(x,y) = \theta^{(-1)}(\theta(x) \cdot \theta(y)).$$

Analyzing the function θ more carefully, it becomes evident that it exactly satisfies the requirements stated in Remark 3.20. The following result about multiplicative generators, which in our case is obtained via pseudo-inverse functions (see Remark 3.20) or in duality to additive generators (see Remark 3.34), can be obtained also completely independently (in fact, this is the case if one generalizes the result in [Schweizer & Sklar 1983, Theorem 5.2.1]). It is presented here (without proof) for the sake of completeness.

3.35 COROLLARY
Let $\theta : [0,1] \longrightarrow [0,1]$ be a strictly increasing function with $\theta(1) = 1$ such that for all $(x,y) \in [0,1]^2$ we have

$$\theta(x) \cdot \theta(y) \in \mathrm{Ran}(\theta) \cup [0, \theta(0^+)].$$

Then the function $T : [0,1]^2 \longrightarrow [0,1]$ given by

$$T(x,y) = \theta^{(-1)}(\theta(x) \cdot \theta(y))$$

is a triangular norm.

3.36 DEFINITION
A *multiplicative generator* of a t-norm T is a strictly increasing function $\theta : [0,1] \longrightarrow [0,1]$ which is right-continuous in 0 and satisfies $\theta(1) = 1$, such that for all $(x,y) \in [0,1]^2$ we have

$$\theta(x) \cdot \theta(y) \in \mathrm{Ran}(\theta) \cup [0, \theta(0)],$$
$$T(x,y) = \theta^{(-1)}(\theta(x) \cdot \theta(y)).$$

Because of the duality between t-norms and t-conorms, additive and multiplicative generators of t-conorms can also be considered. Without presenting technical details and proofs, this can be summarized as follows.

3.37 REMARK
Let T be a t-norm, S the dual t-conorm, $t : [0,1] \longrightarrow [0,\infty]$ an additive generator of T and $\theta : [0,1] \longrightarrow [0,1]$ a multiplicative generator of T. If we define the functions $s : [0,1] \longrightarrow [0,\infty]$ and $\xi : [0,1] \longrightarrow [0,1]$ by

$$s(x) = t(1-x),$$
$$\xi(x) = \theta(1-x),$$

then it is obvious that for all $(x,y) \in [0,1]^2$ we get

$$S(x,y) = s^{(-1)}(s(x) + s(y)),$$
$$S(x,y) = \xi^{(-1)}(\xi(x) \cdot \xi(y)).$$

In complete analogy to Theorem 3.23 and Corollary 3.35 we get the following result for triangular conorms:

3.38 COROLLARY

(i) *Let* $s : [0,1] \longrightarrow [0,\infty]$ *be a strictly increasing function with* $s(0) = 0$ *such that for all* $(x, y) \in [0,1]^2$

$$s(x) + s(y) \in \text{Ran}(s) \cup [s(1^-), \infty].$$

Then the two-place function $S : [0,1]^2 \longrightarrow [0,1]$ *given by*

$$S(x,y) = s^{(-1)}(s(x) + s(y))$$

is a triangular conorm.

(ii) *Let* $\xi : [0,1] \longrightarrow [0,\infty]$ *be a strictly decreasing function with* $\xi(0) = 1$ *such that for all* $(x, y) \in [0,1]^2$

$$\xi(x) \cdot \xi(y) \in \text{Ran}(\xi) \cup [0, \xi(1^-)].$$

Then the two-place function $S : [0,1]^2 \longrightarrow [0,1]$ *given by*

$$S(x,y) = \xi^{(-1)}(\xi(x) \cdot \xi(y))$$

is a triangular conorm.

It is therefore quite natural to define additive and multiplicative generators of t-conorms as follows.

3.39 DEFINITION

(i) An *additive generator* of a t-conorm S is a strictly increasing function $s : [0,1] \longrightarrow [0,\infty]$ which is left-continuous in 1 and satisfies $s(0) = 0$, such that for all $(x, y) \in [0,1]^2$ we have

$$s(x) + s(y) \in \text{Ran}(s) \cup [s(1), \infty],$$
$$S(x,y) = s^{(-1)}(s(x) + s(y)).$$

(ii) A *multiplicative generator* of a t-conorm S is a strictly decreasing function $\xi : [0,1] \longrightarrow [0,1]$ which is left-continuous in 1 and satisfies $\xi(0) = 1$, such that for all $(x, y) \in [0,1]^2$ we have

$$\xi(x) \cdot \xi(y) \in \text{Ran}(\xi) \cup [0, \xi(1)],$$
$$S(x,y) = \xi^{(-1)}(\xi(x) \cdot \xi(y)).$$

The exact relationship between the classes of additive and multiplicative generators of t-norms and t-conorms, respectively, is visualized by the commutative diagram in Figure 3.15 where the operators N, E and L assign to each function $f : [0,1] \longrightarrow [0,\infty]$ the functions $\mathsf{N}f, \mathsf{E}f, \mathsf{L}f : [0,1] \longrightarrow [0,\infty]$ given by

$$\mathsf{N}f(x) = f(1-x), \tag{3.17}$$
$$\mathsf{E}f(x) = e^{-f(x)}, \tag{3.18}$$
$$\mathsf{L}f(x) = -\log(f(x)). \tag{3.19}$$

Note that the same function can be an additive generator for a t-norm and a multiplicative generator for a t-conorm, and vice versa. Put, e.g., $f(x) = 1 - x$ and $g(x) = x$ for $x \in [0,1]$. Then f is an additive generator of $T_{\mathbf{L}}$ and a multiplicative generator of $S_{\mathbf{P}}$, while g is an additive generator of $S_{\mathbf{L}}$ and a multiplicative generator of $T_{\mathbf{P}}$.

3.40 REMARK

(i) The concept of additive generators of t-norms as given in Definition 3.25 can be further generalized (see [Viceník 1998a, 1998b]) in the sense that a strictly decreasing function $f : [0,1] \longrightarrow [0,\infty]$ with $f(1) = 0$ can generate a t-norm T via (3.14), i.e., such that for all $(x,y) \in [0,1]^2$ we have $T(x,y) = f^{(-1)}(f(x) + f(y))$, without satisfying (3.13), i.e., if there exists an $(x,y) \in [0,1]^2$ such that $f(x) + f(y) \notin \mathrm{Ran}(f) \cup [f(0^+),\infty]$. It is even possible to generate t-norms via (3.14) with infinitely many idempotent elements [Viceník 1999a]. However, the complete characterization of the class of all functions $f : [0,1] \longrightarrow [0,\infty]$ generating a t-norm by applying (3.14) is still an open problem.

(ii) If a function $f : [0,1] \longrightarrow [0,1]$ is non-decreasing and satisfies also $f(x) \cdot f(y) \in \mathrm{Ran}(f) \cup [0, f(0^+)]$ for all $(x,y) \in [0,1]^2$, then the function $F : [0,1]^2 \longrightarrow [0,1]$ given by $F(x,y) = f^{(-1)}(f(x) \cdot f(y))$ is a t-subnorm: commutativity (T1) is obvious, associativity (T2) and monotonicity (T3) can be shown using similar arguments as in the proof of Theorem 3.23 (taking into account Remark 3.34), and property (1.6) follows immediately from the monotonicity of $f^{(-1)}$ and from $T_{\mathbf{P}} \leq T_{\mathbf{M}}$. The function f is then called a multiplicative generator of the t-subnorm F. In a dual way it is also possible to consider additive generators of t-subnorms (compare [Jenei 1999]).

3.41 EXAMPLE

(i) Consider the function $f : [0,1] \longrightarrow [0,\infty]$ given by

$$f(x) = \begin{cases} 3 - x & \text{if } x \in [0, 0.5[\,, \\ 1 - x & \text{if } x \in [0.5, 1]\,, \end{cases}$$

then (3.14) leads to the function $T : [0,1]^2 \longrightarrow [0,1]$ given by

$$T(x,y) = \begin{cases} 0.5 & \text{if } (x,y) \in [0.5,1]^2 \text{ and } x + y \in [1, 1.5]\,, \\ T_{\mathbf{L}}(x,y) & \text{otherwise.} \end{cases}$$

In order to see that T is indeed a t-norm, only the associativity (T2) is non-trivial. Since $T|_{[0.5,1]^2} = T^*|_{[0.5,1]^2}$, where T^* is the ordinal sum

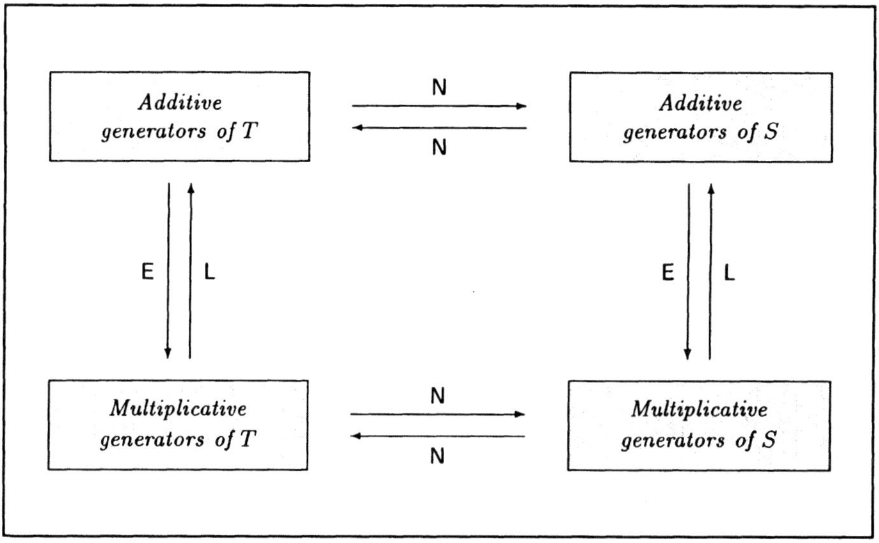

Fig. 3.15. The relationship between additive and multiplicative generators of a t-norm T and its dual t-conorm S: a commutative diagram

$(\langle 0.5, 1, T_{\mathbf{L}}\rangle)$ (see Definition 3.44 below), this proves the associativity in the case $x, y, z \in [0.5, 1]$, whereas in all the other cases we obtain

$$T(x, T(y, z)) = T_{\mathbf{L}}(x, y, z) = T(T(x, y), z).$$

Since 0.5 is a non-trivial idempotent element of T, the T is not Archimedean.

(ii) Consider the non-increasing function $f : [0, 1] \longrightarrow [0, \infty]$ given by

$$f(x) = e^{-\max(3-4x,1)},$$

then for all $(x, y) \in [0, 1]^2$ we have $f(x) \cdot f(y) \in \text{Ran}(f) \cup [0, f(0^+)]$, so f is a multiplicative generator of a t-subnorm F which is given by

$$F(x, y) = \max(\min(x, 0.5) + \min(y, 0.5) - 0.75, 0).$$

Note that the F is continuous but not a t-norm since it violates (T4).

3.3 Ordinal sums

The following method of constructing a new t-norm from a family of given t-norms is based on the results of [Climescu 1946, Clifford 1954, Clifford & Preston 1961] concerning ordinal sums of semigroups (for the case of t-norms and t-conorms see [Schweizer & Sklar 1963, Ling 1965, Frank 1979]). We start with a result drawn from [Clifford 1954].

3.42 THEOREM

Let $(A, \preceq)$ with $A \neq \emptyset$ be a linearly ordered set and $(G_\alpha)_{\alpha \in A}$ with $G_\alpha = (X_\alpha, *_\alpha)$ be a family of semigroups. Assume that for all $\alpha, \beta \in A$ with $\alpha \prec \beta$ the sets X_α and X_β are either disjoint or that $X_\alpha \cap X_\beta = \{x_{\alpha\beta}\}$, where $x_{\alpha\beta}$ is both the unit element of G_α and the annihilator of G_β, and where for each $\gamma \in A$ with $\alpha \prec \gamma \prec \beta$ we have $X_\gamma = \{x_{\alpha\beta}\}$. Put $X = \bigcup_{\alpha \in A} X_\alpha$ and define the binary operation $*$ on X by

$$x * y = \begin{cases} x *_\alpha y & \text{if } (x, y) \in X_\alpha \times X_\alpha, \\ x & \text{if } (x, y) \in X_\alpha \times X_\beta \text{ and } \alpha \prec \beta, \\ y & \text{if } (x, y) \in X_\alpha \times X_\beta \text{ and } \beta \prec \alpha. \end{cases} \tag{3.20}$$

Then $G = (X, *)$ is a semigroup. The semigroup G is commutative if and only if for each $\alpha \in A$ the semigroup G_α is commutative.

Proof: First it is necessary to check that $*$ is well-defined. A problem can only arise if $(x, y) \in X_\alpha \times X_\beta$ and, say, $\alpha \prec \beta$ and $x \in X_\alpha \cap X_\beta$. In this situation the first two cases in (3.20) apply. But then x must be both the unit element of G_α and the annihilator of G_β, i.e., for each $y \in X_\beta$ we get $x *_\beta y = x$, thus producing the same result in either case. This is true even if there exists a $\gamma \in A$ with $\alpha \prec \gamma \prec \beta$ and $X_\gamma = \{x\}$. The situations $y \in X_\alpha \cap X_\beta$ and $\beta \prec \alpha$ are checked in complete analogy.

The associativity of $*$ is easily seen by checking all the possible cases, and the statement about commutativity is obvious. □

Recall that (see Proposition 2.28(i)) $T : [0,1]^2 \longrightarrow [0,1]$ is a t-norm if and only if $([0,1], T, \leq)$ is a fully ordered commutative semigroup with neutral element 1 and annihilator 0. We can therefore apply Theorem 3.42 to construct new t-norms from a given family of t-norms.

3.43 THEOREM

Let $(T_\alpha)_{\alpha \in A}$ be a family of t-norms and $(]a_\alpha, e_\alpha[)_{\alpha \in A}$ be a family of non-empty, pairwise disjoint open subintervals of $[0,1]$. Then the following function $T : [0,1]^2 \longrightarrow [0,1]$ is a t-norm:

$$T(x, y) = \begin{cases} a_\alpha + (e_\alpha - a_\alpha) \cdot T_\alpha\left(\frac{x - a_\alpha}{e_\alpha - a_\alpha}, \frac{y - a_\alpha}{e_\alpha - a_\alpha}\right) & \text{if } (x, y) \in [a_\alpha, e_\alpha]^2, \\ \min(x, y) & \text{otherwise.} \end{cases} \tag{3.21}$$

Proof: The set theoretic complement of $\bigcup_{\alpha \in A}]a_\alpha, e_\alpha[$ with respect to the unit interval $[0,1]$ can always be written as a union $\bigcup_{\beta \in B} [u_\beta, v_\beta]$ of pairwise disjoint closed subintervals of $[0,1]$ (trivial, i.e., one-point intervals are possible) in such a way that the index sets A and B are disjoint. This means that we have $A \cap B = \emptyset$ and also

$$[0,1] \setminus \bigcup_{\alpha \in A}]a_\alpha, e_\alpha[= \bigcup_{\beta \in B} [u_\beta, v_\beta].$$

For each $\alpha \in A$ consider the linear bijection $\varphi_\alpha : [a_\alpha, e_\alpha] \longrightarrow [0,1]$ given by

$$\varphi_\alpha(z) = \frac{z - a_\alpha}{e_\alpha - a_\alpha}. \tag{3.22}$$

Then the commutative semigroup $G_\alpha = ([a_\alpha, e_\alpha], *_\alpha)$, with $*_\alpha$ given by

$$x *_\alpha y = \varphi_\alpha^{-1}(T_\alpha(\varphi_\alpha(x)\varphi_\alpha(y))),$$

is isomorphic to $([0,1], T_\alpha)$; its annihilator is $a_\alpha = \varphi_\alpha^{-1}(0)$ and its neutral element is $e_\alpha = \varphi_\alpha^{-1}(1)$.

For each $\beta \in B$ we construct a commutative semigroup $G_\beta = ([u_\beta, v_\beta], *_\beta)$ with annihilator u_β and neutral element v_β putting

$$x *_\beta y = \min(x, y).$$

Now define a new index set $C = A \cup B$ (which is non-empty even if $A = \emptyset$) and consider the function $s : C \longrightarrow [0,1]$, which assigns the midpoint to each of the intervals under consideration by

$$s(\gamma) = \begin{cases} \frac{a_\gamma + e_\gamma}{2} & \text{if } \gamma \in A, \\ \frac{u_\gamma + v_\gamma}{2} & \text{if } \gamma \in B. \end{cases}$$

Since s is injective and since the intervals $(]a_\alpha, e_\alpha[)_{\alpha \in A}$ cannot be overlapping, we can induce a linear order $\preceq$ on C from the natural order $\leq$ on $[0,1]$ putting

$$\gamma_1 \preceq \gamma_2 \qquad \text{if and only if} \qquad s(\gamma_1) \leq s(\gamma_2).$$

The family $(G_\gamma)_{\gamma \in C}$ fulfills all the conditions of Theorem 3.42. Taking into account the fact that

$$\Big(\bigcup_{\gamma \in A} [a_\gamma, e_\gamma]\Big) \cup \Big(\bigcup_{\gamma \in B} [u_\gamma, v_\gamma]\Big) = [0,1]$$

and that for all $\gamma \in A$ and for all $(x, y) \in [a_\gamma, e_\gamma]^2$

$$x *_\gamma y = a_\gamma + (e_\gamma - a_\gamma) \cdot T_\gamma\big(\tfrac{x - a_\gamma}{e_\gamma - a_\gamma}, \tfrac{y - a_\gamma}{e_\gamma - a_\gamma}\big)$$

and for all $\gamma \in B$ and for all $(x, y) \in [u_\gamma, v_\gamma]^2$

$$x *_\gamma y = \min(x, y),$$

it follows from Theorem 3.42 that $G = ([0,1], T)$ is a commutative semigroup with annihilator 0 and neutral element 1. Since the monotonicity (T3) of T is obvious, T is indeed a t-norm. □

This allows us to adapt the general concept of ordinal sums of abstract semigroups to the case of t-norms as follows:

3.44 DEFINITION

Let $(T_\alpha)_{\alpha\in A}$ be a family of t-norms and $(]a_\alpha, e_\alpha[)_{\alpha\in A}$ be a family of non-empty, pairwise disjoint open subintervals of $[0,1]$. The t-norm T defined by (3.21) is called the *ordinal sum* of the *summands* $\langle a_\alpha, e_\alpha, T_\alpha\rangle$, $\alpha \in A$, and we shall write

$$T = (\langle a_\alpha, e_\alpha, T_\alpha\rangle)_{\alpha\in A}.$$

3.45 REMARK

(i) The concept of ordinal sums introduced in Theorem 3.42 can be applied in a more general setting to a family of groupoids $(X_\alpha, *_\alpha)_{\alpha\in A}$, where $(A, \preceq)$ is a linearly ordered, non-empty index set and each $(X_\alpha, *_\alpha)$ is a groupoid, i.e., a non-empty set X_α equipped with a binary operation $*_\alpha$. It is not difficult to check (compare [Schweizer & Sklar 1983, Definition 5.2.4]) that, if the sets X_α are pairwise disjoint, if $X = \bigcup_{\alpha\in A} X_\alpha$ and if $*$ is defined by (3.20), then $(X, *)$ is a groupoid.

(ii) In the same spirit as in Definition 3.44 we can also introduce the ordinal sum of other binary operations on the unit interval $[0,1]$. Examples for this are the ordinal sum of t-conorms (which is again a t-conorm, see Corollary 3.58), of copulas (yielding a copula, compare Example 9.5(iv)), and of t-subnorms (which always leads to a t-subnorm, sometimes even to a t-norm as in Example 3.61(ii), compare also [Jenei 200x]).

3.46 EXAMPLE

(i) Each t-norm T is a trivial ordinal sum with one summand $\langle 0, 1, T\rangle$, i.e., we have $T = (\langle 0, 1, T\rangle)$.

(ii) The ordinal sum $T = (\langle 0.1, 0.5, T_{\mathbf{P}}\rangle, \langle 0.7, 0.9, T_{\mathbf{L}}\rangle)$ is given by

$$T(x,y) = \begin{cases} 0.1 + 2.5(x-0.1)(y-0.1) & \text{if } (x,y) \in [0.1, 0.5]^2, \\ 0.7 + \max(x+y-1.6, 0) & \text{if } (x,y) \in [0.7, 0.9]^2, \\ \min(x,y) & \text{otherwise.} \end{cases}$$

(iii) An ordinal sum of t-norms may have infinitely many summands. For instance, $T = \left(\left\langle \frac{1}{2^n}, \frac{1}{2^{n-1}}, T_{\mathbf{P}}\right\rangle\right)_{n\in\mathbb{N}}$ means that

$$T(x,y) = \begin{cases} \frac{1}{2^n} + 2^n\left(x - \frac{1}{2^n}\right)\left(y - \frac{1}{2^n}\right) & \text{if } (x,y) \in \left[\frac{1}{2^n}, \frac{1}{2^{n-1}}\right]^2, \\ \min(x,y) & \text{otherwise.} \end{cases}$$

(iv) It is possible to construct families of t-norms using ordinal sums. An example is the family $(T_\lambda^{\mathbf{DP}})_{\lambda\in]0,1[}$ of *Dubois-Prade t-norms* (first introduced in [Dubois & Prade 1980b]) defined by

$$T_\lambda^{\mathbf{DP}}(x,y) = (\langle 0, \lambda, T_{\mathbf{P}}\rangle).$$

The family of Dubois-Prade t-norms $(T_\lambda^{\mathbf{DP}})_{\lambda\in]0,1[}$ is strictly decreasing and continuous with respect to the parameter λ, and the limit cases are $T_{\mathbf{M}}$ and $T_{\mathbf{P}}$, respectively, i.e., for all $\lambda_0 \in]0,1[$

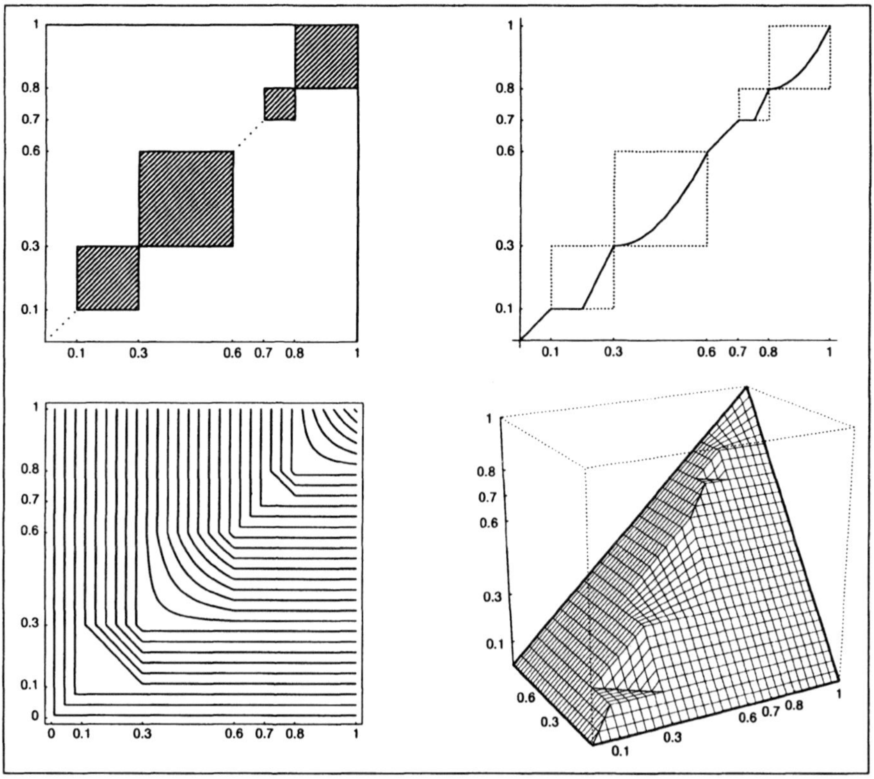

Fig. 3.16. Construction of ordinal sums: the important parts of the domain of $(\langle 0.1, 0.3, T_{\mathbf{L}}\rangle, \langle 0.3, 0.6, T_{\mathbf{P}}\rangle, \langle 0.7, 0.8, T_{\mathbf{L}}\rangle, \langle 0.8, 1, T_{\mathbf{P}}\rangle)$ are marked (top left), top right its diagonal section, bottom left its contour plot, and bottom right its 3D plot

$$\lim_{\lambda \to \lambda_0} T_\lambda^{\mathbf{DP}} = T_{\lambda_0}^{\mathbf{DP}},$$

$$\lim_{\lambda \searrow 0} T_\lambda^{\mathbf{DP}} = T_{\mathbf{M}},$$

$$\lim_{\lambda \nearrow 1} T_\lambda^{\mathbf{DP}} = T_{\mathbf{P}}.$$

All Dubois-Prade t-norms are also copulas (see Section 9.1). Contour plots of two prototypical examples of Dubois-Prade t-norms are given in Figure 3.17.

(v) Another family of t-norms which turn out to be ordinal sums (involving the Łukasiewicz t-norm $T_{\mathbf{L}}$) is the family of Mayor-Torrens t-norms which will be presented in Section 4.9.

(vi) As already mentioned, the t-norms obtained by the constructions, which are described in Examples 3.11, 3.14 and 3.16, are always non-trivial ordinal sums.

3.47 REMARK

Let $T = (\langle a_\alpha, e_\alpha, T_\alpha \rangle)_{\alpha \in A}$ be an ordinal sum of t-norms.

(i) If we have $T_{\alpha_0} = T_{\mathbf{M}}$ for some $\alpha_0 \in A$, then the summand $\langle a_{\alpha_0}, e_{\alpha_0}, T_{\alpha_0} \rangle$ can be omitted, i.e.,

$$(\langle a_\alpha, e_\alpha, T_\alpha \rangle)_{\alpha \in A} = (\langle a_\alpha, e_\alpha, T_\alpha \rangle)_{\alpha \in A \setminus \{\alpha_0\}}.$$

In particular, an empty ordinal sum of t-norms, i.e., an ordinal sum of t-norms with index set $\emptyset$, yields the minimum $T_{\mathbf{M}}$:

$$T_{\mathbf{M}} = (\emptyset) = (\langle a_\alpha, e_\alpha, T_\alpha \rangle)_{\alpha \in \emptyset}.$$

(ii) Each of the intervals $]a_\alpha, e_\alpha[$ is non-empty and, therefore, contains some rational number. Consequently, the cardinality of the index set A of an ordinal sum cannot exceed the cardinality of all rational numbers (in $[0, 1]$), i.e., A must be finite or countably infinite.

(iii) By construction, the set of idempotent elements of T contains the set

$$M = [0, 1] \setminus \bigcup_{\alpha \in A}]a_\alpha, e_\alpha[$$

as a subset, and for each idempotent element a of T with $a \in M$ and for all $x \in [0, 1]$ we have $T(a, x) = \min(a, x)$.

(iv) The set M given in (iv) equals the set of idempotent elements of T if and only if no T_α has a non-trivial idempotent element.

(v) If T is a non-trivial ordinal sum, i.e., if $A \neq \emptyset$ and if no $]a_\alpha, e_\alpha[$ equals $]0, 1[$, then T cannot be Archimedean.

(vi) The ordinal sum T has zero divisors (nilpotent elements) if and only if there is an $\alpha_0 \in A$ such that $a_{\alpha_0} = 0$ and T_{α_0} has zero divisors (nilpotent elements).

(vii) For all $x_1, \ldots, x_n \in [0, 1]$ we get

$$T(x_1, \ldots, x_n) = \begin{cases} a_\alpha + (e_\alpha - a_\alpha) T_\alpha\left(\frac{\min(x_1, e_\alpha) - a_\alpha}{e_\alpha - a_\alpha}, \ldots, \frac{\min(x_n, e_\alpha) - a_\alpha}{e_\alpha - a_\alpha}\right) \\ \qquad\qquad \text{if } \min(x_1, \ldots, x_n) \in]a_\alpha, e_\alpha[, \\ \min(x_1, \ldots, x_n) \qquad\qquad \text{otherwise.} \end{cases}$$

There is a very close relationship between non-trivial idempotent elements and ordinal sums.

3.48 PROPOSITION

Let T be a t-norm and $a \in]0, 1[$ such that $T(a, x) = \min(a, x)$ for all $x \in [0, 1]$. The following are equivalent:

(i) a is a non-trivial idempotent element of T.

(ii) There are t-norms T_1 and T_2 such that $T = (\langle 0, a, T_1 \rangle, \langle a, 1, T_2 \rangle)$.

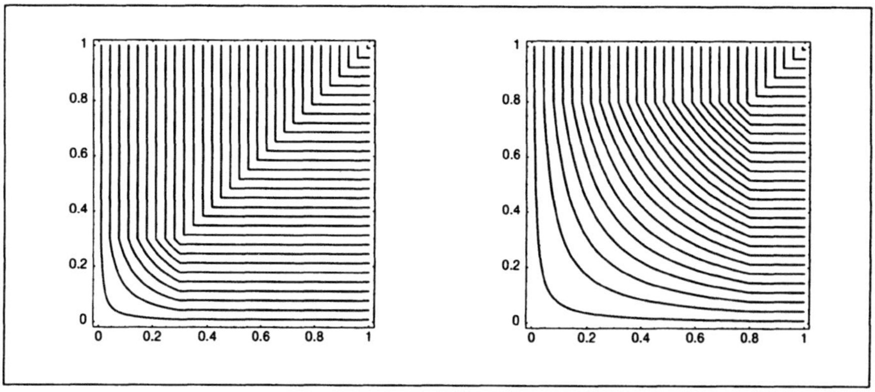

Fig. 3.17. Contour plots of the Dubois-Prade t-norms $T_{0.3}^{\mathbf{DP}}$ (left) and $T_{0.8}^{\mathbf{DP}}$

Proof: Let the t-norm T and $a \in]0,1[$ with $T(a,x) = \min(a,x)$ for all $x \in [0,1]$ be given. If $a \in]0,1[$ is a non-trivial idempotent element of T then consider the strictly increasing bijections $\varphi_1 : [0,a] \longrightarrow [0,1]$ and $\varphi_2 : [a,1] \longrightarrow [0,1]$ given by $\varphi_1(x) = \frac{x}{a}$ and $\varphi_2(x) = \frac{x-a}{1-a}$. Because of Proposition 2.32(ii), the functions $T_1, T_2 : [0,1]^2 \longrightarrow [0,1]$ given by, respectively,

$$T_1(x,y) = \varphi_1(T(\varphi_1^{-1}(x), \varphi_1^{-1}(y))),$$
$$T_2(x,y) = \varphi_2(T(\varphi_2^{-1}(x), \varphi_2^{-1}(y))),$$

are t-norms. By construction, the ordinal sum $(\langle 0,a,T_1\rangle, \langle a,1,T_2\rangle)$ coincides with the given t-norm T on the set $[0,a]^2 \cup [a,1]^2$. If $(x,y) \in [0,a[\times]a,1]$, i.e., $x < a < y$, then $x = T(x,a) \leq T(x,y) \leq x$, i.e., T coincides with the minimum on $[0,a[\times]a,1]$.

The converse follows directly from Remark 3.47(iii). □

The continuity of an ordinal sum of t-norms is equivalent to the continuity of all of its summands.

3.49 PROPOSITION

Let $T = (\langle a_\alpha, e_\alpha, T_\alpha\rangle)_{\alpha\in A}$ be an ordinal sum of t-norms with $A \neq \emptyset$. Then the following are equivalent:

(i) T is continuous.
(ii) For each $\alpha \in A$, T_α is continuous.

Proof: If T is continuous then, because of (3.21) and the continuity of the linear transformations φ_α and φ_α^{-1}, each T_α is continuous too.

To show the converse, observe first that, as a consequence of Proposition 1.19, it suffices to show that for each $x_0 \in [0,1]$ the function $T(x_0, .)$ is continuous. If $x_0 \in [a_\alpha, e_\alpha]$ for some $\alpha \in A$, then the continuity of $T(x_0, .)$ follows from (3.21) together with

$$a_\alpha + (e_\alpha - a_\alpha) \cdot T_\alpha\left(\tfrac{x_0 - a_\alpha}{e_\alpha - a_\alpha}, \tfrac{a_\alpha - a_\alpha}{e_\alpha - a_\alpha}\right) = a_\alpha = \min(x_0, a_\alpha),$$
$$a_\alpha + (e_\alpha - a_\alpha) \cdot T_\alpha\left(\tfrac{x_0 - a_\alpha}{e_\alpha - a_\alpha}, \tfrac{e_\alpha - a_\alpha}{e_\alpha - a_\alpha}\right) = x_0 = \min(x_0, e_\alpha).$$

For $x_0 \in [0,1] \setminus \bigcup_{\alpha \in A} [a_\alpha, e_\alpha]$ we have $T(x_0, .) = \min(x_0, .)$ which is obviously continuous. □

If each t-norm in a summand of an ordinal sum of t-norms has an additive generator, then the ordinal sum also has (partial) additive generators:

3.50 PROPOSITION
Let $T = (\langle a_\alpha, e_\alpha, T_\alpha \rangle)_{\alpha \in A}$ be an ordinal sum of t-norms and assume that for each $\alpha \in A$ the t-norm T_α has an additive generator $t_\alpha : [0,1] \longrightarrow [0,\infty]$. For each $\alpha \in A$ define the function $h_\alpha : [a_\alpha, e_\alpha] \longrightarrow [0,\infty]$ by $h_\alpha = t_\alpha \circ \varphi_\alpha$, where $\varphi_\alpha : [a_\alpha, e_\alpha] \longrightarrow [0,1]$ is defined as in (3.22). Then for all $(x,y) \in [0,1]^2$

$$T(x,y) = \begin{cases} h_\alpha^{(-1)}(h_\alpha(x) + h_\alpha(y)) & \text{if } (x,y) \in [a_\alpha, e_\alpha]^2, \\ \min(x,y) & \text{otherwise.} \end{cases} \tag{3.23}$$

Proof: It is immediately clear that the composite function $h_\alpha = t_\alpha \circ \varphi_\alpha : [a_\alpha, e_\alpha] \longrightarrow [0,\infty]$ is strictly decreasing and satisfies $h_\alpha(e_\alpha) = 0$. Then

$$\begin{aligned} T(x,y) &= \varphi_\alpha^{-1}(T_\alpha(\varphi_\alpha(x), \varphi_\alpha(y))) \\ &= \varphi_\alpha^{-1}(f_\alpha^{(-1)}(t_\alpha \circ \varphi_\alpha(x) + t_\alpha \circ \varphi_\alpha(y))) \\ &= h_\alpha^{(-1)}(h_\alpha(x) + h_\alpha(y)) \end{aligned}$$

holds for each $\alpha \in A$ and for all $x, y \in [a_\alpha, e_\alpha]$. □

3.51 EXAMPLE

(i) For the ordinal sum T considered in Example 3.46(ii) the functions $h_1 : [0.1, 0.5] \longrightarrow [0,\infty]$, $h_2 : [0.7, 0.9] \longrightarrow [0,\infty]$ are given by, respectively,

$$h_1(x) = -\log\left(\tfrac{10x-1}{4}\right),$$
$$h_2(x) = \frac{9}{2} - 5x.$$

(ii) In the case of the ordinal sum T presented in Example 3.46(iii), for each $n \in \mathbb{N}$ the function $h_n : [\frac{1}{2^n}, \frac{1}{2^{n-1}}] \longrightarrow [0,\infty]$ is given by

$$h_n(x) = -\log(2^n x - 1).$$

It is easy to see that the representation of a t-norm as an ordinal sum of t-norms is not unique, in general. For instance, as an immediate consequence of Example 3.46(i) and Remark 3.47(i) we have for each subinterval $[a,e]$ of $[0,1]$

$$T_{\mathbf{M}} = (\emptyset) = (\langle 0, 1, T_{\mathbf{M}} \rangle) = (\langle a, e, T_{\mathbf{M}} \rangle).$$

This gives rise to a natural question: which t-norms T have a unique ordinal sum representation (in which case it necessarily must be the trivial representation $T = (\langle 0, 1, T \rangle)$)?

3.52 DEFINITION

A t-norm T which only has a trivial ordinal sum representation (i.e., there is no ordinal sum representation of T different from $T = (\langle 0, 1, T \rangle)$) is called *ordinally irreducible.*

3.53 PROPOSITION

For each t-norm T the following are equivalent:

(i) *T is not ordinally irreducible.*
(ii) *There is an $x_0 \in\,]0,1[$ such that $T(x_0, y) = \min(x_0, y)$ for all $y \in [0,1]$.*
(iii) *There exists a non-trivial idempotent element x_0 of T such that the vertical section $T(x_0, .) : [0,1] \longrightarrow [0,1]$ is continuous.*

Proof: It is trivial that (i) implies (iii), and the fact that (iii) implies (ii) can be shown using the same reasoning as in the proof of Proposition 2.3.

In order to show that (ii) implies (i) note that $\left([0, x_0] \, , T|_{[0,x_0]^2}, \leq \right)$ and $\left([x_0, 1] \, , T|_{[x_0,1]^2}, \leq \right)$ are fully ordered commutative semigroups with annihilators 0 and x_0 and neutral elements x_0 and 1, respectively. Consequently, the functions $T_1, T_2 : [0,1]^2 \longrightarrow [0,1]$ defined by

$$T_1(x,y) = \frac{T(x_0 x, x_0 y)}{x_0},$$

$$T_2(x,y) = \frac{T(x_0 + (1 - x_0)x, x_0 + (1 - x_0)y) - x_0}{1 - x_0},$$

respectively, are t-norms, and we have $T = (\langle 0, x_0, T_1 \rangle, \langle x_0, 1, T_2 \rangle)$, showing that T is not ordinally irreducible. □

3.54 PROPOSITION

For each t-norm $T \neq T_{\mathbf{M}}$ the following are equivalent:

(i) *T can be uniquely represented as an ordinal sum of ordinally irreducible t-norms.*
(ii) *For all $y \in [0,1]\}$, $M_T = \{x \in [0,1] \mid T(x,y) = \min(x,y)$ is a closed subset of $[0,1]$.*

Proof: Note first that, whenever a t-norm can be written as an ordinal sum of ordinally irreducible t-norms, this representation is clearly unique.

If for $T \neq T_{\mathbf{M}}$ the set M_T is closed then we have $\{0,1\} \subseteq M_T \subset [0,1]$ and, consequently, there is a finite or countable index set A and a family of pairwise disjoint open intervals $(]a_\alpha, e_\alpha[)_{\alpha \in A}$ such that

$$[0,1] \setminus M_T = \bigcup_{\alpha \in A} \,]a_\alpha, e_\alpha[\, .$$

Then, for each $\alpha \in A$, $\left([a_\alpha, e_\alpha] \, , T|_{[a_\alpha,e_\alpha]^2}, \leq \right)$ is a fully ordered commutative semigroup with annihilator a_α and neutral element e_α, and thus the function $T_\alpha : [0,1]^2 \longrightarrow [0,1]$ defined by

$$T_\alpha(x,y) = \frac{T(a_\alpha + (e_\alpha - a_\alpha)x, a_\alpha + (e_\alpha - a_\alpha)y)}{e_\alpha - a_\alpha}$$

is an ordinally irreducible t-norm. Obviously we have $T = (\langle a_\alpha, e_\alpha, T_\alpha \rangle)_{\alpha \in A}$.

Conversely, assume that M_T is not closed, i.e., there exists a sequence $(x_n)_{n\in\mathbb{N}}$ in M_T which converges to $x_0 \in]0,1[\setminus M_T$. Then for each ordinal sum representation $T = (\langle a_\alpha, e_\alpha, T_\alpha \rangle)_{\alpha \in A}$ we have $x_0 \in]a_\alpha, e_\alpha[$ for some $\alpha \in A$ and, subsequently, $x_n \in]a_\alpha, e_\alpha[$ for some $n \in \mathbb{N}$, showing that T_α is not ordinally irreducible. □

3.55 REMARK

(i) Each Archimedean t-norm has only trivial idempotent elements and, therefore, is ordinally irreducible.

(ii) The nilpotent minimum $T^{\mathbf{nM}}$ (see (1.11)) is an example of an ordinally irreducible t-norm with non-trivial idempotent elements.

(iii) Let T be a t-norm and $a, b \in M_T$ (as introduced in Proposition 3.54) with $a < b$. Then the requirements of Proposition 2.32(ii) are satisfied, and therefore $T_{ab} : [0,1]^2 \longrightarrow [0,1]$ defined by

$$T_{ab}(x,y) = \frac{T(a + (b-a)x, a + (b-a)y)}{b-a}$$

is a t-norm and there exists an ordinal sum representation of T such that one of its summands equals $\langle a, b, T_{ab} \rangle$.

(iv) For the t-norm T defined by

$$T(x,y) = \begin{cases} 0 & \text{if } (x,y) \in [0,0.5]^2, \\ 0.5 + \frac{1}{n+2} & \text{if } (x,y) \in \left]0.5 + \frac{1}{n+2}, 0.5 + \frac{1}{n+1}\right[, \\ \min(x,y) & \text{otherwise}, \end{cases}$$

the set $M_T = \{0\} \cup \{0.5 + \frac{1}{n+1} \mid n \in \mathbb{N}\}$ is not closed, showing that T has no ordinal sum representation where the t-norms in the summands are ordinally irreducible.

The following result about ordinal sum representations can be found in [Schweizer & Sklar 1983, Theorem 5.38].

3.56 COROLLARY

Let T be a t-norm satisfying

(i) $x_0 \in [0,1]$ is an idempotent element of T whenever $\lim_{x \searrow x_0} T(x,x) = x_0$.

(ii) If $x_0 \in [0,1]$ is an idempotent element of T then the one-place function $T(x_0, .) : [0,1] \longrightarrow [0,1]$ is continuous.

Then T is uniquely representable as an ordinal sum where in each summand the corresponding t-norm is Archimedean.

Proof: Because of Corollary 2.8 and Proposition 3.54 the t-norm T has a unique ordinal sum representation by means of ordinally irreducible t-norms. Each of these t-norms satisfies properties (i) and (ii) and, consequently, is Archimedean because of Theorem 2.12. □

3.57 EXAMPLE

Consider the non-continuous t-norm $(T_{\mathbf{D}})_{[f_{0.5}]}$ given by

$$(T_{\mathbf{D}})_{[f_{0.5}]}(x,y) = \begin{cases} 0 & \text{if } (x,y) < [0,0.5[^2, \\ \min(x,y) & \text{if } \max(x,y) = 1, \\ \min(x,y,0.5) & \text{otherwise,} \end{cases}$$

which is visualized in Figure 3.2 (right). Then $(T_{\mathbf{D}})_{[f_{0.5}]}$ satisfies the conditions (i) and (ii) in Corollary 3.56, and we obtain

$$(T_{\mathbf{D}})_{[f_{0.5}]} = (\langle 0, 0.5, T_{\mathbf{D}}\rangle, \langle 0.5, 1, T_{\mathbf{D}}\rangle).$$

In the case of t-conorms, a similar construction applying Theorem 3.42 can be used. Since the roles of the neutral element (which is 0 in this case) and the annihilator (which is 1) are interchanged in this case, we obtain max as semigroup operation in the system $(G_\beta)_{\beta \in B}$, when repeating the arguments in the proof of Theorem 3.43. Further, the linear order of the index set C should be reversed, i.e., $\gamma_1 \preceq \gamma_2$ if and only if $s(\gamma_1) \geq s(\gamma_2)$. To be precise, we have the following corollary of Theorem 3.43:

3.58 COROLLARY

Let $(S_\alpha)_{\alpha \in A}$ be a family of t-conorms and $(]a_\alpha, e_\alpha[)_{\alpha \in A}$ be a family of non-empty, pairwise disjoint open subintervals of $[0,1]$. Then the function $S : [0,1]^2 \longrightarrow [0,1]$ defined by

$$S(x,y) = \begin{cases} a_\alpha + (e_\alpha - a_\alpha) \cdot S_\alpha\left(\frac{x-a_\alpha}{e_\alpha - a_\alpha}, \frac{y-a_\alpha}{e_\alpha - a_\alpha}\right) & \text{if } (x,y) \in [a_\alpha, e_\alpha]^2, \\ \max(x,y) & \text{otherwise,} \end{cases} \tag{3.24}$$

is a t-conorm which is called the ordinal sum of the summands $\langle a_\alpha, e_\alpha, S_\alpha\rangle$, $\alpha \in A$, and we shall write $S = (\langle a_\alpha, e_\alpha, S_\alpha\rangle)_{\alpha \in A}$.

All results in this section given for t-norms remain valid for t-conorms with the obvious changes where necessary. In particular, concerning the duality of ordinal sums we get the following:

3.59 REMARK

Let $(\langle a_\alpha, e_\alpha, T_\alpha\rangle)_{\alpha \in A}$ be an ordinal sum of t-norms. Then the dual t-conorm given by (1.9) is just an ordinal sum of t-conorms, i.e., $(\langle 1-e_\alpha, 1-a_\alpha, S_\alpha\rangle)_{\alpha \in A}$, where each t-conorm S_α is the dual of the t-norm T_α. Note, however, that, for each $\alpha \in A$, the t-norm T_α and the t-conorm S_α in general act on different intervals.

Recently, some other methods for constructing new t-norms related to ordinal sums were introduced. We only mention the matrix composition [Drossos & Navara 1996] (see Proposition 3.68) and the annihilation-rotation method [Jenei 200w, 200x, 200y, 200z] (compare Proposition 3.71).

3.4 Constructions of non-continuous t-norms

We already gave examples of several non-continuous t-norms. Most of them are of rather theoretical interest only, e.g., in order to construct counterexamples.

We only mention the drastic product $T_{\mathbf{D}}$, the nilpotent minimum $T^{\mathbf{nM}}$ (Figure 1.5), the t-norms in Figure 2.1, in Example 1.24, or the t-norms defined in (2.10) and (2.11). Also the construction (3.11) in Theorem 3.6 may lead to non-continuous t-norms (see Figures 3.2, 3.4, 3.5, 3.6, 3.7, 3.8 and 3.10). An example of a non-continuous t-norm with an additive generator is given in Figure 3.12, a non-continuous ordinal sum is shown in Figure 3.4 (top left). Here we present some rather general constructions leading to non-continuous t-norms.

Generalizing Proposition 1.6, the following extension of binary operations on subintervals of the half-open unit interval $[0,1[$ to t-norms can be proved. The associativity and the monotonicity of the resulting t-norm can be derived from Theorem 3.42, but for the convenience of the reader we shall provide a complete proof.

3.60 PROPOSITION

Let A be a subinterval of the half-open unit interval $[0,1[$ and let $*: A^2 \longrightarrow A$ be an operation on A which satisfies for all $x,y,z \in A$ the properties (T1)–(T3) and (1.6). Then the function $T: [0,1]^2 \longrightarrow [0,1]$ defined by

$$T(x,y) = \begin{cases} x * y & \text{if } (x,y) \in A^2, \\ \min(x,y) & \text{otherwise,} \end{cases} \tag{3.25}$$

is a triangular norm.

Proof: The commutativity (T1) and the boundary condition (T4) are satisfied by definition. Concerning the associativity (T2), observe that for $x,y,z \in A$ we have $T(T(x,y),z) = T(x,T(y,z))$ as a consequence of the associativity of $*$. If at most one of the values x,y and z is contained in A, then

$$T(T(x,y),z) = \min(x,y,z) = T(x,T(y,z)).$$

Next, assume $x,y \in A$ and $z \notin A$. Then for $x < z$ we obtain

$$T(T(x,y),z) = \min(x*y,z) = x*y = T(x,\min(y,z)) = T(x,T(y,z)),$$

and for $x > z$ we have

$$T(T(x, y), z) = \min(x * y, z) = z = T(x, \min(y, z)) = T(x, T(y, z)).$$

The remaining cases are completely analogous. For the monotonicity, fix $x, y, z \in [0,1]$ with $y \le z$. If $x \notin A$, or if $x \in A$ and both y and z are either elements of A or of $[0,1] \setminus A$, then $T(x, y) \le T(x, z)$ is inherited from the monotonicity of min and $*$. If $x \in A, y \notin A$ and $z \in A$ then clearly $y \le u$ for all $u \in A$, implying $T(x, y) = y \le x * z = T(x, z)$. The last case, namely, $x \in A, y \in A$ and $z \notin A$, is analogous. □

3.61 EXAMPLE

(i) Let $[a, b]$ be a closed subinterval of $[0, 1[$, let $f : [a, b] \longrightarrow [0, \infty]$ be a continuous, non-increasing function with $f(b) = 0$, and define the binary operation $*$ on $[a, b]$ by

$$x * y = f^{(-1)}(f(x) + f(y)),$$

Clearly, $*$ satisfies all the requirements in Proposition 3.60 and, consequently, the function $T : [0,1]^2 \longrightarrow [0,1]$ given by (3.25) is a t-norm.

(ii) If in (i) we put $a = 0$ and $f(x) = 0$ for all $x \in [0, b]$, then the semigroup $([0,1], T)$, where T is given by (3.25), is an ordinal sum of the semigroups $G_1 = ([0, b], *)$ and $G_2 = (]b, 1], \min)$ in the sense of Theorem 3.42, but T is not an ordinal sum of ordinally irreducible t-norms (compare Definition 3.52). Note, however, that T is an ordinal sum of t-subnorms (see Remark 3.45(ii)).

3.62 REMARK

Let A be a subinterval of the half-open unit interval $[0, 1[$ and let $*$ be a binary operation on A satisfying the requirements in Proposition 3.60. Then the t-norm T given by (3.25) is continuous if and only if $*$ is continuous and if for all $x \in A$ we have $\sup\{x * y \mid y \in A\} = x$.

The following t-norms are interesting in the sense that (with the exception of the two extremal cases) they have both nilpotent and non-trivial idempotent elements (in Figure 3.18 a 3D plot of such a t-norm is given).

3.63 PROPOSITION

Let A be a subset of $]0, 1[^2$ with the following properties:

(i) A is symmetric, i.e., $(x, y) \in A$ implies $(y, x) \in A$.
(ii) For all $(x, y) \in A$ we have $]0, x] \times]0, y] \subseteq A$.

Then the following function $T_A : [0,1]^2 \longrightarrow [0,1]$ is a t-norm:

$$T_A(x, y) = \begin{cases} 0 & \text{if } (x, y) \in A, \\ \min(x, y) & \text{otherwise.} \end{cases} \tag{3.26}$$

Proof: When proving the associativity (T2) of T_A, the only non-trivial situation occurs when both (x, y) and (y, z) are elements of $]0,1[^2 \setminus A$. If, in this case, $x \leq y$ and $z \leq y$ then we obtain

$$T_A(T_A(x,y),z) = T_A(x,z) = T_A(x,T_A(y,z)),$$

otherwise the properties of A imply $(x, z) \in\,]0,1[^2 \setminus A$ and, subsequently,

$$T_A(T_A(x,y),z) = \min(x,y,z) = T_A(x,T_A(y,z)).$$

The properties (T1), (T3) and (T4) are obvious. □

3.64 REMARK

Let $A \subseteq\,]0,1[^2$ be a set which satisfies the conditions in Proposition 3.63.

(i) A t-norm T is of the form T_A if and only if we have $T(x,y) \in \{0, \min(x,y)\}$ for all $(x,y) \in [0,1]^2$.

(ii) The t-norm T_A is left-continuous if and only if the set

$$(\{0\} \times [0,1]) \cup ([0,1] \times \{0\}) \cup A$$

is a closed subset of $[0,1]^2$, and it is continuous only in the trivial case $A = \emptyset$ (in which case we obtain $T_A = T_{\mathbf{M}}$).

(iii) If $A \neq \emptyset$, then T_A always has nilpotent elements and zero divisors.

(iv) The only Archimedean case is $A = \,]0,1[^2$ (which means $T_A = T_{\mathbf{D}}$), in all the other cases T_A has non-trivial idempotent elements.

(v) If $f : [0,1] \longrightarrow [0,1]$ is a non-decreasing function, then the set

$$A_f = \{(x,y) \in\,]0,1[^2 \mid f(x) + f(y) \leq 1\}$$

satisfies all the properties in Proposition 3.63, and T_{A_f} is a t-norm (this is also true if we replace the constant 1 by any other real number).

3.65 EXAMPLE

(i) If $f : [0,1] \longrightarrow [0,1]$ is a strictly increasing bijection then the t-norm T_{A_f} given in Remark 3.64(v) is called the *f-nilpotent minimum* (compare [Perny 1992, Perny & Roy 1992, Fodor 1995]).

(ii) In the special case $f = \mathrm{id}_{[0,1]}$ we obtain the nilpotent minimum $T^{\mathbf{nM}}$, which was introduced in Remark 1.21(i) and which is visualized in Figure 1.5.

(iii) If $f : [0,1] \longrightarrow [0,1]$ is a strictly increasing bijection and if T is the nilpotent t-norm defined by

$$T(x,y) = f^{-1}(T_{\mathbf{L}}(f(x), f(y)))$$

(i.e., if f is the unique isomorphism between T and the Łukasiewicz t-norm $T_{\mathbf{L}}$, compare Corollary 5.7(ii) and Remark 5.8), then T_{A_f} is the strongest t-norm which vanishes exactly in the same points of the unit square $[0,1]^2$ as T.

(iv) Also the t-norm visualized on the right-hand side of Figure 3.8 is of the type T_A.

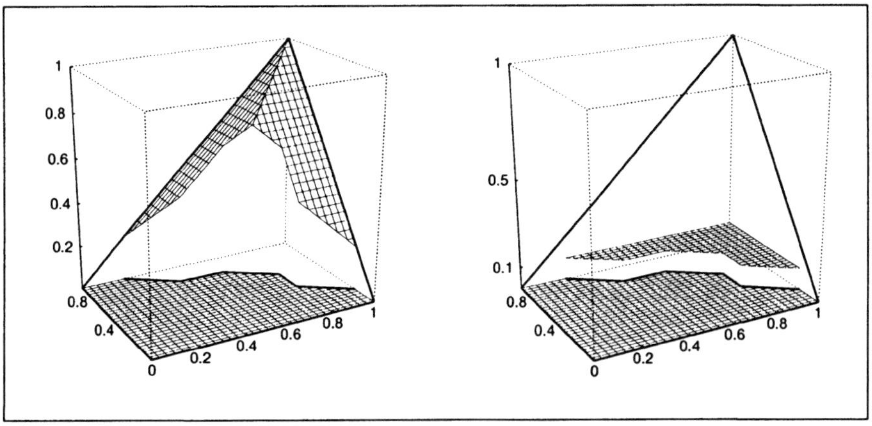

Fig. 3.18. The t-norms T_A (left, see Proposition 3.63) and $T_{A,0.1}$ (see Proposition 3.66); the underlying set A is clearly visible in both cases

The following construction even leads to an Archimedean t-norm (for a 3D visualization see Figure 3.18).

3.66 PROPOSITION
Let A be a subset of $]0,1[^2$ which satisfies the conditions in Proposition 3.63, and assume that $a \in]0,1[$ fulfills $\{a\} \times]0,1[\subseteq A$. Then the following function $T_{A,a} : [0,1]^2 \longrightarrow [0,1]$ is an Archimedean t-norm:

$$T_{A,a}(x,y) = \begin{cases} 0 & \textit{if } (x,y) \in A, \\ a & \textit{if } (x,y) \in]0,1[^2 \setminus A, \\ \min(x,y) & \textit{otherwise,} \end{cases} \tag{3.27}$$

Proof: Since for all $x, y, z \in [0,1]$ we get $T_{A,a}(T_{A,a}(x,y), z) = 0$, $T_{A,a}$ is associative. The properties (T1), (T3) and (T4) are obvious. □

As an immediate consequence of Proposition 3.66 we obtain the following sufficient condition:

3.67 COROLLARY
If $f : [0,1] \longrightarrow [0,1]$ is a non-decreasing function and if $a \in \left]0, f^{(-1)}(1 - f(1))\right[$ then $T_{A_f,a}$ is an Archimedean t-norm.

The method of *matrix composition* presented in [Drossos & Navara 1996] allows generalized t-norms to be constructed on abstract lattices rather than on the unit interval only. Restricted to the case of $[0,1]$, we get the following special situation (the associativity of the resulting t-norm can be checked directly):

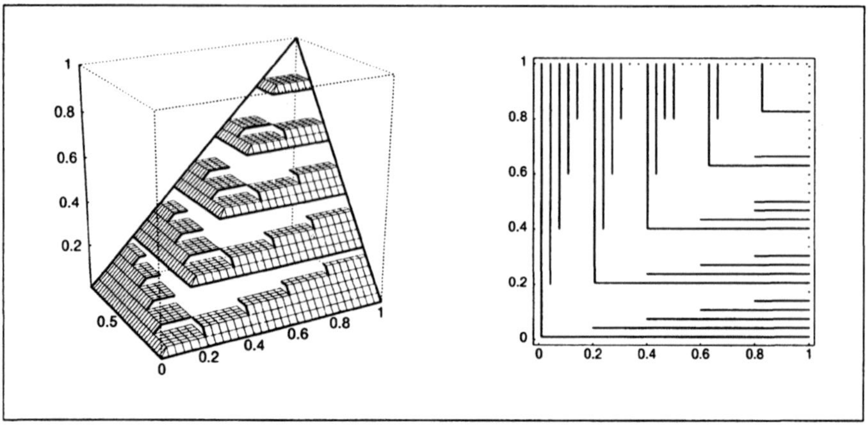

Fig. 3.19. 3D and contour plot of a t-norm obtained via matrix composition based on the chain $\{0, 0.2, 0.4, 0.6, 0.8, 1\}$ (see Proposition 3.68)

3.68 PROPOSITION

Given $n \in \mathbb{N}$, consider a finite subchain $\{\gamma_0, \gamma_1, \dots, \gamma_n\}$ of $[0,1]$ satisfying $0 = \gamma_0 < \gamma_1 < \dots < \gamma_n = 1$ and, for each $x \in [0,1[$, the index $i_x \in \{1, 2, \dots, n\}$ given by $i_x = \min\{j \in \{1, 2, \dots, n\} \mid x < \gamma_j\}$. Further, let $(c_{jk})_{j,k \in \{1,2,\dots,n\}}$ be a symmetric $n \times n$-matrix consisting of numbers in $[0,1]$ such that, for all $j \in \{1, 2, \dots, n\}$, we have $c_{jj} \leq \gamma_j$ and $c_{jk} \leq c_{jl}$ whenever $k \leq l$. If we define the matrix $(T_{jk})_{j,k \in \{1,2,\dots,n\}}$ of t-norms by

$$T_{jk} = \begin{cases} (\langle c_{jk}, 1, T_{\mathbf{D}} \rangle) & \text{if } c_{jk} < 1, \\ T_{\mathbf{M}} & \text{otherwise,} \end{cases}$$

then the function $T : [0,1]^2 \longrightarrow [0,1]$ given by

$$T(x,y) = \begin{cases} T_{i_x i_y}(x,y) & \text{if } (x,y) \in [0,1[^2, \\ \min(x,y) & \text{otherwise,} \end{cases} \tag{3.28}$$

is a t-norm which is non-continuous except for the trivial case $n = 1$ and $c_{11} = 1$. Moreover, for all $j, k \in \{1, 2, \dots, n\}$ the restrictions of T and T_{jk} to $[\gamma_{j-1}, \gamma_j[\times [\gamma_{k-1}, \gamma_k[$ coincide.

Figure 3.19 provides an illustration of a matrix composition based on the chain $\{0, 0.2, 0.4, 0.6, 0.8, 1\}$. In this case the matrix $(c_{jk})_{j,k \in \{1,2,\dots,5\}}$ is computed by the formula $c_{jk} = \frac{1}{30}(5\min(j,k) + \max(j,k) - 5)$, i.e.,

$$(c_{jk})_{j,k \in \{1,2,\dots,5\}} = \frac{1}{30} \begin{pmatrix} 1 & 2 & 3 & 4 & 5 \\ 2 & 7 & 8 & 9 & 10 \\ 3 & 8 & 13 & 14 & 15 \\ 4 & 9 & 14 & 19 & 20 \\ 5 & 10 & 15 & 20 & 25 \end{pmatrix}$$

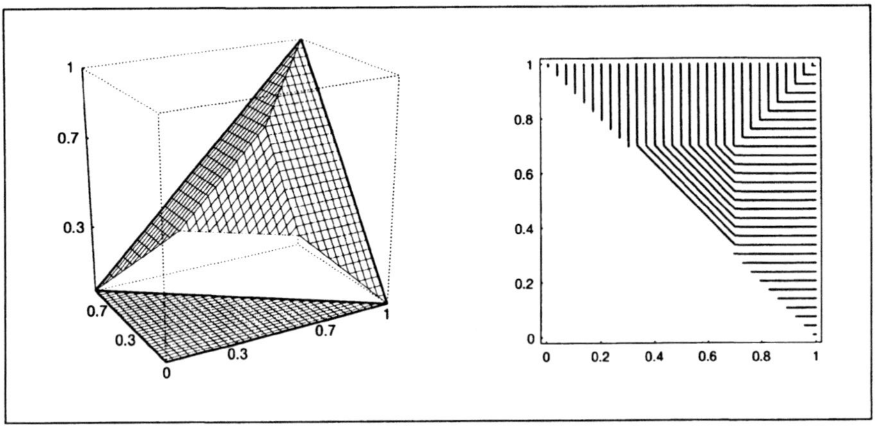

Fig. 3.20. 3D and contour plot of the Jenei t-norm $(T_{\mathbf{L}})^{\mathbf{J}}_{0.3}$(see Example 3.69)

Note that (3.28) not necessarily yields an associative function if one starts from a matrix $(T_{jk})_{j,k\in\{1,2,\ldots,n\}}$ of t-norms different from that in Proposition 3.68.

The following family of non-continuous t-norms is a special subclass of the t-norms described in [Jenei 2000] (where a different but equivalent formula has been used). A member of this family of t-norms is visualized in Figure 3.20.

3.69 EXAMPLE

Fix a nilpotent t-norm T and its unique additive generator $t : [0,1] \longrightarrow [0,\infty]$ satisfying $t(0) = 1$, i.e., $T(x,y) = t^{(-1)}(t(x)+t(y))$ for all $(x,y) \in [0,1]^2$. Then the family $(T^{\mathbf{J}}_\lambda)_{\lambda\in[0,t^{(-1)}(0.5)]}$ of *Jenei t-norms* is given by

$$T^{\mathbf{J}}_\lambda(x,y) = \begin{cases} 0 & \text{if } t(x)+t(y) \geq 1, \\ t^{(-1)}(t(x)+t(y)-1+t(\lambda)) & \\ \qquad \text{if } t(x)+t(y) < 1 \text{ and } (x,y) \in \big[\lambda, t^{(-1)}(1-t(\lambda))\big]^2, & \\ \min(x,y) & \text{otherwise.} \end{cases}$$

3.70 REMARK

Fix a nilpotent t-norm T and its additive generator t as in Example 3.69.

(i) Obviously, we have $T^{\mathbf{J}}_0 = T$. If we define the strictly increasing function $f : [0,1] \longrightarrow [0,1]$ by $f(x) = 1 - t(x)$, then the Jenei t-norm $T^{\mathbf{J}}_{t^{(-1)}(0.5)}$ coincides with the t-norm T_{A_f} discussed in Remark 3.64(v) and Example 3.65(i), i.e., it is an f-nilpotent minimum.

(ii) Each Jenei t-norm $T^{\mathbf{J}}_\lambda$ is left-continuous, but only $T^{\mathbf{J}}_0$ is continuous.

(iii) The family $(T^{\mathbf{J}}_\lambda)_{\lambda\in[0,t^{(-1)}(0.5)]}$ of Jenei t-norms is strictly increasing.

(iv) The family $(T^{\mathbf{J}}_\lambda)_{\lambda\in[0,t^{(-1)}(0.5)]}$ of Jenei t-norms is also continuous with respect to the index λ, i.e., for all $\lambda_0 \in [0, t^{(-1)}(0.5)]$ we immediately get $\lim\limits_{\lambda\to\lambda_0} T^{\mathbf{J}}_\lambda = T^{\mathbf{J}}_{\lambda_0}$.

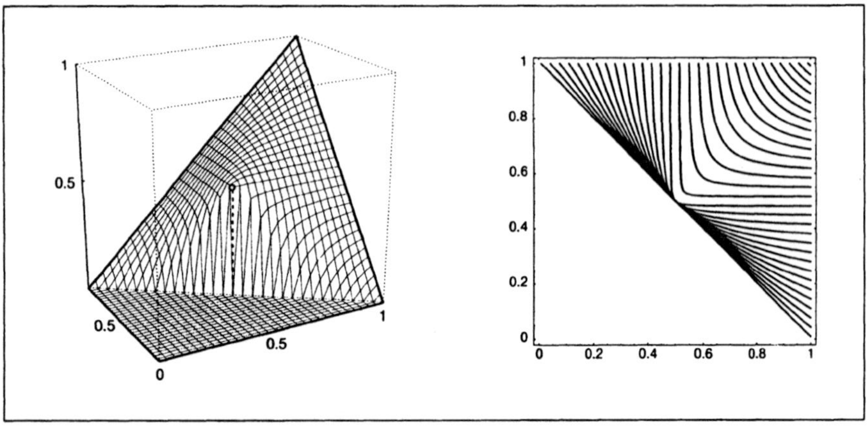

Fig. 3.21. Rotation method: 3D and contour plot of the t-norm $(T_0^{\mathbf{H}})_{\mathrm{rot}}$ (see Example 3.72 and observe the discontinuity at $(0.5, 0.5)$)

(v) For all $\lambda \in [0, t^{(-1)}(0.5)]$ and for all $x, y, z \in [0,1]$ we have

$$T_\lambda^{\mathbf{J}}(x,y) \leq z \quad \text{if and only if} \quad T_\lambda^{\mathbf{J}}\big(t^{(-1)}(1-t(z)), y\big) \leq t^{(-1)}(1-t(x)).$$

One consequence of this is that, in the context of fuzzy logics, for Jenei t-norms the corresponding S-implications (11.4) and R-implications (2.17) coincide (see Section 11.1 and [Fodor 1995]).

Several methods for constructing and characterizing left-continuous t-norms T satisfying $T(x,y) = 0$ if and only if $x + y \leq 1$ have been proposed recently in [Jenei 200x]. Such t-norms are interesting in the context of fuzzy logics (see Section 11.1), since in this case the negation derived from the R-implication (2.17) associated with T equals the standard negation $N_{\mathbf{s}}$ (11.1). An interesting special case is the following *rotation method* (the associativity of the resulting t-norm can be checked directly). Note that in [Jenei 200x] a different but equivalent formula was used.

3.71 PROPOSITION
Let T be a left-continuous t-norm without zero divisors. Then the function $T_{\mathrm{rot}} : [0,1] \longrightarrow [0,1]$ defined by

$$T_{\mathrm{rot}}(x,y) = \begin{cases} \frac{1}{2}\big(1 + T(2x-1, 2y-1)\big) & \textit{if } (x,y) \in \,]0.5,1]^2, \\ \frac{1}{2}\big(1 - \sup\{z \in \,]0.5,1] \mid T(2x-1, 2z-1) \leq 1-2y\}\big) & \\ \qquad \textit{if } (x,y) \in \,]0.5,1] \times [0,0.5] \textit{ and } x+y > 1, & \\ \frac{1}{2}\big(1 - \sup\{z \in \,]0.5,1] \mid T(2y-1, 2z-1) \leq 1-2x\}\big) & \\ \qquad \textit{if } (x,y) \in [0,0.5] \times \,]0.5,1] \textit{ and } x+y > 1, & \\ 0 & \textit{otherwise}, \end{cases}$$

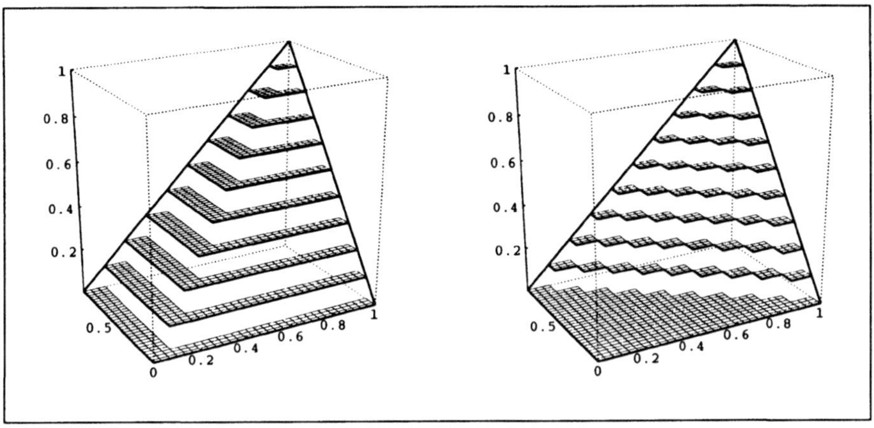

Fig. 3.22. Discretization of $T_{\mathbf{M}}$ and $T_{\mathbf{L}}$: the t-norms $(T_{\mathbf{M}})^{[10]}$ (left) and $(T_{\mathbf{L}})^{[10]}$ (see Example 3.74(ii) and (iii))

is a left-continuous t-norm which is non-continuous in $(0.5, 0.5)$ *and which satisfies* $T_{\text{rot}}(x, y) = 0$ *if and only if* $x + y \leq 1$.

3.72 EXAMPLE

(i) It is not difficult to see that, starting with the minimum $T_{\mathbf{M}}$, we obtain the nilpotent minimum, i.e., $(T_{\mathbf{M}})_{\text{rot}} = T^{\mathbf{nM}}$. If the initial t-norm is the Hamacher product $T_0^{\mathbf{H}}$ (see Section 4.3) then $(T_0^{\mathbf{H}})_{\text{rot}}$ is given by (for a visualization see Figure 3.21)

$$(T_0^{\mathbf{H}})_{\text{rot}}(x, y) = \begin{cases} \frac{x+y-1}{1-(2-2x)(2-2y)} & \text{if } x + y > 1, \\ 0 & \text{otherwise.} \end{cases}$$

(ii) If we start with a strict t-norm T then $(0.5, 0.5)$ is the only point of discontinuity of T_{rot}.

From a practical and computational point of view it is quite often desirable that a t-norm T (restricted to the open unit square $]0, 1[^2$) assumes only finitely many values in the chain $\Gamma_{(n)} = \{0, \frac{1}{n}, \frac{2}{n}, \ldots, \frac{n-1}{n}, 1\}$ for some $n \in \mathbb{N}$. The proof of the following lemma is obvious.

3.73 LEMMA

For each $n \in \mathbb{N}$ *define the function* $f_n : [0, 1] \longrightarrow \Gamma_{(n)}$ *by* $f_n(x) = \frac{1}{n}\lfloor nx \rfloor$. *Then the operation* $T^{[n]} : [0, 1]^2 \longrightarrow [0, 1]$ *given by*

$$T^{[n]}(x, y) = \begin{cases} f_n(T(f_n(x), f_n(y))) & \textit{if } (x, y) \in [0, 1[^2, \\ \min(x, y) & \textit{otherwise,} \end{cases}$$

satisfies the axioms (T1), (T3) and (T4), but not necessarily (T2).

In fact, $T^{[n]}$ is associative (and, subsequently, a t-norm) if and only if $(\varGamma_{(n)}, T^{[n]}|_{\varGamma_{(n)}^2}, \leq)$ is a commutative, fully ordered semigroup with neutral element 1, i.e., if $T^{[n]}|_{\varGamma_{(n)}^2}$ is a discrete t-norm in the sense of Section 7.3 (compare Proposition 7.30).

3.74 EXAMPLE

(i) For each t-norm T we obtain $T^{[1]} = T_{\mathbf{D}}$. However, if the t-norm T is strictly monotone then $T^{[n]}$ is never associative for $n \geq 2$.

(ii) If the t-norm T satisfies $T(x,y) \in \{0, \min(x,y)\}$ for all $(x,y) \in [0,1]^2$ then $T^{[n]}$ is a t-norm for each $n \in \mathbb{N}$. Subsequently, $(T_{\mathbf{M}})^{[n]}$ (see Figure 3.22), $(T_{\mathbf{D}})^{[n]}$, $(T^{\mathbf{nM}})^{[n]}$, and $(T_A)^{[n]}$ (for the definition of T_A see Proposition 3.63) are always t-norms.

(iii) If T is a nilpotent t-norm with additive generator $t : [0,1] \longrightarrow [0,\infty]$ and if for all $i, j \in \{0, 1, \ldots, n\}$ we have

$$t(\tfrac{i}{n}) + t(\tfrac{j}{n}) \in \mathrm{Ran}(t|_{\varGamma_{(n)}}) \cup [t(0), \infty],$$

then $T^{[n]}$ is a t-norm (for a proof see Proposition 7.29). For instance, $(T_{\mathbf{L}})^{[n]}$ is a t-norm for each $n \in \mathbb{N}$ (see Figure 3.22).

(iv) If T is the nilpotent t-norm with additive generator $t : [0,1] \longrightarrow [0,\infty]$

$$t(x) = \begin{cases} 3 - 2x & \text{if } x \in [0, 0.5], \\ 4 - 4x & \text{otherwise,} \end{cases}$$

then $T^{[n]}$ is a t-norm if and only if $n = 1$ or if n is an even number.

Although we have considered already a lot of non-continuous t-norms, they all were Borel measurable functions from the unit square into the unit interval. However, the axioms (T1)–(T4) for t-norms are, analytically speaking, so weak that they do not even imply their Borel measurability (as functions from $[0,1]^2$ into $[0,1]$). A first example of a t-norm which is not Borel measurable was given in [Klement 1982b].

3.75 EXAMPLE

If C is a subset of $]0, 0.5[$ which is not a Borel measurable set (such a set exists since the σ-algebra of Borel subsets of $]0, 0.5[$ is a proper subclass of the power set $\mathcal{P}(]0, 0.5[)$, see [Bauer 1981]), then the sets

$$\begin{aligned} B &= \{x \in \,]0,1[\; \mid x \in C \text{ or } 1 - x \in C\}, \\ A &= \{(x,y) \in \,]0,1[^2 \mid x + y < 1\} \cup \{(x, 1-x) \in \,]0,1[^2 \mid x \in B\}, \end{aligned}$$

are non-Borel subsets of $]0,1[$ and $]0,1[^2$, respectively. Moreover, A satisfies the properties in Proposition 3.63, and the resulting t-norm T_A is not Borel measurable.

4. Families of t-norms

> A good stock of examples, as large as possible, is indispensable for a thorough understanding of any concept, and when I want to learn something new, I make it my first job to build one.
>
> PAUL R. HALMOS

The aim of this chapter is to provide the reader with a collection of parameterized families of t-norms which we think are interesting from various points of view. We have chosen this compact form of presentation in order to simplify the search for specific examples.

To start with, we recall the four basic triangular norms together with their dual t-conorms and summarize some of their most important properties, including their connection to several functional equations.

Then we present (following the order of their introduction) the following families of t-norms and t-conorms: Schweizer-Sklar, Hamacher, Frank, Yager, Dombi, Sugeno-Weber, Aczél-Alsina, and Mayor-Torrens t-norms and t-conorms. We also argue why we feel they are of special significance, either from a theoretical point of view (although some of the results proving this will be presented later) or because of their applications. Even more families of t-norms and t-conorms are listed in, e.g., [Mizumoto 1989a, Lowen 1996].

We shall not give in each single case the proof that the function in question is indeed a t-norm. In fact, in most cases this is quite easily seen: for the continuous Archimedean t-norms (the majority of cases considered) this follows as a classroom exercise from the construction in Theorem 3.23 using additive generators (which are also included here, together with illustrations of some prototypical examples thereof), the remaining t-norms are either basic t-norms or ordinal sums.

We also mention the monotonicity of each of these families and their continuity with respect to the parameter (in those cases where this is not completely obvious, an exact reference to the proof is provided).

Most of the information given in Sections 4.2–4.9 will be summarized in a tabular form in Table 4.1 and also in Appendix A, where also visualizations of some prototypical members of these families of t-norms are given.

4.1 Basic t-norms and t-conorms

The four basic t-norms introduced in Example 1.2, namely, the minimum $T_{\mathbf{M}}$, the product $T_{\mathbf{P}}$, the Łukasiewicz t-norm $T_{\mathbf{L}}$, and the drastic product $T_{\mathbf{D}}$, are very special and interesting t-norms from several points of view.

First of all, the four basic t-norms are special with respect to the order and with respect to important algebraic properties:

4.1 REMARK

(i) In Remark 1.5(i) we have shown that the minimum $T_{\mathbf{M}}$ is the strongest and that the drastic product $T_{\mathbf{D}}$ is the weakest t-norm, i.e., for each t-norm T we have $T_{\mathbf{D}} \leq T \leq T_{\mathbf{M}}$.

(ii) The minimum $T_{\mathbf{M}}$ is the only idempotent t-norm, i.e., the only t-norm whose set of idempotent elements coincides with the unit interval $[0, 1]$ (see Example 2.2(i)).

(iii) The product $T_{\mathbf{P}}$ is a prototypical strict t-norm in the sense that each strict t-norm T is isomorphic to $T_{\mathbf{P}}$ (see Proposition 5.9).

(iv) The Łukasiewicz t-norm $T_{\mathbf{L}}$ is a prototypical nilpotent t-norm in the sense that each nilpotent t-norm T is isomorphic to $T_{\mathbf{L}}$ (see Proposition 5.10).

(v) In Corollary 5.4 we shall see that a function $T : [0,1]^2 \longrightarrow [0,1]$ is a continuous Archimedean t-norm if and only if there exists a continuous, strictly increasing function $\theta : [0,1] \longrightarrow [0,1]$ with $\theta(1) = 1$ (which is uniquely determined up to a positive constant exponent and which is just a multiplicative generator of T) such that for all $(x, y) \in [0,1]^2$ we have

$$T(x,y) = \theta^{(-1)}(T_{\mathbf{P}}(\theta(x), \theta(y))). \tag{4.1}$$

(vi) Admitting also non-continuous functions $\theta : [0,1] \longrightarrow [0,1]$ even the drastic product $T_{\mathbf{D}}$ can be obtained by (4.1): one possibility is to put

$$\theta(x) = \begin{cases} \frac{1+x}{4} & \text{if } x \in [0,1[, \\ 1 & \text{if } x = 1. \end{cases}$$

(vii) In Proposition 2.22(iii) we have seen that $(T_{\mathbf{M}}, S_{\mathbf{M}})$ is the only pair of a t-norm and a t-conorm which are mutually distributive over each other.

The four basic t-norms $T_{\mathbf{M}}$, $T_{\mathbf{P}}$, $T_{\mathbf{L}}$, and $T_{\mathbf{D}}$ are also limit members of several families of t-norms considered in Sections 4.2–4.9 (see in particular Proposition 8.5 and Example 3.32(iii)). Moreover, they can also be characterized by some rather special functional equations.

The basic t-norms $T_{\mathbf{M}}$, $T_{\mathbf{P}}$ and $T_{\mathbf{L}}$ are also copulas (see Section 9.1 and [Nelsen 1999]). The Łukasiewicz t-norm $T_{\mathbf{L}}$ is the weakest copula (Fréchet-Hoeffding lower bound), and the minimum $T_{\mathbf{M}}$ is the strongest copula (Fréchet-Hoeffding upper bound, see Proposition 9.8(i)). The copulas $T_{\mathbf{M}}$, $T_{\mathbf{P}}$ and $T_{\mathbf{L}}$ are also limit cases of some distinguished families of copulas.

4.2 Remark

(i) The minimum $T_{\mathbf{M}}$ is the only t-norm T satisfying for all $x \in [0,1]$ the equation $T(x,x) = x$ (see Proposition 1.9(i)).

(ii) As an immediate consequence of Theorem 5.14, the minimum $T_{\mathbf{M}}$ and the maximum $S_{\mathbf{M}}$ are the only t-norm T and t-conorm S, respectively, such that for all $x, y \in [0,1]$ we have

$$T(x,y) + S(x,y) = x + y,$$
$$T(x,y) \cdot S(x,y) = x \cdot y.$$

(iii) The product $T_{\mathbf{P}}$ is the only continuous t-norm T satisfying for all $x \in [0,1]$ the equations (see Remark 5.13(ii))

$$T(x,x) = x^2,$$
$$T(x,1-x) = x - x^2.$$

(iv) The Łukasiewicz t-norm $T_{\mathbf{L}}$ is the only t-norm T satisfying

$$T(x+z,y) = T(x,y) + z$$

for all $x, y, z \in [0,1]$ such that $x + z \in [0,1]$ and $x + y \geq 1$ (see Remark 5.13(iii)).

(v) The drastic product $T_{\mathbf{D}}$ is the only t-norm T satisfying for all $x \in [0,1[$ the equation $T(x,x) = 0$ (see Proposition 1.9(ii)).

We only mention that the Łukasiewicz t-norm $T_{\mathbf{L}}$, the minimum $T_{\mathbf{M}}$, and the product $T_{\mathbf{P}}$ are the starting points for the Łukasiewicz, the Gödel, and the product fuzzy logic, respectively, as discussed in Section 11.3 (for a more detailed exposition see [Hájek 1998b]).

The minimum $T_{\mathbf{M}}$ obviously plays a key role in a whole branch of mathematics, the so-called idempotent analysis [Kolokoltsov & Maslov 1997].

A particular feature of the Łukasiewicz t-norm $T_{\mathbf{L}}$ is that for $T_{\mathbf{L}}$-measures it is possible to prove, e.g., a Jordan Decomposition Theorem, the Darboux property, and a Liapounoff Theorem (Theorems 14.7 and 14.9, for full proofs consult [Butnariu & Klement 1993]).

4.2 Schweizer-Sklar t-norms

Already in [Schweizer & Sklar 1961] a most interesting family of t-norms was presented, and in [Schweizer & Sklar 1963] the index set was extended to the whole real line (observe that our notation follows the monograph [Schweizer & Sklar 1983], i.e., our index λ corresponds to $-p$ in the original papers). This family of t-norms is remarkable in the sense that it contains all four basic t-norms. Most properties of this family of t-norms are summarized in Appendix A.6, where also 3D and contour plots, as well as plots of diagonal sections, of some representative members are given.

4.3 EXAMPLE

(i) The family $(T_\lambda^{\mathbf{SS}})_{\lambda\in[-\infty,\infty]}$ of *Schweizer-Sklar t-norms* is given by

$$T_\lambda^{\mathbf{SS}}(x,y)=\begin{cases}T_{\mathbf{M}}(x,y) & \text{if } \lambda=-\infty,\\ T_{\mathbf{P}}(x,y) & \text{if } \lambda=0,\\ T_{\mathbf{D}}(x,y) & \text{if } \lambda=\infty,\\ \left(\max\left((x^\lambda+y^\lambda-1),0\right)\right)^{\frac{1}{\lambda}} & \text{if } \lambda\in\,]-\infty,0[\,\cup\,]0,\infty[\,.\end{cases}$$

(ii) The family $(S_\lambda^{\mathbf{SS}})_{\lambda\in[-\infty,\infty]}$ of *Schweizer-Sklar t-conorms* is given by

$$S_\lambda^{\mathbf{SS}}(x,y)=\begin{cases}S_{\mathbf{M}}(x,y) & \text{if } \lambda=-\infty,\\ S_{\mathbf{P}}(x,y) & \text{if } \lambda=0,\\ S_{\mathbf{D}}(x,y) & \text{if } \lambda=\infty,\\ 1-\left(\max\left(((1-x)^\lambda+(1-y)^\lambda-1),0\right)\right)^{\frac{1}{\lambda}} & \\ \qquad\qquad \text{if } \lambda\in\,]-\infty,0[\,\cup\,]0,\infty[\,.\end{cases}$$

4.4 REMARK

(i) Because of $T_1^{\mathbf{SS}}=T_{\mathbf{L}}$ and $S_1^{\mathbf{SS}}=S_{\mathbf{L}}$ the families of Schweizer-Sklar t-norms and t-conorms contain all four basic t-norms and t-conorms, respectively. The subfamily $(T_\lambda^{\mathbf{SS}})_{\lambda\in[-\infty,1]}$ of Schweizer-Sklar t-norms is also a family of copulas (see Section 9.1), in which context (e.g. in [Nelsen 1999]) it is referred to as the family of *Clayton copulas* [Clayton 1978].

(ii) For each $\lambda\in[-\infty,\infty]$ the t-norm $T_\lambda^{\mathbf{SS}}$ and the t-conorm $S_\lambda^{\mathbf{SS}}$ are dual to each other.

(iii) All Schweizer-Sklar t-norms with the exception of $T_\infty^{\mathbf{SS}}$ are continuous. A Schweizer-Sklar t-norm $T_\lambda^{\mathbf{SS}}$ is Archimedean if and only if $\lambda\in\,]-\infty,\infty]$, it is strict if and only if $\lambda\in\,]-\infty,0]$, and it is nilpotent if and only if $\lambda\in\,]0,\infty[$.

(iv) Additive generators $t_\lambda^{\mathbf{SS}}, s_\lambda^{\mathbf{SS}}:[0,1]\longrightarrow[0,\infty]$ of the continuous Archimedean members of the families of Schweizer-Sklar t-norms (see Figure 4.1) and t-conorms are given by, respectively,

$$t_\lambda^{\mathbf{SS}}(x)=\begin{cases}-\log x & \text{if } \lambda=0,\\ \frac{1-x^\lambda}{\lambda} & \text{if } \lambda\in\,]-\infty,0[\,\cup\,]0,\infty[\,,\end{cases}$$

$$s_\lambda^{\mathbf{SS}}(x)=\begin{cases}-\log(1-x) & \text{if } \lambda=0,\\ \frac{1-(1-x)^\lambda}{\lambda} & \text{if } \lambda\in\,]-\infty,0[\,\cup\,]0,\infty[\,.\end{cases}$$

(v) Multiplicative generators $\theta_\lambda^{\mathbf{SS}}, \xi_\lambda^{\mathbf{SS}}:[0,1]\longrightarrow[0,1]$ of the continuous Archimedean members of the families of Schweizer-Sklar t-norms and t-conorms are given by, respectively,

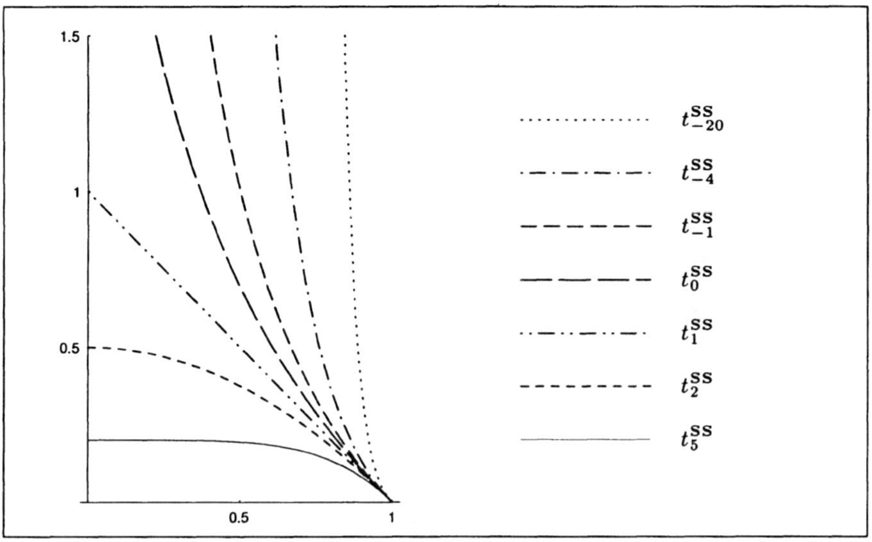

Fig. 4.1. Additive generators of some Schweizer-Sklar t-norms

$$\theta^{\mathbf{SS}}_{\lambda}(x) = \begin{cases} x & \text{if } \lambda = 0, \\ e^{\frac{x^{\lambda}-1}{\lambda}} & \text{if } \lambda \in \,]-\infty, 0[\,\cup\,]0, \infty[\,, \end{cases}$$

$$\xi^{\mathbf{SS}}_{\lambda}(x) = \begin{cases} 1 - x & \text{if } \lambda = 0, \\ e^{\frac{(1-x)^{\lambda}-1}{\lambda}} & \text{if } \lambda \in \,]-\infty, 0[\,\cup\,]0, \infty[\,. \end{cases}$$

(vi) The family of Schweizer-Sklar t-norms is strictly decreasing (see Example 6.6), and the family of Schweizer-Sklar t-conorms is strictly increasing.

(vii) The family of Schweizer-Sklar t-norms is continuous with respect to the parameter λ (see Remark 8.17), i.e., for all $\lambda_0 \in [-\infty, \infty]$ we have $\lim\limits_{\lambda \to \lambda_0} T^{\mathbf{SS}}_{\lambda} = T^{\mathbf{SS}}_{\lambda_0}$.

4.3 Hamacher t-norms

In [Hamacher 1975, 1978] an axiomatic approach for logical connectives conjunction and disjunction, which can be expressed by rational functions, in many-valued logics with $[0, 1]$ as set of truth values (see Section 11.1) was presented. The original axioms for the conjunction $T : [0, 1]^2 \longrightarrow [0, 1]$ include continuity, associativity, strict monotonicity on $]0, 1]^2$ in each component, and $T(1, 1) = 1$. The main result of [Hamacher 1975,1978] can be reformulated in this way: a continuous t-norm T is a rational function (i.e., a quotient of two polynomials) if and only if T belongs to the family $(T^{\mathbf{H}}_{\lambda})_{\lambda \in [0,\infty[}$ (in our presentation we also include the limit case $\lambda = \infty$). A characterization of this family of t-norms by means of a functional equation is given in Proposition 5.19 (this result was first proved in [Fodor & Keresztfalvi 1994]). Most properties of this family of t-norms are summarized in Appendix A.4, where also 3D and contour plots, as well as plots of diagonal sections, of some representative members are given.

4.5 EXAMPLE

(i) The family $(T_\lambda^{\mathbf{H}})_{\lambda\in[0,\infty]}$ of *Hamacher t-norms* is given by

$$T_\lambda^{\mathbf{H}}(x,y) = \begin{cases} T_{\mathbf{D}}(x,y) & \text{if } \lambda = \infty, \\ 0 & \text{if } \lambda = x = y = 0, \\ \dfrac{xy}{\lambda + (1-\lambda)(x+y-xy)} & \text{otherwise.} \end{cases}$$

(ii) The family $(S_\lambda^{\mathbf{H}})_{\lambda\in[0,\infty]}$ of *Hamacher t-conorms* is given by

$$S_\lambda^{\mathbf{H}}(x,y) = \begin{cases} S_{\mathbf{D}}(x,y) & \text{if } \lambda = \infty, \\ 1 & \text{if } \lambda = 0 \text{ and } x = y = 1, \\ \dfrac{x+y-xy-(1-\lambda)xy}{1-(1-\lambda)xy} & \text{otherwise.} \end{cases}$$

4.6 REMARK

(i) It is clear that $T_1^{\mathbf{H}} = T_{\mathbf{P}}$ and $S_1^{\mathbf{H}} = S_{\mathbf{P}}$. The t-norm $T_0^{\mathbf{H}}$ and the t-conorm $S_2^{\mathbf{H}}$, which are given by

$$T_0^{\mathbf{H}}(x,y) = \begin{cases} 0 & \text{if } x = y = 0, \\ \frac{xy}{x+y-xy} & \text{otherwise,} \end{cases}$$

$$S_2^{\mathbf{H}}(x,y) = \frac{x+y}{1+xy},$$

respectively, are sometimes called the *Hamacher product* (note that we have $T_0^{\mathbf{H}} = T_{-1}^{\mathbf{SS}}$) and the *Einstein sum*, respectively. The subfamily $(T_\lambda^{\mathbf{H}})_{\lambda\in[0,2]}$ of Hamacher t-norms is also a family of copulas (see Section 9.1 and [Nelsen 1999]), mentioned first in [Ali *et al.* 1978], in which context it is usually referred to as the family of *Ali-Mikhail-Haq copulas.*

(ii) For each $\lambda \in [0,\infty]$ the t-norm $T_\lambda^{\mathbf{H}}$ and the t-conorm $S_\lambda^{\mathbf{H}}$ are dual to each other.

(iii) All Hamacher t-norms are Archimedean, and all Hamacher t-norms with the exception of $T_\infty^{\mathbf{H}}$ are continuous and strict. There is no nilpotent Hamacher t-norm.

(iv) Here are additive generators $t_\lambda^{\mathbf{H}}, s_\lambda^{\mathbf{H}} : [0,1] \longrightarrow [0,\infty]$ of the strict members of the families of Hamacher t-norms (see Figure 4.2) and t-conorms:

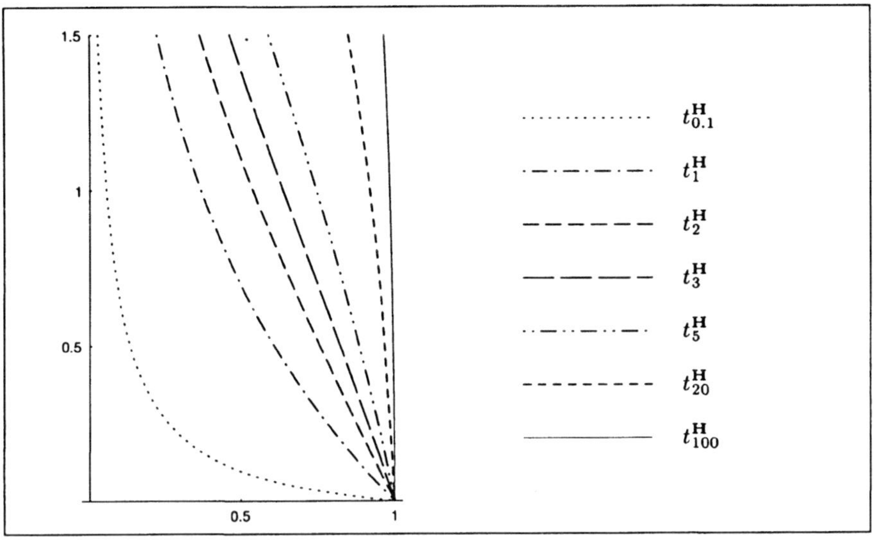

Fig. 4.2. Additive generators of some Hamacher t-norms

$$t^{\mathbf{H}}_{\lambda}(x) = \begin{cases} \frac{1-x}{x} & \text{if } \lambda = 0, \\ \log\left(\frac{\lambda+(1-\lambda)x}{x}\right) & \text{if } \lambda \in \,]0,\infty[\,, \end{cases}$$

$$s^{\mathbf{H}}_{\lambda}(x) = \begin{cases} \frac{x}{1-x} & \text{if } \lambda = 0, \\ \log\left(\frac{\lambda+(1-\lambda)(1-x)}{1-x}\right) & \text{if } \lambda \in \,]0,\infty[\,. \end{cases}$$

(v) Multiplicative generators $\theta^{\mathbf{H}}_{\lambda}, \xi^{\mathbf{H}}_{\lambda} : [0,1] \longrightarrow [0,1]$ of the strict members of the families of Hamacher t-norms and t-conorms are given as follows:

$$\theta^{\mathbf{H}}_{\lambda}(x) = \begin{cases} e^{\frac{x-1}{x}} & \text{if } \lambda = 0, \\ \frac{x}{\lambda+(1-\lambda)x} & \text{if } \lambda \in \,]0,\infty[\,, \end{cases}$$

$$\xi^{\mathbf{H}}_{\lambda}(x) = \begin{cases} e^{\frac{x}{1-x}} & \text{if } \lambda = 0, \\ \frac{1-x}{1-(1-\lambda)x} & \text{if } \lambda \in \,]0,\infty[\,. \end{cases}$$

(vi) The family of Hamacher t-norms is strictly decreasing, and the family of Hamacher t-conorms is strictly increasing (this is easily seen since the inequality $T^{\mathbf{H}}_{\lambda} \geq T^{\mathbf{H}}_{\mu}$, for arbitrary $\lambda, \mu \in [0,\infty[$, is equivalent to $(\lambda-\mu)(1-x)(1-y) \leq 0$ for all $(x,y) \in \,]0,1[^2$, i.e., to $\lambda \leq \mu$).

(vii) The family of Hamacher t-norms is continuous with respect to the parameter λ (see Remark 8.17), i.e., for all $\lambda_0 \in [0,\infty]$ we have $\lim_{\lambda \to \lambda_0} T^{\mathbf{H}}_{\lambda} = T^{\mathbf{H}}_{\lambda_0}$.

4.4 Frank t-norms

The investigations of the associativity of duals of copulas (see Section 9.1) in the framework of distribution functions (see Section 9.3) have led to the following problem: characterize all continuous (or, equivalently, non-decreasing) associative functions $F : [0,1]^2 \longrightarrow [0,1]$, which satisfy for each $x \in [0,1]$ the boundary conditions $F(0,x) = F(x,0)$ and $F(x,1) = F(1,x) = x$, such that the function $G : [0,1]^2 \longrightarrow [0,1]$ given by

$$G(x,y) = x + y - F(x,y)$$

is also associative. In [Frank 1979] it was shown that then F has to be an ordinal sum of members of the following family of t-norms. Most properties of this family of t-norms are summarized in Appendix A.3, where also 3D and contour plots, as well as plots of diagonal sections, of some representative members are given.

4.7 EXAMPLE

(i) The family $(T_\lambda^{\mathbf{F}})_{\lambda\in[0,\infty]}$ of *Frank t-norms* (which were called *fundamental t-norms* in [Butnariu & Klement 1993]) is given by

$$T_\lambda^{\mathbf{F}}(x,y) = \begin{cases} T_{\mathbf{M}}(x,y) & \text{if } \lambda = 0, \\ T_{\mathbf{P}}(x,y) & \text{if } \lambda = 1, \\ T_{\mathbf{L}}(x,y) & \text{if } \lambda = \infty, \\ \log_\lambda\left(1 + \frac{(\lambda^x-1)(\lambda^y-1)}{\lambda-1}\right) & \text{otherwise.} \end{cases}$$

(ii) The family $(S_\lambda^{\mathbf{F}})_{\lambda\in[0,\infty]}$ of *Frank t-conorms* is given by

$$S_\lambda^{\mathbf{F}}(x,y) = \begin{cases} S_{\mathbf{M}}(x,y) & \text{if } \lambda = 0, \\ S_{\mathbf{P}}(x,y) & \text{if } \lambda = 1, \\ S_{\mathbf{L}}(x,y) & \text{if } \lambda = \infty, \\ 1 - \log_\lambda\left(1 + \frac{(\lambda^{1-x}-1)(\lambda^{1-y}-1)}{\lambda-1}\right) & \text{otherwise.} \end{cases}$$

4.8 REMARK

(i) All Frank t-norms are also copulas (see Section 9.1 and [Nelsen 1999]) and have interesting statistical properties in the context of bivariate distributions [Nelsen 1986, Genest 1987].

(ii) For each $\lambda \in [0,\infty]$, the t-norm $T_\lambda^{\mathbf{F}}$ and the t-conorm $S_\lambda^{\mathbf{F}}$ are dual to each other.

(iii) Obviously, all Frank t-norms are continuous. A Frank t-norm $T_\lambda^{\mathbf{F}}$ is Archimedean if and only if $\lambda \in \,]0,\infty]$, and it is strict if and only if $\lambda \in \,]0,\infty[$. The only nilpotent Frank t-norm is $T_\infty^{\mathbf{F}}$.

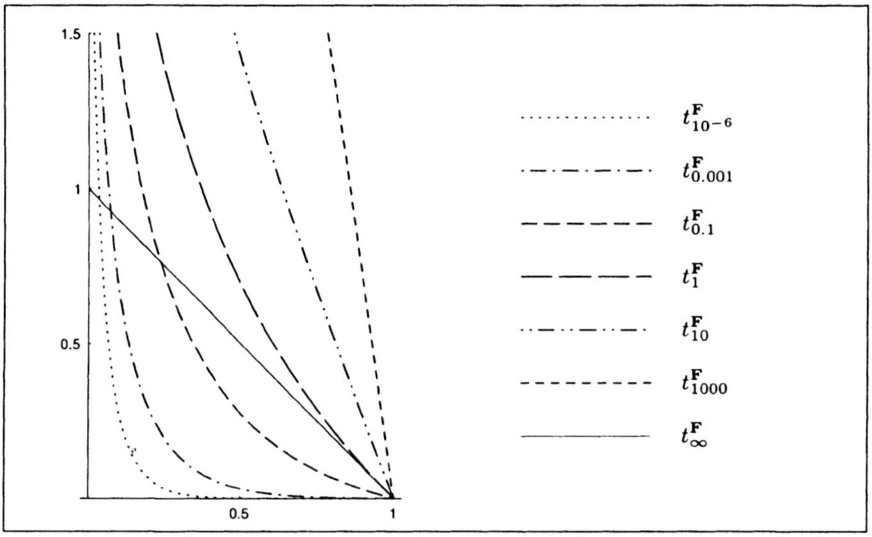

Fig. 4.3. Additive generators of some Frank t-norms

(iv) Additive generators $t^{\mathbf{F}}_\lambda, s^{\mathbf{F}}_\lambda : [0,1] \longrightarrow [0,\infty]$ of the continuous Archimedean members of the families of Frank t-norms (see Figure 4.3) and t-conorms are given by, respectively,

$$t^{\mathbf{F}}_\lambda(x) = \begin{cases} -\log x & \text{if } \lambda = 1, \\ 1 - x & \text{if } \lambda = \infty, \\ \log\left(\frac{\lambda - 1}{\lambda^x - 1}\right) & \text{if } \lambda \in \,]0,1[\,\cup\,]1,\infty[\,, \end{cases}$$

$$s^{\mathbf{F}}_\lambda(x) = \begin{cases} -\log(1 - x) & \text{if } \lambda = 1, \\ x & \text{if } \lambda = \infty, \\ \log\left(\frac{\lambda - 1}{\lambda^{1-x} - 1}\right) & \text{if } \lambda \in \,]0,1[\,\cup\,]1,\infty[\,. \end{cases}$$

(v) The following functions $\theta^{\mathbf{F}}_\lambda, \xi^{\mathbf{F}}_\lambda : [0,1] \longrightarrow [0,1]$ are multiplicative generators of the continuous Archimedean members of the families of Frank t-norms and t-conorms, and they are given by, respectively:

$$\theta^{\mathbf{F}}_\lambda(x) = \begin{cases} x & \text{if } \lambda = 1, \\ e^{x-1} & \text{if } \lambda = \infty, \\ \frac{\lambda^x - 1}{\lambda - 1} & \text{if } \lambda \in \,]0,1[\,\cup\,]1,\infty[\,, \end{cases}$$

$$\xi^{\mathbf{F}}_\lambda(x) = \begin{cases} 1 - x & \text{if } \lambda = 1, \\ e^{-x} & \text{if } \lambda = \infty, \\ \frac{\lambda^{1-x} - 1}{\lambda - 1} & \text{if } \lambda \in \,]0,1[\,\cup\,]1,\infty[\,. \end{cases}$$

(vi) The family of Frank t-norms is strictly decreasing, and the family of Frank t-conorms is strictly increasing (see Proposition 6.8, a first proof of this result was given in [Butnariu & Klement 1993, Proposition 1.12]).

(vii) The family of Frank t-norms is continuous with respect to the parameter λ, i.e., for all $\lambda_0 \in [0, \infty]$ we have $\lim_{\lambda \to \lambda_0} T_\lambda^{\mathbf{F}} = T_{\lambda_0}^{\mathbf{F}}$ (see Remark 8.17).

(viii) The result of [Frank 1979] (presented here as Theorem 5.14) can be reformulated in the sense that a pair (T, S), where T is a continuous t-norm and S is a t-conorm, fulfills the Frank functional equation

$$T(x, y) + S(x, y) = x + y \tag{4.2}$$

for all $(x, y) \in [0, 1]^2$ if and only if T is an ordinal sum of Frank t-norms and S is an ordinal sum of the corresponding dual Frank t-conorms, i.e., if

$$T = (\langle a_\alpha, e_\alpha, T_{\lambda_\alpha}^{\mathbf{F}} \rangle)_{\alpha \in A},$$
$$S = (\langle a_\alpha, e_\alpha, S_{\lambda_\alpha}^{\mathbf{F}} \rangle)_{\alpha \in A}.$$

Note, however, that the t-norm T and the t-conorm S are not necessarily dual to each other since the dual t-conorm of T is given by

$$\big(\langle 1 - e_\alpha, 1 - a_\alpha, S_{\lambda_\alpha}^{\mathbf{F}} \rangle\big)_{\alpha \in A},$$

which coincides with S if and only if for each $\alpha \in A$ there is a $\beta \in A$ such that $\lambda_\alpha = \lambda_\beta$ and $a_\alpha + e_\beta = a_\beta + e_\alpha = 1$.

The family of Frank t-norms plays a key role in the context of fuzzy logics (see Sections 11.3 and 11.4, for full proofs we refer to [Butnariu *et al.* 1995, Hájek 1998b, Klement & Navara 1999b]). Also for T-measures based on Frank t-norms it is possible to prove nice integral representations (see Theorems 14.4–14.6) as well as a Liapounoff Theorem (see Theorems 14.9 and 14.12).

4.5 Yager t-norms

One of the most popular families for modeling the intersection of fuzzy sets (Section 12.1) is the following family of t-norms (which was first introduced in [Yager 1980] for the special case $\lambda \geq 1$ only). The idea was to use the parameter λ as a reciprocal measure for the strength of the logical AND (see Section 11.1). In this context, $\lambda = 1$ expresses the most demanding (i.e., smallest) AND, and $\lambda = \infty$ the least demanding (i.e., largest) AND. Most properties of this family of t-norms are summarized in Appendix A.8, where also 3D and contour plots, as well as plots of diagonal sections, of some representative members are given.

4.9 EXAMPLE

(i) The family $(T_\lambda^{\mathbf{Y}})_{\lambda \in [0,\infty]}$ of *Yager t-norms* is given by

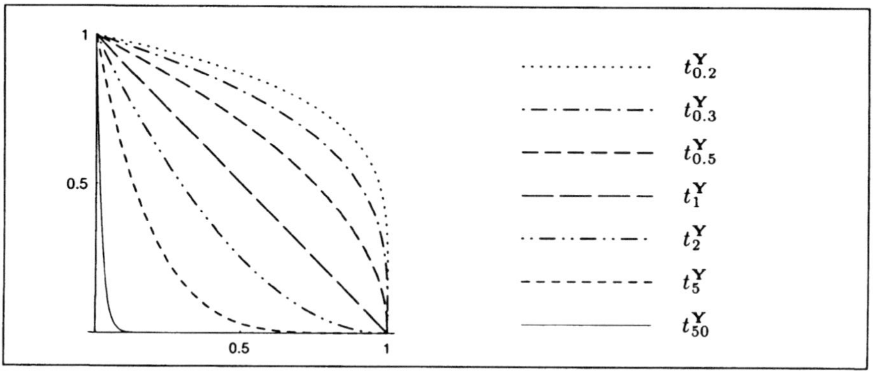

Fig. 4.4. Additive generators of some Yager t-norms

$$T_\lambda^{\mathbf{Y}}(x,y) = \begin{cases} T_{\mathbf{D}}(x,y) & \text{if } \lambda = 0, \\ T_{\mathbf{M}}(x,y) & \text{if } \lambda = \infty, \\ \max\left(1 - \left((1-x)^\lambda + (1-y)^\lambda\right)^{\frac{1}{\lambda}}, 0\right) & \text{otherwise.} \end{cases}$$

(ii) The family $(S_\lambda^{\mathbf{Y}})_{\lambda \in [0,\infty]}$ of *Yager t-conorms* is given by

$$S_\lambda^{\mathbf{Y}}(x,y) = \begin{cases} S_{\mathbf{D}}(x,y) & \text{if } \lambda = 0, \\ S_{\mathbf{M}}(x,y) & \text{if } \lambda = \infty, \\ \min\left(\left(x^\lambda + y^\lambda\right)^{\frac{1}{\lambda}}, 1\right) & \text{otherwise.} \end{cases}$$

4.10 Remark

(i) It is trivial to see that $T_1^{\mathbf{Y}} = T_{\mathbf{L}}$ and $S_1^{\mathbf{Y}} = S_{\mathbf{L}}$. The subfamily $(T_\lambda^{\mathbf{Y}})_{\lambda \in [1,\infty]}$ of Yager t-norms is also a family of copulas (see Section 9.1).

(ii) Obviously, for each $\lambda \in [0,\infty]$, the t-norm $T_\lambda^{\mathbf{Y}}$ and the t-conorm $S_\lambda^{\mathbf{Y}}$ are dual to each other.

(iii) All Yager t-norms with the exception of $T_0^{\mathbf{Y}}$ are continuous. A Yager t-norm $T_\lambda^{\mathbf{Y}}$ is Archimedean if and only if $\lambda \in [0,\infty[$, and it is nilpotent if and only if $\lambda \in]0,\infty[$. There are no strict Yager t-norms.

(iv) Additive generators $t_\lambda^{\mathbf{Y}}, s_\lambda^{\mathbf{Y}} : [0,1] \longrightarrow [0,\infty]$ of the nilpotent members $(T_\lambda^{\mathbf{Y}})_{\lambda \in]0,\infty[}$ and $(S_\lambda^{\mathbf{Y}})_{\lambda \in]0,\infty[}$ of the families of Yager t-norms (see Figure 4.4) and t-conorms are given by, respectively,

$$t_\lambda^{\mathbf{Y}}(x) = (1-x)^\lambda,$$
$$s_\lambda^{\mathbf{Y}}(x) = x^\lambda.$$

(v) Multiplicative generators $\theta_\lambda^{\mathbf{Y}}, \xi_\lambda^{\mathbf{Y}} : [0,1] \longrightarrow [0,1]$ of the nilpotent members $(T_\lambda^{\mathbf{Y}})_{\lambda \in]0,\infty[}$ and $(S_\lambda^{\mathbf{Y}})_{\lambda \in]0,\infty[}$ of the families of Yager t-norms and t-conorms are given by, respectively,

$$\theta_\lambda^{\mathbf{Y}}(x) = e^{-(1-x)^\lambda},$$
$$\xi_\lambda^{\mathbf{Y}}(x) = e^{-x^\lambda}.$$

(vi) The family of Yager t-norms is strictly increasing as a consequence of Examples 3.32(i) and 6.4(i), and the family of Yager t-conorms is strictly decreasing.

(vii) The family of Yager t-norms is continuous with respect to the parameter λ, i.e., for all $\lambda_0 \in [0, \infty]$ we have $\lim\limits_{\lambda \to \lambda_0} T_\lambda^{\mathbf{Y}} = T_{\lambda_0}^{\mathbf{Y}}$ (see Remark 8.17).

The family of Yager t-norms is used in several applications of fuzzy set theory, e.g., in the context of fuzzy numbers (see Section 13.2). In particular, for the addition of fuzzy numbers based on Yager t-norms it has been shown in [Kolesárová 1995] that the sum of piecewise linear fuzzy numbers (compare Example 13.4(ii)) again is a piecewise linear fuzzy number (see Example 13.13(iii)). The Yager t-norms appear also in the investigation of t-norms whose graphs are (partly) ruled surfaces [Alsina & Sklar 1987].

4.6 Dombi t-norms

In an early attempt to study conjunctive and disjunctive operators in fuzzy logic (see Section 11.1), in [Dombi 1982b] the following family of t-norms was introduced. Most properties of this family of t-norms are summarized in Appendix A.2, where also 3D and contour plots, as well as plots of diagonal sections, of some representative members are given.

4.11 EXAMPLE

(i) The family $(T_\lambda^{\mathbf{D}})_{\lambda\in[0,\infty]}$ of *Dombi t-norms* is given by

$$T_\lambda^{\mathbf{D}}(x,y) = \begin{cases} T_{\mathbf{D}}(x,y) & \text{if } \lambda = 0, \\ T_{\mathbf{M}}(x,y) & \text{if } \lambda = \infty, \\ \frac{1}{1+\left(\left(\frac{1-x}{x}\right)^\lambda + \left(\frac{1-y}{y}\right)^\lambda\right)^{\frac{1}{\lambda}}} & \text{otherwise.} \end{cases}$$

(ii) The family $(S_\lambda^{\mathbf{D}})_{\lambda\in[0,\infty]}$ of *Dombi t-conorms* is given by

$$S_\lambda^{\mathbf{D}}(x,y) = \begin{cases} S_{\mathbf{D}}(x,y) & \text{if } \lambda = 0, \\ S_{\mathbf{M}}(x,y) & \text{if } \lambda = \infty, \\ 1 - \frac{1}{1+\left(\left(\frac{x}{1-x}\right)^\lambda + \left(\frac{y}{1-y}\right)^\lambda\right)^{\frac{1}{\lambda}}} & \text{otherwise.} \end{cases}$$

4.12 REMARK

(i) It is easy to see that $T_1^{\mathbf{D}} = T_0^{\mathbf{H}}$ and $S_1^{\mathbf{D}} = S_0^{\mathbf{H}}$. The subfamily $(T_\lambda^{\mathbf{D}})_{\lambda\in[1,\infty]}$ of Dombi t-norms is also a family of copulas (see Section 9.1).

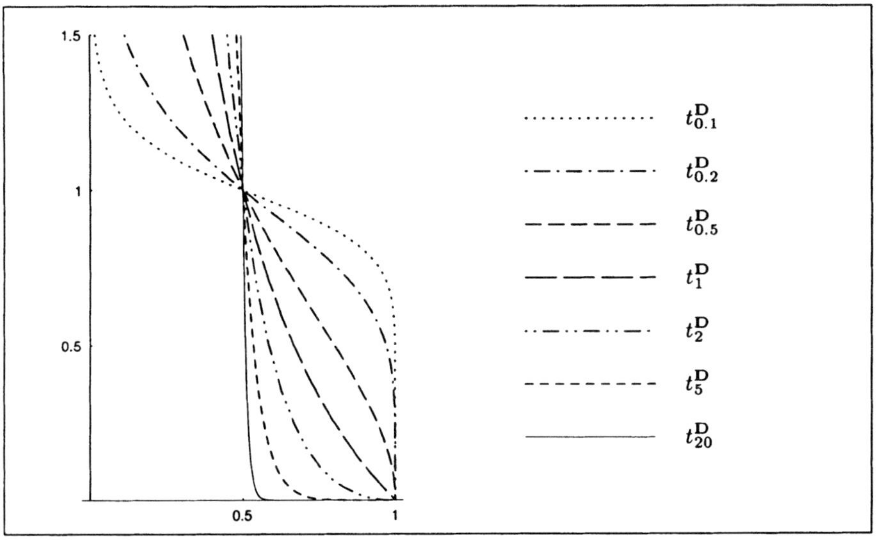

Fig. 4.5. Additive generators of some Dombi t-norms: note that each generator passes through the point $(0.5, 1)$

(ii) For each $\lambda \in [0, \infty]$, the t-norm $T^{\mathbf{D}}_{\lambda}$ and the t-conorm $S^{\mathbf{D}}_{\lambda}$ are dual to each other.

(iii) All Dombi t-norms with the exception of $T^{\mathbf{D}}_{0}$ are continuous. A Dombi t-norm $T^{\mathbf{D}}_{\lambda}$ is Archimedean if and only if $\lambda \in [0, \infty[$, and it is strict if and only if $\lambda \in]0, \infty[$. There are no nilpotent Dombi t-norms.

(iv) Additive generators $t^{\mathbf{D}}_{\lambda}, s^{\mathbf{D}}_{\lambda} : [0,1] \longrightarrow [0, \infty]$ of the strict members $(T^{\mathbf{D}}_{\lambda})_{\lambda \in]0,\infty[}$ and $(S^{\mathbf{D}}_{\lambda})_{\lambda \in]0,\infty[}$ of the families of Dombi t-norms (see Figure 4.5) and t-conorms are given by, respectively,

$$t^{\mathbf{D}}_{\lambda}(x) = \left(\tfrac{1-x}{x}\right)^{\lambda},$$
$$s^{\mathbf{D}}_{\lambda}(x) = \left(\tfrac{x}{1-x}\right)^{\lambda}.$$

(v) Multiplicative generators $\theta^{\mathbf{D}}_{\lambda}, \xi^{\mathbf{D}}_{\lambda} : [0,1] \longrightarrow [0,1]$ of the strict members $(T^{\mathbf{D}}_{\lambda})_{\lambda \in]0,\infty[}$ and $(S^{\mathbf{D}}_{\lambda})_{\lambda \in]0,\infty[}$ of the families of Dombi t-norms and t-conorms are given by, respectively,

$$\theta^{\mathbf{D}}_{\lambda}(x) = e^{-(\frac{1-x}{x})^{\lambda}},$$
$$\xi^{\mathbf{D}}_{\lambda}(x) = e^{-(\frac{x}{1-x})^{\lambda}}.$$

(vi) The family of Dombi t-norms is strictly increasing as a consequence of Examples 3.32(i) and 6.4(i), and the family of Dombi t-conorms is strictly decreasing.

(vii) The family of Dombi t-norms is continuous with respect to the parameter λ, i.e., for all $\lambda_0 \in [0, \infty]$ we have $\lim\limits_{\lambda \to \lambda_0} T^{\mathbf{D}}_{\lambda} = T^{\mathbf{D}}_{\lambda_0}$ (see Remark 8.17).

4.7 Sugeno-Weber t-norms

In [Weber 1983] the use of the families of some special t-norms and t-conorms was suggested in order to model the intersection and union of fuzzy sets (see Section 12.1), respectively. These t-conorms already appeared as possible generalized additions in the context of λ-fuzzy measures in [Sugeno 1974]. Most properties of this family of t-norms are summarized in Appendix A.7, where also 3D and contour plots, as well as plots of diagonal sections, of some representative members are given.

4.13 EXAMPLE

(i) The family $(T_\lambda^{\mathbf{SW}})_{\lambda\in[-1,\infty]}$ of *Sugeno-Weber t-norms* is given by

$$T_\lambda^{\mathbf{SW}}(x,y) = \begin{cases} T_{\mathbf{D}}(x,y) & \text{if } \lambda = -1, \\ T_{\mathbf{P}}(x,y) & \text{if } \lambda = \infty, \\ \max\left(\frac{x+y-1+\lambda xy}{1+\lambda}, 0\right) & \text{otherwise.} \end{cases}$$

(ii) The family $(S_\lambda^{\mathbf{SW}})_{\lambda\in[-1,\infty]}$ of *Sugeno-Weber t-conorms* is given by

$$S_\lambda^{\mathbf{SW}}(x,y) = \begin{cases} S_{\mathbf{P}}(x,y) & \text{if } \lambda = -1, \\ S_{\mathbf{D}}(x,y) & \text{if } \lambda = \infty, \\ \min(x+y+\lambda xy, 1) & \text{otherwise.} \end{cases}$$

4.14 REMARK

(i) Obviously we have $T_0^{\mathbf{SW}} = T_{\mathbf{L}}$ and $S_0^{\mathbf{SW}} = S_{\mathbf{L}}$. Also, the subfamily $(T_\lambda^{\mathbf{SW}})_{\lambda\in[0,\infty]}$ of Sugeno-Weber t-norms is a family of copulas (see Section 9.1).

(ii) A Sugeno-Weber t-norm $T_\lambda^{\mathbf{SW}}$ and a Sugeno-Weber t-conorm $S_\mu^{\mathbf{SW}}$ (for $\lambda, \mu \in \,]-1,\infty[$) are dual to each other if and only if $\mu = -\frac{\lambda}{1+\lambda}$. Of course, $(T_{-1}^{\mathbf{SW}}, S_\infty^{\mathbf{SW}})$ and $(T_\infty^{\mathbf{SW}}, S_{-1}^{\mathbf{SW}})$ are pairs of mutually dual t-norms and t-conorms too.

(iii) All Sugeno-Weber t-norms with the exception of $T_{-1}^{\mathbf{SW}}$ are continuous. Each Sugeno-Weber t-norm $T_\lambda^{\mathbf{SW}}$ is Archimedean, and it is nilpotent if and only if $\lambda \in \,]-1,\infty[$. The only strict Sugeno-Weber t-norm is $T_\infty^{\mathbf{SW}}$.

(iv) Additive generators $t_\lambda^{\mathbf{SW}}, s_\lambda^{\mathbf{SW}} : [0,1] \longrightarrow [0,\infty]$ of the continuous Archimedean members of the families of Sugeno-Weber t-norms (see Figure 4.6) and t-conorms are given by, respectively,

$$t_\lambda^{\mathbf{SW}}(x) = \begin{cases} 1-x & \text{if } \lambda = 0, \\ -\log x & \text{if } \lambda = \infty, \\ 1 - \frac{\log(1+\lambda x)}{\log(1+\lambda)} & \text{if } \lambda \in \,]-1,0[\,\cup\,]0,\infty[, \end{cases}$$

$$s_\lambda^{\mathbf{SW}}(x) = \begin{cases} x & \text{if } \lambda = 0, \\ -\log(1-x) & \text{if } \lambda = -1, \\ \frac{\log(1+\lambda x)}{\log(1+\lambda)} & \text{if } \lambda \in \,]-1,0[\,\cup\,]0,\infty[. \end{cases}$$

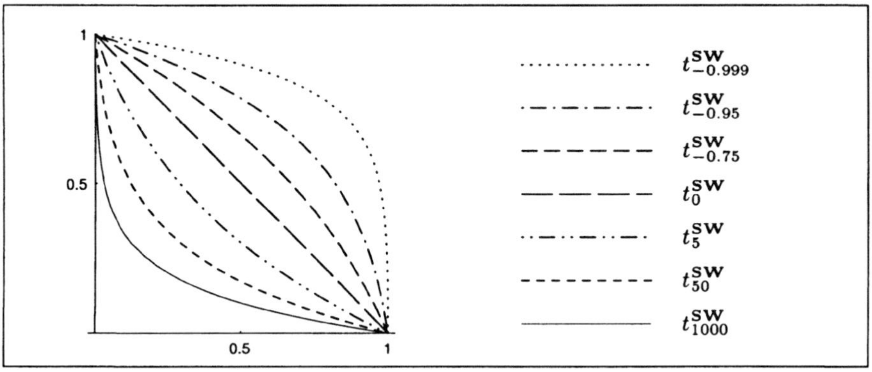

Fig. 4.6. Additive generators of some Sugeno-Weber t-norms

(v) Multiplicative generators $\theta_\lambda^{\mathbf{SW}}, \xi_\lambda^{\mathbf{SW}} : [0,1] \longrightarrow [0,1]$ of the continuous Archimedean members of the families of Sugeno-Weber t-norms and t-conorms are given by, respectively,

$$\theta_\lambda^{\mathbf{SW}}(x) = \begin{cases} e^{x-1} & \text{if } \lambda = 0, \\ x & \text{if } \lambda = \infty, \\ e^{\left(\frac{\log(1+\lambda x)}{\log(1+\lambda)} - 1\right)} & \text{if } \lambda \in \left]-1,0\right[\cup \left]0,\infty\right[, \end{cases}$$

$$\xi_\lambda^{\mathbf{SW}}(x) = \begin{cases} e^{-x} & \text{if } \lambda = 0, \\ 1-x & \text{if } \lambda = -1, \\ e^{-\frac{\log(1+\lambda x)}{\log(1+\lambda)}} & \text{if } \lambda \in \left]-1,0\right[\cup \left]0,\infty\right[. \end{cases}$$

(vi) Obviously, both the family of Sugeno-Weber t-norms and the family of Sugeno-Weber t-conorms are strictly increasing.

(vii) The family of Sugeno-Weber t-norms is continuous with respect to the parameter λ (see Remark 8.17), i.e., for all $\lambda_0 \in [-1,\infty]$ we have $\lim\limits_{\lambda \to \lambda_0} T_\lambda^{\mathbf{SW}} = T_{\lambda_0}^{\mathbf{SW}}$.

(viii) An interesting characterization of a large subfamily of Sugeno-Weber t-norms can be found in [Mayor 1994]: given a t-norm T then we have $T = T_\lambda^{\mathbf{SW}}$ for some $\lambda \in \left]0,\infty\right[$ if and only if T is nilpotent and has an additive generator $t : [0,1] \longrightarrow [0,\infty]$ such that, for each $x_0 \in [0,1]$, the graph of the vertical section $T(x_0, \cdot)$ of T is a straight line segment.

4.8 Aczél-Alsina t-norms

The family of Aczél-Alsina t-norms was formally introduced in [Aczél & Alsina 1984]. The Aczél-Alsina t-norms are the only t-norms such that (see Proposition 5.17 or [Aczél & Alsina 1984]), whenever for $p, q \in \left]0,1\right[\cup \left]0,\infty\right[$ the quotient

$\frac{\log p}{\log q}$ is an irrational number, for all $(x, y) \in [0, 1]^2$ we have

$$T(x^p, y^p) = T(x, y)^p,$$
$$T(x^q, y^q) = T(x, y)^q.$$

Most properties of this family of t-norms are summarized in Appendix A.1, where also 3D and contour plots, as well as plots of diagonal sections, of some representative members are given.

4.15 EXAMPLE

(i) The family $(T_\lambda^{\mathbf{AA}})_{\lambda \in [0,\infty]}$ of *Aczél-Alsina t-norms* is given by

$$T_\lambda^{\mathbf{AA}}(x, y) = \begin{cases} T_{\mathbf{D}}(x, y) & \text{if } \lambda = 0, \\ T_{\mathbf{M}}(x, y) & \text{if } \lambda = \infty, \\ e^{-((-\log x)^\lambda + (-\log y)^\lambda)^{\frac{1}{\lambda}}} & \text{otherwise.} \end{cases}$$

(ii) The family $(S_\lambda^{\mathbf{AA}})_{\lambda \in [0,\infty]}$ of *Aczél-Alsina t-conorms* is given by

$$S_\lambda^{\mathbf{AA}}(x, y) = \begin{cases} S_{\mathbf{D}}(x, y) & \text{if } \lambda = 0, \\ S_{\mathbf{M}}(x, y) & \text{if } \lambda = \infty, \\ 1 - e^{-((-\log(1-x))^\lambda + (-\log(1-y))^\lambda)^{\frac{1}{\lambda}}} & \text{otherwise.} \end{cases}$$

4.16 REMARK

(i) It is easy to see that $T_1^{\mathbf{AA}} = T_{\mathbf{P}}$ and $S_1^{\mathbf{AA}} = S_{\mathbf{P}}$. The members of the subfamily $(T_\lambda^{\mathbf{AA}})_{\lambda \in [1,\infty]}$ of Aczél-Alsina t-norms are also copulas (see Section 9.1 and [Nelsen 1999]), in which context they were discussed in [Gumbel 1960, Hougaard 1986] (according to [Hutchinson & Lai 1990], they are referred to as the family of *Gumbel-Hougaard copulas*). They play an important role in conjunction with bivariate extreme value distributions [Genest & Rivest 1989, Nelsen 1999].

(ii) For each $\lambda \in [0, \infty]$, the t-norm $T_\lambda^{\mathbf{AA}}$ and the t-conorm $S_\lambda^{\mathbf{AA}}$ are dual to each other.

(iii) All Aczél-Alsina t-norms with the exception of $T_0^{\mathbf{AA}}$ are continuous. An Aczél-Alsina t-norm $T_\lambda^{\mathbf{AA}}$ is Archimedean if and only if $\lambda \in [0, \infty[$, and it is strict if and only if $\lambda \in]0, \infty[$. There are no nilpotent Aczél-Alsina t-norms.

(iv) Additive generators $t_\lambda^{\mathbf{AA}}, s_\lambda^{\mathbf{AA}} : [0, 1] \longrightarrow [0, \infty]$ of the strict members $(T_\lambda^{\mathbf{AA}})_{\lambda \in]0,\infty[}$ and $(S_\lambda^{\mathbf{AA}})_{\lambda \in]0,\infty[}$ of the families of Aczél-Alsina t-norms (see Figure 4.7) and t-conorms are given by, respectively,

$$t_\lambda^{\mathbf{AA}}(x) = (-\log x)^\lambda,$$
$$s_\lambda^{\mathbf{AA}}(x) = (-\log(1 - x))^\lambda.$$

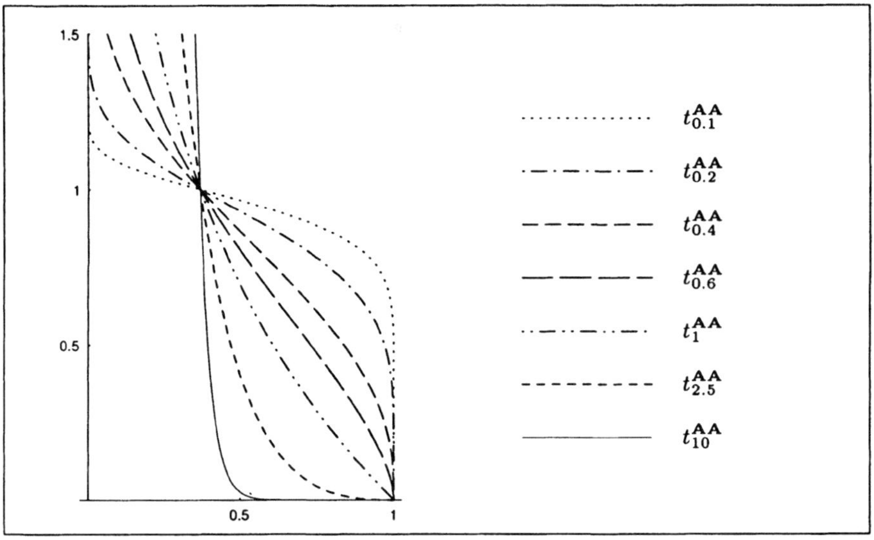

Fig. 4.7. Additive generators of some Aczél-Alsina t-norms: observe that each generator passes through the point $\left(\frac{1}{e}, 1\right)$

(v) Multiplicative generators $\theta_\lambda^{\mathbf{AA}}, \xi_\lambda^{\mathbf{AA}} : [0,1] \longrightarrow [0,1]$ of the strict members $(T_\lambda^{\mathbf{AA}})_{\lambda\in]0,\infty[}$ and $(S_\lambda^{\mathbf{AA}})_{\lambda\in]0,\infty[}$ of the families of Aczél-Alsina t-norms and t-conorms are given by, respectively,

$$\theta_\lambda^{\mathbf{AA}}(x) = e^{-(-\log x)^\lambda},$$
$$\xi_\lambda^{\mathbf{AA}}(x) = e^{-(-\log(1-x))^\lambda}.$$

(vi) The family of Aczél-Alsina t-norms is strictly increasing as a consequence of Examples 3.32(i) and 6.4(i), and the family of Aczél-Alsina t-conorms is strictly decreasing.

(vii) It is trivial to check that the family of Aczél-Alsina t-norms is continuous with respect to the parameter λ, i.e., for all $\lambda_0 \in [0,\infty]$ we have $\lim\limits_{\lambda\to\lambda_0} T_\lambda^{\mathbf{AA}} = T_{\lambda_0}^{\mathbf{AA}}$.

4.9 Mayor-Torrens t-norms

In Proposition 5.18 (for a first proof see [Mayor & Torrens 1991]) the members of the following family will be shown to be the only continuous t-norms satisfying for all $(x,y) \in [0,1]^2$

$$T(x,y) = \max(T(\max(x,y), \max(x,y)) - |x-y|, 0).$$

Most properties of this family of t-norms are summarized in Appendix A.5, where also 3D and contour plots, as well as plots of diagonal sections, of some representative members are given.

4.17 Example

(i) The family $(T_\lambda^{\mathbf{MT}})_{\lambda\in[0,1]}$ of *Mayor-Torrens t-norms* is given by

$$T_\lambda^{\mathbf{MT}}(x,y) = \begin{cases} \max(x+y-\lambda,0) & \text{if } \lambda \in \,]0,1] \text{ and } (x,y) \in [0,\lambda]^2, \\ \min(x,y) & \text{otherwise.} \end{cases}$$

(ii) The family $(S_\lambda^{\mathbf{MT}})_{\lambda\in[0,1]}$ of *Mayor-Torrens t-conorms* is given by

$$S_\lambda^{\mathbf{MT}}(x,y) = \begin{cases} \min(x+y+\lambda-1,1) \\ \qquad\qquad \text{if } \lambda \in \,]0,1] \text{ and } (x,y) \in [1-\lambda,1]^2, \\ \max(x,y) \qquad\qquad\qquad\qquad \text{otherwise.} \end{cases}$$

4.18 Remark

(i) Clearly $T_0^{\mathbf{MT}} = T_{\mathbf{M}}$, $S_0^{\mathbf{MT}} = S_{\mathbf{M}}$, $T_1^{\mathbf{MT}} = T_{\mathbf{L}}$ and $S_1^{\mathbf{MT}} = S_{\mathbf{L}}$. For each $\lambda \in \,]0,1[$, the t-norm $T_\lambda^{\mathbf{MT}}$ and the t-conorm $S_\lambda^{\mathbf{MT}}$ are ordinal sums with one summand, namely, $T_\lambda^{\mathbf{MT}} = (\langle 0,\lambda,T_{\mathbf{L}}\rangle)$ and $S_\lambda^{\mathbf{MT}} = (\langle 1-\lambda,1,S_{\mathbf{L}}\rangle)$. All Mayor-Torrens t-norms are also copulas (see Section 9.1).

(ii) For each $\lambda \in [0,1]$ the t-norm $T_\lambda^{\mathbf{MT}}$ and the t-conorm $S_\lambda^{\mathbf{MT}}$ are dual to each other.

(iii) Each Mayor-Torrens t-norm $T_\lambda^{\mathbf{MT}}$ is continuous. Only $T_1^{\mathbf{MT}}$ is Archimedean and nilpotent. There are no strict Mayor-Torrens t-norms.

(iv) Obviously, the family of Mayor-Torrens t-norms is strictly decreasing, and the family of Mayor-Torrens t-conorms strictly increasing.

(v) The family of Mayor-Torrens t-norms is clearly continuous with respect to the parameter λ, i.e., for all $\lambda_0 \in [0,1]$ we have $\lim\limits_{\lambda\to\lambda_0} T_\lambda^{\mathbf{MT}} = T_{\lambda_0}^{\mathbf{MT}}$.

Family	*continuous*	*Archimedean*	*strict*	*nilpotent*
Schweizer-Sklar t-norms $(T_\lambda^{\mathbf{SS}})_{\lambda\in[-\infty,\infty]}$	$\lambda \in [-\infty, \infty[$	$\lambda \in]-\infty, \infty]$	$\lambda \in]-\infty, 0]$	$\lambda \in]0, \infty[$
Hamacher t-norms $(T_\lambda^{\mathbf{H}})_{\lambda\in[0,\infty]}$	$\lambda \in [0, \infty[$	all	$\lambda \in [0, \infty[$	none
Frank t-norms $(T_\lambda^{\mathbf{F}})_{\lambda\in[0,\infty]}$	all	$\lambda \in]0, \infty]$	$\lambda \in]0, \infty[$	$\lambda = \infty$
Yager t-norms $(T_\lambda^{\mathbf{Y}})_{\lambda\in[0,\infty]}$	$\lambda \in]0, \infty]$	$\lambda \in [0, \infty[$	none	$\lambda \in]0, \infty[$
Aczél-Alsina t-norms $(T_\lambda^{\mathbf{AA}})_{\lambda\in[0,\infty]}$	$\lambda \in]0, \infty]$	$\lambda \in [0, \infty[$	$\lambda \in]0, \infty[$	none
Dombi t-norms $(T_\lambda^{\mathbf{D}})_{\lambda\in[0,\infty]}$	$\lambda \in]0, \infty]$	$\lambda \in [0, \infty[$	$\lambda \in]0, \infty[$	none
Sugeno-Weber t-norms $(T_\lambda^{\mathbf{SW}})_{\lambda\in[-1,\infty]}$	$\lambda \in]-1, \infty]$	all	$\lambda = \infty$	$\lambda \in]-1, \infty[$
Mayor-Torrens t-norms $(T_\lambda^{\mathbf{MT}})_{\lambda\in[0,1]}$	all	$\lambda = 1$	none	$\lambda = 1$

Table 4.1. Some properties of the families of t-norms

5. Representations of t-norms

> In attacking this problem of characterization, it has turned out that the most useful — and intrinsically the most interesting — property is their associativity [...] Now, associativity has been studied in extenso from the algebraic point of view; and even the, by comparison, neglected function-theoretic aspect of associativity [...] can boast a distinguished roster of investigators [Abel 1826, Aczél 1949, Brouwer 1909, Cartan 1930, Hosszú 1954] ...
>
> BERTHOLD SCHWEIZER & ABE SKLAR

For the rather general class of all t-norms, which includes non-continuous t-norms and even t-norms which are not Borel measurable, no universal representation theorems exist so far. In fact, such a characterization of arbitrary t-norms would be closely related to the solution of the famous, still unsolved general associativity functional equation.

We have already seen in Section 3.2 that continuous additive (and multiplicative) generators always give rise to continuous Archimedean t-norms and t-conorms. Now we show that the converse is also true, i.e., for continuous Archimedean t-norms there is a representation in terms of one-place functions. Early traces of this result can be found in [Abel 1826] (compare Theorem 3.22) and [Aczél 1949, Faucett 1955] in the strict case, and in [Mostert & Shields 1957] (in the framework of nilpotent compact topological semigroups). In [Ling 1965] a unique representation of triangular conorms in terms of additive generators was provided (compare also [Schweizer & Sklar 1963]).

In Section 5.4 we prove that each continuous t-norm can be written as an ordinal sum (as considered in Section 3.3) of continuous Archimedean t-norms. This remarkable result also goes back to [Mostert & Shields 1957] (again in the context of compact topological semigroups, compare also [Paalman-de Miranda 1964]).

Some families of t-norms have been shown to solve some functional equations, in particular the families of Frank (Section 4.4), Aczél-Alsina (Section 4.8), Mayor-Torrens (Section 4.9), and Hamacher t-norms (Section 4.3). With the exception of the result of [Frank 1979], where we only give an extended sketch of the exceptionally long proof, these results are presented in Section 5.3 in detail.

5.1 Representation of continuous Archimedean t-norms

In Section 3.2 we have seen that, starting with a continuous additive generator, the construction (3.16) always leads to a continuous Archimedean t-norm. Now we are in the position to show that the converse also holds true.

5.1 THEOREM
For a function $T : [0,1]^2 \longrightarrow [0,1]$ the following are equivalent:

(i) T is a continuous Archimedean t-norm.
(ii) T has a continuous additive generator, i.e., there exists a continuous, strictly decreasing function $t : [0,1] \longrightarrow [0,\infty]$ with $t(1) = 0$, which is uniquely determined up to a positive multiplicative constant, such that for all $(x,y) \in [0,1]^2$ (3.16) holds, i.e., $T(x,y) = t^{(-1)}(t(x) + t(y))$.

Proof: As an immediate consequence of Theorem 3.23 and Propositions 3.26 and 3.29, (ii) implies (i).

To prove the converse, let T be a continuous Archimedean t-norm. If, for some $x \in [0,1]$ and some $n \in \mathbb{N} \cup \{0\}$, we have $x_T^{(n)} = x_T^{(n+1)}$ then, in the standard way by induction, we obtain

$$x_T^{(n)} = x_T^{(2n)} = \left(x_T^{(n)}\right)_T^{(2)}$$

and, since T is continuous Archimedean, $x_T^{(n)} \in \{0,1\}$. This means that we have $x_T^{(n)} > x_T^{(n+1)}$ whenever $x_T^{(n)} \in]0,1[$.

Now choose and fix an arbitrary element $a \in]0,1[$, and define the function $h : \mathbb{Q} \cap [0,\infty[\longrightarrow [0,1]$ by $h(r) = a_T^{(r)}$ (for the notion $a_T^{(r)}$ see (3.7)). Since T is continuous and since (3.5) holds, h is a continuous function. As a consequence of (3.8) we obtain

$$h(r+s) = a_T^{(r+s)} = T(a_T^{(r)}, a_T^{(s)}) \leq a_T^{(r)} = h(r),$$

i.e., h is also non-increasing. The function h is even strictly decreasing on the preimage of $]0,1]$ since for all $\frac{m}{n}, \frac{p}{q} \in \mathbb{Q} \cap [0,\infty[$ with $h(\frac{m}{n}) > 0$ we get

$$h(\tfrac{m}{n} + \tfrac{p}{q}) \leq h\left(\tfrac{mq+1}{nq}\right) = \left(a_T^{(\frac{1}{nq})}\right)_T^{(mq+1)} < \left(a_T^{(\frac{1}{nq})}\right)_T^{(mq)} = h(\tfrac{m}{n}).$$

The monotonicity and continuity of h on $\mathbb{Q} \cap [0,\infty[$ allows us to extend it uniquely to a function $\overline{h} : [0,\infty] \longrightarrow [0,1]$ via

$$\overline{h}(x) = \inf\{h(r) \mid r \in \mathbb{Q} \cap [0,x]\}.$$

Then $\overline{h}$ is continuous and non-increasing, and we have for all $x, y \in [0,\infty]$

$$\overline{h}(x+y) = T(\overline{h}(x), \overline{h}(y)).$$

Moreover, $\overline{h}$ is strictly decreasing on the preimage of $]0,1]$. Denote the quasi-inverse of $\overline{h}$ by t, then t is a function from $[0,1]$ to $[0,\infty]$ which is continuous, strictly decreasing, and satisfies $t(1) = 0$. A combination of all the arguments so far yields that t is indeed a continuous additive generator of T since for each $(x,y) \in [0,1]^2$

$$T(x,y) = T(\overline{h}(t(x)), \overline{h}(t(y))) = \overline{h}(t(x) + t(y)) = t^{(-1)}(t(x) + t(y)).$$

To show that the continuous additive generator t of T constructed above is unique up to a positive multiplicative constant, assume that the two functions $t_1, t_2 : [0,1] \longrightarrow [0,\infty]$ are both continuous additive generators of T, i.e., we have for each $(x,y) \in [0,1]^2$ the equality

$$t_1^{(-1)}(t_1(x) + t_1(y)) = t_2^{(-1)}(t_2(x) + t_2(y)).$$

Substituting $u = t_2(x)$ and $v = t_2(y)$, we obtain that, for all $u, v \in [0, t_2(0)]$ satisfying $u + v \in [0, t_2(0)[$,

$$t_1 \circ t_2^{(-1)}(u) + t_1 \circ t_2^{(-1)}(v) = t_1 \circ t_2^{(-1)}(u+v). \tag{5.1}$$

Then from the continuity of t_1 and $t_2^{(-1)}$ it follows that (5.1) holds for all $u, v \in [0, t_2(0)]$ with $u + v \in [0, t_2(0)]$.

Equation (5.1) is a Cauchy functional equation (see [Aczél 1966]), whose continuous, strictly increasing solutions $t_1 \circ t_2^{(-1)} : [0, t_2(0)] \longrightarrow [0,\infty]$ must satisfy $t_1 \circ t_2^{(-1)} = b \cdot \mathrm{id}_{[0,t_2(0)]}$ for some $b \in]0,\infty[$. As a consequence, we get $t_1 = b \cdot t_2$ for some $b \in]0,\infty[$, thus completing the proof. □

Because of Theorem 2.43, in Theorem 5.1(i) we may replace the continuous Archimedean t-norm by an Archimedean I-semigroup on $[0,1]$ with neutral element 1, i.e., the commutativity (T1) and the monotonicity (T3) need not be required. The Archimedean property, however, cannot be omitted as a consequence of Proposition 3.29.

In [Krause 1981] it was shown that, in Theorem 5.1, the commutativity (T1) can be dropped and the continuity assumption can be relaxed. Another generalization of Theorem 5.1, where the continuity of the operation T is required, but the commutativity (T1) is dropped and the boundary condition (T4), associativity (T2) and monotonicity (T3) are relaxed, was proved in [Bacchelli 1986].

Based on the proof of Theorem 5.1, it is possible to give some constructive way to obtain additive generators of continuous Archimedean t-norms. As an illustrating example, we include the following result of [Craigen & Páles 1989] (compare also [Aczél 1949, Pi-Calleja 1954, Alsina 1992]) for the case of strict t-norms which can be derived in a straightforward manner from the proof of Theorem 5.1.

5.2 COROLLARY

Let T be a strict t-norm. Fix an arbitrary element $x_0 \in \,]0,1[$, and define the function $t : [0,1] \longrightarrow [0,\infty]$ by

$$t(x) = \inf\left\{ \tfrac{m-n}{k} \;\middle|\; m,n,k \in \mathbb{N} \text{ and } (x_0)_T^{(m)} < T\big((x_0)_T^{(n)}, x_T^{(k)}\big) \right\}.$$

Then t is an additive generator of T.

5.3 EXAMPLE

If we consider the Hamacher product, i.e., $T = T_0^{\mathbf{H}}$ (see Remark 4.6(i)), observe that we get (taking into account $\frac{1}{\infty} = 0$ and $\frac{1}{0} = \infty$) for all $(x,y) \in [0,1]^2$

$$T(x,y) = \frac{1}{\frac{1}{x} + \frac{1}{y} - 1}$$

and, for each $x \in [0,1]$ and each $n \in \mathbb{N}$

$$x_T^{(n)} = \frac{1}{\frac{n}{x} - n + 1}.$$

For an arbitrary fixed $x_0 \in \,]0,1[$ the inequality $(x_0)_T^{(m)} < T\big((x_0)_T^{(n)}, x_T^{(k)}\big)$ is easily seen to be equivalent to $(m-n)\big(\frac{1}{x_0} - 1\big) > k\big(\frac{1}{x} - 1\big)$, yielding the additive generator $t : [0,1] \longrightarrow [0,\infty]$ of T specified by

$$t(x) = \inf\left\{ \tfrac{m-n}{k} \;\middle|\; m,n,k \in \mathbb{N} \text{ and } \tfrac{m-n}{k} > \frac{\frac{1}{x}-1}{\frac{1}{x_0}-1} \right\} = \frac{x_0}{1-x_0} \cdot \frac{1-x}{x}.$$

The representation of continuous Archimedean t-norms given in Theorem 5.1 is based on the addition on the interval $[0,\infty]$. There is a completely analogous representation based on the multiplication on $[0,1]$, which is an immediate consequence of the fact that there is a one-to-one correspondence between additive and multiplicative generators expressed by the operators E (defined in (3.18)) and L (given in (3.19), see Figure 3.15). Note that the following representation of continuous Archimedean t-norms by means of multiplicative generators can be derived directly from the results for I-semigroups in [Mostert & Shields 1957] which are equivalent because of Theorem 2.43.

5.4 COROLLARY

For a function $T : [0,1]^2 \longrightarrow [0,1]$ the following are equivalent:

(i) T is a continuous Archimedean t-norm.

(ii) T has a continuous multiplicative generator, i.e., there exists a continuous, strictly increasing function $\theta : [0,1] \longrightarrow [0,1]$ with $\theta(1) = 1$, which is uniquely determined up to a positive constant exponent, such that for all $(x,y) \in [0,1]^2$ we have $T(x,y) = \theta^{(-1)}(\theta(x) \cdot \theta(y))$.

Moreover, T is a strict t-norm if and only if each continuous multiplicative generator θ of T satisfies $\theta(0) = 0$, and T is a nilpotent t-norm if and only if each continuous multiplicative generator θ of T satisfies $\theta(0) > 0$.

Because of the operator N (given in (3.17), see Figure 3.15) we have the following representation of continuous Archimedean t-conorms in terms of continuous additive and multiplicative generators.

5.5 COROLLARY
For a function $S : [0,1]^2 \longrightarrow [0,1]$ *the following are equivalent:*

(i) S *is a continuous Archimedean t-conorm.*
(ii) S *has a continuous additive generator, i.e., there exists a continuous, strictly increasing function* $s : [0,1] \longrightarrow [0,\infty]$ *with* $s(0) = 0$, *which is uniquely determined up to a positive multiplicative constant, such that for all* $(x,y) \in [0,1]^2$ *we have* $S(x,y) = s^{(-1)}(s(x) + s(y))$.
(iii) S *has a continuous multiplicative generator, i.e., there exists a continuous, strictly decreasing function* $\sigma : [0,1] \longrightarrow [0,1]$ *with* $\sigma(0) = 1$, *which is uniquely determined up to a positive constant exponent, such that for all* $(x,y) \in [0,1]^2$ *we have* $S(x,y) = \sigma^{(-1)}(\sigma(x) \cdot \sigma(y))$.

It is straightforward that each strictly increasing bijection $\varphi : [0,1] \longrightarrow [0,1]$ preserves the strictness or nilpotency of t-norms. Now we are in the position to prove that all strict and all nilpotent t-norms are mutually isomorphic.

5.6 PROPOSITION
Let T_1 *and* T_2 *be two continuous Archimedean t-norms. The following are equivalent:*

(i) T_1 *and* T_2 *are isomorphic.*
(ii) Either both T_1 *and* T_2 *are strict or both* T_1 *and* T_2 *are nilpotent.*

Proof: If T_1 and T_2 are two strict or two nilpotent t-norms with continuous additive generators t_1 and t_2, respectively, such that $t_1(0) = t_2(0)$ (observe that this condition is always satisfied in the case of continuous additive generators of strict t-norms, and that it is possible to choose appropriate generators satisfying it in the case of nilpotent t-norms), then in both cases $\varphi : [0,1] \longrightarrow [0,1]$ defined by $\varphi = t_1^{(-1)} \circ t_2$ is an isomorphism between T_2 and T_1. □

An immediate consequence of Propositions 3.31 and 5.6 is that the product $T_{\mathbf{P}}$ and the Łukasiewicz t-norm $T_{\mathbf{L}}$ are not only prototypical examples of strict and nilpotent t-norms, respectively, but that each continuous Archimedean t-norm is isomorphic either to $T_{\mathbf{P}}$ or to $T_{\mathbf{L}}$.

5.7 COROLLARY
(i) A function $T : [0,1]^2 \longrightarrow [0,1]$ *is a strict t-norm if and only if it is isomorphic to the product* $T_{\mathbf{P}}$.

(ii) *A function* $T : [0,1]^2 \longrightarrow [0,1]$ *is a nilpotent t-norm if and only if it is isomorphic to the Łukasiewicz t-norm* $T_{\mathbf{L}}$.

5.8 REMARK

As a consequence of Proposition 3.31 and Theorem 5.1, a strictly increasing isomorphism $\varphi : [0,1] \longrightarrow [0,1]$ satisfies $(T_{\mathbf{P}})_\varphi = T_{\mathbf{P}}$ if and only if $\varphi(x) = x^b$ for some $b \in]0,\infty[$, and it satisfies $(T_{\mathbf{L}})_\varphi = T_{\mathbf{L}}$ if and only if $\varphi = \mathrm{id}_{[0,1]}$. This means that there are infinitely many isomorphisms between a given strict t-norm and $T_{\mathbf{P}}$, whereas the isomorphism between a nilpotent t-norm and $T_{\mathbf{L}}$ is uniquely determined (for an alternative construction of this isomorphism see Proposition 5.10 in the next section).

5.2 Strict and nilpotent t-norms

In Section 5.1, the assertions of Corollary 5.7 have been derived mainly as consequences of Theorem 5.1. However, it is possible to show these results in a more direct way, which will be presented in this section. This follows the historical development since results leading to representations of strict t-norms have been established in [Abel 1826, Aczél 1949, Faucett 1955, Schweizer & Sklar 1961], and characterizations of nilpotent I-semigroups were proved in [Mostert & Shields 1957, Paalman-de Miranda 1964, Ling 1965].

5.9 PROPOSITION

For a function $T : [0,1]^2 \longrightarrow [0,1]$ *the following are equivalent:*

(i) T *is a strict t-norm.*

(ii) T *is isomorphic to the product* $T_{\mathbf{P}}$, *i.e., there is a strictly increasing bijection* $\varphi : [0,1] \longrightarrow [0,1]$ *such that for all* $(x,y) \in [0,1]^2$

$$T(x,y) = \varphi^{-1}(T_{\mathbf{P}}(\varphi(x),\varphi(y))) = \varphi^{-1}(\varphi(x)\cdot\varphi(y)). \tag{5.2}$$

Proof: From Remark 2.30(ii) we know that each function $T : [0,1]^2 \longrightarrow [0,1]$ which is isomorphic to $T_{\mathbf{P}}$ must be a strict t-norm, so (ii) implies (i).

Conversely, let T be a strict t-norm. Because of (3.5), for each $x \in]0,1[$

$$\lim_{n\to\infty} x_T^{(\frac{1}{2^n})} = 1.$$

Now fix an arbitrary element $a \in]0,1[$ and consider the set

$$D = \{a_T^{(\frac{m}{2^n})} \mid m \in \mathbb{N}, n \in \mathbb{N}\cup\{0\}\} \subseteq [0,1].$$

Of course, $(D, T|_{D^2})$ is a commutative, fully ordered subsemigroup of $([0,1],T)$.

We claim that D is a dense subset of $[0,1]$. Indeed, assuming that $\mathrm{cl}(D)$, the topological closure of D, is a proper subset of $[0,1]$ implies that there is a non-empty open interval $]u,v[\subseteq [0,1] \setminus \mathrm{cl}(D)$ with $u,v \in \mathrm{cl}(D)$. Then the sequence

$$\left(T\left(a_T^{(\frac{1}{2^n})}, v\right)\right)_{n \in \mathbb{N}}$$

converges to the value $T(1, v) = v$, and the cancellation law (CL) (which necessarily holds for the strict t-norm T) implies for all $n \in \mathbb{N}$

$$T\left(a_T^{(\frac{1}{2^n})}, v\right) < v.$$

Hence, for a sufficiently large $n \in \mathbb{N}$ we must have

$$T\left(a_T^{(\frac{1}{2^n})}, v\right) \in \left]u, v\right[\subseteq [0,1] \setminus \mathrm{cl}(D),$$

which is a contradiction to the fact that for all $n \in \mathbb{N}$ we also must have $T\left(a_T^{(\frac{1}{2^n})}, v\right) \in \mathrm{cl}(D)$, as a consequence of $a_T^{(\frac{1}{2^n})} \in \mathrm{cl}(D)$ and $v \in \mathrm{cl}(D)$.

Now define the function $\psi : D \longrightarrow [0,1]$ by $\psi\left(a_T^{(r)}\right) = a^r$, where a^r is the usual rth power of the real number a (with respect to the multiplication). It is clear that ψ is a strictly increasing injection and that $\psi(D)$, the image of D under ψ, is a dense subset of $[0,1]$. Moreover, for all $m, p \in \mathbb{N}$ and for all $n, q \in \mathbb{N} \cup \{0\}$ we have

$$\psi\left(T\left(a_T^{(\frac{m}{2^n})}, a_T^{(\frac{p}{2^q})}\right)\right) = \psi\left(a_T^{(\frac{m}{2^n} + \frac{p}{2^q})}\right) = a^{\frac{m}{2^n} + \frac{p}{2^q}} = a^{\frac{m}{2^n}} \cdot a^{\frac{p}{2^q}}.$$

Thus, $\psi : D \longrightarrow [0,1]$ can be extended to a strictly increasing bijection with domain $\mathrm{cl}(D) = [0,1]$, i.e., to a function $\varphi : [0,1] \longrightarrow [0,1]$, which satisfies (5.2) for all $(x, y) \in [0,1]^2$. □

5.10 PROPOSITION

For a function $T : [0,1]^2 \longrightarrow [0,1]$ the following are equivalent:

(i) T is a nilpotent t-norm.

(ii) T is isomorphic to the Łukasiewicz t-norm $T_{\mathbf{L}}$, i.e., there is a strictly increasing bijection $\varphi : [0,1] \longrightarrow [0,1]$ such that for all $(x,y) \in [0,1]^2$

$$T(x, y) = \varphi^{-1}(T_{\mathbf{L}}(\varphi(x), \varphi(y))) = \varphi^{-1}(\max(\varphi(x) + \varphi(y) - 1, 0)).$$

Proof: From Remark 2.30(ii) we know that each function $T : [0,1]^2 \longrightarrow [0,1]$ which is isomorphic to the nilpotent t-norm $T_{\mathbf{L}}$ must be a nilpotent t-norm, so (ii) implies (i).

Conversely, if T is a nilpotent t-norm then put

$$a = \sup\{x \in \left]0,1\right[\mid x_T^{(2)} = 0\} > 0,$$

from which we immediately infer

$$a_T^{(\frac{p}{2^q})} > a_T^{(\frac{m}{2^n})} > 0 \qquad \text{whenever } \tfrac{p}{2^q} < \frac{m}{2^n} < 2, \tag{5.3}$$

$$a_T^{(\frac{m}{2^n})} = 0 \qquad \text{whenever } \tfrac{m}{2^n} \geq 2. \tag{5.4}$$

As shown in the proof of Proposition 5.9, the following set D is a dense subset of $[0,1]$, i.e., $\mathrm{cl}(D) = [0,1]$:

$$D = \{a_T^{(\frac{m}{2^n})} \mid m \in \mathbb{N}, n \in \mathbb{N} \cup \{0\}\}.$$

The function $\psi : D \longrightarrow [0.5, 1]$ given by

$$\psi(a_T^{(\frac{m}{2^n})}) = \max(2^{-\frac{m}{2^n+1}}, 0.5)$$

is well-defined because of $2^{-\frac{m}{2^n+1}} \leq 0.5$ for all $m \in \mathbb{N}$ and all $n \in \mathbb{N} \cup \{0\}$ with $\frac{m}{2^n} \geq 2$, as a consequence of (5.4). Because of (5.3), ψ is strictly increasing, and $\psi(D)$, the image of D under ψ, is a dense subset of $[0.5, 1]$. Therefore, ψ has a unique strictly increasing extension $\overline{\psi} : [0,1] \to [0.5, 1]$ which is an isomorphism between the two semigroups $([0,1], T)$ and $([0.5,1], *)$ with $x * y = \max(xy, 0.5)$ (see Example 2.26(i)). Since $([0.5,1], *)$ is isomorphic to $([0,1], T_{\mathbf{L}})$, as pointed out in Example 2.29(i), this completes the proof. □

5.3 Representation of continuous t-norms

The following representation of continuous t-norms by means of ordinal sums can be derived from results in [Mostert & Shields 1957] in the context of I-semigroups. It also follows from Corollary 3.56 (originally proved in [Schweizer & Sklar 1983, Theorem 5.38]), but for the convenience of the reader we present a self-contained proof of this important result.

5.11 THEOREM
For a function $T : [0,1]^2 \longrightarrow [0,1]$ the following are equivalent:

(i) T is a continuous t-norm.
(ii) T is uniquely representable as an ordinal sum of continuous Archimedean t-norms, i.e., there exists a uniquely determined (finite or countably infinite) index set A, a family of uniquely determined pairwise disjoint open subintervals $(]a_\alpha, e_\alpha[)_{\alpha \in A}$ of $[0,1]$ and a family of uniquely determined continuous Archimedean t-norms $(T_\alpha)_{\alpha \in A}$ such that

$$T = (\langle a_\alpha, e_\alpha, T_\alpha \rangle)_{\alpha \in A}.$$

Proof: Because of Proposition 3.49, each ordinal sum of continuous t-norms is continuous, so (ii) implies (i).

In order to prove the converse, recall that from Corollary 2.8 we know that the set of idempotent elements of the continuous t-norm T is closed. Therefore, the set of non-idempotent elements of T is either empty (in which case we have $T = T_{\mathbf{M}}$) or it can be written as the (non-trivial) union of a finite or countably infinite family of pairwise disjoint open subintervals $(]a_\alpha, e_\alpha[)_{\alpha \in A}$ where, of course, each a_α and each e_α (but no element in $]a_\alpha, e_\alpha[$) is an idempotent element of T.

For the time being, assume that $A \neq \emptyset$ and fix an arbitrary $\alpha \in A$. Then the monotonicity of T implies that for all $(x, y) \in [a_\alpha, e_\alpha]^2$

$$a_\alpha = T(a_\alpha, a_\alpha) \leq T(x, y) \leq T(e_\alpha, e_\alpha) = e_\alpha$$

and for all $x \in [a_\alpha, 1]$

$$a_\alpha = T(a_\alpha, a_\alpha) \leq T(x, a_\alpha) \leq T(1, a_\alpha) = a_\alpha,$$

showing that $([a_\alpha, e_\alpha], T|_{[a_\alpha,e_\alpha]^2})$ is a semigroup with annihilator a_α and with trivial idempotent elements only (actually, a_α acts as annihilator on $[a_\alpha, 1]$). Because of the monotonicity and continuity of T we also have for each $\alpha \in A$

$$\{T(z, e_\alpha) \mid z \in [0, 1]\} = [0, e_\alpha],$$

which means that each $x \in [0, e_\alpha]$ can be written as $x = T(z, e_\alpha)$ for some $z \in [0, 1]$. This, together with the associativity of T, implies that

$$T(x, e_\alpha) = T(T(z, e_\alpha), e_\alpha) = T(z, T(e_\alpha, e_\alpha)) = T(z, e_\alpha) = x,$$

showing that e_α acts as neutral element on $[0, e_\alpha]$ and, subsequently, in the I-semigroup $([a_\alpha, e_\alpha], T|_{[a_\alpha,e_\alpha]^2})$.

Let $\varphi_\alpha : [0, 1] \longrightarrow [a_\alpha, e_\alpha]$ be the strictly increasing bijection given by

$$\varphi_\alpha(x) = a_\alpha + (e_\alpha - a_\alpha) \cdot x,$$

then for each $\alpha \in A$ the function $T_\alpha : [0, 1]^2 \longrightarrow [0, 1]$ defined by

$$T_\alpha(x, y) = \varphi_\alpha^{-1}(T(\varphi_\alpha(x), \varphi_\alpha(y))$$

is a continuous t-norm which has only trivial idempotent elements, and which is also Archimedean because of Proposition 2.15(i). A simple computation verifies that for all $\alpha \in A$ and for all $(x, y) \in [a_\alpha, e_\alpha]^2$ we have

$$T(x, y) = a_\alpha + (e_\alpha - a_\alpha) \cdot T_\alpha\left(\tfrac{x - a_\alpha}{e_\alpha - a_\alpha}, \tfrac{y - a_\alpha}{e_\alpha - a_\alpha}\right).$$

If $(x, y) \in [0, 1]^2$ (without loss of generality we may assume $x \leq y$) is contained in none of the squares $[a_\alpha, e_\alpha]^2$ then there exists some idempotent element $b \in [x, y]$ which acts as neutral element on $[0, b]$ and as annihilator on $[b, 1]$, and we have

$$T(x, y) = T(T(x, b), y) = T(x, T(b, y)) = T(x, b) = x = \min(x, y),$$

completing the proof that $T = (\langle a_\alpha, e_\alpha, T_\alpha \rangle)_{\alpha \in A}$.

The uniqueness of the representation of T is an immediate consequence of the one-to-one correspondence between the set of idempotent elements of T and the family of intervals $(]a_\alpha, e_\alpha[)_{\alpha \in A}$. □

By duality, we have a completely analogous representation for continuous t-conorms:

5.12 COROLLARY
For a function $S : [0,1]^2 \longrightarrow [0,1]$ the following are equivalent:

(i) S is a continuous t-conorm.
(ii) S is uniquely representable as an ordinal sum of continuous Archimedean t-conorms, i.e., there exists a uniquely determined (finite or countably infinite) index set A, a family of uniquely determined pairwise disjoint open subintervals $(]a_\alpha, e_\alpha[)_{\alpha\in A}$ of $[0,1]$ and a family of uniquely determined continuous Archimedean t-conorms $(S_\alpha)_{\alpha\in A}$ such that $S = (\langle a_\alpha, e_\alpha, S_\alpha\rangle)_{\alpha\in A}$.

5.4 Functional equations

The four basic triangular norms can be characterized by functional equations.

5.13 REMARK
(i) The minimum $T_{\mathbf{M}}$ is the only t-norm T solving the functional equation

$$T(x,x) = x$$

for all $x \in [0,1]$ (see Proposition 1.9(i)).
(ii) The product t-norm $T_{\mathbf{P}}$ is the only continuous t-norm T solving

$$T(x,x) = x^2,$$
$$T(x,1-x) = x - x^2,$$

for all $x \in [0,1]$. Because of the continuity of T and the representation in Theorem 5.11, either one of the functional equations above implies that T is strict, and the result follows from Proposition 7.13.
(iii) The Łukasiewicz t-norm $T_{\mathbf{L}}$ is the only t-norm T solving

$$T(x+z,y) = T(x,y) + z$$

for all $x,y,z \in [0,1]$ such that $x+z \in [0,1]$ and $x+y \geq 1$. Indeed, from $y = T(1-y+y,y) = T(1-y,y)+y$ and the monotonicity (T3) we conclude $T(x,y) = 0$ whenever $x+y \leq 1$. In the case $x+y > 1$ we obtain $T(x,y) = T(1-y+x+y-1,y) = x+y-1$, showing that $T = T_{\mathbf{L}}$.
(iv) The drastic product $T_{\mathbf{D}}$ is the only t-norm T solving $T(x,x) = 0$ for all $x \in [0,1[$ (see Proposition 1.9(ii)).

Frank t-norms
The most famous and important result in the context of t-norms and functional equations states that only the Frank t-norms and t-conorms (together with

ordinal sums thereof) solve the Frank functional equation (4.2) [Frank 1979] (compare also [Alsina 1985b]).

Recall that a function $f : I \longrightarrow \mathbb{R}$, where $I \subseteq \mathbb{R}$ is an arbitrary non-empty interval, is called a *convex* or a *concave function* if for all $x, y \in I$ we have

$$f\left(\tfrac{x+y}{2}\right) \le \tfrac{1}{2}(f(x) + f(y)), \tag{5.5}$$
$$f\left(\tfrac{x+y}{2}\right) \ge \tfrac{1}{2}(f(x) + f(y)), \tag{5.6}$$

respectively. Trivially, a function $f : I \longrightarrow \mathbb{R}$ is convex if and only if the set

$$\{(x, y) \mid x \in I \text{ and } y \ge f(x)\}$$

is a *convex set*, i.e., a set which, together with any two elements thereof, contains also the straight line segment joining them. Moreover, $f : I \longrightarrow \mathbb{R}$ is a concave function if and only if $-f$ is a convex function.

The original proof of the subsequent theorem is rather long and involves a lot of technical details, therefore we can present here an extended sketch only (another sketch of the same proof can be found in [Flegel 1998]).

5.14 THEOREM

(i) A pair (T, S), where T is an Archimedean t-norm and S is a t-conorm, fulfills the Frank functional equation (4.2), i.e., the equation

$$T(x, y) + S(x, y) = x + y$$

for all $(x, y) \in [0, 1]^2$ if and only if there exists a $\lambda \in \,]0, \infty]$ such that $T = T^{\mathbf{F}}_\lambda$ and $S = S^{\mathbf{F}}_\lambda$.

(ii) A pair (T, S), where T is a t-norm and S is a t-conorm, fulfills the Frank functional equation (4.2) for all $(x, y) \in [0, 1]^2$ if and only if there exists a family $(]a_\alpha, e_\alpha[)_{\alpha \in A}$ of pairwise disjoint open subintervals of $[0, 1]$ and a family of indexes $(\lambda_\alpha)_{\alpha \in A}$ in $[0, \infty[$ such that

$$T = (\langle a_\alpha, e_\alpha, T^{\mathbf{F}}_{\lambda_\alpha} \rangle)_{\alpha \in A}, \tag{5.7}$$
$$S = (\langle a_\alpha, e_\alpha, S^{\mathbf{F}}_{\lambda_\alpha} \rangle)_{\alpha \in A}. \tag{5.8}$$

Sketch of the proof: Obviously, the pairs $(T_{\mathbf{M}}, S_{\mathbf{M}})$, $(T_{\mathbf{P}}, S_{\mathbf{P}})$ and $(T_{\mathbf{L}}, S_{\mathbf{L}})$ satisfy (4.2), and for each $\lambda \in \,]0, 1[\, \cup \,]1, \infty[$ and for all $(x, y) \in [0, 1]^2$ we obtain

$$\begin{aligned} T^{\mathbf{F}}_\lambda(x, y) + S^{\mathbf{F}}_\lambda(x, y) &= \log_\lambda\left(1 + \tfrac{(\lambda^x - 1)(\lambda^y - 1)}{\lambda - 1}\right) \\ &\quad + 1 - \log_\lambda\left(1 + \tfrac{(\lambda^{1-x} - 1)(\lambda^{1-y} - 1)}{\lambda - 1}\right) \\ &= 1 + \log_\lambda\left(\tfrac{\lambda - 1 + (\lambda^x - 1)(\lambda^y - 1)}{\lambda - 1 + (\lambda^{1-x} - 1)(\lambda^{1-y} - 1)}\right) \\ &= 1 + \log_\lambda(\lambda^{x+y-1}) \\ &= x + y. \end{aligned}$$

But then also each pair of ordinal sums of the form given in (5.7) and (5.8) fulfills (4.2).

Conversely, we first show that for each solution (T,S) of (4.2) both the t-norm T and the t-conorm S necessarily are continuous. If in (4.2) we fix one coordinate, say, $y_0 \in [0,1]$, then we obtain

$$T(\cdot,y_0) + S(\cdot,y_0) = \mathrm{id}_{[0,1]}(\cdot) + y_0.$$

Since the function on the right-hand side is obviously continuous and since both $T(\cdot,y_0), S(\cdot,y_0) : [0,1] \longrightarrow [0,1]$ are non-decreasing functions because of (T3), the only possibility is that both $T(\cdot,y_0)$ and $S(\cdot,y_0)$ are continuous, thus implying the continuity of T and S as a consequence of Proposition 1.19.

Next, consider first a solution (T,S) of (4.2), where T is a nilpotent t-norm, and let $t : [0,1] \longrightarrow [0,\infty]$ be a continuous additive generator of T with $t(0) = 1$. The generator t is a convex function, as a consequence of [Frank 1979, Lemma 2.1], [Moynihan 1978b, Theorem 3.2] and [Schweizer & Sklar 1983, Theorem 6.3.3]. Writing $x_0 = t^{(-1)}(\frac{1}{2})$, we obtain $T(x_0,v) = 0$ if and only if $v \in [0,x_0]$. The convexity of t implies $x_0 \le \frac{1}{2}$ and $t(\frac{x_0}{2}) + t(\frac{3x_0}{2}) \ge 2t(x_0) = 1$, from which we obtain $T(\frac{x_0}{2},\frac{3x_0}{2}) = 0$. From the associativity of S we infer $S(S(x_0,\frac{x_0}{2}),\frac{3x_0}{2}) = S(x_0,S(\frac{x_0}{2},\frac{3x_0}{2}))$, i.e.,

$$\begin{aligned} S(\tfrac{3x_0}{2},\tfrac{3x_0}{2}) &= S(x_0 + \tfrac{x_0}{2} - T(x_0,\tfrac{x_0}{2}),\tfrac{3x_0}{2}) \\ &= S(x_0,\tfrac{x_0}{2} + \tfrac{3x_0}{2} - T(\tfrac{x_0}{2},\tfrac{3x_0}{2})) \\ &= S(x_0,2x_0) \end{aligned}$$

and, consequently, $T(\frac{3x_0}{2},\frac{3x_0}{2}) = T(x_0,2x_0) > 0$, i.e., $2t(\frac{3x_0}{2}) = t(x_0) + t(2x_0)$. Then, because t is a convex function, $t|_{[x_0,2x_0]}$ must be linear, i.e., for each $x \in [0,x_0]$ we have $t(x_0 + x) = t(x_0) + k \cdot x$. Taking into account that, for each $x \in \,]0,x_0[$, we have $T(x_0,x_0+x) > 0$ and using similar arguments as before, we obtain $T(2x_0,x) = T(x_0,x_0+x) > 0$, i.e., $t(2x_0) + t(x) = t(x_0) + t(x_0+x)$ or, equivalently, $t(x) = t(x_0) + t(x_0+x) - t(2x_0) = t(x_0) + k\cdot(x - x_0)$, which, together with the continuity of t, shows that $t|_{[0,2x_0]}$ is linear. From $t(0) = 1$ and $t(x_0) = \frac{1}{2}$ we get $t(2x_0) = 0$, i.e., $2x_0 = 1$, and $t(x) = 1 - x$ for all $x \in [0,1]$, i.e., $T = T_{\mathbf{L}}$ and, subsequently, $S = S_{\mathbf{L}}$.

If (T,S) is a solution of (4.2) with a strict t-norm T the situation is much more complicated. In [Frank 1979, Lemma 3.14] it was shown in full detail that in this case we must have $T = T_\lambda^{\mathbf{F}}$ and $S = S_\lambda^{\mathbf{F}}$ for some $\lambda \in \,]0,\infty[$.

For the remaining part of the proof of (ii), let (T,S) be a solution of (4.2), where T is a continuous t-norm. Observe first that S must be a continuous t-conorm, and that T and S have the same idempotent elements. Therefore, Theorem 5.11 implies $T = (\langle a_\alpha, e_\alpha, T_\alpha\rangle)_{\alpha\in A}$ and $S = (\langle a_\alpha, e_\alpha, S_\alpha\rangle)_{\alpha\in A})$ with continuous Archimedean t-norms T_α and t-conorms S_α, respectively. Because (4.2) obviously holds for each $(x,y) \in [a_\alpha,e_\alpha]^2$, it is a matter of computation only to verify that each pair (T_α, S_α) must be a solution of (4.2), i.e., $(T_\alpha, S_\alpha) = (T^{\mathbf{F}}_{\lambda_\alpha}, S^{\mathbf{F}}_{\lambda_\alpha})$ for some $\lambda_\alpha \in \,]0,\infty]$. □

Observe that the original result in [Frank 1979] was formulated in the context of I-semigroups on $[0,1]$. However, the operations in such I-semigroups are t-norms or t-conorms because of Theorem 2.43.

It is interesting to note that, because of [Fodor & Jenei 1999], Theorem 5.14 remains valid if we fix an arbitrary strictly decreasing bijection $f : [0,1] \longrightarrow [0,1]$ and require the validity of (4.2), i.e., $T(x,y) + S(x,y) = x + y$ only for those $(x,y) \in [0,1]^2$ satisfying $f(x) + f(y) > 1$.

The following result was proved in [Fodor & Roubens 1994, Theorem 3.3]. For the notions T_φ and S_φ see Proposition 2.28(iv), in particular (2.8) and (2.9).

5.15 THEOREM

Let T_1 and T_2 be two continuous t-norms and S be a continuous t-conorm. The following are equivalent:

(i) For all $(x,y) \in [0,1]^2$, the triple (T_1, T_2, S) fulfills the functional equation

$$S(T_1(x, 1-y), T_2(x,y)) = x. \tag{5.9}$$

(ii) There exist a strictly increasing bijection $\varphi : [0,1] \longrightarrow [0,1]$ satisfying $\varphi(x) + \varphi(1-x) = 1$ for all $x \in [0,1]$ and a number $\lambda \in [0,\infty]$ such that $T_1 = \left(T^{\mathbf{F}}_{\lambda}\right)_\varphi$, $T_2 = \left(T^{\mathbf{F}}_{\frac{1}{\lambda}}\right)_\varphi$ and $S = (S_{\mathbf{L}})_\varphi$.

Proof: A simple computation shows that, for each $\lambda \in [0,\infty]$ and for each strictly increasing bijection $\varphi : [0,1] \longrightarrow [0,1]$ satisfying $\varphi(x) + \varphi(1-x) = 1$ for all $x \in [0,1]$, the t-norms $T_1 = \left(T^{\mathbf{F}}_{\lambda}\right)_\varphi$ and $T_2 = \left(T^{\mathbf{F}}_{\frac{1}{\lambda}}\right)_\varphi$ and the t-conorm $S = (S_{\mathbf{L}})_\varphi$ satisfy (5.9).

Conversely, assume that the two continuous t-norms T_1 and T_2 and the continuous t-conorm S fulfill (5.9). Putting $x = 1$, we get $S(y, 1-y) = 1$ for all $y \in [0,1]$. Then the validity of (5.9) together with the continuity of T_1, T_2 and S yields $S(x,y) < 1$ whenever $x + y < 1$. As a consequence of Corollaries 5.5 and 5.12, S is a nilpotent t-conorm generated by an additive generator $s : [0,1] \longrightarrow [0,\infty]$ satisfying $s(1) = 1$. Then $S(y, 1-y) = 1$ is equivalent to $s(y) + s(1-y) \geq 1$, and $S(x,y) < 1$ is equivalent to $s(x) + s(y) < 1$, from which we get $s(x) + s(1-x) = 1$ for all $x \in [0,1]$ because of the continuity of s. In particular, s is a strictly increasing bijection from $[0,1]$ to $[0,1]$, so we may rename it $\varphi : [0,1] \longrightarrow [0,1]$ given by $\varphi(x) = s(x)$. Therefore, the functional equation (5.9) transforms into

$$\varphi(T_1(x, 1-y)) + \varphi(T_2(x,y)) = \varphi(x),$$

and, writing $\varphi(x) = u$ and $\varphi(y) = v$, and defining the two continuous t-norms $T_1^* = (T_1)_{\varphi^{-1}}$ and $T_2^* = (T_2)_{\varphi^{-1}}$, we get for all $(u,v) \in [0,1]^2$

$$T_1^*(u, 1-v) + T_2^*(u, v) = u. \tag{5.10}$$

In particular, we obtain the two equations $T_1^*(u,v) + T_2^*(u, 1-v) = u$ and $T_1^*(1-v, 1-u) + T_2^*(1-v, u) = 1-v$. Their subtraction yields

$$T_1^*(u,v) + 1 - T_1^*(1-u, 1-v) = u + v$$

or, equivalently, $T_1^*(u,v) + S_1^*(u,v) = u+v$, where S_1^* is the t-conorm which is dual to T_1^*. This means, however, that the pair (T_1^*, S_1^*) is a solution of the Frank functional equation (4.2), and, because of Theorem 5.14, T_1^* must be either a Frank t-norm or an ordinal sum of Frank t-norms.

In a completely analogous way it is possible to show that also T_2^* must be either a Frank t-norm or an ordinal sum of Frank t-norms.

If T_1^* has no non-trivial idempotent elements, then the only possibility is $T_1^* = T_\lambda^{\mathbf{F}}$ for some $\lambda \in \,]0, \infty]$. In the case $T_1^* = T_\infty^{\mathbf{F}} = T_{\mathbf{L}}$ equation (5.10) leads to $T_2^* = T_{\mathbf{M}} = T_0^{\mathbf{F}}$, and, if $T_1^* = T_\lambda^{\mathbf{F}}$ for some $\lambda \in \,]0, \infty[$, then a straightforward computation shows that (5.10) implies for all $(u,v) \in [0,1]^2$

$$T_2^*(u,v) = u - T_\lambda^{\mathbf{F}}(u, 1-v) = T_{\frac{1}{\lambda}}^{\mathbf{F}}(u,v).$$

If $a \in \,]0,1[$ is an idempotent element of T_1^* then, because of the monotonicity (T3) and the boundary condition (T4), for all $b \in [a,1]$ we have $T_1^*(a,b) = a$ and, as a consequence of (5.10), $T_2^*(a, 1-b) = 0$ and, in particular, $T_2^*(a, 1-a) = 0$. But the only Frank t-norm or ordinal sum of Frank t-norms satisfying this latter equality is $T_{\mathbf{L}}$, and (5.10) implies that then only $T_1^* = T_{\mathbf{M}}$ is possible, i.e., we have $T_1^* = T_{\mathbf{M}} = T_0^{\mathbf{F}}$ and $T_2^* = T_{\mathbf{L}} = T_\infty^{\mathbf{F}}$.

Recalling the definitions of T_1^* and T_2^* and Remark 2.30(ii), this completes the proof. □

If we put $T = T_1 = T_2$ then the only solutions of the functional equation (5.9) are a t-norm T and a t-conorm S such that there exists a multiplicative generator of T which is simultaneously an additive generator of S (this result was originally proved in [Alsina 1985a]).

5.16 COROLLARY

Let T be a continuous t-norm and S be a continuous t-conorm. The following are equivalent:

(i) For all $(x,y) \in [0,1]^2$, the pair (T,S) fulfills the functional equation

$$S(T(x, 1-y), T(x,y)) = x.$$

(ii) There exists a strictly increasing bijection $\varphi : [0,1] \longrightarrow [0,1]$ satisfying $\varphi(x) + \varphi(1-x) = 1$ for all $x \in [0,1]$ such that $T = (T_{\mathbf{P}})_\varphi$ and $S = (S_{\mathbf{L}})_\varphi$.

Aczél-Alsina t-norms

The following characterization of Aczél-Alsina t-norms (see Section 4.8) is based on [Aczél & Alsina 1984].

5.17 PROPOSITION

Let $p, q \in \,]0,1[\, \cup \,]0,\infty[$ such that $\frac{\log p}{\log q}$ is an irrational number, and let T be a t-norm. The following are equivalent:

(i) For all $(x, y) \in [0,1]^2$, T is a solution of the functional equations

$$T(x^p, y^p) = T(x,y)^p, \tag{5.11}$$
$$T(x^q, y^q) = T(x,y)^q. \tag{5.12}$$

(ii) There exists a $\lambda \in [0,\infty]$ such that $T = T_\lambda^{\mathbf{AA}}$.

Proof: It is easily checked that each $T_\lambda^{\mathbf{AA}}$ solves both (5.11) and (5.12).

Conversely, let T be a t-norm satisfying equations (5.11) and (5.12) for all $(x,y) \in [0,1]^2$, and consider the set $M = \{p^n q^m \mid m, n \in \mathbb{Z}\} \subseteq \,]0,\infty[$. From the irrationality of $\frac{\log p}{\log q}$ we conclude that M is a dense subset of $]0,\infty[$. By induction, for all $z \in M$ we get $T(x^z, y^z) = T(x,y)^z$.

If T is a continuous t-norm satisfying equation (5.11), we obtain for all $(x,y) \in [0,1]^2$ and for all $z \in \,]0,\infty[$ that

$$T(x^z, y^z) = T(x,y)^z. \tag{5.13}$$

If T is a non-Archimedean solution of (5.11) and (5.12), i.e., $T(a,a) = a$ for some $a \in \,]0,1[$, then for all $x \in \,]0,1[$

$$T(x,x) = T\left(a^{\log_a x}, a^{\log_a x}\right) = T(a,a)^{\log_a x} = a^{\log_a x} = x,$$

implying $T = T_{\mathbf{M}} = T_\infty^{\mathbf{AA}}$.

If T is a nilpotent solution of (5.11) and (5.12), then $T(a,a) = a$ for some $a \in \,]0,1[$. The same reasoning as in the non-Archimedean case shows that $T(x,x) = 0$ for all $x \in \,]0,1[$ which would imply $T = T_{\mathbf{D}}$, thus contradicting the continuity of T.

If T is a strict solution of (5.11) and (5.12) with continuous additive generator $t : [0,1] \longrightarrow [0,\infty]$, then (5.13) can be rewritten as

$$t^{-1}\left(2t(x^z)\right) = \left(t^{-1}(2t(x))\right)^z. \tag{5.14}$$

Defining the function $h : [0,\infty] \longrightarrow [0,\infty]$ by $h(x) = t(e^{-x})$, the functional equation (5.14) can be transformed into the equivalent equation $2h(x) = h(kx)$ for some $k \in \,]1,\infty[$ and for all $x \in \,]0,\infty[$. Each solution of this Cauchy functional equation [Aczél 1966] has the form $h(x) = h(1)x^\lambda$, where $\lambda = \frac{\log 2}{\log k} > 0$. Consequently, we get $t(x) = h(1)(-\log x)^\lambda$ with $\lambda \in \,]0,\infty[$, i.e., $T = T_\lambda^{\mathbf{AA}}$.

Combining these three cases, $T = T_\lambda^{\mathbf{AA}}$ for some $\lambda \in \,]0,\infty]$ is a necessary condition for a continuous t-norm T to solve equations (5.11) and (5.12).

Turning to non-continuous solutions T of (5.11) and (5.12), the fact that M is a dense subset of $]0,\infty[$ implies that T must be continuous on $[0,1[^2$, i.e., the only points of discontinuity are of the form $(a,1)$ or $(1,a)$ with $a \in \,]0,1]$.

If T is discontinuous in the point $(1,1)$, i.e., if

$$x_0 = \sup\{T(x,y) \mid (x,y) \in [0,1[^2\} < 1,$$

then $x_0 = 0$, i.e., $T = T_{\mathbf{D}} = T_0^{\mathbf{AA}}$. To see this, assume $x_0 \in \,]0,1[$ and put $r = \min(p, \frac{1}{p}) \in M$. Then there is an $\varepsilon \in \,]0, x_0[$ such that $x_0 < (x_0 - \varepsilon)^r$ and a point $(x,y) \in \,]0,1[^2$ such that $x_0 - \varepsilon < T(x,y)$, leading to the contradiction $x_0 < (x_0 - \varepsilon)^r < T(x,y)^r = T(x^r, y^r) \leq x_0$.

If T is continuous in $(1,1)$ and discontinuous in $(a,1)$ with $a \in \,]0,1[$, i.e., if

$$a_0 = \sup\{T(x,y) \mid (x,y) \in [0,a] \times [0,1[\} < a,$$

then the case $a_0 = 0$ is impossible since this would imply $T(x^s, y^s) = 0$ for all $s \in M$ and all $(x,y) \in [0,a] \times [0,1[$, i.e., $T = T_{\mathbf{D}}$, thus violating the continuity in $(1,1)$. For $a_0 \in \,]0,a[$ put $b = \frac{\log a_0}{\log a} > 1$, which means that $a_0 = a^b$. Then for each $\varepsilon \in \,\big]0, a_0 - (a_0)^b\big[$ there is some point $(x,y) \in [0,a] \times [0,1[$ with $(a_0)^b < a_0 - \varepsilon < T(x,y) \leq a_0$ and, since T is continuous in $(1,1)$, there is a $z \in [y,1[$ such that $y \leq T(z,z)$. But then we would obtain (recall that T is continuous on $[0,1[^2$)

$$T(T(x,z),z) \leq T\big(a^b, \big(z^{\frac{1}{b}}\big)^b\big) = T\big(a, z^{\frac{1}{b}}\big)^b < a_0 - \varepsilon \leq T(x,y) \leq T(x, T(z,z)),$$

thus violating (T2) and showing that this case is impossible too. □

Mayor-Torrens t-norms

A characterization of Mayor-Torrens t-norms (see Section 4.9) proving that they are completely determined by their values on $\{(x,x) \mid x \in [0,1]\}$, i.e., on the diagonal of $[0,1]^2$, was presented in [Mayor & Torrens 1991].

5.18 PROPOSITION

Let T be a continuous t-norm. The following are equivalent.

(i) For all $(x,y) \in [0,1]^2$, T satisfies the functional equation

$$T(x,y) = \max(T(\max(x,y), \max(x,y)) - |x-y|, 0). \tag{5.15}$$

(ii) There exists a $\lambda \in [0,1]$ such that $T = T_\lambda^{\mathbf{MT}}$.

Proof: It is easy to check that each $T_\lambda^{\mathbf{MT}}$ with $\lambda \in [0,1]$ is a solution of (5.15). Conversely, consider a continuous t-norm T which fulfills equation (5.15) for all $(x,y) \in [0,1]^2$.

If T has no zero divisors then, for all $(x,y) \in \,]0,1]^2$ with $x < y$, we have $0 < T(x,y) = T(y,y) - y + x$. From the continuity of T we conclude $T(y,y) = y$ for each $y \in \,]0,1]$, i.e., $T = T_{\mathbf{M}} = T_0^{\mathbf{MT}}$ (implying that there are no strict t-norms solving (5.15)).

Let T be a nilpotent solution of (5.15) and $t : [0,1] \longrightarrow [0,\infty]$ be a continuous additive generator of T with $t(0) = 1$. Fix $x, y \in [0,1]$ with, say, $y \geq x$ and assume $T(x,y) > 0$, i.e., $t(x) + t(y) < 1$. Then the equation (5.15) is transformed into $T(x,y) = T(y,y) - y + x$. From the continuity of t we immediately get $y - T(y,y) = t^{(-1)}(1 - t(y))$ for all $y \in \left[t^{(-1)}(0.5), 1\right]$. Consequently, taking into account that T is a solution of (5.15), we obtain for all $y \in [t^{(-1)}(0.5), 1]$

$$T\left(t^{(-1)}(0.5), y\right) = t^{(-1)}(0.5 + t(y)) = t^{(-1)}(0.5) - t^{(-1)}(1 - t(y)).$$

In [Mayor & Torrens 1991] it was shown in detail that the only solution of this functional equation is given by $t(x) = 1 - x$ for all $x \in [0, 0.5]$ and, by induction, for all $x \in [0,1]$, i.e., $T_{\mathbf{L}}$ is the only Archimedean solution of (5.15).

Let T be a non-Archimedean solution of (5.15) different from $T_{\mathbf{M}}$, i.e., T is a non-trivial ordinal sum with some non-trivial idempotent element a of T. Then for all $(x,y) \in \,]0,1]^2$ with $x < a \leq y$ we get $0 < x = T(x,y) = T(y,y) - y + x$, i.e,. $T(y,y) = y$. This means that the set of idempotent element of T is of the form $\{0\} \cup [\lambda, 1]$ for some $\lambda \in \,]0,1[$, which implies $T = (\langle 0, \lambda, T_{\mathbf{L}} \rangle) = T_\lambda^{\mathbf{MT}}$. □

Hamacher t-norms

The following characterization of Hamacher t-norms (see Section 4.3) can be found in [Fodor & Keresztfalvi 1994].

5.19 PROPOSITION

Let T be a t-norm and $\theta : [0,1] \longrightarrow [0,1]$ a continuous multiplicative generator of T with $\theta'(0^+) \in \,]0,\infty[$. The following are equivalent:

(i) For all $(x,y) \in \,]0,1]^2$, T satisfies the functional equation

$$T\left(x, 1 - \tfrac{T(x,1-y)}{x}\right) = xy. \tag{5.16}$$

(ii) $T = T^{\mathbf{H}}_{\frac{1}{\theta'(0^+)}}$.

Proof: A simple computation shows that each $T_\lambda^{\mathbf{H}}$ with $\lambda \in \,]0,\infty[$ solves (5.16).

Conversely, consider a t-norm T having a continuous multiplicative generator $\theta : [0,1] \longrightarrow [0,1]$ with $\theta'(0^+) \in \,]0,\infty[$, which fulfills equation (5.16) for all $(x,y) \in [0,1]^2$. Obviously, T must be continuous and Archimedean.

If T has zero divisors, i.e., if there are $x, y \in \,]0,1[$ with $T(x,y) = 0$, then

$$x > x(1-y) = T\left(x, 1 - \tfrac{T(x,y)}{x}\right) = T(x,1) = x,$$

which is again a contradiction. Therefore, T must be strict, i.e., $\theta(0) = 0$. Equation (5.16) can now be rewritten as

$$\theta(x) \cdot \theta\left(1 - \tfrac{\theta^{-1}(\theta(x)\cdot\theta(1-y))}{x}\right) = \theta(x \cdot y). \tag{5.17}$$

The differentiability of the solutions of this equation was shown in [Fodor & Keresztfalvi 1994]. Taking the partial derivative of (5.17) with respect to y and taking the limit if y approaches 1 from the left, we get the differential equation

$$\theta'(x) = \theta'(1^-) \cdot \left(\tfrac{\theta(x)}{x}\right)^2,$$

whose solution is given by

$$\theta(x) = \frac{x}{\theta'(1^-) + (1 - \theta'(1^-)) \cdot x},$$

i.e., $T = T^{\mathbf{H}}_{\theta'(1^-)}$. A straightforward computation yields $\theta'(1^-) = \frac{1}{\theta'(0^+)}$. □

Each continuous t-norm T which solves (5.16) must be Archimedean: the existence of a non-trivial idempotent element a would imply

$$0 < a(1-a) = T\big(a, 1 - \tfrac{T(a,a)}{a}\big) = T(a,0) = 0.$$

Since in the proof we excluded nilpotent solutions, each continuous solution T of (5.16) must be strict. Observe also that, if a strict t-norm T has a multiplicative generator θ with $\theta'(0^+) \in \,]0,\infty[$ then no other multiplicative generator of T (which must be of the form θ^c for some $c \in \,]0,\infty[$ according to Corollary 5.4) has this property (we have $(\theta^c)'(0^+) = 0$ for $c > 1$ and $(\theta^c)'(0^+) = \infty$ for $c < 1$). On the other hand, there are t-norms (e.g., $T^{\mathbf{H}}_0$) solving (5.16) without having a multiplicative generator θ with $\theta'(0^+) \in \,]0,\infty[$.

Conditional distributivity

In Definition 2.21 the distributivity of t-norms and t-conorms was considered. A relaxed version of this concept was discussed, e.g., in [Bertoluzza 1993]. It will prove to be useful in the context of integrals based on t-norms and t-conorms (see Section 14.3).

5.20 DEFINITION

Let T be a t-norm and S be a t-conorm. Then we say that T *is conditionally distributive over* S if for all $x, y, z \in [0,1]$

$$\text{(CD)} \qquad T(x, S(y,z)) = S(T(x,y), T(x,z)) \quad \text{whenever } S(y,z) < 1.$$

It is possible to characterize all pairs of continuous t-norms and t-conorms where the t-norm is conditionally distributive over the t-conorm.

5.21 THEOREM

A continuous t-norm T is conditionally distributive over a continuous t-conorm S if and only if we have either one of the following cases:

(i) $S = S_{\mathbf{M}}$.

(ii) *There is a strict t-norm T^* and a nilpotent t-conorm S^* such that the additive generator s of S^* satisfying $s(1) = 1$ is also a multiplicative generator of T^*, and there is an $a \in [0,1[$ such that, for some continuous t-norm T^{**}, we have $T = (\langle 0, a, T^{**}\rangle, \langle a, 1, T^*\rangle)$ and $S = (\langle a, 1, S^*\rangle)$.*

Proof: If, for a continuous t-norm T and a continuous t-conorm S, we have case (i) or (ii), then obviously T is conditionally distributive over S.

Conversely, assume that for a continuous t-norm T and a continuous t-conorm S (CD) holds. If $b \in [0,1[$ is an idempotent element of S, then, for each $x \in [0,1]$, $T(x,b)$ is an idempotent element of S, and, since T is continuous, each element in $[0,b]$ is an idempotent element of S. Therefore, from Corollary 5.12 it follows that either $S = S_{\mathbf{M}}$ (in which case we are done) or $S = (\langle a, 1, S^*\rangle)$, where S^* is a continuous Archimedean t-conorm and $a \in [0,1[$.

If $S = (\langle a, 1, S^*\rangle)$ for some continuous Archimedean t-conorm S^* and some $a \in [0,1[$, then a is an idempotent element not only of S, but also of T. This can be seen as follows: for all $x \in]a,1[$ with $S(x,x) < 1$ we have

$$T\big(a, x_S^{(2)}\big) = T(a, S(x,x)) = S(T(a,x), T(a,x)) = T(a,x),$$

and, consequently, for all $x \in]a,1[$ we get $T(a,x) = T\big(a, x_S^{(\frac{1}{2})}\big)$ (see Remark 3.5). If S^* is a strict t-conorm, then fix $c = \frac{a+1}{2}$, and by induction we obtain for all $n \in \mathbb{N}$

$$T(a,c) = T\big(a, c_S^{(2^n)}\big) = T\big(a, c_S^{(2^{-n})}\big).$$

Since $\lim_{n\to\infty} c_S^{(2^n)} = 1$ and $\lim_{n\to\infty} c_S^{(2^{-n})} = a$, this, together with the continuity of T, implies $a = T(a,1) = T(a,a)$ in this case. If the t-conorm S^* is nilpotent, fix $c = \inf\{x \in [0,1] \mid x_S^{(2)} = 1\}$. Then $T(a,x) = T\big(a, x_S^{(2)}\big)$ for all $x \in [0,c[$ and, because of the continuity of both T and S, also

$$T(a,c) = T\big(a, c_S^{(2)}\big) = T(a,1) = a.$$

By induction, we get $T(a,c) = T\big(a, c_S^{(2^{-n})}\big)$ for all $n \in \mathbb{N}$, implying $T(a,a) = a$ in this case too. Therefore, also T can be written as an ordinal sum (see Theorem 5.11), one of its summands being $\langle a, 1, T^*\rangle$, where T^* is some continuous t-norm. Since T is conditionally distributive over S, T^* must be conditionally distributive over the Archimedean t-conorm S^*.

Next we show that T^* is also Archimedean. Indeed, the existence of a non-trivial idempotent element d of T^* would imply the existence of $x \in [0,1]$ with $x < d < S^*(x,x) < 1$, leading to the contradiction

$$d = T^*(d, S^*(x,x)) = S^*(T^*(d,x), T^*(d,x)) = S^*(x,x).$$

Moreover, T^* is not nilpotent: if $0 < e = \sup\{x \in [0,1] \mid T^*(x,x) = 0\}$ then there exists a $y \in [0,1]$ with $y < e < S^*(y,y) < 1$, leading to the contradiction

$$0 < T^*(e, S^*(y,y)) = S^*(T^*(e,y), T^*(e,y)) = 0.$$

Now fix a multiplicative generator θ of the strict t-norm T^* and an additive generator s of the continuous Archimedean t-conorm S^* and observe that we have $S(y,z) = s^{-1}(s(y)+s(z))$ for all $(y,z) \in [0,1]^2$ with $S(y,z) < 1$. Define the continuous, strictly increasing function $f : [0, s(1)] \longrightarrow [0,1]$ by $f = \theta \circ s^{-1}$ and note that $f(s(1)) = 1$. Writing $u = s(x)$, $v = s(y)$ and $w = s(z)$, the fact that T^* is conditionally distributive over S^* can be rewritten as

$$f(u) \cdot f(v+w) = f(f^{-1}(f(u) \cdot f(v)) + f^{-1}(f(u) \cdot f(w))) \tag{5.18}$$

for all $u, v, w \in [0, s(1)]$ and $v + w < s(1)$. Fixing $u \in \,]0, s(1)]$, define the continuous, strictly increasing function $g_u : [0, s(1)] \longrightarrow [0, s(1)]$ by $g_u(x) = f^{-1}(f(u) \cdot f(x))$ and observe that $g_u(0) = 0$ and $g_u(s(1)) = u$. Then (5.18) transforms into the Cauchy functional equation

$$g_u(v+w) = g_u(v) + g_u(w)$$

for all $v, w \in [0, s(1)]$ and $v + w \leq s(1)$, where the case $v + w = s(1)$ follows from the continuity of g_u. If $s(1) = \infty$ then this equation has no solution, so S^* must be a nilpotent t-conorm. From [Aczél 1966, Section 2.1, Theorem 3] it follows that $g_u(x) = \frac{u}{s(1)} x$, i.e.,

$$f\left(\tfrac{u}{s(1)} x\right) = f(u) \cdot f(x)$$

for all $x, u \in [0, s(1)]$ (the case $u = 0$ follows from $f(0) = 0$). The only solutions of this modified Cauchy functional equation are given by $f(x) = \left(\frac{x}{s(1)}\right)^p$, where $p \in \,]0, \infty[$. Since $s(1)$ can be chosen arbitrarily in $]0, \infty[$ we may take $s(1) = 1$, leading to $f(x) = x^p$ and, consequently, we have $\theta = s^p$, i.e., s is a multiplicative generator of T^*. □

Examples of pairs (T, S) of (not necessarily continuous) t-norms and t-conorms such that T is conditionally distributive over S are, for instance, $(T_{\mathbf{D}}, S)$ for t-conorms S satisfying $S(x,y) < 1$ if and only if $(x,y) \in [0,1[^2$, and $(T, S_{\mathbf{D}})$ for an arbitrary t-norm T. A full characterization of all pairs (T, S) of t-norms and t-conorms satisfying (CD) is still an open problem.

6. Comparison of t-norms

> The fundamental importance of the subject of order may be inferred from the fact that all the concepts required in geometry can be expressed in terms of the concept of order alone.
>
> EDWARD V. HUNTINGTON

We have already seen that there is a strict (pointwise) order relationship (1.5) between the four basic t-norms $T_{\mathbf{M}}$, $T_{\mathbf{P}}$, $T_{\mathbf{L}}$, and $T_{\mathbf{D}}$, and that each t-norm lies between the two extremes $T_{\mathbf{M}}$ and $T_{\mathbf{D}}$ (see (1.4)). It is also clear that the pointwise comparison of two t-norms is a partial order. For two arbitrary t-norms, however, it may be difficult to decide whether they are comparable or not. A well-known example was the family of Frank t-norms (see Section 4.4) whose monotonicity was shown first in [Butnariu & Klement 1993].

We start our discussion with some examples of continuous Archimedean t-norms whose diagonal sections are incomparable, thus implying the incomparability of the t-norms themselves and showing that we are not dealing with a linear order. We also give examples of t-norms which are incomparable in spite of the fact that their diagonal sections coincide.

The main result for the pointwise comparison of continuous Archimedean t-norms, involving their additive generators, is due to [Schweizer & Sklar 1983, Lemma 5.5.8]. We also give a sufficient condition which may be easier to check in some cases. These results allow us to show the monotonicity of most of the families of t-norms considered in Chapter 4. We also mention that, for any two continuous Archimedean t-norms, there is always an upper and lower bound within the class of continuous Archimedean t-norms.

Turning our attention to general continuous t-norms, we exploit the fact that they can be written as ordinal sums of continuous Archimedean t-norms and present a necessary and sufficient condition for their comparability [Klement *et al.* 1997].

Finally, we consider the domination relation between t-norms which turns out to be stronger than the pointwise comparability and which plays a crucial role in the construction of Cartesian products, e.g., of probabilistic metric spaces (see Section 9.3) [Tardiff 1976, Schweizer & Sklar 1983] and of fuzzy equivalence and ordering relations (see Section 12.2) [De Baets & Mesiar 1998, Bodenhofer 1999]).

6.1 Comparison of continuous Archimedean t-norms

We first want to study the pointwise comparability of two t-norms T_1 and T_2 in the sense of Definition 1.4, i.e., whether for all $(x, y) \in [0,1]^2$ we have $T_1(x, y) \leq T_2(x, y)$. Obviously, the relation $\leq$ is a partial order on the class of all t-norms.

Although we have $T_{\mathbf{D}} < T_{\mathbf{L}} < T_{\mathbf{P}} < T_{\mathbf{M}}$ and, for an arbitrary t-norm T, $T_{\mathbf{D}} \leq T \leq T_{\mathbf{M}}$ (see (1.5) and (1.4)), $\leq$ is not a linear order, i.e., there are incomparable t-norms. It is easy to find examples of strict and nilpotent t-norms which are incomparable, as well as pairs of incomparable strict and nilpotent t-norms, respectively. As illustrated in Figure 6.1, the incomparability of the t-norms sometimes can be seen from the incomparability of their respective diagonal sections. Unfortunately, we cannot restrict ourselves to the comparison of diagonal sections only. Example 6.1 and Figure 6.2 give two pairs of strict and nilpotent t-norms which coincide on the diagonal (in the nilpotent case also on the opposite diagonal), but are nevertheless incomparable.

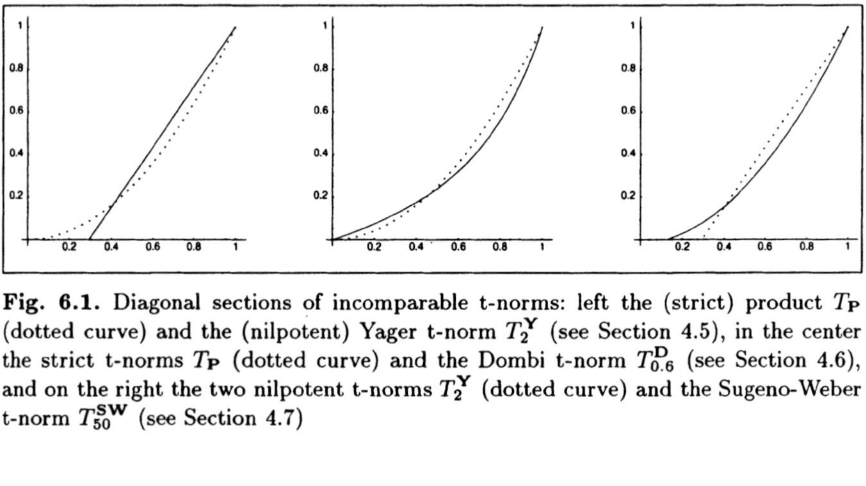

Fig. 6.1. Diagonal sections of incomparable t-norms: left the (strict) product $T_{\mathbf{P}}$ (dotted curve) and the (nilpotent) Yager t-norm $T_2^{\mathbf{Y}}$ (see Section 4.5), in the center the strict t-norms $T_{\mathbf{P}}$ (dotted curve) and the Dombi t-norm $T_{0.6}^{\mathbf{D}}$ (see Section 4.6), and on the right the two nilpotent t-norms $T_2^{\mathbf{Y}}$ (dotted curve) and the Sugeno-Weber t-norm $T_{50}^{\mathbf{SW}}$ (see Section 4.7)

However, the incomparability of the diagonal sections is sufficient for the incomparability of the corresponding t-norms, but by no means necessary.

6.1 EXAMPLE

(i) Consider the function $t_1 : [0,1] \longrightarrow [0,\infty]$ defined by (the index n may be any number in $\mathbb{Z}$)

$$t_1(x) = \begin{cases} \infty & \text{if } x = 0, \\ 2^n\left(2 - \left(4x^{\frac{1}{2^n}} - 1\right)^2\right) & \text{if } x \in \left[\frac{1}{2^{2^{n+1}}}, \frac{1}{2^{2^n}}\right[, \\ 0 & \text{if } x = 1, \end{cases}$$

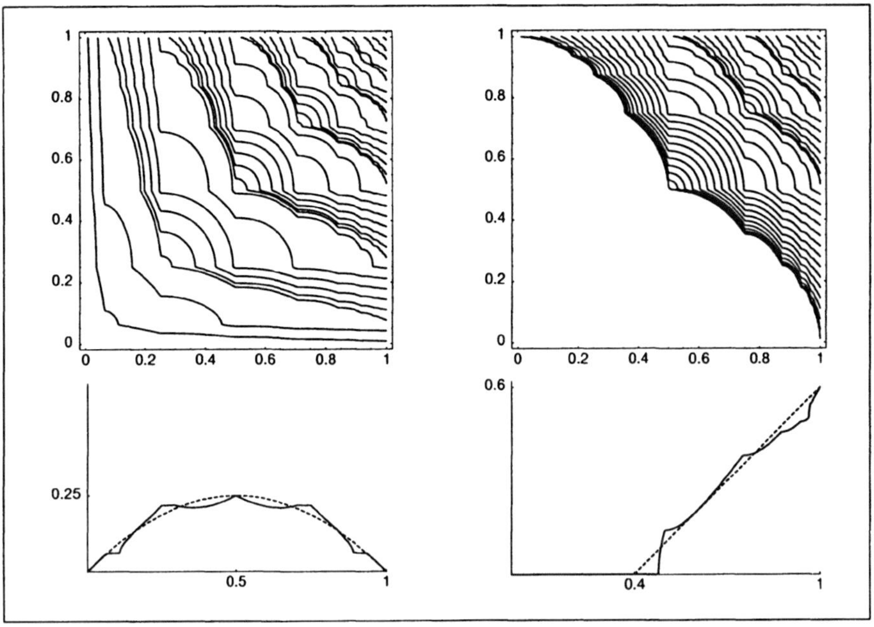

Fig. 6.2. Contour plots of the strict t-norm T_1 (top left) and the nilpotent t-norm T_2 (top right) considered in Example 6.1 together with the incomparable opposite diagonal sections of T_1 and $T_{\mathbf{P}}$ (bottom left), and the incomparable vertical sections $T_2(0.6, .)$ and $T_{\mathbf{L}}(0.6, .)$ (bottom right)

then t_1 is an additive generator of some strict t-norm T_1. A simple computation shows that the diagonal sections of T_1 and $T_{\mathbf{P}}$ coincide, but the opposite diagonal sections $d_{T_1}, d_{T_{\mathbf{P}}} : [0,1] \longrightarrow [0,1]$ given by $d_{T_1}(x) = T_1(x, 1-x)$ and $d_{T_{\mathbf{P}}}(x) = T_{\mathbf{P}}(x, 1-x)$ are incomparable (see Figure 6.2).

(ii) The function $t_2 : [0,1] \longrightarrow [0,\infty]$ defined by (this time we have $n \in \mathbb{N}$)

$$t_2(x) = \begin{cases} \frac{1-2(2^{n-1}x-2^{n-1}+1)^2}{2^{n-1}} & \text{if } x \in \left[1-\frac{1}{2^{n-1}}, 1-\frac{1}{2^n}\right[, \\ 0 & \text{if } x = 1, \end{cases}$$

is an additive generator of some nilpotent t-norm T_2. Now the diagonal sections and the opposite diagonal sections of T_2 and $T_{\mathbf{L}}$ coincide, but the vertical sections $T_2(0.6, .), T_{\mathbf{L}}(0.6, .) : [0,1] \longrightarrow [0, 0.6]$ are incomparable (see Figure 6.2).

Turning now to the special case of continuous Archimedean t-norms, remember that their additive generators contain the full information about them because of Theorem 5.1. Therefore, also the pointwise comparability of continuous Archimedean t-norms should be derivable from appropriate properties of their respective additive generators. Recall also that each additive generator t of a continuous Archimedean t-norm, viewed as a function from $[0,1]$ to $[0, t(0)]$, is a strictly decreasing bijection, and that, subsequently, the inverse function $t^{-1} : [0, t(0)] \longrightarrow [0,1]$ always exists.

The following necessary and sufficient condition for the comparison of continuous Archimedean t-norms can be found in [Schweizer & Sklar 1983, Lemma 5.5.8], for the special case of strict t-norms it was proved first in [Schweizer & Sklar 1961] (see also [Alsina & Gimenez 1984]).

6.2 THEOREM
Let T_1 and T_2 be two continuous Archimedean t-norms with additive generators $t_1, t_2 : [0,1] \longrightarrow [0,\infty]$, respectively. The following are equivalent:

(i) $T_1 \leq T_2$.

(ii) The function $t_1 \circ t_2^{-1} : [0, t_2(0)] \longrightarrow [0,\infty]$ *is subadditive, i.e., for all* $u, v \in [0, t_2(0)]$ *with* $u + v \in [0, t_2(0)]$ *we have*

$$t_1 \circ t_2^{-1}(u+v) \leq t_1 \circ t_2^{-1}(u) + t_1 \circ t_2^{-1}(v). \tag{6.1}$$

Proof: Note first that $T_1 \leq T_2$ if and only if for all $x, y \in [0,1]$

$$t_1^{-1}(\min(t_1(0), t_1(x) + t_1(y))) \leq t_2^{-1}(\min(t_2(0), t_2(x) + t_2(y))).$$

Fixing $x, y \in [0,1]$, writing $t_2(x) = u$ and $t_2(y) = v$ and taking into account that t_1 is strictly decreasing, this is equivalent to the fact that for all $u, v \in [0, t_2(0)]$

$$\begin{aligned} &\min(t_1(0), t_1 \circ t_2^{-1}(u) + t_1 \circ t_2^{-1}(v)) \\ &\qquad\qquad \geq \min(t_1(0), t_1 \circ t_2^{-1}(\min(t_2(0), u+v))). \end{aligned} \tag{6.2}$$

If $t_1 \circ t_2^{-1}$ is subadditive then (6.2) trivially holds for all $u, v \in [0, t_2(0)]$, i.e., we have $T_1 \leq T_2$.

Conversely, if $u, v \in [0, t_2(0)]$ such that $u + v < t_2(0)$ then $T_1 \leq T_2$ implies

$$\min(t_1(0), t_1 \circ t_2^{-1}(u+v)) = t_1 \circ t_2^{-1}(u+v) < t_1(0)$$

and, as a consequence of (6.2), we have (6.1). Now the continuity of $t_1 \circ t_2^{-1}$ implies that (6.1) must hold also for $u, v \in [0, t_2(0)]$ with $u + v = t_2(0)$, i.e., $t_1 \circ t_2^{-1}$ is subadditive. □

There exist criteria (some of which are only sufficient) for the comparability of continuous Archimedean t-norms which sometimes are easier to check than the subadditivity in Theorem 6.2. The following one can be derived easily from Theorem 6.2 (recall the definition of a concave function (5.6)).

6.3 COROLLARY
Let T_1 and T_2 be two continuous Archimedean t-norms with additive generators $t_1, t_2 : [0,1] \longrightarrow [0,\infty]$, respectively. If the function $t_1 \circ t_2^{-1} : [0, t_2(0)] \longrightarrow [0,\infty]$ is concave then we have $T_1 \leq T_2$.

Proof: It is not difficult to check that, for each $a \in \,]0,\infty]$, each concave function $f : [0,a] \longrightarrow [0,\infty]$ satisfying $f(0) = 0$ is necessarily subadditive. □

We only mention that, in general, the subadditivity of a function does not imply its concavity. Here is an example of a continuous, strictly increasing function $f : [0,\infty] \longrightarrow [0,\infty]$ which is subadditive but obviously not concave:

$$f(x) = \begin{cases} 4x & \text{if } x \in [0,4]\,, \\ 2x+8 & \text{if } x \in \,]4,6]\,, \\ 3x+2 & \text{otherwise.} \end{cases}$$

6.4 EXAMPLE

(i) In Example 3.32(i) we have seen that for each continuous Archimedean t-norm T with additive generator $t : [0,1] \longrightarrow [0,\infty]$, and for each $\lambda \in \,]0,\infty[$, the function $t^\lambda : [0,1] \longrightarrow [0,\infty]$ is an additive generator of a continuous Archimedean t-norm which was denoted there $T^{(\lambda)}$. Moreover, the family $(T^{(\lambda)})_{\lambda\in]0,\infty[}$ is strictly increasing with respect to the parameter λ. Indeed, for $\lambda, \mu \in \,]0,\infty[$ the composite function $t^\lambda \circ (t^\mu)^{-1} : [0, t(0)^\mu] \longrightarrow [0,\infty]$ is given by

$$t^\lambda \circ (t^\mu)^{-1}(x) = x^{\frac{\lambda}{\mu}},$$

and it is concave whenever $\lambda \leq \mu$, showing that $(T^{(\lambda)})_{\lambda\in]0,\infty[}$ is a strictly increasing family of t-norms. Consequently, the families $(T_\lambda^{\mathbf{Y}})_{\lambda\in[0,\infty]}$ of Yager t-norms (see Section 4.5), $(T_\lambda^{\mathbf{AA}})_{\lambda\in[0,\infty]}$ of Aczél-Alsina t-norms (see Section 4.8), and $(T_\lambda^{\mathbf{D}})_{\lambda\in[0,\infty]}$ of Dombi t-norms (see Section 4.6) are strictly increasing families of t-norms.

(ii) In Example 3.32(ii) we have seen that for each continuous Archimedean t-norm T with additive generator $t : [0,1] \longrightarrow [0,\infty]$, and for each $\lambda \in \,]0,\infty[$, the function $t_{(\lambda)} : [0,1] \longrightarrow [0,\infty]$ given by $t_{(\lambda)}(x) = t(x^\lambda)$ is an additive generator of a continuous Archimedean t-norm which was denoted there $T_{(\lambda)}$. However, the family $(T_{(\lambda)})_{\lambda\in]0,\infty[}$ is not necessarily monotone: consider the additive generator $t : [0,1] \longrightarrow [0,\infty]$ defined by

$$t(x) = \begin{cases} \frac{24}{25}(3-10x)^2 & \text{if } x \in [0,0.2], \\ 1-x^2 & \text{otherwise,} \end{cases}$$

and let $T = T_{(1)}$ be the t-norm generated by t. Then we obtain

$$T_{(1)}(0.5,0.5) = 0.175 < 0.400307 = T_{(2)}(0.5,0.5),$$
$$T_{(1)}(0.2,0.2) = \tfrac{3-\sqrt{2}}{10} > 0 = T_{(2)}(0.2,0.2),$$

showing that $T_{(1)}$ and $T_{(2)}$ are incomparable.

6.5 COROLLARY

Let T_1 and T_2 be two continuous Archimedean t-norms with differentiable additive generators $t_1, t_2 : [0,1] \longrightarrow [0,\infty]$, respectively. If the function

$$\frac{t_1'}{t_2'} :]0,1[\longrightarrow [0,\infty[$$

is non-decreasing, then we have $T_1 \leq T_2$.

Proof: Suppose that $\frac{t_1'}{t_2'}$ is non-decreasing and observe that for all $x \in]0, t_2(0)[$

$$\frac{t_1'}{t_2'}(t_2^{-1}(x)) = \left(t_1 \circ t_2^{-1}\right)'(x).$$

Taking into account that $t_2^{-1} : [0, t_2(0)]$ is strictly decreasing, this means that the function $\left(t_1 \circ t_2^{-1}\right)'$ is non-increasing. Subsequently, $t_1 \circ t_2^{-1}$ is concave, and we have $T_1 \leq T_2$ because of Corollary 6.3. □

6.6 EXAMPLE

The family of Schweizer-Sklar t-norms $(T_\lambda^{\mathbf{SS}})_{\lambda\in[-\infty,\infty]}$ (see Section 4.2) is a strictly decreasing family of t-norms [Sherwood 1984]. From (1.4) it is clear that for each $\lambda \in]-\infty,\infty[$ we have

$$T_{-\infty}^{\mathbf{SS}} = T_{\mathbf{M}} \geq T_\lambda^{\mathbf{SS}} \geq T_{\mathbf{D}} = T_\infty^{\mathbf{SS}}.$$

Recalling the additive generators $(t_\lambda^{\mathbf{SS}})_{\lambda\in]-\infty,\infty[}$ of the continuous Archimedean Schweizer-Sklar t-norms given in Remark 4.4(iv), it is easy to see that, for each $x \in]0,1[$ and for all $\lambda, \mu \in]-\infty,\infty[$ with $\lambda < \mu$, we obtain

$$\frac{(t_\mu^{\mathbf{SS}})'}{(t_\lambda^{\mathbf{SS}})'}(x) = x^{\mu-\lambda},$$

which means that $\frac{(t_\mu^{\mathbf{SS}})'}{(t_\lambda^{\mathbf{SS}})'}$ is a non-decreasing function, and $(T_\lambda^{\mathbf{SS}})_{\lambda\in[-\infty,\infty]}$ is strictly decreasing because of Corollary 6.5.

From [Hardy *et al.* 1952, (103)] we immediately get the following sufficient criterion for the comparability of continuous Archimedean t-norms.

6.7 COROLLARY

Let T_1 and T_2 be two continuous Archimedean t-norms with additive generators $t_1, t_2 : [0,1] \longrightarrow [0,\infty]$, respectively. If $f :]0, t_2(0)] \longrightarrow [0,\infty]$ defined by

$$f(x) = \frac{(t_1 \circ t_2^{-1})(x)}{x}$$

is a non-increasing function, then we have $T_1 \leq T_2$.

To prove the monotonicity of the family of Frank t-norms $(T_\lambda^{\mathbf{F}})_{\lambda\in[0,\infty]}$ (see Section 4.4) was a non-trivial task (a first proof thereof, based on [Takacs 1991], was published in [Butnariu & Klement 1993, Proposition 1.12(ii)]). We present here a shorter proof given in [Klement *et al.* 1997]).

6.8 PROPOSITION
The family $(T_\lambda^{\mathbf{F}})_{\lambda\in[0,\infty]}$ of Frank t-norms is strictly decreasing.

Proof: Trivially we have $T_0^{\mathbf{F}} = T_{\mathbf{M}} > T_\lambda^{\mathbf{F}}$ for all $\lambda \in\,]0,\infty]$. Recalling the additive generators $(t_\lambda^{\mathbf{F}})_{\lambda\in]0,\infty]}$ of the continuous Archimedean Frank t-norms given in Remark 4.8(iv), then from

$$\frac{(t_\infty^{\mathbf{F}})'}{(t_\lambda^{\mathbf{F}})'}(x) = \begin{cases} x & \text{if } \lambda = 1, \\ \frac{1}{\log\lambda}(1-\lambda^{-x}) & \text{if } \lambda \in\,]0,1[\,\cup\,]1,\infty[\,, \end{cases}$$

it follows that for each $\lambda \in\,]0,\infty[$ the function $\frac{(t_\infty^{\mathbf{F}})'}{(t_\lambda^{\mathbf{F}})'}$ is non-decreasing, implying $T_\infty^{\mathbf{F}} \le T_\lambda^{\mathbf{F}}$ and, since $T_\infty^{\mathbf{F}}$ is nilpotent and $T_\lambda^{\mathbf{F}}$ is strict, even $T_\infty^{\mathbf{F}} < T_\lambda^{\mathbf{F}}$.

Now let us show that $T_\mu^{\mathbf{F}} \le T_\lambda^{\mathbf{F}}$ whenever $1 < \lambda < \mu < \infty$. Observe that for all $x \in\,]0,1[$ we get

$$\frac{(t_\mu^{\mathbf{F}})'}{(t_\lambda^{\mathbf{F}})'}(x) = \frac{\mu^x(\lambda^x-1)\log\mu}{\lambda^x(\mu^x-1)\log\lambda} = \frac{\log\mu}{\log\lambda}\cdot\frac{1-\left(\frac{1}{\lambda}\right)^x}{1-\left(\frac{1}{\mu}\right)^x}.$$

Then $\frac{(t_\mu^{\mathbf{F}})'}{(t_\lambda^{\mathbf{F}})'}$ is non-decreasing on $]0,1[$ if and only if

$$\left(1-\left(\tfrac{1}{\mu}\right)^x\right)\left(\tfrac{1}{\lambda}\right)^x\log\tfrac{1}{\lambda} \le \left(1-\left(\tfrac{1}{\lambda}\right)^x\right)\left(\tfrac{1}{\mu}\right)^x\log\tfrac{1}{\mu},$$

i.e., if and only if we have the inequality

$$\frac{\left(\frac{1}{\lambda}\right)^x\log\frac{1}{\lambda}}{\left(\frac{1}{\mu}\right)^x\log\frac{1}{\mu}} \ge \frac{1-\left(\frac{1}{\lambda}\right)^x}{1-\left(\frac{1}{\mu}\right)^x}. \tag{6.3}$$

Consider now the functions $f, g :]0,1[\, \longrightarrow [0,\infty[$ defined by $f(x) = 1-\left(\frac{1}{\lambda}\right)^x$ and $g(x) = 1-\left(\frac{1}{\mu}\right)^x$. Then, by the Cauchy Mean Value Theorem, for each $x \in\,]0,1[$ there exists a $y \in\,]0,x[$ such that

$$\frac{1-\left(\frac{1}{\lambda}\right)^x}{1-\left(\frac{1}{\mu}\right)^x} = \frac{f(x)-f(0)}{g(x)-g(0)} = \frac{f'(y)}{g'(y)} = \frac{\left(\frac{1}{\lambda}\right)^y\log\frac{1}{\lambda}}{\left(\frac{1}{\mu}\right)^y\log\frac{1}{\mu}} < \frac{\left(\frac{1}{\lambda}\right)^x\log\frac{1}{\lambda}}{\left(\frac{1}{\mu}\right)^x\log\frac{1}{\mu}}.$$

This proves inequality (6.3) and, consequently, the function $\frac{(t_\mu^{\mathbf{F}})'}{(t_\lambda^{\mathbf{F}})'}$ is non-decreasing, i.e., $T_\mu^{\mathbf{F}} \le T_\lambda^{\mathbf{F}}$ and, because of $T_\mu^{\mathbf{F}} \ne T_\lambda^{\mathbf{F}}$, even $T_\mu^{\mathbf{F}} < T_\lambda^{\mathbf{F}}$ in this case. Similarly we can show $T_1^{\mathbf{F}} < T_\lambda^{\mathbf{F}}$ for all $\lambda \in\,]1,\infty[$.

The case $0 < \lambda < \mu \leq 1$ can be transformed into $1 \leq \frac{1}{\mu} < \frac{1}{\lambda} < \infty$, and the case $0 < \lambda < 1 < \mu < \infty$ is proved combining the two latter cases. □

Given two continuous Archimedean t-norms T_1 and T_2, then obviously the minimum $T_{\mathbf{M}}$ is an upper bound of T_1 and T_2, i.e., $T_{\mathbf{M}} \geq \max(T_1, T_2)$. A much less trivial problem is to find an upper and lower bound of T_1 and T_2 which is also a continuous Archimedean t-norm.

6.9 THEOREM
Let T_1 and T_2 be two strict t-norms. Then there exists a strict t-norm T such that $T \geq \max(T_1, T_2)$.

Proof: Fix two additive generators $t_1, t_2 : [0,1] \longrightarrow [0,\infty]$ of T_1 and T_2, respectively, satisfying $t_1(0.5) = t_2(0.5) = 1$, and consider the strictly decreasing bijection $\varphi : [0,1] \longrightarrow [0,\infty]$ given by $\varphi(x) = \frac{1-x}{x}$. To simplify the notations, we shall use the two strictly increasing bijections $f_1, f_2 : [0,\infty] \longrightarrow [0,\infty]$ given by $f_1 = \varphi \circ t_1^{-1}$ and $f_2 = \varphi \circ t_2^{-1}$.

Consider the set $M = \{(x,y) \in [0,\infty]^2 \mid x < y\}$ and, for each $(a,b) \in M$ and $n \in \mathbb{N}$, the set

$$P_{ab}^{(n)} = \{(x_0, x_1, \ldots, x_n) \in [0,\infty]^{n+1} \mid a = x_0 < x_1 < \cdots < x_n = b\}.$$

Define the functions $p : \bigcup_{n\in\mathbb{N}} P_{ab}^{(n)} \longrightarrow [0,\infty]$ and $q : M \longrightarrow [0,\infty]$ by

$$p(x_0, x_1, \ldots, x_n) = \prod_{i=1}^{n} \min\left(\frac{f_1^{-1}(x_{i-1})}{f_1^{-1}(x_i)}, \frac{f_2^{-1}(x_{i-1})}{f_2^{-1}(x_i)}\right),$$

$$q(x,y) = \inf\left\{p(x_0, x_1, \ldots, x_n) \,\middle|\, (x_0, x_1, \ldots, x_n) \in P_{xy}^{(n)}, n \in \mathbb{N}\right\}.$$

For all $(x,y) \in M$ and for all $z \in \,]x,y[$ we have $q(x,y) = q(x,z) \cdot q(z,y)$ and

$$\frac{f_1^{-1}(x) \cdot f_2^{-1}(x)}{f_1^{-1}(y) \cdot f_2^{-1}(y)} \leq q(x,y) \leq \min\left(\frac{f_1^{-1}(x)}{f_1^{-1}(y)}, \frac{f_2^{-1}(x)}{f_2^{-1}(y)}\right).$$

Define the strictly increasing bijection $g : [0,\infty] \longrightarrow [0,\infty]$ by

$$g(x) = \begin{cases} q(x,1) & \text{if } x \in [0,1[\,, \\ 1 & \text{if } x = 1, \\ \frac{1}{q(1,x)} & \text{if } x \in \,]1,\infty]\,, \end{cases}$$

and observe that, for all $(x,y) \in M$, we get $g(x) = q(x,y) \cdot g(y)$. This implies for all $(x,y) \in [0,\infty]^2$

$$\begin{aligned} g \circ f_1(x) &= q(f_1(x), f_1(x+y)) \cdot (g \circ f_1)(x+y) \\ &\leq \frac{f_1^{-1} \circ f_1(x)}{f_1^{-1} \circ f_1(x+y)} \cdot (g \circ f_1)(x+y) \\ &\leq \frac{x}{x+y} \cdot (g \circ f_1)(x+y), \end{aligned}$$

yielding $g \circ f_1(x + y) \geq g \circ f_1(x) + g \circ f_1(y)$, i.e., $g \circ f_1$ is superadditive. The superadditivity of $g \circ f_2$ is shown analogously.

Defining the function $t : [0, 1] \longrightarrow [0, \infty]$ by $t = g \circ \varphi$ it is immediately clear that this is a strictly decreasing bijection with $t(0.5) = 1$, i.e., an additive generator of some strict t-norm T. Because of the superadditivity of the function $g \circ f_1 = t \circ \varphi^{-1} \circ \varphi \circ t_1^{-1} = t \circ t_1^{-1}$ and Theorem 6.2, we obtain $T \geq T_1$ and, in complete analogy, $T \geq T_2$. □

The proof of the corresponding result for nilpotent t-norms is very similar to the proof of Theorem 6.9 (compare also [Marko & Mesiar 1999]).

6.10 COROLLARY
Let T_1 and T_2 be two nilpotent t-norms. Then there exists a nilpotent t-norm T such that $T \geq \max(T_1, T_2)$.

Proof: For additive generators $t_1, t_2 : [0, 1] \longrightarrow [0, \infty]$ of T_1 and T_2, respectively, satisfying $t_1(0) = t_2(0) = 1$ define the functions $f_1, f_2 : [0, 1] \longrightarrow [0, 1]$ by $f_1(x) = 1 - t_1(x)$ and $f_2(x) = 1 - t_2(x)$. Repeating the construction of $g : [0, 1] \longrightarrow [0, 1]$ exactly as in the proof of Theorem 6.9, and defining the function $t : [0, 1] \longrightarrow [0, \infty]$ by $t(x) = 1 - g(x)$, it turns out that t is an additive generator of some nilpotent t-norm T which satisfies $T \geq \max(T_1, T_2)$. □

Now the result for continuous Archimedean t-norms is a simple combination of Theorem 6.9 and Corollary 6.10.

6.11 COROLLARY
Let T_1 and T_2 be two continuous Archimedean t-norms. Then there exists a continuous Archimedean t-norm T such that $T \geq \max(T_1, T_2)$.

Proof: If the t-norms T_1 and T_2 are both strict or both nilpotent, then the result follows directly from Theorem 6.9 and Corollary 6.10.

If one t-norm, say T_1, is nilpotent and the other one strict then Proposition 5.10 implies that there is a strictly increasing bijection $\varphi : [0, 1] \longrightarrow [0, 1]$ such that $T_1 = (T_{\mathbf{L}})_\varphi$. Then $T^* = (T_{\mathbf{P}})_\varphi$ is a strict t-norm with $T^* \geq T_1$, and the result follows again from Theorem 6.9. □

The proof of the existence of a lower bound for two continuous Archimedean t-norms T_1 and T_2, which itself is a continuous Archimedean t-norm, is much more complicated (see [Marko & Mesiar 200x]).

6.2 Comparison of continuous t-norms

When comparing general continuous t-norms recall first that, because of Theorem 5.11, each continuous t-norm T is uniquely representable as an ordinal sum of continuous Archimedean t-norms, i.e., there exists a uniquely determined index set A, a family of uniquely determined pairwise disjoint open subintervals $(]a_\alpha, e_\alpha[)_{\alpha\in A}$ of $[0,1]$ and a family of uniquely determined continuous Archimedean t-norms $(T_\alpha)_{\alpha\in A}$ such that $T = (\langle a_\alpha, e_\alpha, T_\alpha\rangle)_{\alpha\in A}$.

If T_1 and T_2 are two continuous t-norms which have the respective, uniquely determined ordinal sum representations

$$T_1 = (\langle a_\alpha^{(1)}, e_\alpha^{(1)}, T_\alpha^{(1)}\rangle)_{\alpha\in A_1},$$
$$T_2 = (\langle a_\alpha^{(2)}, e_\alpha^{(2)}, T_\alpha^{(2)}\rangle)_{\alpha\in A_2},$$

and if the t-norm T_1 has an idempotent element x_0 with $x_0 \in \left]a_\alpha^{(2)}, e_\alpha^{(2)}\right[$ for some $\alpha \in A_2$, then $T_1 \le T_2$ is impossible: since $T_\alpha^{(2)}$ is Archimedean, the assumption $T_1 \le T_2$ immediately leads to the contradiction

$$T_2(x_0, x_0) < x_0 = T_1(x_0, x_0).$$

This argument explains the relationship between the ordinal sum representations of T_1 and T_2 required in the following result due to [Klement *et al.* 1997].

6.12 THEOREM
Let T_1 and T_2 be two continuous t-norms which have the respective, uniquely determined ordinal sum representations

$$T_1 = (\langle a_\alpha^{(1)}, e_\alpha^{(1)}, T_\alpha^{(1)}\rangle)_{\alpha\in A_1},$$
$$T_2 = (\langle a_\alpha^{(2)}, e_\alpha^{(2)}, T_\alpha^{(2)}\rangle)_{\alpha\in A_2},$$

such that for each $\alpha_2 \in A_2$ there is a unique $\alpha_1 \in A_1$ with

$$\left]a_{\alpha_2}^{(2)}, e_{\alpha_2}^{(2)}\right[\subseteq \left]a_{\alpha_1}^{(1)}, e_{\alpha_1}^{(1)}\right[, \tag{6.4}$$

and consider for each such pair $(\alpha_1, \alpha_2) \in A_1 \times A_2$ the affine transformation $f_{\alpha_1,\alpha_2} : \mathbb{R} \longrightarrow \mathbb{R}$ defined by

$$f_{\alpha_1,\alpha_2}(x) = \frac{a_{\alpha_1}^{(1)} - a_{\alpha_2}^{(2)}}{e_{\alpha_2}^{(2)} - a_{\alpha_2}^{(2)}} + \frac{e_{\alpha_1}^{(1)} - a_{\alpha_1}^{(1)}}{e_{\alpha_2}^{(2)} - a_{\alpha_2}^{(2)}} \cdot x.$$

Further, for $i \in \{1,2\}$ and $\alpha \in A_1$ or $\alpha \in A_2$, let $t_\alpha^{(i)} : [0,1] \longrightarrow [0,\infty]$ be an additive generator of the continuous Archimedean t-norm $T_\alpha^{(i)}$. Then the following are equivalent:

(i) $T_1 \le T_2$.

(ii) For each $\alpha_2 \in A_2$ and the corresponding $\alpha_1 \in A_1$, the function

$$h_{\alpha_2} = t_{\alpha_1}^{(1)} \circ f_{\alpha_1,\alpha_2}^{-1} \circ (t_{\alpha_2}^{(2)})^{-1} : [0, t_{\alpha_2}^{(2)}(0)] \longrightarrow [0,\infty]$$

satisfies the inequality

$$h_{\alpha_2}(u+v) \le h_{\alpha_2}(u) + h_{\alpha_2}(v) \tag{6.5}$$

for all $(u,v) \in [0, t_{\alpha_2}^{(2)}(0)]^2$ with

$$h_{\alpha_2}(u) + h_{\alpha_2}(v) < t_{\alpha_1}^{(1)}\left(\frac{a_{\alpha_2}^{(2)} - a_{\alpha_1}^{(1)}}{e_{\alpha_1}^{(1)} - a_{\alpha_1}^{(1)}}\right). \tag{6.6}$$

Proof: Because of $T \leq T_{\mathbf{M}}$ for each t-norm T, the inequality $T_1 \leq T_2$ is equivalent to $T_1(x,y) \leq T_2(x,y)$ for all $\alpha_2 \in A_2$ and all $(x,y) \in \left]a_{\alpha_2}^{(2)}, e_{\alpha_2}^{(2)}\right[^2$. Now fix $\alpha_2 \in A_2$, $(x,y) \in \left]a_{\alpha_2}^{(2)}, e_{\alpha_2}^{(2)}\right[^2$ and choose the unique $\alpha_1 \in A_1$ such that (6.4) holds. Then $T_1(x,y) \leq T_2(x,y)$ is equivalent to

$$a_{\alpha_1}^{(1)} + (e_{\alpha_1}^{(1)} - a_{\alpha_1}^{(1)}) \cdot T_{\alpha_1}^{(1)}\left(\frac{x-a_{\alpha_1}^{(1)}}{e_{\alpha_1}^{(1)}-a_{\alpha_1}^{(1)}}, \frac{y-a_{\alpha_1}^{(1)}}{e_{\alpha_1}^{(1)}-a_{\alpha_1}^{(1)}}\right)$$
$$\leq a_{\alpha_2}^{(2)} + (e_{\alpha_2}^{(2)} - a_{\alpha_2}^{(2)}) \cdot T_{\alpha_2}^{(2)}\left(\frac{x-a_{\alpha_2}^{(2)}}{e_{\alpha_2}^{(2)}-a_{\alpha_2}^{(2)}}, \frac{y-a_{\alpha_2}^{(2)}}{e_{\alpha_2}^{(2)}-a_{\alpha_2}^{(2)}}\right). \quad (6.7)$$

Putting $a = \frac{x-a_{\alpha_2}^{(2)}}{e_{\alpha_2}^{(2)}-a_{\alpha_2}^{(2)}}$ and $b = \frac{y-a_{\alpha_2}^{(2)}}{e_{\alpha_2}^{(2)}-a_{\alpha_2}^{(2)}}$, then inequality (6.7) is equivalent to

$$f_{\alpha_1,\alpha_2}\big(T_{\alpha_1}^{(1)}(f_{\alpha_1,\alpha_2}^{-1}(a), f_{\alpha_1,\alpha_2}^{-1}(b))\big) \leq T_{\alpha_2}^{(2)}(a,b). \quad (6.8)$$

Next, write $u = t_{\alpha_2}^{(2)}(a)$ and $v = t_{\alpha_2}^{(2)}(b)$. Then either the left-hand side of (6.8) is non-positive, i.e., (6.6) is violated, or it is positive, i.e., (6.6) holds and we have

$$h_{\alpha_2}(u) + h_{\alpha_2}(v) \geq h_{\alpha_2}\big(\min(t_{\alpha_2}^{(2)}(0), u+v)\big).$$

This latter inequality, however, is equivalent to (6.5) for all $(u,v) \in [0, t_{\alpha_2}^{(2)}(0)]^2$ satisfying (6.6). □

Similar as in Section 6.1, there is a sufficient condition for the comparison of continuous t-norms which sometimes may be easier to check.

6.13 COROLLARY

Let T_1 and T_2 be two continuous t-norms. If, keeping the notations of Theorem 6.12, for each $\alpha_2 \in A_2$ the function $g_{\alpha_2} :]0,1[\longrightarrow]-\infty,\infty[$ defined by

$$g_{\alpha_2}(x) = \frac{(t_{\alpha_1}^{(1)})'(f_{\alpha_1,\alpha_2}(x))}{(t_{\alpha_2}^{(2)})'(x)}$$

is non-decreasing, then we have $T_1 \leq T_2$.

6.14 REMARK

In [Gottwald 1995] left-continuous t-norms stronger than $T_{\mathbf{L}}$ are considered. Restricting ourselves to the case of continuous t-norms, we can apply Theorem 6.2. Recall that the function $t : [0,1] \longrightarrow [0,\infty]$ defined by $t(x) = 1 - x$ is an additive generator of $T_{\mathbf{L}}$. Then we have $T_{\mathbf{L}} \leq T$, where $T = (\langle a_\alpha, e_\alpha, T_\alpha \rangle)_{\alpha \in A}$ is an ordinal sum of continuous Archimedean t-norms and where, for each $\alpha \in A$, $t_\alpha : [0,1] \longrightarrow [0,\infty]$ is an additive generator of T_α, if and only if for each $\alpha \in A$ and for all $x, y \in [0, f_\alpha(0)]$

$$t_\alpha^{-1}(x) + t_\alpha^{-1}(y) - t_\alpha^{-1}(x+y) \leq \frac{1-a_\alpha}{e_\alpha - a_\alpha}.$$

The latter condition is obviously fulfilled if the t-norm T_α is stronger than $T_{\mathbf{L}}$ or if $e_\alpha \leq \frac{1+a_\alpha}{2}$.

6.15 EXAMPLE
Let T be a (not necessarily continuous) t-norm which has an additive generator $t : [0,1] \longrightarrow [0,\infty]$ such that the function $\mathsf{N}t : [0,1] \longrightarrow [0,\infty]$ (which is given by $\mathsf{N}t(x) = t(1-x)$, see (3.17) is subadditive. Then it is easy to see that $T \leq T_{\mathbf{L}}$.

As a consequence of Theorem 6.2, the converse is true for continuous t-norms T, i.e., we have $T \leq T_{\mathbf{L}}$ if and only if T is Archimedean and if, for each additive generator $t : [0,1] \longrightarrow [0,\infty]$ of T, the function g_t is subadditive.

Note that, whenever the additive generator t is concave (5.6) then the function g_t is necessarily subadditive.

Furthermore, each continuous t-norm weaker than $T_{\mathbf{L}}$ is nilpotent. This means that neither a strict t-norm nor a t-norm with a non-trivial idempotent element can be weaker than $T_{\mathbf{L}}$.

6.3 Domination of t-norms

When constructing the Cartesian product of probabilistic metric spaces (Section 9.3, for more details see [Tardiff 1976] and [Schweizer & Sklar 1983, Section 12.7]) and of fuzzy equivalence and ordering relations (Section 12.2, compare also [De Baets & Mesiar 1998, Bodenhofer 1999]), another relation between t-norms is needed.

6.16 DEFINITION
Let T_1 and T_2 be two t-norms. Then we say that T_1 *dominates* T_2 (in symbols $T_1 \gg T_2$) if for all $x, y, u, v \in [0,1]$

$$T_1(T_2(x,y), T_2(u,v)) \geq T_2(T_1(x,u), T_1(y,v)). \tag{6.9}$$

6.17 EXAMPLE

(i) Trivially, for each t-norm T we have $T_{\mathbf{M}} \gg T$ and $T \gg T_{\mathbf{D}}$.

(ii) If for two t-norms T_1 and T_2 we have $T_1 \gg T_2$ then, putting $y = u = 1$ in (6.9), we immediately see that also $T_1 \geq T_2$ holds. Consequently, the two t-norms $T_{(1)}$ and $T_{(2)}$ considered in Example 6.4(ii) are also incomparable with respect to $\gg$. However, for the t-norm T introduced in (2.3) which satisfies $T(x,y) = \frac{xy}{2}$ whenever $(x,y) \in [0,1[^2$, we have $T_{\mathbf{P}} > T$ but not $T_{\mathbf{P}} \gg T$. In Example 6.20 we shall see that, even for strict t-norms, $T_1 \geq T_2$ does not imply $T_1 \gg T_2$.

(iii) The relation $\gg$ on the set of all t-norms is reflexive and antisymmetric: from the commutativity (T1) and the associativity (T2) we obtain for each t-norm T and all $x, y, u, v \in [0,1]$ the equality

$$T(T(x,y), T(u,v)) = T(T(x,u), T(y,v)),$$

i.e., $T \gg T$, and the assumptions $T_1 \gg T_2$ and $T_2 \gg T_1$ imply, as a consequence of (ii), $T_1 = T_2$. However, it is still an open problem whether the domination is also transitive, i.e., an order relation on the set of all t-norms.

(iv) Consider again the family of Schweizer-Sklar t-norms $(T^{\mathbf{SS}}_\lambda)_{\lambda\in[-\infty,\infty]}$ (see Section 4.2). Here we have $T^{\mathbf{SS}}_\lambda \gg T^{\mathbf{SS}}_\mu$ if and only if $\lambda \geq \mu$, i.e., $\gg$ is an order relation on the set $\{T^{\mathbf{SS}}_\lambda \mid \lambda \in [-\infty,\infty]\}$: because of Example 6.6 and (ii) we know that $T^{\mathbf{SS}}_\lambda \gg T^{\mathbf{SS}}_\mu$ implies $\lambda \geq \mu$, the (rather long) proof of the opposite direction can be found in [Sherwood 1984].

(v) In Example 3.32(i) we have seen that for each continuous Archimedean t-norm T with additive generator $t : [0,1] \longrightarrow [0,\infty]$, and for each index $\lambda \in\ ]0,\infty[$, the function $t^\lambda : [0,1] \longrightarrow [0,\infty]$ is an additive generator of a continuous Archimedean t-norm which was denoted there $T^{(\lambda)}$. Now we claim that for $\lambda,\mu \in\ ]0,\infty[$ with $\lambda > \mu$ we always have $T^{(\lambda)} \gg T^{(\mu)}$.
To see this, apply the Minkowski inequality (see [Taylor 1958]), which states that for all $a,b,c,d \in \mathbb{R}$ and for all $p \in [1,\infty[$

$$\left((a+b)^p + (c+d)^p\right)^{\frac{1}{p}} \leq \left(a^p + c^p\right)^{\frac{1}{p}} + \left(b^p + d^p\right)^{\frac{1}{p}},$$

yielding for all $x,y,u,v \in [0,1]$

$$\left((t^\mu(x)+t^\mu(y))^{\frac{\lambda}{\mu}} + (t^\mu(u)+t^\mu(v))^{\frac{\lambda}{\mu}}\right)^{\frac{\mu}{\lambda}} \leq \left(t^\lambda(x)+t^\lambda(u)\right)^{\frac{\mu}{\lambda}} + \left(t^\lambda(y)+t^\lambda(v)\right)^{\frac{\mu}{\lambda}}$$

or, equivalently,

$$\left((t^\mu(x)+t^\mu(y))^{\frac{\lambda}{\mu}} + (t^\mu(u)+t^\mu(v))^{\frac{\lambda}{\mu}}\right)^{\frac{1}{\lambda}} \leq \left(\left((t^\lambda(x)+t^\lambda(u))^{\frac{1}{\lambda}}\right)^\mu + \left((t^\lambda(y)+t^\lambda(v))^{\frac{1}{\lambda}}\right)^\mu\right)^{\frac{1}{\mu}}.$$

If T is a strict t-norm then we immediately get

$$\begin{aligned} T^{(\lambda)}(T^{(\mu)}(x,y)T^{(\mu)}(u,v)) &= t^{-1}\left(\left((t^\mu(x)+t^\mu(y))^{\frac{\lambda}{\mu}} + (t^\mu(u)+t^\mu(v))^{\frac{\lambda}{\mu}}\right)^{\frac{1}{\lambda}}\right) \\ &\geq t^{-1}\left(\left(\left((t^\lambda(x)+t^\lambda(u))^{\frac{1}{\lambda}}\right)^\mu + \left((t^\lambda(y)+t^\lambda(v))^{\frac{1}{\lambda}}\right)^\mu\right)^{\frac{1}{\mu}}\right) \\ &= T^{(\mu)}(T^{(\lambda)}(x,u),T^{(\lambda)}(y,v)), \end{aligned}$$

i.e., $T^{(\lambda)} \gg T^{(\mu)}$. For a nilpotent t-norm T, if

$$\left((t^\mu(x)+t^\mu(y))^{\frac{\lambda}{\mu}} + (t^\mu(u)+t^\mu(v))^{\frac{\lambda}{\mu}}\right)^{\frac{\mu}{\lambda}} \geq t(0),$$

then trivially

$$T^{(\lambda)}(T^{(\mu)}(x,y),T^{(\mu)}(u,v)) \geq 0 = T^{(\mu)}(T^{(\lambda)}(x,u),T^{(\lambda)}(y,v)).$$

Finally, if $\left((t^\mu(x)+t^\mu(y))^{\frac{\lambda}{\mu}}+(t^\mu(u)+t^\mu(v))^{\frac{\lambda}{\mu}}\right)^{\frac{\mu}{\lambda}} < t(0)$, then all arguments of the pseudo-inverse $t^{(-1)}$ are strictly smaller than $t(0)$, and the same reasoning as in the case of a strict t-norm T leads to $T^{(\lambda)} \gg T^{(\mu)}$.

The following result gives a necessary and sufficient condition for the domination between strict t-norms.

6.18 THEOREM

Let T_1 and T_2 be strict t-norms with additive generators $t_1, t_2 : [0,1] \longrightarrow [0,\infty]$, respectively. The following are equivalent:

(i) $T_1 \gg T_2$.

(ii) For $h = t_1 \circ t_2^{-1} : [0,\infty] \longrightarrow [0,\infty]$ *and for all* $a,b,c,d \in [0,\infty]$ *we have*

$$h^{-1}(h(a+b) + h(c+d)) \leq h^{-1}(h(a) + h(c)) + h^{-1}(h(b) + h(d)) \quad (6.10)$$

Proof: Since $t_2 : [0,1] \longrightarrow [0,\infty]$ is a bijection, there is a one-to one correspondence between the quadruples $(x,y,u,v) \in [0,1]^4$ and the quadruples $(a,b,c,d) \in [0,\infty]^4$ given by $a = t_2(x)$, $b = t_2(y)$, $c = t_2(u)$, and $d = t_2(v)$. Fix a quadruple $(x,y,u,v) \in [0,1]^4$. Then (6.9) is equivalent to

$$\begin{aligned} T_1(T_2(t_2^{-1}(a), t_2^{-1}(b)), T_2(t_2^{-1}(c), t_2^{-1}(d))) \\ \geq T_2(T_1(t_2^{-1}(a), t_2^{-1}(c)), T_1(t_2^{-1}(b), t_2^{-1}(d))), \end{aligned}$$

which, if we rewrite the t-norms T_1 and T_2 by means of their additive generators t_1 and t_2, respectively (use Theorem 5.1 and take again into account that t_1 and t_2 are bijections), is transformed into

$$\begin{aligned} t_1^{-1}((t_1 \circ t_2^{-1})(a+b) &+ (t_1 \circ t_2^{-1})(c+d)) \\ &\geq t_2^{-1}((t_2 \circ t_1^{-1})((t_1 \circ t_2^{-1})(a) + (t_1 \circ t_2^{-1})(c)) \\ &\quad + (t_2 \circ t_1^{-1})((t_1 \circ t_2^{-1})(b) + (t_1 \circ t_2^{-1})(d))). \end{aligned}$$

Applying on both sides the strictly decreasing bijection t_2 leads to

$$\begin{aligned} (t_2 \circ t_1^{-1})((t_1 \circ t_2^{-1})(a+b) &+ (t_1 \circ t_2^{-1})(c+d)) \\ &\leq (t_2 \circ t_1^{-1})((t_1 \circ t_2^{-1})(a) + (t_1 \circ t_2^{-1})(c)) \\ &\quad + (t_2 \circ t_1^{-1})((t_1 \circ t_2^{-1})(b) + (t_1 \circ t_2^{-1})(d)), \end{aligned}$$

which, because of $h = t_1 \circ t_2^{-1}$ and $h^{-1} = t_2 \circ t_1^{-1}$, is equivalent to (6.10). □

A simple computation of the function h used in Theorem 6.18 shows that the following result holds.

6.19 COROLLARY

If T and T^* are two strict t-norms and if, for $\lambda \in\,]0,\infty[$, $T_{(T^*,\lambda)}$ is the strict t-norm introduced in Example 3.32(ii), then the following are equivalent:

(i) $T \gg T^*$.

(ii) For some $\lambda \in\,]0,\infty[$ *we have* $T_{(T^*,\lambda)} \gg T^*$.

(iii) For all $\lambda \in\,]0,\infty[$ *we have* $T_{(T^*,\lambda)} \gg T^*$.

The following example which is due to [Tardiff 1976] shows that, even for strict t-norms, $T_1 \geq T_2$ does not imply $T_1 \gg T_2$.

6.20 EXAMPLE
Consider the strictly decreasing functions $t_1, t_2 : [0,1] \longrightarrow [0,\infty]$ defined by

$$t_1(x) = \frac{1}{x} - 1,$$
$$t_2(x) = \begin{cases} \frac{1}{2x} & \text{if } x \in [0, 0.5], \\ \frac{1}{x} - 1 & \text{otherwise,} \end{cases}$$

and let T_1 and T_2 be the strict t-norms generated by t_1 and t_2, respectively, via (3.11). Then the function $h = t_1 \circ t_2^{-1} : [0,\infty] \longrightarrow [0,\infty]$ is given by

$$h(x) = \begin{cases} x & \text{if } x \in [0,1], \\ 2x - 1 & \text{otherwise,} \end{cases}$$

and we obtain, in violation of (6.10),

$$h^{-1}(h(1+1) + h(0.5+1)) = 3,$$
$$h^{-1}(h(1) + h(1)) + h^{-1}(h(0.5) + h(1)) = 2.75.$$

Therefore, as a consequence of Theorem 6.18, we do not have $T_1 \gg T_2$.

However, $T_1 \geq T_2$ holds because of Corollary 6.3: $h^{-1} = t_2 \circ t_1^{-1}$ is a concave function (see (5.6)) and satisfies $h^{-1}(0) = 0$, implying that h^{-1} is subadditive.

When comparing strict t-norms with respect to $\gg$, some sufficient conditions for functions to satisfy (6.10) in Theorem 6.18 are helpful. For this purpose, recall the definition of a convex function (5.5).

6.21 LEMMA
If a strictly increasing, continuous function $h : [0,\infty] \longrightarrow [0,\infty]$ satisfies (6.10) and $h(0) = 0$, then it is convex and superadditive.

Proof: Fix two arbitrary numbers $p, q \in [0,\infty]$ with $p \leq q$ and put $a = h^{-1}(p)$, $b = h^{-1}(\frac{p+q}{2}) - h^{-1}(p)$, $c = h^{-1}(\frac{q-p}{2})$, and $d = 0$. Inserting these values into (6.10) we obtain the inequality

$$\frac{h^{-1}(p) + h^{-1}(q)}{2} \leq h^{-1}\left(\frac{p+q}{2}\right).$$

Therefore, h^{-1} is a concave function, and h is a convex function satisfying $h(0) = 0$, implying that h is superadditive. □

The function $h : [0,\infty] \longrightarrow [0,\infty]$ in Example 6.20 shows that the converse is not true, i.e., that there are strictly increasing, continuous convex functions which satisfy $h(0) = 0$, but not (6.10).

The following result, the proof of which can be found in [Tardiff 1984], gives a sufficient condition for a strictly increasing, continuous convex function $h : [0,\infty] \longrightarrow [0,\infty]$ with $h(0) = 0$, to satisfy (6.10).

6.22 LEMMA

Let $h : [0, \infty] \longrightarrow [0, \infty]$ be continuous strictly increasing, everywhere differentiable convex function with $h(0) = 0$. If the function $g : \mathbb{R} \longrightarrow \mathbb{R}$ given by $g(x) = \log(h'(e^x))$ is convex, then h satisfies (6.10).

This result is crucial in the following construction (again due to [Tardiff 1984]) of a strict t-norm which dominates a given strict t- norm.

6.23 PROPOSITION

Let T_1 be a strict t-norm and $t_1 : [0, 1] \longrightarrow [0, \infty]$ be an additive generator of T_1. Further, let $g : \mathbb{R} \longrightarrow \mathbb{R}$ be a non-decreasing, continuous convex function with

$$\int_0^\infty e^{g(\log u)}\, du = \infty.$$

Then the function $t_2 : [0, 1] \longrightarrow [0, \infty]$ defined by

$$t_2(x) = \int_0^{t_1(x)} e^{g(\log u)}\, du$$

is a strictly decreasing bijection, i.e., an additive generator of some strict t-norm T_2, and we have $T_2 \gg T_1$.

7. Values and discretization of t-norms

> A key reason for the growth in the importance of discrete mathematics is that information is stored and manipulated by computing machines in a discrete fashion.
>
> KENNETH H. ROSEN

The full information about the values of a t-norm is given by the preimages of one-point sets $\{\beta\}$ with $\beta \in [0,1]$ under the t-norm T or, equivalently, by the sets of solutions of the equations $T(x,y) = \beta$. For continuous t-norms these preimages are completely described in Section 7.1. For arbitrary t-norms, a complete characterization of the preimages of $\{0\}$ is given. Also, for continuous t-norms T, the unique solvability of the equation $T(a,x) = b$ with $(a,b) \in [0,1[^2$ and $b \leq a$ is investigated.

The focus in Section 7.2 is on the identification of small subsets of the unit square which uniquely determine a continuous Archimedean t-norm [Kimberling 1973, Schweizer & Sklar 1983]. First we give necessary and sufficient conditions for the diagonal sections of continuous Archimedean t-norms [Kimberling 1973, Mesiar & Navara 1999]. The main results are that a strict t-norm is uniquely determined by its diagonal section and the section along the graph of a strictly decreasing bijection on the unit interval (in [Darsow & Frank 1984, Bézivin & Tomás 1993] these requirements are even weakened considerably), and that a nilpotent t-norm is uniquely determined by its diagonal section and its preimage of $\{0\}$ (see [Burgués 1981]). A by-product is that continuous t-norms with identical diagonal sections are either identical or incomparable.

Motivated by the fact that in most practical situations it is necessary to discretize the unit interval, in Section 7.3 we consider discrete t-norms, i.e., fully ordered commutative semigroups with neutral element 1 on finite subchains of the infinite chain $[0,1]$ (compare [De Baets 1998b, De Baets & Mesiar 1999]). The proper equivalent of the continuity of t-norms turns out to be the divisibility of discrete t-norms [Birkhoff 1973, Godo & Sierra 1988]. In full analogy to Theorem 5.11, there is a representation of divisible discrete t-norms as ordinal sums of Archimedean discrete t-norms [Mayor & Torrens 1993]. For t-norms with additive generators we give necessary and sufficient conditions for the restriction to be a discrete t-norm. Finally, each discrete t-norm can be extended to a t-norm (which is continuous if we start from a divisible discrete t-norm).

7.1 Values and preimages

We now want to study the set of those elements of the unit square which are mapped by a t-norm T to a given number $\beta \in [0,1]$ or, equivalently, the solutions of the equation $T(x,y) = \beta$.

To make things precise, if T is a t-norm and β an element of $[0,1]$, then we are interested in the preimage of the one-point set $\{\beta\}$ with respect to T:

$$T^{-1}(\{\beta\}) = \{(x,y) \in [0,1]^2 \mid T(x,y) = \beta\}.$$

7.1 EXAMPLE

For the four basic t-norms $T_{\mathbf{M}}$, $T_{\mathbf{P}}$, $T_{\mathbf{L}}$ and $T_{\mathbf{D}}$ we have (compare the contour lines in the contour plots of $T_{\mathbf{M}}$, $T_{\mathbf{P}}$ and $T_{\mathbf{L}}$ in Figure 1.2)

$$T_{\mathbf{M}}{}^{-1}(\{\beta\}) = (\{\beta\} \times [\beta,1]) \cup ([\beta,1] \times \{\beta\}),$$

$$T_{\mathbf{P}}{}^{-1}(\{\beta\}) = \begin{cases} (\{0\} \times [0,1]) \cup ([0,1] \times \{0\}) & \text{if } \beta = 0, \\ \{(x, \frac{\beta}{x}) \mid x \in [\beta,1]\} & \text{otherwise,} \end{cases}$$

$$T_{\mathbf{L}}{}^{-1}(\{\beta\}) = \begin{cases} \{(x,y) \in [0,1]^2 \mid x+y \leq 1\} & \text{if } \beta = 0, \\ \{(x, 1-x+\beta) \mid x \in [\beta,1]\} & \text{otherwise,} \end{cases}$$

$$T_{\mathbf{D}}{}^{-1}(\{\beta\}) = \begin{cases} [0,1[^2 & \text{if } \beta = 0, \\ \{(\beta,1),(1,\beta)\} & \text{otherwise.} \end{cases}$$

In general, the boundary condition (T4) and the commutativity (T1) of a t-norm T imply that the set $T^{-1}(\{\beta\})$ contains the trivial points $(\beta, 1)$ and $(1, \beta)$ and that it is symmetric, i.e., for all $(x,y) \in [0,1]^2$ we have

$$(x,y) \in T^{-1}(\{\beta\}) \quad \text{if and only if} \quad (y,x) \in T^{-1}(\{\beta\}).$$

Because of $T \leq T_{\mathbf{M}}$ for each t-norm T, we immediately get $T^{-1}(\{\beta\}) \subseteq [\beta,1]^2$. In particular, for $\beta = 1$ the preimage is a singleton, i.e., for each t-norm T we have $T^{-1}(\{1\}) = \{(1,1)\}$.

Turning to the opposite case $\beta = 0$, the set $T^{-1}(\{0\})$ always contains the lower left boundary of the unit square $[0,1]^2$, i.e., for each t-norm

$$(\{0\} \times [0,1]) \cup ([0,1] \times \{0\}) \subseteq T^{-1}(\{0\}). \tag{7.1}$$

It is also clear that $(x,y) \in T^{-1}(\{0\})$ if and only if (x,y) belongs to the lower left boundary of $[0,1]^2$ (i.e., if either $x = 0$ or $y = 0$) or if x and y are zero divisors of T (see Definition 2.1). In other words, in (7.1) the equality holds if and only if T has no zero divisors.

For $\beta \in]0,1[$ there may also be non-trivial elements of $T^{-1}(\{\beta\})$ (see, e.g., the continuous basic t-norms $T_{\mathbf{M}}$, $T_{\mathbf{P}}$ and $T_{\mathbf{L}}$ above). To be precise, we have $T^{-1}(\{\beta\}) = \{(\beta,1),(1,\beta)\}$ if and only if $\beta \notin \operatorname{Ran}(T|_{]0,1[^2})$ (which, of course, can only be possible for non-continuous t-norms T).

When constructing a new t-norm by means of some isomorphism (see Proposition 2.28(iv)), with the help of an additive generator (see Theorem 3.23), or as an ordinal sum of a family of t-norms (see Theorem 3.43), then we can determine $T^{-1}(\{\beta\})$ from the components used in the construction.

7.2 PROPOSITION
Let T be a t-norm and let $\varphi : [0,1] \longrightarrow [0,1]$ be a strictly increasing bijection. Then for the t-norm T_φ (as defined in (2.8)) and for each $\beta \in [0,1]$ we have

$$T_\varphi^{-1}(\{\beta\}) = \varphi^{-1}(T^{-1}(\{\varphi(\beta)\})),$$

where, as usual, for $A \subseteq [0,1]^2$ the generalized preimage $\varphi^{-1}(A)$ is given by

$$\varphi^{-1}(A) = \{(x,y) \in [0,1]^2 \mid (\varphi(x), \varphi(y)) \in A\}.$$

Proof: This follows immediately from the fact that the equality $T_\varphi(x,y) = \beta$ holds if and only if $T(\varphi(x), \varphi(y)) = \varphi(\beta)$. □

The following result for continuous Archimedean t-norms can either be derived directly or from Propositions 7.2, 5.9 and 5.10 together with Example 7.1.

7.3 COROLLARY
Let T be a continuous Archimedean t-norm and $t : [0,1] \longrightarrow [0,\infty]$ an additive generator of T. Then we have

$$T^{-1}(\{\beta\}) = \begin{cases} \{(x,y) \in [0,1]^2 \mid t(x) + t(y) \geq t(0)\} & \text{if } \beta = 0, \\ \{(x,y) \in [0,1]^2 \mid t(x) + t(y) = t(\beta)\} & \text{otherwise.} \end{cases}$$

An immediate consequence is that, for a continuous Archimedean t-norm T with additive generator $t : [0,1] \longrightarrow [0,\infty]$ and for $\beta \in]0,1]$, the set $T^{-1}(\{\beta\})$ coincides with the graph of the function $g_{T,\beta} : [\beta,1] \longrightarrow [\beta,1]$ given by

$$g_{T,\beta}(x) = t^{-1}(t(\beta) - t(x))$$

(observe that this function is well-defined since $t|_{[\beta,1]} : [\beta,1] \longrightarrow [0,t(\beta)]$ is a bijection).

Moreover, we have the following relationship between the shapes of t the generator $t : [0,1] \longrightarrow [0,\infty]$ and the function $g_{T,\beta} : [\beta,1] \longrightarrow [\beta,1]$ in this case.

7.4 PROPOSITION
Let T be a continuous Archimedean t-norm, $t : [0,1] \longrightarrow [0,\infty]$ be an additive generator of T, and let $\beta \in]0,1[$.

(i) If t is a convex function, so is $g_{T,\beta}$.
(ii) If t is a concave function, so is $g_{T,\beta}$.

Proof: Assume that the additive generator t is a convex function, and fix $\beta \in]0,1[$. Then Corollary 7.3 implies for all $x, y \in [\beta, 1]$ and for all $\gamma \in [0,1]$

$$\begin{aligned} g_{T,\beta}(\gamma x + (1-\gamma)y) &= t^{-1}(t(\beta) - t(\gamma x + (1-\gamma)y)) \\ &\leq t^{-1}(\gamma(t(\beta) - t(x)) + (1-\gamma)(t(\beta) - t(y))) \\ &\leq \gamma g_{T,\beta}(x) + (1-\gamma) g_{T,\beta}(y), \end{aligned}$$

i.e., also $g_{T,\beta}$ is a convex function. The proof of (ii) is completely dual. □

7.5 Example

(i) Continuous Archimedean t-norms with convex additive generators are, for example, the Schweizer-Sklar t-norms $T_\lambda^{\mathbf{SS}}$ for $\lambda \in]-\infty, 1]$ (see Figure 4.1), the Hamacher t-norms $T_\lambda^{\mathbf{H}}$ for $\lambda \in [0,1]$ (see Figure 4.2), the Frank t-norms $T_\lambda^{\mathbf{F}}$ for $\lambda \in]0,\infty]$ (see Figure 4.3), the Yager t-norms $T_\lambda^{\mathbf{Y}}$ for $\lambda \in [1,\infty[$ (see Figure 4.4), the Aczél-Alsina t-norms $T_\lambda^{\mathbf{AA}}$ for $\lambda \in [1,\infty[$ (see Figure 4.7), the Dombi t-norms $T_\lambda^{\mathbf{D}}$ for $\lambda \in [1,\infty[$ (see Figure 4.5), and the Sugeno-Weber t-norms $T_\lambda^{\mathbf{SW}}$ for $\lambda \in [0,\infty]$ (see Figure 4.6). Observe that all these t-norms are copulas because of Theorem 9.12.

(ii) Examples of continuous Archimedean t-norms with concave additive generators are the Schweizer-Sklar t-norms $T_\lambda^{\mathbf{SS}}$ for $\lambda \in [1,\infty[$, the Yager t-norms $T_\lambda^{\mathbf{Y}}$ for $\lambda \in]0,1]$, and the Sugeno-Weber t-norms $T_\lambda^{\mathbf{SW}}$ for $\lambda \in]-1,0]$. None of these t-norms is a copula (see Theorem 9.12).

(iii) Examples of continuous Archimedean t-norms whose additive generators are neither convex nor concave are (for $\lambda \in]1,\infty[$) the Hamacher t-norms $T_\lambda^{\mathbf{H}}$, the Aczél-Alsina t-norms $T_\lambda^{\mathbf{AA}}$ for $\lambda \in]0,1[$, the Dombi t-norms $T_\lambda^{\mathbf{D}}$ for $\lambda \in]0,1[$, and the incomparable t-norms $T_\lambda^{\mathbf{I}}$ for $\lambda \in]0,0.5[$ (see Example 3.33 and Appendix B.2). Again, none of these t-norms is a copula because of Theorem 9.12.

For the preimages of the ordinal sum of a family of t-norms we obtain the following characterization which is an immediate consequence of the construction of ordinal sums given in Theorem 3.43:

7.6 Corollary

Let $(T_\alpha)_{\alpha\in A}$ be a family of t-norms, $(]a_\alpha, e_\alpha[)_{\alpha\in A}$ be a family of non-empty, pairwise disjoint open subintervals of $[0,1]$, and $(\varphi_\alpha : [a_\alpha, e_\alpha] \longrightarrow [0,1])_{\alpha\in A}$ the family of linear strictly increasing bijections given by (3.22). Then for the ordinal sum $T = (\langle a_\alpha, e_\alpha, T_\alpha\rangle)_{\alpha\in A}$ we have:

$$T^{-1}(\{\beta\}) = \begin{cases} (\{\beta\} \times [e_\alpha, 1]) \cup ([e_\alpha, 1] \times \{\beta\}) \cup \varphi_\alpha^{-1}(T_\alpha^{-1}(\{\varphi_\alpha(\beta)\})) \\ \qquad\qquad\qquad\qquad\qquad\qquad\qquad\qquad \textit{if } \beta \in [a_\alpha, e_\alpha[, \\ (\{\beta\} \times [\beta, 1]) \cup ([\beta, 1] \times \{\beta\}) \qquad\qquad\qquad \textit{otherwise.} \end{cases}$$

Combining the statements of Corollaries 7.6 and 7.3 and our knowledge about the preimages of the minimum $T_{\mathbf{M}}$, the subsequent characterization of preimages of continuous t-norms follows readily. Recall that, because of Theorem 5.11, each continuous t-norm T is uniquely representable as an ordinal sum of continuous Archimedean t-norms.

7.7 COROLLARY

Let T be a continuous t-norm, and let $(]a_\alpha, e_\alpha[)_{\alpha\in A}$ and $(T_\alpha)_{\alpha\in A}$ be the uniquely determined families of non-empty, pairwise disjoint open subintervals and continuous Archimedean t-norms, respectively, such that $T = (\langle a_\alpha, e_\alpha, T_\alpha\rangle)_{\alpha\in A}$. For each $\alpha \in A$, let $t_\alpha : [0,1] \longrightarrow [0,\infty]$ be an additive generator of the t-norm T_α, $\varphi_\alpha : [a_\alpha, e_\alpha] \longrightarrow [0,1]$ be the linear strictly increasing bijection given by (3.22), and define the strictly decreasing function $h_\alpha : [a_\alpha, e_\alpha] \longrightarrow [0,\infty]$ by $h_\alpha = t_\alpha \circ \varphi_\alpha$. Then we have:

$$T^{-1}(\{\beta\}) = \begin{cases} (\{\beta\}\times[e_\alpha,1])\cup([e_\alpha,1]\times\{\beta\}) \\ \qquad \cup\{(x,y)\in[a_\alpha,e_\alpha]^2 \mid h_\alpha(x)+h_\alpha(y)=h_\alpha(\beta)\} \\ \qquad\qquad\qquad\qquad\qquad\qquad\qquad \text{if } \beta\in\,]a_\alpha,e_\alpha[, \\ (\{\beta\}\times[e_\alpha,1])\cup([e_\alpha,1]\times\{\beta\}) \\ \qquad \cup\{(x,y)\in[a_\alpha,e_\alpha]^2 \mid h_\alpha(x)+h_\alpha(y)\geq h_\alpha(\beta)\} \\ \qquad\qquad\qquad\qquad\qquad\qquad\qquad \text{if } \beta = a_\alpha, \\ (\{\beta\}\times[\beta,1])\cup([\beta,1]\times\{\beta\}) \qquad\qquad \text{otherwise.} \end{cases}$$

We only mention that this implies that, for each continuous t-norm T and for each $\beta \in [0,1]$, the preimage $T^{-1}(\{\beta\})$ is a compact and connected subset of the unit square $[0,1]^2$ which has a non-zero Lebesgue measure if and only if, taking the unique representation as an ordinal sum of continuous Archimedean t-norms existing because of Theorem 5.11, $\beta = a_\alpha$ and T_α is nilpotent (see Figure 7.1).

In particular, if T is a continuous t-norm such that $T^{-1}(\{0\})$ has a non-zero Lebesgue measure, then we put

$$x_0 = \sup\{x \in\,]0,1[\; \mid T(x,y) = 0 \text{ for some } y \in\,]0,1[\}$$

(observe that the interval $]0, x_0[$ coincides with the set of zero divisors of T). A consequence of Corollary 7.7 is that the following function $\varphi : [0, x_0] \longrightarrow [0, x_0]$ (recall that $\sup\emptyset = 0$ in this case) is an involution:

$$\varphi(x) = \sup\{y \in\,]0, x_0[\; \mid T(x,y) = 0\}.$$

Turning our attention to arbitrary t-norms, we first study the case $\beta = 0$.

7.8 PROPOSITION

Let A be a subset of $]0,1[^2$. Then there exists a t-norm T such that

$$T^{-1}(\{0\}) = A \cup (\{0\}\times[0,1]) \cup ([0,1]\times\{0\})$$

if and only if A satisfies the following properties:

(i) A is symmetric, i.e., $(x, y) \in A$ implies $(y, x) \in A$.
(ii) For all $(x, y) \in A$ we have $]0, x] \times]0, y] \subseteq A$.

Proof: Obviously, the properties (i) and (ii) are necessary because of the commutativity (T1) and the monotonicity (T3). Conversely, the t-norm T_A introduced in Proposition 3.63 obviously satisfies $T_A^{-1}(\{0\}) = A$. □

Consider now an arbitrary t-norm T which has a non-trivial idempotent element $\beta \in]0, 1[$. Then the restriction of T to $[\beta, 1]^2$ coincides with the restriction of some ordinal sum $(\langle \beta, 1, T^* \rangle)$ to $[\beta, 1]^2$. In this case, we can combine the results of Proposition 7.8 and Corollary 7.6 in a straightforward way to obtain the following result.

7.9 COROLLARY
Let $\beta \in]0, 1[$ and let A be a subset of $]\beta, 1[^2$. Then there exists a t-norm T such that β is an idempotent element of T and

$$T^{-1}(\{\beta\}) = A \cup (\{\beta\} \times [\beta, 1]) \cup ([\beta, 1] \times \{\beta\})$$

if and only if A has the following properties:

(i) A is symmetric, i.e., $(x, y) \in A$ implies $(y, x) \in A$.
(ii) For all $(x, y) \in A$ we have $]\beta, x] \times]\beta, y] \subseteq A$.

If $\beta \in]0, 1[$ is not an idempotent element of the t-norm T, then we know only that $T^{-1}(\{\beta\})$ is symmetric and that the sets $T^{-1}(\{\beta\})$ and $[0, \beta]^2$ are disjoint subsets of $[0, 1]^2$.

It is immediately clear that the equation $T(a, x) = b$ has a solution for each pair $(a, b) \in [0, 1]^2$ with $b \le a$ if and only if T is continuous (or, if the integral l-monoid $([0, 1], \le, T)$ is divisible (see Section 2.4 and [Birkhoff 1973]). This solution is unique in the case $a = 1$. The following result shows under which conditions the equation $T(a, x) = b$ with $(a, b) \in [0, 1]^2$ and $b \le a < 1$ has a unique solution.

7.10 PROPOSITION
Let T be a continuous t-norm and $(a, b) \in [0, 1[^2$ with $b \le a$. Then the following are equivalent:

(i) The equation $T(a, x) = b$ has a unique solution.
(ii) If $a_T^{(n)} = b$ for some $n \in \mathbb{N}$ then b is not an idempotent element of T and, if $a = b$, then no element of $[b, 1[$ is an idempotent element of T.

Proof: For a continuous Archimedean t-norm T, because of Corollary 7.3, the equation $T(a, x) = b$ has more than one solution if and only if $b = 0$ and if either $a = 0$ or T is nilpotent. In either case we then have $a_T^{(n)} = b = 0$ for some $n \in \mathbb{N}$ and $b \in [b, 1[$ is an idempotent element of T. This shows that (ii) implies (i), the converse being obvious for continuous Archimedean t-norms. For continuous t-norms the claim then follows from Corollary 7.6. □

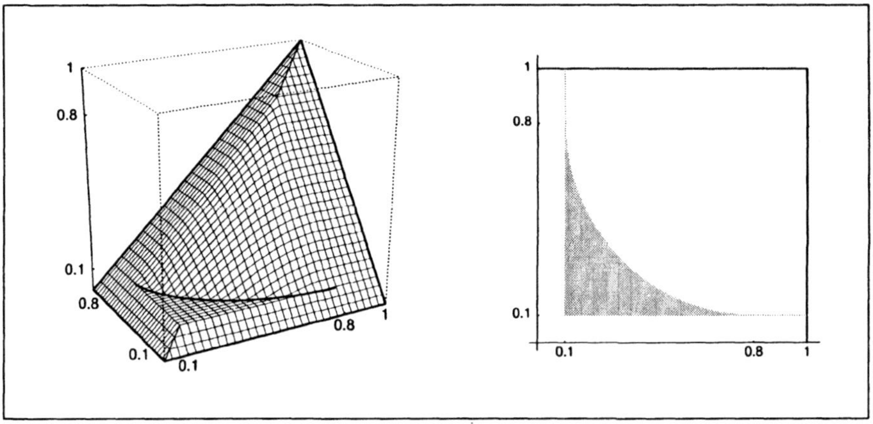

Fig. 7.1. The t-norm $T = (\langle 0.1, 0.8, T^{\mathbf{Y}}_{0.2}\rangle)$ (left) and the set $T^{-1}(\{0.1\})$ (right)

In fact, more can be said about the uniqueness of solutions of $T(a, x) = b$ under the hypotheses of Proposition 7.10. If $a = b$ then $T(a, x) = b$ has a unique solution if and only if the ordinal sum decomposition of T according to Theorem 5.11 contains a summand of the form $\langle c, 1, T^*\rangle$ with $c \in [0, b[$. If $a > b$ then the uniqueness of the solution of $T(a, x) = b$ is violated only if there is a summand of the form $\langle b, c, T^*\rangle$, where $c \in]a, 1]$ and T^* is nilpotent.

If T is left-continuous and if $T(a, x) = b$ has a solution for $(a, b) \in [0, 1]^2$ with $b \leq a$, then there exists also a maximal solution of $T(a, x) = b$, and it coincides with $a \to_T b = \sup\{x \in [0, 1] \mid T(a, x) \leq b\}$ (for a more detailed discussion of the T-residuum $\to_T$ see (2.17)). Note that, in a more general context, the maximal solution of $T(a, x) = b$ for $(a, b) \in [0, 1]^2$ with $b \leq a$ coincides with $g(b)$, where the one-place functions $T(a, .) : [0, 1] \longrightarrow [0, a]$ and $g : [0, a] \longrightarrow [0, 1]$ form a Galois connection [Gierz *et al.* 1980].

7.2 Sections and segments

In Remark 1.3(i) we have seen that, as a consequence of the axioms (T1)–(T4), all t-norms coincide on the boundary of the unit square $[0, 1]^2$. Therefore, when specifying a t-norm T, there is only some freedom on the open unit square $]0, 1[^2$. This freedom is further restricted, especially by the commutativity (T1) (forcing that the graph of T be symmetric), by the associativity (T2), the monotonicity (T3), and by the fact that we always have the inequality $T \leq T_{\mathbf{M}}$ (i.e., $T(x, y) \leq \min(x, y)$ must hold for all $(x, y) \in [0, 1]^2$). In fact, Proposition 3.60 shows that each binary operation $* : A^2 \longrightarrow A$, where A is a subinterval of $[0, 1[$ and $*$ satisfies the properties mentioned above, can be extended (not necessarily in a unique way) to a t-norm.

We shall study now the values of t-norms on some specific subsets of the unit square $[0,1]^2$ and, in some cases, investigate the existence of a t-norm extension for functions defined on such subsets of $[0,1]^2$.

A prominent subset of $[0,1]^2$ is the diagonal $\{(x,x) \mid x \in [0,1]\}$. To simplify notations, for a given t-norm T its diagonal section will be denoted $\delta_T : [0,1] \longrightarrow [0,1]$, specified by

$$\delta_T(x) = T(x,x) \quad \bigl(= x_T^{(2)} \bigr).$$

We already know that several algebraic properties of a t-norm T are indeed properties of its diagonal section δ_T. By definition, an element of $[0,1]$ is an idempotent element of T if and only if it is a fixed point of δ_T. Also, the Archimedean property of T as well as the existence of zero divisors or nilpotent elements of T can be described by corresponding properties of δ_T (see Theorem 2.12 and Remark 2.4).

It is evident that, for each t-norm T, its diagonal section δ_T is a non-decreasing function, satisfying also $\delta_T(0) = 0$, $\delta_T(1) = 1$, and $\delta_T \leq \mathrm{id}_{[0,1]}$. Obviously, δ_T is continuous whenever T is a continuous t-norm. Let us write, for convenience,

$$\mathcal{D} = \{\delta \in [0,1]^{[0,1]} \mid \delta \text{ is non-decreasing and } \delta \leq \mathrm{id}_{[0,1]} \text{ and } \delta(1) = 1\}.$$

The diagonal sections of strict t-norms can be completely characterized as follows [Kimberling 1973, Mesiar & Navara 1999].

7.11 PROPOSITION

For $\delta \in \mathcal{D}$ the following are equivalent:

(i) There exists a strict t-norm T with $\delta_T = \delta$.

(ii) δ is a strictly increasing bijection such that $\delta(x) < x$ for all $x \in \,]0,1[$.

Proof: Since each strict t-norm is Archimedean and has a strictly increasing diagonal section, (i) obviously implies (ii).

Conversely, if δ satisfies (ii) it suffices to find an additive generator, i.e., a strictly decreasing bijection $t : [0,1] \longrightarrow [0,\infty]$, such that for all $x \in [0,1]$ we have $\delta(x) = t^{-1}(2t(x))$ or, equivalently,

$$t \circ \delta = 2t \tag{7.2}$$

(note that this is the so-called Schröder functional equation [Kuczma 1963]). Without loss of generality we may put $t(0.5) = 1$. Then, if t solves (7.2), for each $n \in \mathbb{Z}$ we must necessarily have $t\bigl(\delta^{(n)}(0.5)\bigr) = 2^n$, where $\delta^{(0)} = \mathrm{id}_{[0,1]}$, $\delta^{(n)} = \delta \circ \delta^{(n-1)}$ whenever $n \in \mathbb{N}$, and $\delta^{(n)} = \bigl(\delta^{(-n)}\bigr)^{-1}$ whenever $-n \in \mathbb{N}$. The strict monotonicity, the continuity of δ and the fact that $\delta(x) < x$ for all $x \in \,]0,1[$ imply $\lim_{n\to\infty} \delta^{(n)}(0.5) = 0$ and $\lim_{n\to-\infty} \delta^{(n)}(0.5) = 1$. Now choose an arbitrary strictly decreasing bijection $f : [\delta(0.5), 0.5] \longrightarrow [1,2]$, and define the function $t : [0,1] \longrightarrow [0,\infty]$ by (observe that n can be any value in the set of integers $\mathbb{Z}$)

$$t(x) = \begin{cases} \infty & \text{if } x = 0, \\ 2^n \cdot f\left(\delta^{(-n)}(x)\right) & \text{if } x \in \left]\delta^{(n+1)}(0.5), \delta^{(n)}(0.5)\right], \\ 0 & \text{if } x = 1. \end{cases} \tag{7.3}$$

It is only a matter of computation to check that t is a strictly decreasing bijection, i.e., an additive generator of some strict t-norm T, which solves (7.2), i.e., $\delta_T = \delta$ holds. □

A generalization of Proposition 7.11, showing that the diagonal section δ_T may be replaced by the function $x \longmapsto x_T^{(n)}$ for some fixed $n > 2$, was proved in [Kimberling 1973].

A consequence of the proof of Proposition 7.11 is that different strict t-norms may have the same diagonal section and that, for a given strictly increasing bijection $\delta \in \mathcal{D}$ with $\delta(x) < x$ for all $x \in]0,1[$, the generator of each strict t-norm T with $\delta_T = \delta$ must have the form (7.3), where $f : [\delta(0.5), 0.5] \longrightarrow [1,2]$ is some strictly decreasing bijection.

7.12 PROPOSITION
Let T_1 and T_2 be two strict t-norms with $\delta_{T_1} = \delta_{T_2}$. Then either $T_1 = T_2$ or T_1 and T_2 are incomparable.

Proof: From the construction in the proof of Proposition 7.11 we know that the additive generators t_1 and t_2 of T_1 and T_2, respectively, must be of the form (7.3) with appropriate strictly decreasing bijections $f_1, f_2 : [\delta(0.5), 0.5] \longrightarrow [1,2]$. Then for all $x \in [0,\infty]$ we get

$$t_1 \circ t_2^{-1}(x) = \begin{cases} 0 & \text{if } x = 0, \\ 2^n f_1 \circ f_2^{-1}(2^{-n}x) & \text{if } x \in \left[2^n, 2^{n+1}\right[, \\ \infty & \text{if } x = \infty, \end{cases}$$

and either $f_1 \circ f_2^{-1} = \mathrm{id}_{[1,2]}$, i.e., $t_1 = t_2$ and $T_1 = T_2$, or $t_1 \circ t_2^{-1}$ is neither subadditive nor superadditive, i.e., T_1 and T_2 are incomparable as a consequence of Theorem 6.2. □

In [Darsow & Frank 1984, Bézivin & Tomás 1993] it was shown that a strict t-norm T is uniquely determined by its diagonal section δ_T together with some other appropriate section of T (compare Proposition 7.12).

7.13 PROPOSITION
Let $f : [0,1] \longrightarrow [0,1]$ be a strictly decreasing bijection. If T_1 and T_2 are two strict t-norms with the same diagonal section, i.e., $\delta_{T_1} = \delta_{T_2}$, and if

$$T_1(x, f(x)) = T_2(x, f(x)) \tag{7.4}$$

holds for all $x \in [0,1]$, then $T_1 = T_2$.

Proof: If $t_1, t_2 : [0,1] \longrightarrow [0,\infty]$ are additive generators of T_1 and T_2, respectively, define the functions $\varphi, \psi : [0,\infty] \longrightarrow [0,\infty]$ and $\zeta :]0,\infty[\longrightarrow]0,\infty[$ by

$$\varphi = t_1 \circ t_2^{-1}, \qquad \psi = t_2 \circ f \circ t_2^{-1}, \qquad \text{and} \qquad \zeta(x) = \frac{\varphi(x)}{x}.$$

Obviously, φ is a strictly increasing bijection, ψ is a strictly decreasing bijection, and ζ is a continuous function. Moreover, the property $\delta_{T_1} = \delta_{T_2}$ is equivalent to $\varphi(2x) = 2 \cdot \varphi(x)$ for all $x \in [0,\infty]$, and condition (7.4) transforms into

$$\varphi(x + \psi(x)) = \varphi(x) + \varphi(\psi(x)),$$

and, subsequently, we get for all $x \in]0,\infty[$

$$\zeta(2x) = \zeta(x), \tag{7.5}$$

$$\zeta(x + \psi(x)) = \frac{x\zeta(x) + \psi(x)\zeta(\psi(x))}{x + \psi(x)}. \tag{7.6}$$

From (7.5) we see that ζ is completely determined by its values on the interval $[1,2]$. From the continuity of ζ on $[1,2]$ it follows that

$$m = \sup\{\zeta(x) \mid x \in [1,2]\} < \infty$$

and that there exists some $x_0 \in [1,2]$ with $\zeta(x_0) = m$. We shall show that for each $x \in]0,\infty[$ we have $\zeta(x) = m$, i.e., $t_1 = m \cdot t_2$, which will imply $T_1 = T_2$.

We shall consider the closed subset $A = \{x \in]0,\infty[\mid \zeta(x) = m\}$ of $]0,\infty[$, and we shall show that $A =]0,\infty[$. If we suppose that the open set $]0,\infty[\setminus A$ is non-empty, then there exists an open subinterval $]a,b[$ of $]0,\infty[\setminus A$ (i.e., $A \cap]a,b[= \emptyset$) with $a, b \in A$. The fact $0 + \psi(0) = \infty + \psi(\infty) = \infty$ and the continuity of ψ imply the existence of $x_1 \in]0,\infty[$ satisfying

$$x_1 + \psi(x_1) = \inf\{x + \psi(x) \mid x \in]0,\infty[\}.$$

We choose an integer $n \in \mathbb{N}$ such that $x_1 < 2^n(b-a)$ and $x_1 + \psi(x_1) < 2^n b$. From $b \in A$ and (7.5) we get $2^n b \in A$ and, subsequently, $x_2 + \psi(x_2) = 2^n b$ for some $x_2 \in]0, x_1[$. Then from (7.6) we obtain

$$m = \zeta(2^n b) = \frac{x_2\zeta(x_2) + \psi(x_2)\zeta(\psi(x_2))}{x_2 + \psi(x_2)}$$

implying, together with $\zeta(x_2) \leq m$ and $\zeta(\psi(x_2)) \leq m$, $\zeta(x_2) = \zeta(\psi(x_2)) = m$ and, because of the construction of A, $\psi(x_2) \in A$. Then, for $x_3 = \frac{\psi(x_2)}{2^n}$ we get $x_3 \in A$ as a consequence of (7.5). But we also have

$$a < b - \frac{x_1}{2^n} < b - \frac{x_2}{2^n} = x_3 < b,$$

i.e., $x_3 \in]a,b[$. Therefore we cannot have $A \cap]a,b[= \emptyset$, implying $A =]0,\infty[$ and, subsequently, $T_1 = T_2$. □

The perhaps simplest special case of Proposition 7.13 is that a strict t-norm is uniquely determined by its values on the two diagonals $\{(x,x) \mid x \in [0,1]\}$ and $\{(x,1-x) \mid x \in [0,1]\}$ of the unit square. In fact, in [Bézivin & Tomás 1993] it was shown that it suffices to know the values of a strict t-norm on some appropriate subset of the two diagonals, e.g, on $\{(x,x) \mid x \in [0,1]\}$ and $\{(x,1-x) \mid x \in [0,\varepsilon]\}$, for some $\varepsilon \in [0,1]$ (see Figure 7.2 (top left)).

Turning our attention to nilpotent t-norms, a result similar to that given in Proposition 7.11 holds:

7.14 PROPOSITION
For $\delta \in \mathcal{D}$ the following are equivalent:

(i) There exists a nilpotent t-norm T with $\delta_T = \delta$.
(ii) δ is continuous, $\delta|_{\delta^{-1}(]0,1])}$ is strictly increasing, $\delta^{-1}(\{0\}) \neq \{0\}$, and $\delta(x) < x$ for all $x \in]0,1[$.

Proof: Again the necessity is obvious. In order to show the sufficiency, denote $x_0 = \sup\{x \in [0,1] \mid \delta(x) = 0\}$, observe $x_0 \in]0,1[$ and choose an arbitrary strictly decreasing bijection $f : [0,x_0] \longrightarrow [0.5,1]$. Then, similarly as in the proof of Proposition 7.11, the function $t : [0,1] \longrightarrow [0,\infty]$ by (this time we have $n \in \mathbb{N}$)

$$t(x) = \begin{cases} f(x) & \text{if } x \in [0,x_0]\,, \\ 2^{-n} \cdot f\left(\delta^{(n)}(x)\right) & \text{if } x \in \left]\delta^{(1-n)}(x_0), \delta^{(-n)}(x_0)\right]\,, \\ 0 & \text{if } x = 1, \end{cases} \tag{7.7}$$

is an additive generator of a nilpotent t-norm T with $\delta_T = \delta$. □

An immediate consequence of the proof of Proposition 7.14 is that different nilpotent t-norms may have the same diagonal section and that, given $\delta \in \mathcal{D}$ satisfying the requirements of Proposition 7.14, the generator of each nilpotent t-norm T with $\delta_T = \delta$ must have the form (7.7), where

$$x_0 = \sup\{x \in [0,1] \mid \delta(x) = 0\}$$

and $f : [0,x_0] \longrightarrow [0.5,1]$ is some strictly decreasing bijection.

Using the same arguments as in Proposition 7.12, we can show an analogous result for nilpotent t-norms with the same diagonal section.

7.15 PROPOSITION
Let T_1 and T_2 be two nilpotent t-norms with $\delta_{T_1} = \delta_{T_2}$. Then either $T_1 = T_2$ or T_1 and T_2 are incomparable.

In [Burgués 1981] it was shown that a nilpotent t-norm T is uniquely determined by its diagonal section δ_T and the preimage of $\{0\}$ (compare Proposition 7.15). However, there is no direct analogue of Proposition 7.13 in the case of nilpotent t-norms, as can be seen from the t-norms T_2 from Example 6.1(ii) and $T_{\mathbf{L}}$ (compare also Figure 6.2).

7.16 PROPOSITION
If T_1 and T_2 are two nilpotent t-norms with $\delta_{T_1} = \delta_{T_2}$ and $T_1^{-1}(\{0\}) = T_2^{-1}(\{0\})$ then $T_1 = T_2$.

Proof: Let t_1 and t_2 be continuous additive generators of T_1 and T_2, respectively, with $t_1(0) = t_2(0) = 1$, and denote

$$x_0 = \sup\{x \in [0,1] \mid \delta_{T_1}(x) = 0\} = \sup\{x \in [0,1] \mid \delta_{T_2}(x) = 0\}.$$

Then $\delta_{T_1} = \delta_{T_2}$ is equivalent to

$$t_1^{(-1)}(2t_1(x)) = t_2^{(-1)}(2t_2(x)) \tag{7.8}$$

for all $x \in [x_0, 1]$, and $T_1^{-1}(\{0\}) = T_2^{-1}(\{0\})$ means that $t_1(x) + t_1(y) \geq 1$ if and only if $t_2(x) + t_2(y) \geq 1$, i.e.,

$$t_1^{(-1)}(1 - t_1(x)) = t_2^{(-1)}(1 - t_2(x)) \tag{7.9}$$

for all $x \in [0,1]$. Putting $t = t_2 \circ t_1^{(-1)}$ and substituting $z = t_1(x)$, then (7.8) is equivalent to $t(2z) = 2t(z)$ for all $z \in [0, 0.5]$, and (7.9) is equivalent to $t(1-z) = 1 - t(z)$ for all $z \in [0,1]$. Now it is evident that $t(0) = 0$, $t(1) = 1$ and $t(0.5) = 0.5$, and by induction we get $t(r) = r$ for each dyadic rational number in $[0,1]$. Because of the continuity of t this means $t = \mathrm{id}_{[0,1]}$, implying $t_1 = t_2$ and, subsequently, $T_1 = T_2$. □

The only continuous t-norm which is uniquely determined by its diagonal section is the minimum $T_{\mathbf{M}}$ (see Proposition 1.9). Combining the results of Propositions 7.11 and 7.14 and the general representation of continuous t-norms given in Theorem 5.11, we immediately obtain the following characterization of diagonal sections of continuous t-norm (compare also [Mesiar & Navara 1999, Tkadlec 1999]). The diagonal sections of the drastic product $T_{\mathbf{D}}$ (compare again Proposition 1.9) and of some other non-continuous t-norms are discussed in [Tkadlec 1999, Viceník 1999b].

7.17 PROPOSITION
For $\delta \in \mathcal{D}$ the following are equivalent:

(i) There exists a continuous t-norm T with $\delta_T = \delta$.
(ii) δ is continuous and the restriction $\delta|_{[0,1]\setminus\delta^{-1}(\{x\in[0,1[\mid\delta(x)=x\})}$ is strictly increasing.

A straightforward consequence of Theorem 5.11 and Propositions 7.12 and 7.15 is the following:

7.18 COROLLARY
Let T_1 and T_2 be two continuous t-norms with $\delta_{T_1} = \delta_{T_2}$. Then either $T_1 = T_2$ or T_1 and T_2 are incomparable.

7.19 EXAMPLE
For non-continuous t-norms this is no longer true: consider t-norms T_A and T_B (as introduced in Proposition 3.63), where A is a proper subset of B and $A \cap \{(x,x) \mid x \in [0,1]\} = B \cap \{(x,x) \mid x \in [0,1]\}$, then we have $\delta_{T_A} = \delta_{T_B}$, but T_A is strictly stronger than T_B.

7.20 REMARK
(i) In [Schweizer & Sklar 1983, Problem 5.8.1] the following question can be found: is an Archimedean t-norm T with a continuous diagonal section δ_T necessarily a continuous t-norm? This problem has a negative answer [Krause 200x], and the Krause t-norm $T^{\mathbf{K}}$ is an example of a non-continuous Archimedean t-norm with a continuous diagonal section (details are given in Appendix B.1). However, a characterization of the class of all t-norms with continuous diagonal sections has not been given so far. Note that there are left-continuous (but non-continuous) t-norms with continuous diagonal sections (see Remark B.2).
(ii) Another related problem is still unsolved: can each (continuous) function in $\mathcal{D}$ be extended to a t-norm, i.e., do we have $\mathcal{D} = \{\delta_T \mid T \text{ is a t-norm}\}$ (compare [Kimberling 1973])? For instance, it is unknown whether there exists a (necessarily Archimedean and non-continuous) t-norm T whose continuous diagonal section δ_T is given by

$$\delta_T(x) = \begin{cases} \frac{x}{2} & \text{if } x \in [0, 0.5]\,, \\ 0.25 & \text{if } x \in \,]0.5, 0.75]\,, \\ 3x - 2 & \text{otherwise.} \end{cases}$$

Because of the commutativity (T1) of t-norms, each t-norm is uniquely determined by its values on the triangle whose edges are $(0,0)$, $(1,0)$ and $(1,1)$. Continuous Archimedean t-norms, however, are uniquely determined by their values on an appropriate triangle with arbitrarily small surface (see [Kimberling 1973]).

7.21 PROPOSITION
Let T be a continuous Archimedean t-norm and $\varepsilon \in \,]0,1]$. Then T is completely determined by its values on the triangle $\Delta_\varepsilon \subseteq [0,1]^2$, whose edges are the points $(1-\varepsilon, 0)$, $(1,0)$ and $(1,1)$.

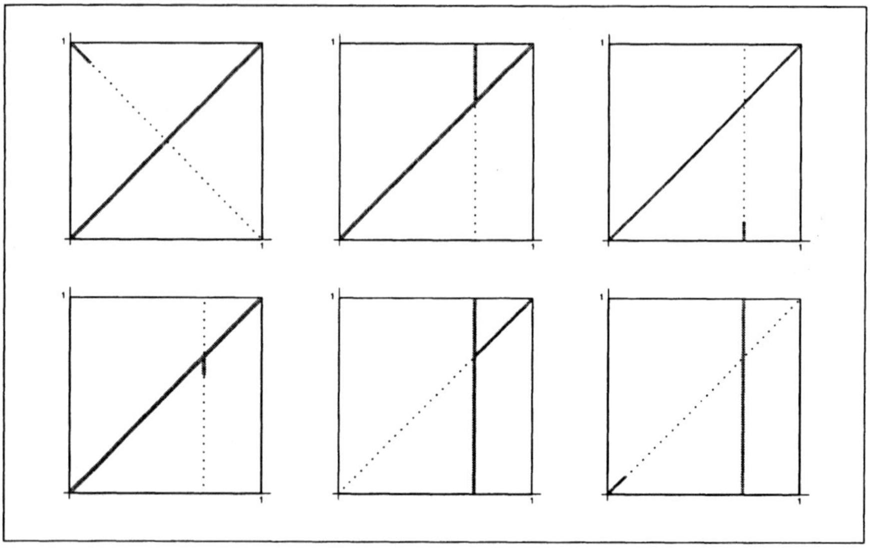

Fig. 7.2. Some typical subsets (bold lines) of the unit square which determine a strict t-norm according to [Bézivin & Tomás 1993] (top left) and [Darsow & Frank 1984] (others)

Proof: Because of the commutativity and the boundary conditions of t-norms, it suffices to fix a point $(x_0, y_0) \in]0,1[^2 \setminus \Delta_\varepsilon$ with $y_0 \leq x_0$. Because of the equalities $T(1-\varepsilon, 0) = 0$, $T(1,1) = 1$ and the continuity of T, there exists a point (u, v) on the straight line segment connecting the points $(1-\varepsilon, 0)$ and $(1,1)$ (i.e., $(u,v) \in \Delta_\varepsilon$) satisfying $T(u,v) = x_0$. Note that, since T is continuous Archimedean, there is exactly one such (u, v). Now put $x_1 = v$ and $y_1 = T(u, y_0)$ and observe that the associativity (T2) implies $T(x_0, y_0) = T(x_1, y_1)$. Moreover, we have $x_1 > x_0$ and $y_1 < y_0$. Iterating this procedure, we either reach a boundary point (x_{n_0}, y_{n_0}) of the unit square with $T(x_0, y_0) = T(x_{n_0}, y_{n_0})$, or we obtain a sequence of points $(x_n, y_n)_{n \in \mathbb{N}}$, where the sequence $(x_n)_{n \in \mathbb{N}}$ is strictly increasing and converges to 1 and the sequence $(y_n)_{n \in \mathbb{N}}$ is strictly decreasing and converges to 0, such that $T(x_0, y_0) = T(x_n, y_n)$ for all $n \in \mathbb{N}$. In the latter case there exists an $n_0 \in \mathbb{N}$ such that $(x_{n_0}, y_{n_0}) \in \Delta_\varepsilon$, and in both cases we may put $T(x_0, y_0) = T(x_{n_0}, y_{n_0})$. □

Similar results as for diagonals can also be obtained for vertical (or horizontal) sections of continuous t-norms, in the case of a strict t-norm T involving the solution of the so-called Abel functional equation [Aczél 1966]

$$t \circ T(x_0, .) = t + 1,$$

where $T(x_0, .) : [0,1] \longrightarrow [0,1]$ is the vertical section of the t-norm T for some $x_0 \in]0,1[$, and the unknown $t : [0,1] \longrightarrow [0,\infty]$ is the additive generator of T satisfying $t(x_0) = 1$ (see [Darsow & Frank 1984]).

In Figure 7.2 some other typical subsets of the unit square with the property that the knowledge of the values of a strict t-norm on these sets uniquely determines the t-norm on the whole unit square are visualized (see again [Darsow & Frank 1984]).

7.3 Discrete t-norms

Denote, for each $n \in \mathbb{N}$, by $\Gamma_{(n)}$ the finite subchain $\{0, \frac{1}{n}, \frac{2}{n}, \ldots, \frac{n-1}{n}, 1\}$ of the infinite chain $[0,1]$ (both with respect to the usual order $\leq$, a fact we usually will not mention in the sequel). Since each finite chain $\{\gamma_0, \gamma_1, \ldots, \gamma_n\}$ with

$$0 = \gamma_0 < \gamma_1 < \cdots < \gamma_n = 1$$

obviously is isomorphic to $\Gamma_{(n)}$, we can restrict ourselves to to the case $\Gamma_{(n)}$.

We now want to discuss the relationship between triangular norms and appropriate operations on the finite chains $\Gamma_{(n)}$. In this context, the notions of semigroups (see Section 2.2) and l-monoids (see Section 2.4) will be helpful.

7.22 DEFINITION
Given a number $n \in \mathbb{N}$, a function $D : \Gamma_{(n)}^2 \longrightarrow \Gamma_{(n)}$ is called a *discrete t-norm* if $(\Gamma_{(n)}, D, \leq)$ is a fully ordered commutative semigroup with neutral element 1 or, equivalently, a finite commutative integral l-monoid.

The notions of idempotent element, nilpotent element or Archimedean property of a discrete t-norm $D : \Gamma_{(n)}^2 \longrightarrow \Gamma_{(n)}$ are, of course, just the respective notions in the semigroup $(\Gamma_{(n)}, D, \leq)$. For $n \geq 2$ the class of Archimedean discrete t-norms coincides with the class of discrete t-norms for which $\frac{n-1}{n}$ is a nilpotent element (compare Remark 2.4(iv)). Finally note that, if $D : \Gamma_{(n)}^2 \longrightarrow \Gamma_{(n)}$ is a discrete t-norm then the l-monoid $(\Gamma_{(n)}, D, \leq)$ is always residuated (see Example 2.46(i)).

7.23 EXAMPLE

(i) There is only one discrete t-norm on the trivial chain $\Gamma_{(1)} = \{0,1\}$. By definition it is both idempotent and Archimedean, and it coincides with the conjunction in Boolean logic.

(ii) For each $n \in \mathbb{N}$ the following operations $D_{\mathbf{M}}^{(n)}, D_{\mathbf{L}}^{(n)}, D_{\mathbf{D}}^{(n)} : \Gamma_{(n)}^2 \longrightarrow \Gamma_{(n)}$ are discrete t-norms:

$$D_{\mathbf{M}}^{(n)}(\tfrac{i}{n}, \tfrac{j}{n}) = \tfrac{1}{n}\min(i,j),$$
$$D_{\mathbf{L}}^{(n)}(\tfrac{i}{n}, \tfrac{j}{n}) = \tfrac{1}{n}\max(i+j-n, 0),$$
$$D_{\mathbf{D}}^{(n)}(\tfrac{i}{n}, \tfrac{j}{n}) = \begin{cases} 0 & \text{if } \max(i,j) < n, \\ \frac{1}{n}\min(i,j) & \text{otherwise.} \end{cases}$$

Obviously, $D_{\mathbf{M}}^{(n)}$ is the strongest and the only idempotent discrete t-norm, and $D_{\mathbf{D}}^{(n)}$ is Archimedean and the weakest discrete t-norm on $\Gamma_{(n)}$. We only mention that $D_{\mathbf{L}}^{(n)}$ is the only divisible Archimedean discrete t-norm (see Definition 2.49) on $\Gamma_{(n)}$, as will be shown in Theorem 7.26.

(iii) The number of discrete t-norms on $\Gamma_{(n)}$ is illustrated in Table 7.1 (compare [De Baets 1998a, De Baets & Mesiar 1999]).

n	Discrete t-norms	n	Discrete t-norms	n	Discrete t-norms
1	1	5	94	9	86.417
2	2	6	451	10	590.489
3	6	7	2.386	11	4.446.029
4	22	8	13.775	12	37.869.449

Table 7.1. The number of discrete t-norms on $\Gamma_{(n)}$

Also ordinal sums of finitely many discrete t-norms can be introduced, and they are again discrete t-norms. They are obtained by induction from the following result.

7.24 PROPOSITION

Let D_1 and D_2 be discrete t-norms on $\Gamma_{(m)}$ and $\Gamma_{(n)}$, respectively. Then the function $D : \Gamma_{(m+n)}^2 \longrightarrow \Gamma_{(m+n)}$ defined by

$$D\left(\tfrac{i}{m+n}, \tfrac{j}{m+n}\right) = \begin{cases} \frac{m}{m+n} \cdot D_1\left(\frac{i}{m}, \frac{j}{m}\right) & \text{if } (i,j) \in \{0,1,\ldots,m\}^2, \\ \frac{m}{m+n} + \frac{n}{m+n} \cdot D_2\left(\frac{i}{n}, \frac{j}{n}\right) & \\ \qquad\qquad \text{if } (i,j) \in \{m, m+1, \ldots, m+n\}^2, & \\ \frac{1}{m+n} \min(i,j) & \text{otherwise}, \end{cases}$$

is a discrete t-norm on $\Gamma_{(m+n)}$, and it is called the ordinal sum of D_1 and D_2.

Proof: This is an immediate consequence of Theorem 3.42. □

The proper equivalent to the continuity of t-norms on $[0,1]$ in the framework of discrete t-norms is their divisibility as given in Definition 2.49 [Birkhoff 1973, Höhle 1995] (in [Godo & Sierra 1988] divisible discrete t-norms were called smooth). The divisibility of discrete t-norms can be characterized by means of some Lipschitz property (a fact which is not true in the context of continuous t-norms).

7.25 PROPOSITION

Given $n \in \mathbb{N}$, then a discrete t-norm $D : \Gamma_{(n)}^2 \longrightarrow \Gamma_{(n)}$ is divisible if and only if it satisfies the Lipschitz property with constant 1, i.e., if for all integers $i, j, k, l \in \{0, 1, \ldots, n\}$

$$\left|D\left(\tfrac{i}{n}, \tfrac{j}{n}\right) - D\left(\tfrac{k}{n}, \tfrac{l}{n}\right)\right| \le \frac{1}{n}(|i-k| + |j-l|). \tag{7.10}$$

Proof: If D is a divisible discrete t-norm then for all $i, j \in \{0, 1, \ldots, n\}$ with $i \geq 1$ the monotonicity of D yields $D(\frac{i-1}{n}, \frac{j}{n}) \leq D(\frac{i}{n}, \frac{j}{n})$, which, together with the divisibility of D, implies $\left|D(\frac{i-1}{n}, \frac{j}{n}) - D(\frac{i}{n}, \frac{j}{n})\right| \leq \frac{1}{n}$. The validity of (7.10) for all $i, j, k, l \in \{0, 1, \ldots, n\}$ then follows by induction.

Conversely, fix $i, j \in \{0, 1, \ldots, n\}$ with $j \leq i$, and put $k_l = n \cdot D(\frac{i}{n}, \frac{l}{n})$ for each $l \in \{0, 1, \ldots, n\}$. Then we have $0 = k_0 \leq k_1 \leq \cdots \leq k_n = i$ and, because of (7.10), $k_l - k_{l-1} \leq 1$, implying that there must exist an index $k \in \{0, 1, \ldots, n\}$ with $D(\frac{i}{n}, \frac{k}{n}) = \frac{j}{n}$. □

The following representation theorem of divisible discrete t-norms was first proved in [Mayor & Torrens 1993].

7.26 THEOREM

Let $n \in \mathbb{N}$ and $D : \Gamma_{(n)}^2 \longrightarrow \Gamma_{(n)}$ be a divisible discrete t-norm. Then we have:

(i) If D is Archimedean then $D = D_{\mathbf{L}}^{(n)}$.

(ii) If D is not Archimedean then it can be represented as an ordinal sum, where all summands are Archimedean discrete t-norms.

Proof: For $n = 1$ assertion (i) holds because of Example 7.23(i). If $n \geq 2$ and if D is Archimedean, then $\gamma_0 = \frac{n-1}{n}$ is a nilpotent element of D, and, consequently, the sequence $\left((\gamma_0)_D^{(k)}\right)_{k\in\mathbb{N}}$ is strictly decreasing until it reaches the minimal value 0. Because of Proposition 7.25 we obtain

$$\begin{aligned}
D(1, \gamma_0) &= \gamma_0 &&= (\gamma_0)_D^{(1)}, \\
D(\tfrac{n-1}{n}, \gamma_0) &= \tfrac{n-2}{n} &&= (\gamma_0)_D^{(2)}, \\
D(\tfrac{n-2}{n}, \gamma_0) &= \tfrac{n-3}{n} &&= (\gamma_0)_D^{(3)}, \\
&\ \vdots && \ \vdots \qquad \vdots \\
D(\tfrac{2}{n}, \gamma_0) &= \tfrac{1}{n} &&= (\gamma_0)_D^{(n-1)}, \\
D(\tfrac{1}{n}, \gamma_0) &= 0 &&= (\gamma_0)_D^{(n)},
\end{aligned}$$

i.e., for each $i \in \{0, 1, \ldots, n\}$ we have $(\gamma_0)_D^{(n-i)} = \frac{i}{n}$. Then, because of the associativity of D, we get for all $i, j \in \{0, 1, \ldots, n\}$

$$\begin{aligned}
D(\tfrac{i}{n}, \tfrac{j}{n}) &= D\big((\gamma_0)_D^{(n-i)}, (\gamma_0)_D^{(n-j)}\big) \\
&= (\gamma_0)_D^{(2n-i-j)} \\
&= (\gamma_0)_D^{(\min(2n-i-j,n))} \\
&= \tfrac{1}{n} \max(i + j - n, 0),
\end{aligned}$$

showing that indeed $D = D_{\mathbf{L}}^{(n)}$.

In order to prove (ii), assume that D is not Archimedean, i.e., that it has a non-trivial idempotent element $\frac{k_0}{n}$ with $k_0 \in \{1, 2, \ldots, n-1\}$. Then the monotonicity of D implies that $D(\frac{j}{n}, \frac{k_0}{n}) = \frac{k_0}{n}$ for each $j \in \{k_0, k_0+1, \ldots, n\}$. For $j \in \{0, 1, \ldots, k_0\}$ Proposition 7.25 infers

$$\frac{k_0}{n} - D(\tfrac{j}{n}, \tfrac{k_0}{n}) \le \frac{k_0}{n} - \frac{j}{n},$$

i.e., together with $D(\frac{j}{n}, \frac{k_0}{n}) \le \frac{j}{n}$ we get $D(\frac{j}{n}, \frac{k_0}{n}) = \frac{j}{n}$. This means that for all $j \in \{0, 1, \ldots, n\}$ we have $D(\frac{j}{n}, \frac{k_0}{n}) = \min(\frac{j}{n}, \frac{k_0}{n})$. It is easy to check that the operations $D_1 : \varGamma^2_{(k_0)} \longrightarrow \varGamma_{(k_0)}$ and $D_2 : \varGamma^2_{(n-k_0)} \longrightarrow \varGamma_{(n-k_0)}$ given by

$$D_1(\tfrac{i}{k_0}, \tfrac{j}{k_0}) = \tfrac{n}{k_0} D(\tfrac{i}{n}, \tfrac{j}{n}),$$
$$D_2(\tfrac{i}{n-k_0}, \tfrac{j}{n-k_0}) = \tfrac{n}{n-k_0} D(\tfrac{k_0+i}{n}, \tfrac{k_0+j}{n}),$$

respectively, are discrete t-norms and that D is just the ordinal sum of D_1 and D_2. Continuing this process until no summand possesses a non-trivial idempotent element, D can be written as an ordinal sum of Archimedean discrete t-norms. □

A direct consequence of Theorem 7.26 is that each divisible discrete t-norm is uniquely determined by its idempotent elements. Consequently, on $\varGamma_{(n)}$ there are exactly 2^{n-1} different divisible discrete t-norms.

Given a t-norm T (which acts on $[0,1]$), it is quite natural to ask under which conditions the restriction to $\varGamma_{(n)}$ leads to a submonoid of $([0,1], T, \le)$ or, equivalently, under which conditions $T|_{\varGamma^2_{(n)}}$ is a discrete t-norm.

First we list several assertions which can be verified in a straightforward manner.

7.27 Remark

(i) If T is a t-norm and $n \in \mathbb{N}$ then $T|_{\varGamma^2_{(n)}}$ is a discrete t-norm if and only if $\varGamma_{(n)}$ is closed under T, i.e., if for all $i, j \in \{0, 1, \ldots, n\}$

$$T(\tfrac{i}{n}, \tfrac{j}{n}) \in \{0, \tfrac{1}{n}, \ldots, \tfrac{1}{n}\min(i,j)\}.$$

(ii) For each t-norm T the restriction $T|_{\varGamma^2_{(1)}}$ equals the unique discrete t-norm on $\varGamma_{(1)} = \{0, 1\}$ (compare Example 7.23(i)).

(iii) The restrictions of the basic t-norms $T_{\mathbf{M}}$, $T_{\mathbf{L}}$ and $T_{\mathbf{D}}$, for each $n \in \mathbb{N}$, are given by $T_{\mathbf{M}}|_{\varGamma^2_{(n)}} = D^{(n)}_{\mathbf{M}}$, $T_{\mathbf{L}}|_{\varGamma^2_{(n)}} = D^{(n)}_{\mathbf{L}}$, and $T_{\mathbf{D}}|_{\varGamma^2_{(n)}} = D^{(n)}_{\mathbf{D}}$.

(iv) If T is a t-norm and $T|_{\varGamma^2_{(n)}}$ a discrete t-norm then an element of $\varGamma_{(n)}$ is an idempotent or nilpotent element of T if and only if it is an idempotent or nilpotent element, respectively, of $T|_{\varGamma^2_{(n)}}$.

(v) If T is an Archimedean t-norm and if $T|_{\varGamma^2_{(n)}}$ is a discrete t-norm then also $T|_{\varGamma^2_{(n)}}$ is Archimedean.

7.28 Example

(i) The class of t-norms T_A, where $A \subseteq \,]0,1[^2$ satisfies conditions (i) and (ii) in Proposition 3.63, has the property $T(x,y) \in \{0, \min(x,y)\}$ for all $(x,y) \in [0,1]^2$ (see Remark 3.64(i)). Therefore, for each $n \in \mathbb{N}$ the restriction $T|_{\varGamma^2_{(n)}}$ is a discrete t-norm.

(ii) Another class of t-norms with the property that each restriction to $\Gamma^2_{(n)}$ is a discrete t-norm is the family of Jenei t-norms $((T_{\mathbf{L}})^{\mathbf{J}}_{\lambda})_{\lambda\in[0,0.5]}$ (based on the Łukasiewicz t-norm $T_{\mathbf{L}}$, see Example 3.69).

7.29 PROPOSITION
Let T be a t-norm and $t : [0,1] \longrightarrow [0,\infty]$ an additive generator of T.

(i) If $t(0) = \infty$, i.e., T is strictly monotone, then only the restriction $T|_{\Gamma^2_{(1)}}$ is a discrete t-norm.

(ii) If $t(0) < \infty$ then for $n \in \mathbb{N}$ the restriction $T|_{\Gamma^2_{(n)}}$ is a discrete t-norm if and only if for all $i, j \in \{0, 1, \ldots, n\}$ we have

$$t(\tfrac{i}{n}) + t(\tfrac{j}{n}) \in \mathrm{Ran}(t|_{\Gamma_{(n)}}) \cup [t(0), \infty].$$

Proof: Let $t(0) = \infty$ and $n \geq 2$. Then the sequence $\left((\frac{1}{n})^{(k)}_T\right)_{k\in\mathbb{N}}$ is a strictly decreasing sequence, so the finite chain $\Gamma_{(n)}$ cannot be closed under T.

In order to show (ii) it is equivalent to observe that the closedness of $\Gamma_{(n)}$ under T is equivalent to the fact that, for all $i, j \in \{0, 1, \ldots, n\}$, there is either an element $k \in \{0, 1, \ldots, n\}$ with $t(\frac{i}{n}) + t(\frac{j}{n}) = t(\frac{k}{n})$ (in which case we get $T(\frac{i}{n}, \frac{j}{n}) = \frac{k}{n}$), or we have $t(\frac{i}{n}) + t(\frac{j}{n}) > t(0)$ (in which case we necessarily get $T(\frac{i}{n}, \frac{j}{n}) = 0$). □

Now let us consider the inverse problem: given a discrete t-norm D on $\Gamma_{(n)}$, is there a t-norm T which extends D, i.e., such that $T|_{\Gamma^2_{(n)}} = D$? Recall that for $x \in \mathbb{R}$ the term $\lfloor x \rfloor$ denotes the floor of x, i.e., the largest integer which is smaller than or equal to x.

7.30 PROPOSITION
Let $n \in \mathbb{N}$ and let D be a discrete t-norm on $\Gamma_{(n)}$. Then the following function $T : [0,1]^2 \longrightarrow [0,1]$ is a t-norm which satisfies $T|_{\Gamma^2} = D$:

$$T(x,y) = \begin{cases} D\left(\frac{\lfloor nx \rfloor}{n}, \frac{\lfloor ny \rfloor}{n}\right) & \text{if } (x,y) \in [0,1[^2, \\ \min(x,y) & \text{otherwise.} \end{cases}$$

Proof: Obviously, T satisfies (T1), (T3) and (T4), and we have $T|_{\Gamma^2_{(n)}} = D$. For the associativity, the only non-trivial situation is if $x, y, z \in [0,1[$. But in this case the associativity of T follows from

$$\frac{1}{n}\left\lfloor nD\left(\frac{\lfloor nx \rfloor}{n}, \frac{\lfloor ny \rfloor}{n}\right)\right\rfloor = D\left(\frac{\lfloor nx \rfloor}{n}, \frac{\lfloor ny \rfloor}{n}\right)$$

and the associativity of D. □

The extension of a discrete t-norm given in Proposition 7.30 is always right-continuous, but never continuous. In particular, such an extension of the conjunction in Boolean (two-valued) logic is just the drastic product $T_{\mathbf{D}}$. Moreover, these extensions are just the t-norms $T^{[n]}$ considered in Lemma 3.73 and Example 3.74.

In general, a continuous t-norm which is an extension of a given discrete t-norm does not necessarily exist.

7.31 EXAMPLE

Consider the function $D : \mathit{\Gamma}^2_{(4)} \longrightarrow \mathit{\Gamma}_{(4)}$ given by

$$D(\tfrac{i}{4}, \tfrac{j}{4}) = \begin{cases} \frac{1}{4} & \text{if } (i,j) \in \{2,3\}^2, \\ \min(x,y) & \text{if } \max(i,j) = 4, \\ 0 & \text{otherwise.} \end{cases}$$

Then D is an Archimedean discrete t-norm. A continuous extension T of D would have to satisfy $T(\frac{2}{4}, \frac{2}{4}) = T(\frac{2}{4}, \frac{3}{4}) = \frac{1}{4}$. However, this is possible only if there is a number $c \in \left]\frac{3}{4}, 1\right]$ and a nilpotent t-norm T^* such that T is an ordinal sum and $\langle \frac{1}{4}, c, T^* \rangle$ is a summand thereof, implying that $\frac{1}{4}$ is an idempotent element of T and, consequently, of D. Since this is obviously not the case, there is no continuous t-norm T which is an extension of D.

However, for a divisible discrete t-norm D there is always a continuous t-norm which is an extension of D.

7.32 THEOREM

For each divisible discrete t-norm D on $\mathit{\Gamma}_{(n)}$ there exists a continuous t-norm T such that $T|_{\mathit{\Gamma}^2_{(n)}} = D$.

Proof: If D is an Archimedean discrete divisible t-norm on $\mathit{\Gamma}_{(n)}$ it is sufficient to consider a continuous, strictly decreasing function $t : [0,1] \longrightarrow [0,\infty]$ which satisfies $t(\frac{i}{n}) = n - i$. Obviously, t is an additive generator of some nilpotent t-norm T, and we have $T|_{\mathit{\Gamma}^2_{(n)}} = D$.

If D is not Archimedean, it can be written as an ordinal sum of Archimedean discrete t-norms according to Theorem 7.26(ii). Applying the construction of the additive generator above to each Archimedean summand of D, then the ordinal sum of the corresponding t-norms is a continuous t-norm which extends the discrete t-norm D. □

A consequence of this result is that, for each divisible discrete t-norm D on $\mathit{\Gamma}_{(n)}$, there exists a continuous t-norm T which is an extension of D and which can be written as an ordinal sum, where the t-norm in each summand equals the Łukasiewicz t-norm $T_{\mathbf{L}}$.

8. Convergence of t-norms

> An exact description of any real situation is virtually impossible. This is a fact we have had to accept and adjust to. As a result, one of the major problems in description (essential to communication, decision making, and, in a broader sense, to any human activity) is to reduce the necessary imprecision to a level of relative unimportance. We must balance the needs for exactness and simplicity, and reduce complexity without oversimplification in order to match the level of detail at each step with the problem we face.
>
> RICHARD E. BELLMAN

When we introduced several important families of t-norms in Chapter 4, we mentioned already (without proof) that all these families are continuous with respect to the parameter, i.e., that we have pointwise convergence of the t-norms if the corresponding parameters converge (some of these statements are trivial, some of them follow directly from [Dombi 1982, 1986]).

When studying the problem whether an arbitrary (usually continuous) t-norm can be approximated with arbitrary precision by a strict or nilpotent t-norm, it first turns out that the pointwise convergence of a sequence of t-norms (which may be non-continuous) to a continuous t-norm is sufficient for the uniform convergence. Then, proving a series of lemmas, we give an affirmative answer to our initial problem (compare also [Nguyen *et al.* 1998]), first in the case of continuous Archimedean t-norms, and the also for arbitrary continuous t-norms.

In the subsequent Section 8.2 we investigate the relationship between the (pointwise) convergence of continuous Archimedean t-norms and the convergence of their respective multiplicative generators. The main result is that we have convergence of the t-norms if and only if the corresponding (uniquely determined) multiplicative generators, which pass through the same point of the open unit square, converge [Jenei 1998, Mesiar 1998b].

We also give necessary and sufficient conditions for the convergence to the extreme t-norms $T_{\mathbf{M}}$ and $T_{\mathbf{D}}$.

As a by-product of the results in this chapter, we obtain proofs for the continuity (with respect to the underlying parameter) of most of the families of t-norms considered in Chapter 4.

8.1 Approximation of continuous t-norms

We shall now investigate under which conditions a (usually continuous) t-norm T can be approximated with arbitrary precision by an appropriate strict or nilpotent t-norm. Of course this is equivalent to asking whether we can find a sequence of strict or nilpotent t-norms which converges uniformly to T. In order to do so, we shall work with the following notion.

8.1 DEFINITION

For each $\varepsilon > 0$ we say that two t-norms T_1 and T_2 are *ε-close* to each other if for all points $(x, y) \in [0,1]^2$ we have $|T_1(x,y) - T_2(x,y)| \leq \varepsilon$.

8.2 REMARK

From the monotonicity (T3) of t-norms we can derive immediately:

(i) The t-norms T and $T_{\mathbf{M}}$ are ε-close if and only if, for each $x \in [0,1]$, we have $T(x,x) > x - \varepsilon$.

(ii) The t-norms T and $T_{\mathbf{D}}$ are ε-close if and only if, for each $x \in [0,1[$, we have $T(x,x) < \varepsilon$.

Observe that this means that ε-closeness of a t-norm T to $T_{\mathbf{M}}$ or $T_{\mathbf{D}}$, respectively, is equivalent to the ε-closeness of the diagonal sections of T and $T_{\mathbf{M}}$ or $T_{\mathbf{D}}$, respectively.

It is well-known from calculus that, for a sequence of continuous real functions defined on a compact set, the pointwise convergence to a continuous function is equivalent to the uniform convergence of these functions. However, in the case of t-norms the pointwise convergence of a sequence $(T_n)_{n\in\mathbb{N}}$ to a continuous t-norm T is sufficient for the uniform convergence of $(T_n)_{n\in\mathbb{N}}$, regardless whether the t-norms T_n are continuous or not [Jenei & Fodor 1998].

8.3 PROPOSITION

If T is a continuous t-norm and if $(T_n)_{n\in\mathbb{N}}$ is a sequence of t-norms which converges pointwise to T, then $(T_n)_{n\in\mathbb{N}}$ converges uniformly to T, i.e., for each $\varepsilon > 0$ there exists an $N_\varepsilon \in \mathbb{N}$ such that for all $n \geq N_\varepsilon$ the t-norms T and T_n are ε-close.

Proof: Choose an $\varepsilon > 0$ and $(x_0, y_0) \in [0,1]^2$. Because of the continuity of T, there exists a $\delta_{x_0y_0} > 0$ (which also depends on ε) such that for all $(x, y) \in [0,1]^2 \cap ([x_0 - \delta_{x_0y_0}, x_0 + \delta_{x_0y_0}] \times [y_0 - \delta_{x_0y_0}, y_0 + \delta_{x_0y_0}])$

$$|T(x,y) - T(x_0,y_0)| < \tfrac{\varepsilon}{3}.$$

Writing $\underline{x_0} = \max(x_0 - \delta_{x_0y_0}, 0)$, $\underline{y_0} = \max(y_0 - \delta_{x_0y_0}, 0)$, $\overline{x_0} = \min(x_0 + \delta_{x_0y_0}, 1)$, and $\overline{y_0} = \min(y_0 + \delta_{x_0y_0}, 1)$, we immediately get

$$T(\overline{x_0}, \overline{y_0}) - T(\underline{x_0}, \underline{y_0}) < \tfrac{2\varepsilon}{3}.$$

Next, from the pointwise convergence of $(T_n)_{n\in\mathbb{N}}$ to T it follows that there is an $N \in \mathbb{N}$ such that for all $n \geq N$

$$\begin{aligned}|T_n(\underline{x_0}, \underline{y_0}) - T(\underline{x_0}, \underline{y_0})| &< \tfrac{\varepsilon}{3},\\ |T_n(\overline{x_0}, \overline{y_0}) - T(\overline{x_0}, \overline{y_0})| &< \tfrac{\varepsilon}{3},\end{aligned}$$

and for all $(x, y) \in [0,1]^2 \cap ([x_0 - \delta_{x_0y_0}, x_0 + \delta_{x_0y_0}] \times [y_0 - \delta_{x_0y_0}, y_0 + \delta_{x_0y_0}])$ with $T_n(x, y) < T(x, y)$ we have

$$\begin{aligned}T(x,y) - T_n(x,y) &\leq T(\overline{x_0}, \overline{y_0}) - T_n(\underline{x_0}, \underline{y_0})\\ &\leq T(\overline{x_0}, \overline{y_0}) - T(\underline{x_0}, \underline{y_0}) + |T(\underline{x_0}, \underline{y_0}) - T_n(\underline{x_0}, \underline{y_0})|\\ &< \varepsilon.\end{aligned}$$

For all $(x, y) \in [0,1]^2 \cap ([x_0 - \delta_{x_0y_0}, x_0 + \delta_{x_0y_0}] \times [y_0 - \delta_{x_0y_0}, y_0 + \delta_{x_0y_0}])$ with $T_n(x, y) \geq T(x, y)$ we get, in complete analogy,

$$\begin{aligned}T_n(x,y) - T(x,y) &\leq T_n(\overline{x_0}, \overline{y_0}) - T(\underline{x_0}, \underline{y_0})\\ &\leq |T_n(\overline{x_0}, \overline{y_0}) - T(\overline{x_0}, \overline{y_0})| + T(\overline{x_0}, \overline{y_0}) - T(\underline{x_0}, \underline{y_0})\\ &< \varepsilon.\end{aligned}$$

Since, as a compact set, the unit square $[0,1]^2$ can be covered by a finite number of rectangles of the form $[0,1]^2 \cap (]x - \frac{\delta_{xy}}{2}, x + \frac{\delta_{xy}}{2}[\times]y - \frac{\delta_{xy}}{2}, y + \frac{\delta_{xy}}{2}[)$, this implies that there exists an $N_\varepsilon \in \mathbb{N}$ such that for all $n \geq N_\varepsilon$ the t-norms T and T_n are ε-close. □

The following result can be obtained as an immediate consequence of Proposition 8.3.

8.4 COROLLARY

Let $(T_n)_{n\in\mathbb{N}}$ be a sequence of t-norms which converges to a continuous t-norm T, and let $(x_n)_{n\in\mathbb{N}}$ and $(y_n)_{n\in\mathbb{N}}$ be two convergent sequences of elements of $[0,1]$. Then we have $\lim\limits_{n\to\infty} T_n(x_n, y_n) = T\big(\lim\limits_{n\to\infty} x_n, \lim\limits_{n\to\infty} y_n\big)$.

Next we can show that, starting from an (almost) arbitrary continuous Archimedean or nilpotent t-norm T, it is possible to approximate each of the three continuous basic t-norms, i.e., the minimum $T_{\mathbf{M}}$, the product $T_{\mathbf{P}}$, and the Łukasiewicz t-norm $T_{\mathbf{L}}$, in a uniform way. The (non-continuous) drastic product $T_{\mathbf{D}}$, however, can only be obtained as pointwise limit of continuous Archimedean t-norms (it is well-known from calculus that the uniform limit of a sequence of continuous functions is necessarily continuous too).

For this purpose, recall Example 3.32: if T is a continuous Archimedean t-norm with additive generator $t : [0,1] \longrightarrow [0,\infty]$, then for each $\lambda \in\,]0,\infty[$ the function $t^\lambda : [0,1] \longrightarrow [0,\infty]$ defined by $t^\lambda(x) = (t(x))^\lambda$ is an additive generator of a continuous Archimedean t-norm which was denoted there by $T^{(\lambda)}$. Similarly, the function $t_{(\lambda)} : [0,1] \longrightarrow [0,\infty]$ defined by $t_{(\lambda)}(x) = t(x^\lambda)$ is an additive generator of a continuous Archimedean t-norm denoted $T_{(\lambda)}$. Assertions (i) and (ii) of the following result go back to [Dombi 1982, 1986].

8.5 PROPOSITION

Let T be a continuous Archimedean t-norm and let $t : [0,1] \longrightarrow [0,\infty]$ be an additive generator of T.

(i) For each $\varepsilon > 0$ there exists a $\lambda_0 \in \,]0,\infty[$ such that $T^{(\lambda_0)}$ and $T_{\mathbf{M}}$ are ε-close.

(ii) $\lim\limits_{\lambda \searrow 0} T^{(\lambda)} = T_{\mathbf{D}}$.

(iii) If $t'(1^-) \in \,]-\infty,0[$ then for each $\varepsilon > 0$ there is a $\lambda_0 \in \,]0,\infty[$ such that $T_{(\lambda_0)}$ and $T_{\mathbf{P}}$ are ε-close.

(iv) If T is nilpotent then $\lim\limits_{\lambda \to \infty} T_{(\lambda)} = T_{\mathbf{D}}$.

(v) If T is nilpotent and $t'(0^+) \in \,]-\infty,0[$ and if for each $\lambda \in \,]0,1]$ the function $t_{[\lambda]} : [0,1] \longrightarrow [0,\infty]$ is given by $t_{[\lambda]}(x) = t(\lambda x) - t(\lambda)$, then, for each $\lambda \in \,]0,1]$, $t_{[\lambda]}$ is the additive generator of some nilpotent t-norm $T_{[\lambda]}$, and for each $\varepsilon > 0$ there is a λ_0 such that $T_{[\lambda_0]}$ and $T_{\mathbf{L}}$ are ε-close.

Proof: Choose a pair $(x,y) \in \,]0,1[^2$ with $x \leq y$. Then the strict monotonicity of t implies $0 < \frac{t(y)}{t(x)} \leq 1$, and for all $\lambda \in \,]0,\infty[$ we get

$$T^{(\lambda)}(x,y) = (t^\lambda)^{(-1)}\left(t^\lambda(x) + t^\lambda(y)\right) = t^{(-1)}\Big(t(x) \cdot \big(1 + \big(\tfrac{t(y)}{t(x)}\big)^\lambda\big)^{\frac{1}{\lambda}}\Big).$$

From $\lim\limits_{\lambda \to \infty} (1 + c^\lambda)^{\frac{1}{\lambda}} = 1$ and $\lim\limits_{\lambda \searrow 0} (1 + c^\lambda)^{\frac{1}{\lambda}} = \infty$ for all $c \in \,]0,1]$ we deduce

$$\lim_{\lambda \to \infty} T^{(\lambda)}(x,y) = t^{(-1)}(t(x) \cdot 1) = x = T_{\mathbf{M}}(x,y),$$
$$\lim_{\lambda \to 0^+} T^{(\lambda)}(x,y) = t^{(-1)}(t(x) \cdot \infty) = 0 = T_{\mathbf{D}}(x,y).$$

From the commutativity (T1) and from the fact that all t-norms coincide on the boundary of the unit square $[0,1]^2$ we obtain pointwise convergence to $T_{\mathbf{M}}$ and $T_{\mathbf{D}}$, respectively, and Proposition 8.3 tells us that the convergence in (i) is uniform.

Next, assume $t'(1^-) = \lim_{z \nearrow 1} \frac{t(z)}{z-1} \in \,]-\infty,0[$. Observe that for each pair $(x,y) \in \,]0,1[^2$ there exists a $\lambda_1 \in \,]0,\infty[$ such that $t(x^{\lambda_1}) + t(y^{\lambda_1}) < t(0)$. Consequently, we get

$$\begin{aligned}
\lim_{\lambda \searrow 0} T_{(\lambda)}(x,y) &= \lim_{\lambda \searrow 0} \left(t^{(-1)}\left(t(x^\lambda) + t(y^\lambda)\right)\right)^{\frac{1}{\lambda}} \\
&= \lim_{\lambda \searrow 0} e^{\frac{\log(t^{(-1)}(t(x^\lambda)+t(y^\lambda)))}{\lambda}} \\
&= \lim_{\lambda \searrow 0} e^{\frac{\log(x^\lambda+y^\lambda-1)}{\lambda}} \\
&= xy,
\end{aligned}$$

and assertion (iii) follows applying Proposition 8.3.

In order to show (iv), choose $(x,y) \in \,]0,1[^2$. Because of $t(0) < \infty$ there exists a $\lambda_0 \in \,]0,\infty[$ such that $t(x^{\lambda_0}) + t(y^{\lambda_0}) > t(0)$, i.e., $T_{(\lambda_0)}(x,y) = 0$.

Finally, it is obvious that each $t_{[\lambda]}$ is a continuous, strictly decreasing function with $t_{[\lambda]}(0) < \infty$ and $t_{[\lambda]}(1) = 0$ and, therefore, the additive generator of some nilpotent t-norm $T_{[\lambda]}$. For each $(x,y) \in \,]0,1[^2$ we have

$$T_{[\lambda]}(x,y) = \frac{t^{-1}(\min(t(0), t(\lambda x) + t(\lambda y) - t(\lambda)))}{\lambda}.$$

If $x + y > 1$ then direct computation yields

$$\lim_{\lambda \searrow 0} T_{[\lambda]}(x,y) = \lim_{\lambda \searrow 0} \frac{t^{-1}(\lambda(x+y-1)t'(0^+) + t(0))}{\lambda} = x + y - 1.$$

If $x + y < 1$ then for a sufficiently small $\lambda \in \,]0,1]$ we get

$$t(\lambda x) + t(\lambda y) - t(\lambda) > t(0),$$

i.e., $T_{[\lambda]}(x,y) = 0$. Together with the monotonicity of the t-norms $T_{[\lambda]}$ this means $\lim_{\lambda \searrow 0} T_{[\lambda]} = T_{\mathbf{L}}$. By virtue of Proposition 8.3, the proof is complete. □

8.6 Remark

(i) If we start from a strict t-norm T, then for each $\lambda \in \,]0,\infty[$ also $T^{(\lambda)}$ is a strict t-norm, Proposition 8.5(i) tells us that there exist strict t-norms which are ε-close to $T_{\mathbf{M}}$. Similarly, there are nilpotent t-norms which are ε-close to $T_{\mathbf{M}}$.

(ii) As a consequence of Proposition 8.5, each of the following families of t-norms is continuous with respect to the parameter λ (compare Example 3.32(i)): the family $(T_\lambda^{\mathbf{SS}})_{\lambda \in [-\infty,\infty]}$ of Schweizer-Sklar t-norms (see Section 4.2), the family $(T_\lambda^{\mathbf{Y}})_{\lambda \in [0,\infty]}$ of Yager t-norms (see Section 4.5), the family $(T_\lambda^{\mathbf{AA}})_{\lambda \in [0,\infty]}$ of Aczél-Alsina t-norms (see Section 4.8), and the family $(T_\lambda^{\mathbf{D}})_{\lambda \in [0,\infty]}$ of Dombi t-norms (see Section 4.6).

The main result of this section is that each continuous t-norm can be approximated in a uniform way by a strict t-norm (see [Jenei & Fodor 1998, Mesiar 1998a, Nguyen *et al.* 1998]). This will be prepared by a sequence of lemmas.

In the first step we show that two continuous Archimedean t-norms, whose additive generators coincide on a the closed subinterval $[\varepsilon, 1]$ of $[0,1]$, are necessarily ε-close.

8.7 Lemma

Let T_1 and T_2 be continuous Archimedean t-norms with additive generators $t_1, t_2 : [0,1] \longrightarrow [0,\infty]$, respectively. If for some $\varepsilon \in \,]0,1[$ the restrictions $t_1|_{[\varepsilon,1]}$ and $t_2|_{[\varepsilon,1]}$ coincide, then T_1 and T_2 are ε-close.

Proof: Fix an arbitrary $\varepsilon > 0$. If $(x, y) \in [0,1]^2$ with $T_1(x,y) \geq \varepsilon$, then necessarily $x \geq \varepsilon$, $y \geq \varepsilon$ and

$$\begin{aligned} T_1(x,y) &= t_1^{(-1)}(t_1(x) + t_1(y)) \\ &= t_1^{(-1)}(t_2(x) + t_2(y)) \\ &= \sup\{z \in [0,1] \mid t_1(z) > t_2(x) + t_2(y)\} \\ &= \sup\{z \in [\varepsilon,1] \mid t_1(z) > t_2(x) + t_2(y)\} \\ &= \sup\{z \in [\varepsilon,1] \mid t_2(z) > t_2(x) + t_2(y)\} \\ &= t_2^{(-1)}(t_2(x) + t_2(y)) \\ &= T_2(x,y). \end{aligned}$$

Similarly we obtain $T_1(x,y) < \varepsilon$ if and only if $T_2(x,y) < \varepsilon$. Consequently, T_1 and T_2 are ε-close. □

From this result we can derive that each strict t-norm is ε-close to an appropriate nilpotent t-norm, and vice versa.

8.8 LEMMA

(i) Let T be a strict t-norm. For each $\varepsilon > 0$ there is a nilpotent t-norm T_ε such that T and T_ε are ε-close.

(ii) Let T be a nilpotent t-norm. For each $\varepsilon > 0$ there is a strict t-norm T_ε such that T and T_ε are ε-close.

Proof: If $t : [0,1] \longrightarrow [0,\infty]$ is an additive generator of some strict t-norm T, i.e., $t(0) = \infty$, then for each $\varepsilon > 0$ the function $t_\varepsilon : [0,1] \longrightarrow [0,\infty]$ given by

$$t_\varepsilon(x) = \begin{cases} 2t(\varepsilon) - \frac{t(\varepsilon)}{\varepsilon}x & \text{if } x \in [0,\varepsilon[\,, \\ t(x) & \text{otherwise,} \end{cases}$$

is a continuous, strictly decreasing function with $t_\varepsilon(0) < \infty$. This means that t_ε is an additive generator of some nilpotent t-norm T_ε, and T and T_ε are ε-close because of Lemma 8.7.

The proof of (ii) is similar: if $t : [0,1] \longrightarrow [0,\infty]$ is an additive generator of some nilpotent t-norm T then we put

$$t_\varepsilon(x) = \begin{cases} \frac{t(\varepsilon)}{\log \varepsilon} \log x & \text{if } x \in [0,\varepsilon[\,, \\ t(x) & \text{otherwise,} \end{cases}$$

and the corresponding strict t-norm T_ε and T are ε-close. □

We already know from Theorem 5.11 that each continuous t-norm T can be written as an ordinal sum of continuous Archimedean t-norms. Now we construct an ordinal sum of continuous Archimedean t-norms with finitely many summands which is ε-close to T.

8.9 LEMMA

Let T be a continuous t-norm. For each $\varepsilon > 0$ there is a finite index set A with $\text{card}(A) < \frac{1}{\varepsilon} + 1$, *a family $(T_\alpha)_{\alpha \in A}$ of continuous Archimedean t-norms and a family $(]a_\alpha, e_\alpha[)_{\alpha \in A}$ of pairwise disjoint open subintervals of $[0, 1]$ such that T and the ordinal sum $(\langle a_\alpha, e_\alpha, T_\alpha \rangle)_{\alpha \in A}$ are ε-close.*

Proof: Because of Theorem 5.11 we may write $T = (\langle a_\alpha, e_\alpha, T_\alpha \rangle)_{\alpha \in B}$ where each summand T_α is continuous Archimedean. Fix $\varepsilon > 0$ and define the subset A of B by $A = \{\alpha \in B \mid e_\alpha - a_\alpha \geq \varepsilon\}$, which means that A is a finite set such that $\text{card}(A) < \frac{1}{\varepsilon} + 1$. Next we put

$$T^*_\alpha = \begin{cases} T_\alpha & \text{if } \alpha \in A, \\ T_\mathbf{M} & \text{if } \alpha \in B \setminus A. \end{cases}$$

Then the ordinal sums $(\langle a_\alpha, e_\alpha, T_\alpha \rangle)_{\alpha \in B}$ and $(\langle a_\alpha, e_\alpha, T^*_\alpha \rangle)_{\alpha \in B}$ are ε-close, since for all $\alpha \in B \setminus A$ and for all $(x, y) \in [a_\alpha, e_\alpha]$ we have $e_\alpha - a_\alpha < \varepsilon$ and

$$a_\alpha \leq T(x, y) \leq T_\mathbf{M}(x, y) \leq e_\alpha.$$

Because of Remark 3.47(i) and the construction of T^*_α, however, we have $(\langle a_\alpha, e_\alpha, T^*_\alpha \rangle)_{\alpha \in B} = (\langle a_\alpha, e_\alpha, T_\alpha \rangle)_{\alpha \in A}$, implying that T and $(\langle a_\alpha, e_\alpha, T_\alpha \rangle)_{\alpha \in A}$ are ε-close. □

Now we show that each continuous t-norm is ε-close to some ordinal sum of strict t-norms with finitely many summands.

8.10 LEMMA

Let T be a continuous t-norm. For each $\varepsilon > 0$ there exist a number $n \in \mathbb{N}$, strict t-norms $T_1, T_2, \ldots, T_n$ and numbers $b_0, b_1, \ldots, b_n \in [0, 1]$ with

$$0 = b_0 < b_1 < \cdots < b_n = 1$$

such that T and the ordinal sum $(\langle b_{i-1}, b_i, T_i \rangle)_{i \in \{1,2,\ldots,n\}}$ are ε-close.

Proof: First note that, if $(\langle a_\alpha, e_\alpha, T_\alpha \rangle)_{\alpha \in A}$ and $(\langle a_\alpha, e_\alpha, T^*_\alpha \rangle)_{\alpha \in A}$ are two ordinal sums such that, for each $\alpha \in A$, the summands T^*_α and T_α are ε-close, and if $\varepsilon^* = \sup\{(e_\alpha - a_\alpha)\varepsilon \mid \alpha \in A\}$, then the ordinal sums $(\langle a_\alpha, e_\alpha, T_\alpha \rangle)_{\alpha \in A}$ and $(\langle a_\alpha, e_\alpha, T^*_\alpha \rangle)_{\alpha \in A}$ are ε^*-close and, subsequently, ε-close.

Fixing a continuous t-norm T and $\varepsilon > 0$, Lemma 8.9 guarantees the existence of an ordinal sum $(\langle a_\alpha, e_\alpha, T_\alpha \rangle)_{\alpha \in A}$ with a finite index set A such that T and $(\langle a_\alpha, e_\alpha, T_\alpha \rangle)_{\alpha \in A}$ are $\frac{\varepsilon}{2}$-close. Then, putting the numbers a_α and e_α in their natural order and relabeling them, we obtain a finite chain, i.e., numbers $b_0, b_1, \ldots, b_n \in [0, 1]$ with $0 = b_0 < b_1 < \cdots < b_n = 1$. Defining $T^*_i = T_\alpha$ if $[b_{i-1}, b_i] = [a_\alpha, e_\alpha]$, and $T^*_i = T_\mathbf{M}$ otherwise, then Remark 3.47(i) yields

$$(\langle a_\alpha, e_\alpha, T_\alpha \rangle)_{\alpha \in A} = (\langle b_{i-1}, b_i, T^*_i \rangle)_{i \in \{1,2,\ldots,n\}}.$$

By construction and Theorem 2.18, each T_i^* is either strict or nilpotent or equal to $T_{\mathbf{M}}$. Now we replace each non-strict T_i^* by a strict t-norm T_i such that T_i and T_i^* are $\frac{\varepsilon}{2}$-close (this is possible because of Proposition 8.5(i) and Lemma 8.8(ii)), and we put $T_i = T_i^*$ whenever T_i^* is strict. Then $(\langle b_{i-1}, b_i, T_i \rangle)_{i \in \{1,2,\ldots,n\}}$ and $(\langle b_{i-1}, b_i, T_i^* \rangle)_{i \in \{1,2,\ldots,n\}}$ are $\frac{\varepsilon}{2}$-close, implying that T and $(\langle b_{i-1}, b_i, T_i \rangle)_{i \in \{1,2,\ldots,n\}}$ are ε-close. □

Finally, we prove that each ordinal sum of two strict t-norms is ε-close to some strict t-norm.

8.11 LEMMA

Let T_1 and T_2 be strict t-norms, $c \in \,]0,1[$ and denote the ordinal sum $(\langle 0, c, T_1 \rangle, \langle c, 1, T_2 \rangle)$ by T. For each $\varepsilon > 0$ there exists a strict t-norm T_ε such that T_ε and T are ε-close.

Proof: Let $t_1, t_2 : [0,1] \longrightarrow [0,\infty]$ be additive generators of the strict t-norms T_1 and T_2, respectively, and define the strictly decreasing bijections $h_1 : [0,c] \longrightarrow [0,\infty]$ and $h_2 : [c,1] \longrightarrow [0,\infty]$ by

$$h_1(x) = t_1\left(\tfrac{x}{c}\right) \qquad \text{and} \qquad h_2(x) = t_2\left(\tfrac{x-c}{1-c}\right).$$

Then Proposition 3.50 tells us that for all $(x,y) \in [0,1]^2$

$$T(x,y) = \begin{cases} h_1^{-1}(h_1(x) + h_1(y)) & \text{if } (x,y) \in [0,c]^2, \\ h_2^{-1}(h_2(x) + h_2(y)) & \text{if } (x,y) \in [c,1]^2, \\ \min(x,y) & \text{otherwise.} \end{cases}$$

Fix an $\varepsilon \in \,]0, \min(c, 1-c)[$, put

$$\begin{aligned} a_\varepsilon &= \inf\left\{ h_1\left(x - \tfrac{\varepsilon}{2}\right) - h_1(x) \mid x \in \left]\tfrac{\varepsilon}{2}, c\right] \right\}, \\ \delta &= \min\left(c - h_1^{-1}(a_\varepsilon), \tfrac{\varepsilon}{2}\right), \end{aligned}$$

and observe that for all $x \in [0,c]$ we have $0 \le x - T(x, c-\delta) \le \frac{\varepsilon}{2} < \varepsilon$. Now we define the strictly decreasing bijection $h_\varepsilon : [0,1] \longrightarrow [0,\infty]$, i.e., an additive generator of some strict t-norm T_ε, by

$$h_\varepsilon(x) = \begin{cases} h_1(x) & \text{if } x \in [0, c-\delta], \\ \frac{h_1(c-\delta)}{5}\left(1 + \left(\frac{x-c-\delta}{\delta}\right)^2\right) & \text{if } x \in \,]c-\delta, c+\delta[, \\ \frac{h_1(c-\delta)}{5 h_2(c+\delta)} h_2(x) & \text{otherwise.} \end{cases}$$

Fix a pair $(x_0, y_0) \in [0,1]^2$ and assume without loss of generality $x_0 \le y_0$. In order to complete the proof we have to show

$$|T_\varepsilon(x_0, y_0) - T(x_0, y_0)| \le \varepsilon. \tag{8.1}$$

This will be done by checking the subsequent six cases.

Case 1: $(x_0, y_0) \in [0, c-\delta]^2$.
Since h_ε coincides with h_1 on $[0, c-\delta]$ we have $T_\varepsilon(x_0, y_0) = T(x_0, y_0)$.

Case 2: $(x_0, y_0) \in [0, c-\delta] \times]c-\delta, 1]$.
Here we can derive (8.1) from the two inequalities

$$x_0 \geq T(x_0, y_0) \geq T(x_0, c-\delta) \geq T\big(x_0, h_1^{-1}(a_\varepsilon)\big) \geq h_1^{-1}\big(h_1(x_0 - \tfrac{\varepsilon}{2})\big) \geq x_0 - \tfrac{\varepsilon}{2},$$
$$x_0 \geq T_\varepsilon(x_0, y_0) \geq T_\varepsilon(x_0, c-\delta) \geq T(x_0, c-\delta) \geq x_0 - \tfrac{\varepsilon}{2}.$$

Case 3: $x_0 \in]c-\delta, c]$.
Here (8.1) is implied by $x_0 \geq T(x_0, y_0) \geq T(c-\delta, c-\delta) \geq x_0 - \varepsilon$ and

$$x_0 \geq T_\varepsilon(x_0, y_0) \geq T_\varepsilon(c-\delta, c-\delta) \geq T(c-\delta, c-\delta) \geq x_0 - \varepsilon.$$

Case 4: $x_0 \in]c, c+\delta]$.
Here (8.1) follows from $c + \frac{\varepsilon}{2} \geq c + \delta \geq x_0 \geq T(x_0, y_0) > c$ and

$$c + \tfrac{\varepsilon}{2} \geq x_0 \geq T_\varepsilon(x_0, y_0) > T_\varepsilon(c, c) = h_\varepsilon^{-1}\big(\tfrac{4}{5} h_1(c-\delta)\big) > c - \delta \geq c - \tfrac{\varepsilon}{2}.$$

Case 5: $x_0 \in]c+\delta, 1]$ and $T(x_0, y_0) < c + \delta$.
In this case we obtain (8.1) from $c + \frac{\varepsilon}{2} \geq c + \delta > T(x_0, y_0) > c$ and from $h_2(x_0) + h_2(y_0) > h_2(c+\delta)$, i.e., from $h_\varepsilon(x_0) + h_\varepsilon(y_0) > h_\varepsilon(c+\delta)$, which in turn implies

$$c + \tfrac{\varepsilon}{2} \geq c + \delta > T_\varepsilon(x_0, y_0) > T_\varepsilon(c+\delta, c+\delta) = h_\varepsilon^{-1}\big(\tfrac{2}{5} h_1(c-\delta)\big) = c.$$

Case 6: $x_0 \in]c+\delta, 1]$ and $T(x_0, y_0) \geq c + \delta$.
By construction of h_ε we obtain $T_\varepsilon(x_0, y_0) = T(x_0, y_0)$ in this case. □

Now it is not difficult to combine the assertions of Lemmas 8.10 and 8.11 and to prove that each continuous t-norm can be approximated uniformly by a strict t-norm.

8.12 THEOREM
Let T be a continuous t-norm. For each $\varepsilon > 0$ there exists a strict t-norm T_ε such that T and T_ε are ε-close.

Proof: Fix a continuous t-norm T and $\varepsilon > 0$. Because of Lemma 8.10 there exists an ordinal sum of strict t-norms

$$(\langle 0, b_1, T_1\rangle, \langle b_1, b_2, T_2\rangle, \ldots, \langle b_{n-1}, 1, T_n\rangle) \tag{8.2}$$

which is $\frac{\varepsilon}{2}$-close to T.

Picking the first two summands of (8.2) and applying Lemma 8.11 we can construct a strict t-norm $T^{(1)}$ which is $\frac{\varepsilon}{2n}$-close to the ordinal sum $(\langle 0, \frac{b_1}{b_2}, T_1\rangle, \langle \frac{b_1}{b_2}, 1, T_2\rangle)$. This implies that T is $\frac{(n+1)\varepsilon}{2n}$-close to the ordinal sum

$$(\langle 0, b_2, T^{(1)}\rangle, \langle b_2, b_3, T_3\rangle, \ldots, \langle b_{n-1}, 1, T_n\rangle). \tag{8.3}$$

Applying again Lemma 8.11 to the first two summands of (8.3) and repeating this procedure, after k steps we obtain a strict t-norm $T^{(k)}$ such that T is $\frac{(n+k)\varepsilon}{2n}$-close to the ordinal sum (which has $n-k$ summands)

$$(\langle 0, b_{k+1}, T^{(k)}\rangle, \langle b_{k+1}, b_{k+2}, T_{k+2}\rangle, \ldots, \langle b_{n-1}, 1, T_n\rangle).$$

Finally, after $n-1$ steps we reach a strict t-norm $T_\varepsilon = T^{(n-1)}$ such that T and T_ε are ε-close. □

As an immediate consequence of Theorem 8.12 and Lemma 8.8(i) there is a completely analogous result for the approximation by nilpotent t-norms.

8.13 COROLLARY
Let T be a continuous t-norm. For each $\varepsilon > 0$ there exists a nilpotent t-norm T_ε such that T and T_ε are ε-close.

8.2 Convergence of t-norms and their generators

For continuous Archimedean t-norms, all the information is contained in their additive generators (because of Theorem 5.1) or, equivalently, in their multiplicative generators (see Corollary 5.4). It therefore is to be expected that the convergence of continuous Archimedean t-norms is closely related to the convergence of their generators.

We shall use multiplicative generators throughout this section, exploiting the fact that they can be considered as uniformly continuous functions from $[0,1]$ into $[0,1]$. However, since convergence of multiplicative generators is equivalent to the convergence of additive generators, all results of this section can be easily restated for additive generators.

In Corollary 5.4 we also have seen that, given a continuous Archimedean t-norm T, there are infinitely many multiplicative generators of T (each positive power of a multiplicative generator of T is a multiplicative generator of the same T).

Therefore, given an arbitrary sequence $(T_n)_{n\in\mathbb{N}}$ of continuous Archimedean t-norms, it is always possible to find a convergent sequence of corresponding multiplicative generators whose limit vanishes in $[0,1[$ and, consequently, cannot be a multiplicative generator of some continuous Archimedean t-norm. So, the mere fact that a sequence of multiplicative generators converges is not sufficient to derive the convergence of the corresponding t-norms.

However, fixing some $x_0 \in\,]0,1[$ and some $a \in\,]0,1[$, there is exactly one multiplicative generator $\theta : [0,1] \longrightarrow [0,1]$ of T such that $\theta(x_0) = a$. Without loss of generality, we may put $x_0 = a = 0.5$.

If we denote by $\mathcal{G}$ the class of all strictly increasing continuous functions from $\theta : [0,1] \longrightarrow [0,1]$ such that $\theta(0.5) = 0.5$ and $\theta(1) = 1$, then we have a natural one-to-one correspondence between $\mathcal{G}$ and the class of continuous Archimedean t-norms in the sense that each continuous Archimedean t-norm corresponds to its multiplicative generator in $\mathcal{G}$.

Now we can prove the following close relationship between the convergence of continuous Archimedean t-norms and the convergence of their multiplicative generators contained in $\mathcal{G}$, first given in [Jenei 1998] (see also [Mesiar 1998b]).

8.14 THEOREM

Let $(T_n)_{n\in\mathbb{N}}$ be a sequence of continuous Archimedean t-norms with multiplicative generators $(\theta_n : [0,1] \longrightarrow [0,1])_{n\in\mathbb{N}}$ in $\mathcal{G}$, and let T be a continuous Archimedean t-norm with multiplicative generator $\theta : [0,1] \longrightarrow [0,1]$ in $\mathcal{G}$. Then the following are equivalent:

(i) $\lim_{n\to\infty} T_n = T$.

(ii) For each $x \in \,]0,1]$ we have $\lim_{n\to\infty} \theta_n(x) = \theta(x)$.

Proof: In order to show that (i) implies (ii), assume that $(T_n)_{n\in\mathbb{N}}$ converges to T. It is evident that for each $n \in \mathbb{N}$ we have $\theta_n^{(-1)}(0.5) = \theta^{(-1)}(0.5) = 0.5$. The convergence of $(T_n)_{n\in\mathbb{N}}$ to T and Corollary 8.4 imply that for each $k \geq 2$

$$\lim_{n\to\infty} \theta_n^{(-1)}(0.5^k) = \lim_{n\to\infty} \theta_n^{(-1)}\left(\theta_n(0.5)^k\right) = \theta^{(-1)}\left(\theta(0.5)^k\right) = \theta^{(-1)}(0.5^k). \quad (8.4)$$

Next we claim that for each $k \in \mathbb{N}$

$$\lim_{n\to\infty} \theta_n^{(-1)}(0.5^{\frac{1}{k}}) = \theta^{(-1)}(0.5^{\frac{1}{k}}). \quad (8.5)$$

Write $a_{n,k} = \theta_n^{(-1)}\left(0.5^{\frac{1}{k}}\right)$ and $a_k = \theta^{(-1)}\left(0.5^{\frac{1}{k}}\right)$, and assume first that

$$\limsup_{n\to\infty} a_{n,k} = b > a_k + \varepsilon$$

for some $\varepsilon > 0$. Then there exists a subsequence $(a_{n_i,k})_{i\in\mathbb{N}}$ of $(a_{n,k})_{n\in\mathbb{N}}$ which converges to b, and because of Corollary 8.4 we obtain the contradiction

$$\begin{aligned} 0.5 &= \lim_{i\to\infty} \theta_{n_i,k}^{(-1)}(\theta_{n_i,k}(a_{n_i,k})^k) \\ &= \lim_{i\to\infty} (a_{n_i,k})_{T_{n_i}}^{(k)} \\ &= (b)_T^{(k)} \\ &= \theta^{(-1)}\left(\theta(b)^k\right) \\ &> \theta^{(-1)}\left(\theta(a_k)^k\right) \\ &= 0.5. \end{aligned}$$

Also the assumption $\liminf_{n\to\infty} a_{n,k} < a_k - \varepsilon$ for some $\varepsilon > 0$ leads to a contradiction, so (8.5) must hold. Combining (8.4), (8.5) and Corollary 8.4, we get for each $r \in \mathbb{Q} \cap \,]0,\infty[$, i.e., for each positive rational number,

$$\lim_{n\to\infty} \theta_n^{(-1)}(0.5^r) = \theta^{(-1)}(0.5^r).$$

Taking into account the continuity of the pseudo-inverses $\theta_n^{(-1)}$ and $\theta^{(-1)}$ as well as the fact that $\theta_n^{(-1)}(1) = \theta^{(-1)}(1) = 1$, it follows that

$$\lim_{n\to\infty} \theta_n^{(-1)} = \theta^{(-1)}.$$

Assume now that there is an $x_0 \in]0,1]$ such that $\liminf_{n\to\infty} \theta_n(x_0) < \theta(x_0)$. Then there exist an $\varepsilon > 0$ and a subsequence $(\theta_{n_j})_{j\in\mathbb{N}}$ of $(\theta_n)_{n\in\mathbb{N}}$ such that $0 < \theta_{n_j}(x_0) < \theta(x_0) - \varepsilon$ and, subsequently, we obtain the contradiction

$$x_0 = \lim_{j\to\infty} \theta_{n_j}^{(-1)}(\theta_{n_j}(x_0)) \leq \lim_{j\to\infty} \theta_{n_j}^{(-1)}(\theta(x_0) - \varepsilon) = \theta^{(-1)}(\theta(x_0) - \varepsilon) < x_0,$$

Also $\limsup_{n\to\infty} \theta_n(x_0) > \theta(x_0)$ for some $x_0 \in]0,1]$ leads to a contradiction, showing that (ii) must be true.

To show the converse, assume $\lim_{n\to\infty} \theta_n(x) = \theta(x)$ for each $x \in]0,1]$. We first claim that the sequence of pseudo-inverses $(\theta_n^{(-1)})_{n\in\mathbb{N}}$ converges to $\theta^{(-1)}$ (observe that the convergence then is automatically uniform). If for some $x_0 \in]0,1[$ (in 0 and in 1 we always have convergence) we have

$$\limsup_{n\to\infty} \theta_n^{(-1)}(x_0) > \theta^{(-1)}(x_0),$$

then there exist an $\varepsilon > 0$ and a subsequence $(\theta_{n_i}^{(-1)})_{i\in\mathbb{N}}$ such that

$$\theta_{n_i}^{(-1)}(x_0) > \theta^{(-1)}(x_0)s + \varepsilon > 0$$

for all $i \in \mathbb{N}$. This leads to the contradiction

$$x_0 = \lim_{i\to\infty} \theta_{n_i}(\theta_{n_i}^{(-1)}(x_0)) \geq \lim_{i\to\infty} \theta_{n_i}(\theta^{(-1)}(x_0) + \varepsilon) = \theta(\theta^{(-1)}(x_0) + \varepsilon) > x_0.$$

In complete analogy, the assumption

$$0 < \liminf_{n\to\infty} \theta_n^{(-1)}(x_0) < \theta^{(-1)}(x_0)$$

for some $x_0 \in]0,1[$ also yields a contradiction, and in the remaining case $\liminf_{n\to\infty} \theta_n^{(-1)}(x_0) = 0$ we necessarily have $\theta^{(-1)}(x_0) = 0$. Therefore, for each $\varepsilon > 0$ there is an $n_0 \in \mathbb{N}$ such that for all $x \in [0,1]$ and for all $n \geq n_0$

$$|\theta_n^{(-1)}(x) - \theta^{(-1)}(x)| < \varepsilon.$$

Since the pseudo-inverse $\theta^{(-1)}$ (as a continuous function on the compact unit interval) is even uniformly continuous, there exists some $\delta > 0$ such that for all $u, v \in [0,1]$ with $|u - v| < \delta$ we have

$$|\theta^{(-1)}(u) - \theta^{(-1)}(v)| < \varepsilon.$$

Now, fix two arbitrary elements $x_0, y_0 \in \,]0,1]$ (observe that for $x_0 = 0$ or $y_0 = 0$ we have $T_n(x_0, y_0) = 0 = T(x_0, y_0)$ for all $n \in \mathbb{N}$). Then, because of the convergence of the sequences $(\theta_n(x_0))_{n\in\mathbb{N}}$ and $(\theta_n(y_0))_{n\in\mathbb{N}}$, there is some $n_1 \geq n_0$ such that for all $n \geq n_1$ we have

$$|\theta_n(x_0) + \theta_n(y_0) - \theta(x_0) - \theta(y_0)| < \delta,$$

and for all $n \geq n_1$ we obtain

$$\begin{aligned}|T_n(x_0, y_0) - T(x_0, y_0)| &= |\theta_n^{(-1)}(\theta_n(x_0) + \theta_n(y_0)) - \theta^{(-1)}(\theta(x_0) + \theta(y_0))| \\ &\leq |\theta_n^{(-1)}(\theta_n(x_0) + \theta_n(y_0)) - \theta^{(-1)}(\theta_n(x_0) + \theta_n(y_0))| \\ &\quad + |\theta^{(-1)}(\theta_n(x_0) + \theta_n(y_0)) - \theta^{(-1)}(\theta(x_0) + \theta(y_0))| \\ &< 2\varepsilon,\end{aligned}$$

showing that $\lim_{n\to\infty} T_n = T$. □

Obviously, if the limit t-norm T in Theorem 8.14 is strict, then the multiplicative generators $(\theta_n)_{n\in\mathbb{N}}$ converge on the whole interval $[0,1]$ to θ (implying that the convergence is uniform).

8.15 Example

Consider the family of Frank t-norms $(T_\lambda^{\mathbf{F}})_{\lambda\in[0,\infty]}$ (see Section 4.4) and recall the multiplicative generators $(\theta_\lambda^{\mathbf{F}})_{\lambda\in]0,\infty]}$ given in Remark 4.8(v). Then we obtain $\lim_{n\to\infty} T_n^{\mathbf{F}} = T_{\mathbf{L}}$ and $\lim_{n\to\infty} T_{\frac{1}{n}}^{\mathbf{F}} = T_{\mathbf{M}}$, but also

$$\lim_{n\to\infty} \theta_n^{\mathbf{F}} = \mathbf{1}_{\{1\}} \qquad \text{and} \qquad \lim_{n\to\infty} \theta_{\frac{1}{n}}^{\mathbf{F}} = \mathbf{1}_{]0,1]},$$

where $\mathbf{1}_A : [0,1] \longrightarrow [0,1]$ is the characteristic function of the set $A \subseteq [0,1]$. However, neither $\mathbf{1}_{\{1\}}$ nor $\mathbf{1}_{]0,1]}$ can be a multiplicative generator of a t-norm.

If we replace $\theta_\lambda^{\mathbf{F}}$ by $(\theta_\lambda^{\mathbf{F}})^{\frac{1}{\log_2(1+\sqrt{\lambda})}}$ (which is contained in $\mathcal{G}$ and is also a multiplicative generator of $T_\lambda^{\mathbf{F}}$ for $\lambda \in \,]0,1[\,\cup\,]1,\infty[$), then for each $x \in \,]0,1]$

$$\lim_{n\to\infty} (\theta_n^{\mathbf{F}})^{\frac{1}{\log_2(1+\sqrt{n})}}(x) = 4^{x-1}.$$

This means that on $]0,1]$ the limit $\lim_{n\to\infty}(\theta_n^{\mathbf{F}})^{\frac{1}{\log_2(1+\sqrt{n})}}$ coincides with the multiplicative generator $\theta : [0,1] \longrightarrow [0,1]$ of the limit t-norm $T_{\mathbf{L}}$ which is given by $\theta(x) = 4^{x-1}$ (which also belongs to $\mathcal{G}$). However, since all Frank t-norms $T_n^{\mathbf{F}}$ are strict and the limit $T_{\mathbf{L}}$ is nilpotent, it is impossible to have convergence of the corresponding multiplicative generators of $T_n^{\mathbf{F}}$ (all of which map 0 to 0) to a multiplicative generator of $T_{\mathbf{L}}$ (which has a positive value in 0) in the point 0.

It may also happen that a sequence of continuous Archimedean t-norms converges to a t-norm which is not Archimedean (e.g., $\lim_{n\to\infty} T_{\frac{1}{n}}^{\mathbf{F}} = T_{\mathbf{M}}$, see Section 4.4) or which is not continuous (e.g., $\lim_{n\to\infty} T_{\frac{1}{n}}^{\mathbf{Y}} = T_{\mathbf{D}}$, see Section 4.5). The following is a consequence of Remark 8.2:

8.16 COROLLARY
Let $(T_n)_{n\in\mathbb{N}}$ be a sequence of continuous Archimedean t-norms with multiplicative generators $(\theta_n : [0,1] \longrightarrow [0,1])_{n\in\mathbb{N}}$. Then we have:

(i) $\lim_{n\to\infty} T_n = T_{\mathbf{M}}$ *if and only if* $\lim_{n\to\infty} \theta_n^{(-1)}(\theta_n(x)^2) = x$ *for all* $x \in [0,1]$.

(ii) $\lim_{n\to\infty} T_n = T_{\mathbf{D}}$ *if and only if* $\lim_{n\to\infty} \theta_n^{(-1)}(\theta_n(x)^2) = 0$ *for all* $x \in [0,1[$.

8.17 REMARK
From Theorem 8.14 and Corollary 8.16 it follows as an easy exercise that the families of Schweizer-Sklar t-norms (see Section 4.2), of Hamacher t-norms (see Section 4.3), of Frank t-norms (see Section 4.4), of Yager t-norms (see Section 4.5), of Aczél-Alsina t-norms (see Section 4.8), of Dombi t-norms (see Section 4.4), and of Sugeno-Weber t-norms (see Section 4.4) are continuous with respect to the parameter.

The conditions in Corollary 8.16 are not always easy to check, so the following more intuitive necessary and sufficient conditions may be helpful.

8.18 PROPOSITION
Let $(T_n)_{n\in\mathbb{N}}$ be a sequence of continuous Archimedean t-norms with multiplicative generators $(\theta_n : [0,1] \longrightarrow [0,1])_{n\in\mathbb{N}}$ in $\mathcal{G}$.

(i) If $\lim_{n\to\infty} T_n = T_{\mathbf{M}}$ *then we have*

$$\lim_{n\to\infty} \theta_n(x) = \begin{cases} 0 & \text{if } x \in [0,0.5[\,, \\ 0.5 & \text{if } x = 0.5, \\ 1 & \text{if } x \in \,]0.5,1]\,. \end{cases}$$

(ii) If the sequence $(\theta_n)_{n\in\mathbb{N}}$ satisfies

$$\lim_{n\to\infty} \theta_n(x) = \begin{cases} 0 & \text{if } x = 0, \\ 0.5 & \text{if } x \in \,]0,1[\,, \\ 1 & \text{if } x = 1, \end{cases}$$

then we have $\lim_{n\to\infty} T_n = T_{\mathbf{D}}$.

Proof: Assume first $\liminf_{n\to\infty} \theta_n(x_0) = a < 1$ for some $x_0 \in \,]0.5,1[$. Then there is a subsequence $(\theta_{n_i})_{i\in\mathbb{N}}$ of $(\theta_n)_{n\in\mathbb{N}}$ such that $\theta_{n_i}(x_0) < \frac{1+a}{2}$ for all $i \in \mathbb{N}$. Since there exists a number $k \in \mathbb{N}$ with $(\frac{1+a}{2})^k \leq 0.5$, we have $(x_0)_{T_{n_i}}^{(k)} \leq 0.5$ for all $i \in \mathbb{N}$ and, subsequently,

$$\liminf_{i\to\infty} (x_0)_{T_{n_i}}^{(k)} \leq 0.5 < x_0,$$

contradicting the assumption $\lim_{n\to\infty} T_n = T_{\mathbf{M}}$. A similar contradiction is obtained assuming $\limsup_{n\to\infty} \theta_n(x_0) = a > 0$ for some $x_0 \in [0,0.5[$, thus proving (i).

If we fix an $x_0 \in \,]0.5, 1[$ then, because of the uniform convergence of $(\theta_n : [0,1] \longrightarrow [0,1])_{n\in\mathbb{N}}$ on the interval $[1-x_0, x_0]$, there is an $n_0 \in \mathbb{N}$ such that for all $n \geq n_0$ and for all $x \in [1-x_0, x_0]$ we have $|\theta_n(x) - 0.5| < 0.1$. Then $\theta_n(x_0) < 0.6$ and $\theta_n(x_0)^2 < 0.36 < 0.4 < \theta_n(1-x_0)$ and, subsequently, $\theta^{(-1)}(\theta_n(x_0)^2) \leq 1 - x_0$ for all $n \geq n_0$. Therefore for each $x \in \,]0,1[$ we have $\limsup_{n\to\infty} \delta_{T_n}(x) < 1 - x$ and, because of the monotonicity of the diagonal sections δ_{T_n}, $\limsup_{n\to\infty} \delta_{T_n} = 0$, i.e., $\lim_{n\to\infty} T_n = T_{\mathbf{D}}$ as a consequence of Remark 8.2. □

8.19 EXAMPLE

(i) Let $(\theta_n : [0,1] \longrightarrow [0,1])_{n\in\{3,4,\dots\}}$ be the sequence of multiplicative generators given by

$$\theta_n(x) = \begin{cases} \frac{2}{n-2}x & \text{if } x \in \left[0, 0.5 - \frac{1}{n}\right[, \\ \frac{n-2}{2}x - \frac{n-4}{4} & \text{if } x \in \left[0.5 - \frac{1}{n}, 0.5 + \frac{1}{n}\right[, \\ \frac{(n-1)(2nx-2-n)}{n^2(0.6n-2)} + 1 - \frac{1}{n} & \text{if } x \in \left[0.5 + \frac{1}{n}, 0.8\right[, \\ \left(1 - \frac{1}{n^2} + \sqrt{1 - \frac{1}{n^2}}\right)(10x - 8) + 1 - \frac{1}{n^2} & \text{if } x \in [0.8, 0.9[, \\ 1 - \left(\sqrt{1 - \frac{1}{n^2}} - 1\right)(10x - 10) & \text{if } x \in [0.9, 1]. \end{cases}$$

Then $\lim_{n\to\infty} \theta_n$ is as required in Proposition 8.18(i), but for the corresponding t-norms $(T_n : [0,1] \longrightarrow [0,1])_{n\in\{3,4,\dots\}}$ we have

$$\lim_{n\to\infty} T_n(0.9, 0.9) = 0.8 < T_{\mathbf{M}}(0.9, 0.9),$$

so the condition in Proposition 8.18(i) is only necessary for the convergence to $T_{\mathbf{M}}$, but not sufficient.

(ii) Let $(\theta_n : [0,1] \longrightarrow [0,1])_{n\in\{3,4,\dots\}}$ be the sequence of multiplicative generators given by

$$\theta_n(x) = \begin{cases} 0.4nx & \text{if } x \in \left[0, \frac{1}{n}\right[, \\ \frac{n(0.4+0.2x)-1}{n-2} & \text{if } x \in \left[\frac{1}{n}, 1 - \frac{1}{n}\right[, \\ 1 - 0.4n(1-x) & \text{if } x \in \left[1 - \frac{1}{n}, 1\right]. \end{cases}$$

Then for the corresponding t-norms $(T_n : [0,1] \longrightarrow [0,1])_{n\in\{3,4,\dots\}}$ we have $\lim_{n\to\infty} T_n = T_{\mathbf{D}}$, but $\lim_{n\to\infty} \theta_n(x) = 0.4 + 0.2x$ for all $x \in \,]0,1[$, so the condition in Proposition 8.18(ii) is sufficient for $\lim_{n\to\infty} T_n = T_{\mathbf{D}}$, but not necessary.

It is even possible that a sequence of continuous Archimedean t-norms converges pointwise to a function which is not a t-norm at all (e.g., the sequence of t-norms $\left(T^{\mathbf{I}}_{0.5-\frac{1}{n}}\right)_{n\in\mathbb{N}}$, see Appendix B.2).

The requirement in Theorem 8.14 that all multiplicative generators belong to $\mathcal{G}$ may be modified as follows.

8.20 PROPOSITION

Let $(T_n)_{n\in\mathbb{N}}$ be a sequence of continuous Archimedean t-norms and let T be a continuous Archimedean t-norm. Then the following are equivalent:

(i) $\lim_{n\to\infty} T_n = T$.

(ii) There is a sequence of multiplicative generators $(\theta_n : [0,1] \longrightarrow [0,1])_{n\in\mathbb{N}}$ *of* $(T_n)_{n\in\mathbb{N}}$ *such that the restriction*

$$\big(\lim_{n\to\infty} \theta_n\big)|_{]0,1]}$$

coincides with the restriction of a multiplicative generator of T to $]0,1]$.

Proof: The fact that (i) implies (ii) is an immediate consequence of Theorem 8.14.

If $(\theta_n)_{n\in\mathbb{N}}$ is a sequence of multiplicative generators of $(T_n)_{n\in\mathbb{N}}$ and if θ is a multiplicative generator of T satisfying (ii), then there is a sequence $(\lambda_n)_{n\in\mathbb{N}}$ in $]0,\infty[$ and a $\lambda \in]0,\infty[$ such that all $(\theta_n)^{\lambda_n}$ and θ^λ are contained in $\mathcal{G}$. It is easily seen that then $\lim_{n\to\infty} \lambda_n = \lambda$ and, subsequently, $\lim_{n\to\infty}(\theta_n)^{\lambda_n}(x) = \theta^\lambda(x)$ for all $x \in]0,1]$. Therefore, $\lim_{n\to\infty} T_n = T$ because of Theorem 8.14. □

The duality between additive and multiplicative generators of continuous Archimedean t-norms, which is expressed by the operators E and L (given in (3.18) and (3.19), respectively, see also Figure 3.15), allows us to formulate the result of Proposition 8.20 with the help of additive generators.

8.21 COROLLARY

Let $(T_n)_{n\in\mathbb{N}}$ be a sequence of continuous Archimedean t-norms and let T be a continuous Archimedean t-norm. Then the following are equivalent:

(i) $\lim_{n\to\infty} T_n = T$.

(ii) There exists a sequence of additive generators $(t_n : [0,1] \longrightarrow [0,\infty])_{n\in\mathbb{N}}$ *of* $(T_n)_{n\in\mathbb{N}}$ *such that the restriction*

$$\big(\lim_{n\to\infty} t_n\big)|_{]0,1]}$$

coincides with the restriction of an additive generator of T to $]0,1]$.

Note that, whenever in Corollary 8.21 the limit t-norm T is strict, then $\lim_{n\to\infty} t_n$ is an additive generator of T.

PART II

9. Distribution functions

> ... leads to the concept of a space in which a distribution function rather than a definite number is associated with every pair of elements. [...] The distribution function associated with two elements of a statistical metric space might be said to give, for every x, the probability that the distance between the two points in question does not exceed x. Such a statistical generalization of metric spaces appears to be well adapted for the investigation of physical quantities and physiological thresholds. The idealization of the local behavior of rods and boards, implied by this statistical approach, differs radically from that of Euclid. In spite of this fact, or perhaps just because of it, the statistical approach may provide a useful means for geometrizing the physics of the microcosm.
>
> KARL MENGER

One basic idea in probability theory is to replace numbers by distribution functions, taking into account the fact that in many measurements a smaller or larger amount of uncertainty cannot be avoided. In [Schweizer & Sklar 1983, page 123] distribution functions were even expected to be "numbers of the future". Obviously, the revolutionary developments in physics and chemistry (in particular relativity and quantum theory, and statistical mechanics) in the first decades of the twentieth century provided a strong motivation for such considerations.

When working with distribution functions, one interesting aspect is the connection between joint distribution functions and their marginals. According to a result now bearing the name of Abe Sklar, this connection is completely described by so-called copulas [Sklar 1959, Nelsen 1999], which are the topic of Section 9.1. Copulas are also operations on the unit interval, and many important copulas are also t-norms, as many t-norms turn out to be copulas. We discuss this relationship between t-norms and copulas in detail, including several examples.

Next we consider the so-called triangle functions [Šerstnev 1962, 1964] which are operations on distribution functions in the same sense as t-norms are operations on the unit interval, and which are proper generalizations of t-norms. Again, our exposition concentrates on the description of the close relationship between t-norms and triangle functions.

Then we turn our attention to probabilistic metric spaces, where distances are described by distribution functions rather than by numbers. The first definition of a statistical metric space (as it was called originally, the change of the adjective was only made in 1964 in order to conform to the by then customary usage of these terms) was given in [Menger 1942]. When generalizing the classical triangle inequality, in this paper Karl Menger also coined the term triangular norm (still without the associativity (T2) and with weaker boundary conditions, so triangular conorms were included as well). Shortly after, an alternative definition of a statistical metric space (with a different triangle inequality based on the convolution of distribution functions) was given in [Wald 1943]. Karl Menger continued his work in [Menger 1951a, 1951b, 1951c]. A good overview of his ideas in many other mathematical fields can be found in [Menger 1979b], and [Menger 1979a] is of particular interest in the context of geometrical aspects of probabilistic metric spaces.

The formulation of the actual set of axioms for t-norms by Berthold Schweizer and Abe Sklar in [Schweizer & Sklar 1958, 1960, 1961] and the definition of statistical metric spaces given in [Šerstnev 1962] (containing both the Menger and the Wald spaces as special cases) marked the starting point for a remarkable and fast development of the field of probabilistic metric spaces (compare, among many others, [Schweizer & Sklar 1963, Šerstnev 1964, Sherwood 1971, Sherwood & Taylor 1974, Moynihan 1975, Schweizer 1975, Tardiff 1976, Alsina 1978], [Moynihan 1978a, 1978b] and [Moynihan & Schweizer 1979, Tardiff 1980, Alsina 1983, Alsina *et al.* 1993b]). In [Höhle 1978a, 1978b, 1978c, 1982b] an interesting connection with fuzzy set theory (see Section 12.1), in particular with fuzzy topology, was established. Most of these results are contained in the monograph [Schweizer & Sklar 1983], which now forms the classical source for researchers in the field.

In Section 9.3, we start with the definition of a probabilistic metric space in the sense of [Šerstnev 1962], paying particular attention to Menger spaces [Menger 1942] and Wald spaces [Wald 1943]. Again we focus on aspects where the role of triangle functions (and, implicitly, t-norms) is most crucial. Readers more deeply interested in probabilistic metric spaces, or looking for a complete description of the many facets thereof, are once again referred to [Schweizer & Sklar 1983]. We only discuss here some remarkable results concerning probabilistic contraction mappings and fixed point theorems in probabilistic metric spaces (compare, e.g., [Sehgal & Bharucha-Reid 1972], [Hadžić 1978, 1982] and [Hicks 1983, Radu 1985, Schweizer *et al.* 1988, Tardiff 1992]).

Finally, we would like to mention that the fuzzy reals as considered in Section 13.2 are in a one-to-one correspondence with a natural class of distribution functions, and that they are also used as generalizations of real numbers. Some elementary properties of the addition of these fuzzy reals (based on Zadeh's extension principle and involving various t-norms) will be discussed in some detail in Section 13.2.

9.1 Copulas

In the context of the construction of a joint distribution function from two given marginal distribution functions, the notion of a copula was introduced in [Sklar 1959], following some pioneering work in [Hoeffding 1940, 1941] and, later on, in [Fréchet 1951, Féron 1956]. This work was continued in, e.g., [Sklar 1973, Moynihan 1978b, Schweizer & Sklar 1983]. More recent references concerning copulas are [Dall'Aglio *et al.* 1991, Nelsen 1991, Schweizer 1991, Nelsen 1995, Sklar 1996], particularly recommended is the monograph [Nelsen 1999].

Let us first recall and fix the notions of a distribution function and a joint distribution function [Bauer 1981].

9.1 DEFINITION

(i) A *distribution function* (on $[-\infty, \infty]$) is a function $F : [-\infty, \infty] \longrightarrow [0, 1]$ satisfying the subsequent properties:

(DF1) F is left-continuous on $]\infty, \infty[$.

(DF2) F is non-decreasing.

(DF3) $F(-\infty) = 0$.

(DF4) $F(\infty) = 1$.

The class of all distribution functions on $[-\infty, \infty]$ will be denoted by Δ.

(ii) The natural order on $[-\infty, \infty]$ is carried over to Δ pointwise, i.e., for $F, G \in \Delta$ we have $F \leq G$ if and only if $F(x) \leq G(x)$ for all $x \in [-\infty, \infty]$.

9.2 EXAMPLE

Special distribution functions in Δ are the so-called *unit steps* (or *Dirac distributions*) $\varepsilon_a : [-\infty, \infty] \longrightarrow [0, 1]$, which are defined for $a \in [-\infty, \infty[$ by

$$\varepsilon_a(x) = \begin{cases} 0 & \text{if } x \in [-\infty, a], \\ 1 & \text{if } x \in]a, \infty], \end{cases}$$

while $\varepsilon_\infty : [-\infty, \infty] \longrightarrow [0, 1]$ is given by

$$\varepsilon_\infty(x) = \begin{cases} 0 & \text{if } x \in [-\infty, \infty[, \\ 1 & \text{if } x = \infty. \end{cases}$$

Obviously we have $\varepsilon_a \leq \varepsilon_b$ if and only if $a \geq b$. Moreover, $\varepsilon_{-\infty}$ is the maximal element of Δ, and ε_∞ is the minimal element of the complete lattice $(\Delta, \leq)$. This means that the function $\varphi : [-\infty, \infty] \longrightarrow \Delta$ given by $\varphi(x) = \varepsilon_x$ provides a canonical, order reversing embedding from $[-\infty, \infty]$ into Δ.

Although we shall usually restrict ourselves to distribution functions, let us mention here the close connection with random variables (for more details see, e.g., [Bauer 1981]): if $(\Omega, \mathcal{A}, P)$ is a probability space (i.e., $\mathcal{A}$ is a σ-algebra of subsets of Ω and $P : \mathcal{A} \longrightarrow [0, 1]$ is a probability measure), if $X : \Omega \longrightarrow \mathbb{R}$ is an $\mathcal{A}$-measurable function, a so-called *random variable*, and if we write briefly

$$P(X < x) = P(\{\omega \in \Omega \mid X(\omega) < x\}),$$

then the function $F_X : [-\infty, \infty] \longrightarrow [0, 1]$ defined by

$$F_X(x) = P(X < x)$$

is a distribution function (called the *distribution of the random variable* X) which satisfies $\inf\{F_X(x) \mid x \in \mathbb{R}\} = 0$ and $\sup\{F_X(x) \mid x \in \mathbb{R}\} = 1$.

9.3 DEFINITION

(i) A *joint distribution function* (on the extended real plane $[-\infty, \infty]^2$) is a function $H : [-\infty, \infty]^2 \longrightarrow [0, 1]$ such that for all $x, x^*, y, y^* \in [-\infty, \infty]$ with $x \le x^*$ and $y \le y^*$:

(JDF1) H is left-continuous on $]\infty, \infty[^2$.

(JDF2) $H(x, y) + H(x^*, y^*) \ge H(x, y^*) + H(x^*, y)$.

(JDF3) $H(x, -\infty) = H(-\infty, x) = 0$.

(JDF4) $H(\infty, \infty) = 1$.

(ii) If $H : [-\infty, \infty]^2 \longrightarrow [0, 1]$ is a joint distribution function then the *marginal distribution functions* $F, G \in \Delta$ are defined by, respectively:

$$F(x) = H(x, \infty),$$
$$G(x) = H(\infty, x).$$

There is a natural interpretation of a joint distribution function and its marginals in terms of a *random vector* (X, Y) (whose components X and Y are random variables as above): the function $H_{XY} : [-\infty, \infty]^2 \longrightarrow [0, 1]$ given by

$$H_{XY}(x, y) = P(X < x \text{ and } Y < y)$$

is a joint distribution function (the *joint distribution of the random vector* (X, Y)), and its marginal distributions coincide with the distributions F_X and F_Y of X and Y, respectively.

After these preliminaries on distribution functions we are ready to give the definition of a copula. We shall start with a two-dimensional copula, and we only will briefly mention the higher-dimensional case later on.

9.4 DEFINITION

A *(two-dimensional) copula* (briefly a 2-*copula* or, simply, a *copula*) is a binary operation C on the unit interval $[0, 1]$, i.e., a function $C : [0, 1]^2 \longrightarrow [0, 1]$, such that for all $x, x^*, y, y^* \in [0, 1]$ with $x \le x^*$ and $y \le y^*$ we have:

(CO1) $C(x, y) + C(x^*, y^*) \ge C(x, y^*) + C(x^*, y)$.

(CO2) $C(x, 0) = C(0, x) = 0$.

(CO3) $C(x, 1) = C(1, x) = x$.

We only mention that, for an arbitrary function $F : [0,1]^2 \longrightarrow \mathbb{R}$, the axiom (CO1) is logically independent from the monotonicity (1.3): if we put $F(x,y) = (2x-1)(2y-1)$ then (CO1) holds and (1.3) is violated, while, on the other hand, the maximum $S_{\mathbf{M}}$ satisfies (1.3) but not (CO1).

Here is a list of examples of copulas which is by no means exhaustive. They were selected according to their relationship to t-norms, on the one hand, and to their importance for subsequent results and remarks, on the other hand. Many additional examples of copulas can be found in [Nelsen 1999].

9.5 EXAMPLE

(i) Among the four basic t-norms, the minimum $T_{\mathbf{M}}$ (also called the *Fréchet-Hoeffding upper bound*), the product $T_{\mathbf{P}}$ and the Łukasiewicz t-norm $T_{\mathbf{L}}$ (also called the *Fréchet-Hoeffding lower bound*) are associative and commutative copulas. The drastic product $T_{\mathbf{D}}$ violates (CO1) and is not a copula.

(ii) The only copulas C satisfying either $C(x,x) = x$ or $C(x, 1-x) = 0$ for each $x \in [0,1]$ are $C = T_{\mathbf{M}}$ and $C = T_{\mathbf{L}}$, respectively.

(iii) It is easy to see that each convex combination of copulas is again a copula. For each copula C, the following functions $C_{0,1}, C_{1,0}, \hat{C} : [0,1]^2 \longrightarrow [0,1]$ are easily seen to be copulas too:

$$\begin{aligned} C_{0,1}(x,y) &= x - C(x, 1-y), \\ C_{1,0}(x,y) &= y - C(1-x, y), \\ \hat{C}(x,y) &= x + y - 1 + C(1-x, 1-y). \end{aligned}$$

Because of its natural applications in reliability theory, $\hat{C}$ is often called a *survival copula* (see [Nelsen 1999]). Obviously, for each copula C we get

$$\begin{aligned} \hat{C} &= (C_{0,1})_{1,0} = (C_{1,0})_{0,1}, \\ (\hat{C})_{0,1} &= C_{1,0}, \\ (\hat{C})_{1,0} &= C_{0,1}. \end{aligned}$$

Moreover, from Theorem 5.15 it follows that the only associative copula C satisfying $C = C_{0,1}$ (or, equivalently, $C = C_{1,0}$) is the product t-norm $T_{\mathbf{P}}$.

(iv) Starting with a family $(]a_\alpha, e_\alpha[)_{\alpha \in A}$ of pairwise disjoint open subintervals of $[0,1]$ and a family of copulas $(C_\alpha)_{\alpha \in A}$, construction (3.21) yields again a copula (which is called an ordinal sum of the copulas $(C_\alpha)_{\alpha \in A}$). In fact, each copula can be written as an ordinal sum of copulas without non-trivial idempotent elements. Being ordinal sums of copulas, all Mayor-Torrens t-norms $T_\lambda^{\mathbf{MT}}$ (see Section 4.9) and all Dubois-Prade t-norms $T_\lambda^{\mathbf{DP}}$ (see Example 3.46(iv)) are copulas.

(v) The following function $C : [0,1]^2 \longrightarrow [0,1]$ is a non-commutative copula and, therefore, not a t-norm:

$$C(x,y) = xy + x^2 y(1-x)(1-y).$$

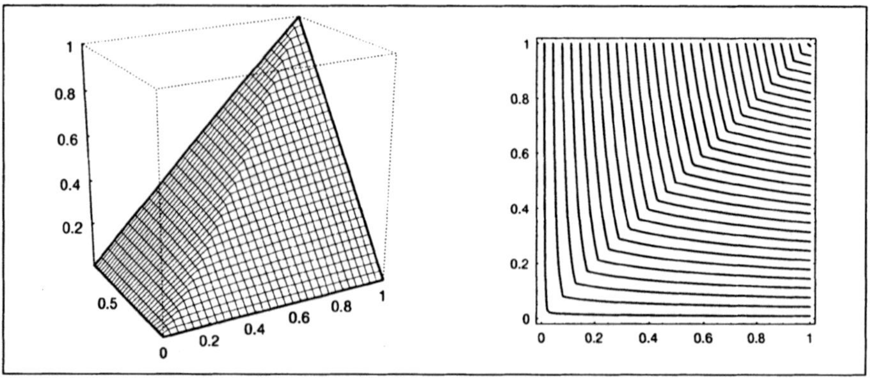

Fig. 9.1. 3D and contour plots of the Cuadras-Augé copula $C_{0.75}^{\mathbf{CA}}$

(vi) The family $(C_\lambda^{\mathbf{FGM}})_{\lambda\in[-1,1]}$ of *Farlie-Gumbel-Morgenstern copulas* is given by (see [Morgenstern 1956, Gumbel 1958, Farlie 1960], compare also [Eyraud 1938] and [Schweizer & Sklar 1983, page 87])

$$C_\lambda^{\mathbf{FGM}}(x,y) = xy + \lambda xy(1-x)(1-y).$$

With the exception of $C_0^{\mathbf{FGM}} = T_{\mathbf{P}}$, the Farlie-Gumbel-Morgenstern copulas $C_\lambda^{\mathbf{FGM}}$ are non-associative copulas and, therefore, no t-norms. Note also that each $C_\lambda^{\mathbf{FGM}}$ is a convex combination of the two extremal members of the family because of

$$C_\lambda^{\mathbf{FGM}} = \tfrac{1-\lambda}{2} \cdot C_{-1}^{\mathbf{FGM}} + \tfrac{1+\lambda}{2} \cdot C_1^{\mathbf{FGM}},$$

and that we have $C_\lambda^{\mathbf{FGM}} = (C_\lambda^{\mathbf{FGM}})^{\frown}$, i.e., each Farlie-Gumbel-Morgenstern copula coincides with the corresponding survival copula. Moreover, we also get for each $\lambda \in [-1,1]$

$$(C_\lambda^{\mathbf{FGM}})_{0,1} = (C_\lambda^{\mathbf{FGM}})_{1,0} = C_{-\lambda}^{\mathbf{FGM}}.$$

(vii) The family $(C_\lambda^{\mathbf{CA}})_{\lambda\in[0,1]}$ of *Cuadras-Augé copulas* is given by

$$C_\lambda^{\mathbf{CA}} = {T_{\mathbf{M}}}^{\lambda} \cdot {T_{\mathbf{P}}}^{1-\lambda}$$

(see [Cuadras & Augé 1981]). Observe that this is just the family of weighted geometric means of $T_{\mathbf{M}}$ and $T_{\mathbf{P}}$, and that we have $C_0^{\mathbf{CA}} = T_{\mathbf{P}}$ and $C_1^{\mathbf{CA}} = T_{\mathbf{M}}$. Except for these limit cases, $C_\lambda^{\mathbf{CA}}$ is not associative and, subsequently, not a t-norm. The family of Cuadras-Augé copulas is strictly increasing with respect to the parameter λ. For a visualization of $C_{0.75}^{\mathbf{CA}}$ see Figure 9.1.

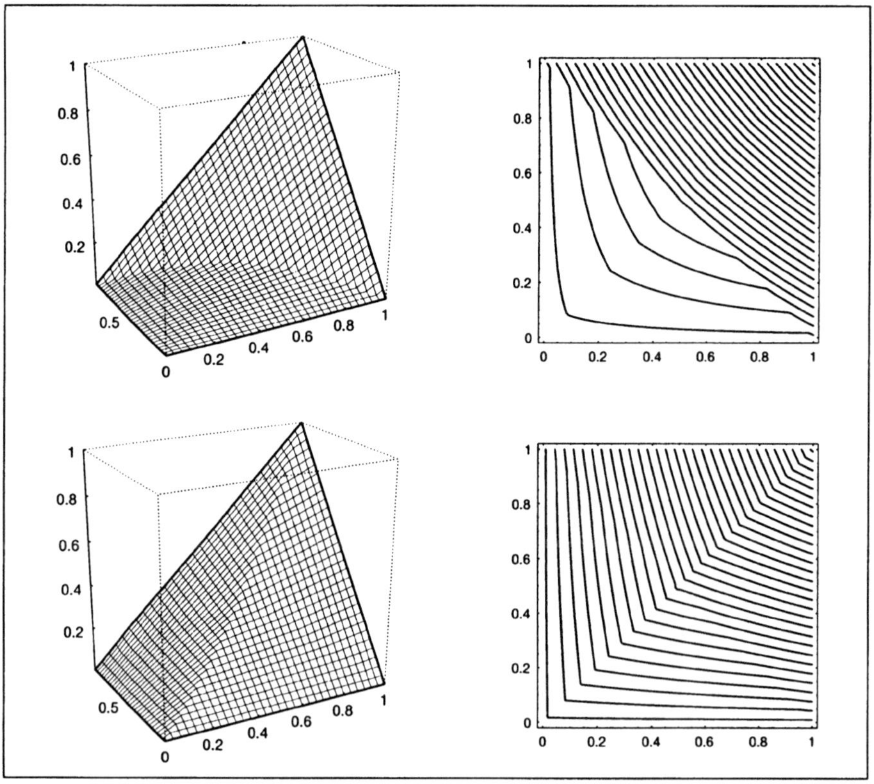

Fig. 9.2. 3D and contour plots of the Mardia copulas $C^{\mathbf{M}}_{-0.75}$ (top) and $C^{\mathbf{M}}_{0.75}$

(viii) The family of $(C^{\mathbf{M}}_{\lambda})_{\lambda\in[-1,1]}$ of *Mardia copulas* is given by [Mardia 1970]

$$C^{\mathbf{M}}_{\lambda} = \tfrac{\lambda^2(1+\lambda)}{2} \cdot T_{\mathbf{M}} + (1-\lambda^2)\cdot T_{\mathbf{P}} + \tfrac{\lambda^2(1-\lambda)}{2} \cdot T_{\mathbf{L}}.$$

As a convex combination of $T_{\mathbf{M}}$, $T_{\mathbf{P}}$ and $T_{\mathbf{L}}$, each $C^{\mathbf{M}}_{\lambda}$ is trivially a copula, and we have $C^{\mathbf{M}}_{-1} = T_{\mathbf{L}}$, $C^{\mathbf{M}}_{0} = T_{\mathbf{P}}$ and $C^{\mathbf{M}}_{1} = T_{\mathbf{M}}$. Except for these special cases, $C^{\mathbf{M}}_{\lambda}$ is never associative and, subsequently, not a t-norm. The family of Mardia copulas is not monotone with respect to the parameter λ. For a visualization of $C^{\mathbf{M}}_{-0.75}$ and $C^{\mathbf{M}}_{0.75}$ and see Figure 9.2.

The importance of copulas in probability and statistics comes from the following result (appearing first in [Sklar 1959], see also [Nelsen 1999, Theorem 2.4.1]), which is usually referred to as *Sklar's Theorem* (we state it here in the language of random variables, equivalently it can be formulated in terms of distribution functions). It shows that the joint distribution of a random vector and the corresponding marginal distributions are linked by a copula.

9.6 THEOREM

Let (X, Y) be a random vector, $F_X, F_Y : [-\infty, \infty] \longrightarrow [0, 1]$ its marginal distribution functions, and $H_{XY} : [-\infty, \infty]^2 \longrightarrow [0, 1]$ a function. The following are equivalent:

(i) H_{XY} is a joint distribution of (X, Y).

(ii) There is a 2-copula C_{XY} such that for all $(x, y) \in [-\infty, \infty]^2$ we have

$$H_{XY}(x, y) = C_{XY}(F_X(x), F_Y(y)).$$

If F_X and F_Y are continuous then C_{XY} is unique, otherwise, C_{XY} is uniquely determined only on $\text{Ran}(F_X) \times \text{Ran}(F_Y)$.

This connection between copulas, joint and marginal distributions provides a natural interpretation of axiom (CO1): it simply makes sure that the probability of the rectangle $[x, x^*] \times [y, y^*]$ is always non-negative.

As a consequence of Sklar's Theorem, the copula $C_{XY} = T_{\mathbf{P}}$ characterizes independence of random variables X and Y, provided the corresponding marginal distribution functions F_X and F_Y are both continuous. Moreover, if in this case ρ_{XY} denotes the correlation coefficient of X and Y, then $\rho_{XY} = -1$ corresponds to $C_{XY} = T_{\mathbf{L}}$, and $\rho_{XY} = 1$ corresponds to $C_{XY} = T_{\mathbf{M}}$.

Given two copulas C_1 and C_2, their *$*$-product* $C_1 * C_2 : [0, 1]^2 \longrightarrow [0, 1]$ [Darsow *et al.* 1992] is defined by

$$C_1 * C_2(x, y) = \int_0^1 \frac{\partial C_1(x, t)}{\partial t} \cdot \frac{\partial C_2(t, y)}{\partial t}\, dt.$$

The function $C_1 * C_2$ is well-defined since the partial derivatives exist almost everywhere, and it is always a copula, i.e., the $*$-product is an operation on the set of all copulas. Moreover, the set of all copulas, equipped with the $*$-product, is a non-commutative semigroup whose annihilator is the product t-norm $T_{\mathbf{P}}$ and whose neutral element is the minimum t-norm $T_{\mathbf{M}}$ [Li *et al.* 1997]. We also have $T_{\mathbf{L}} * T_{\mathbf{L}} = T_{\mathbf{M}}$, and for each copula C it is possible to show that $T_{\mathbf{L}} * C * T_{\mathbf{L}} = \hat{C}$, and that $C = \hat{C}$ if and only if $T_{\mathbf{L}} * C = C * T_{\mathbf{L}}$. Both the family of Farlie-Gumbel-Morgenstern copulas $(C_\lambda^{\mathbf{FGM}})_{\lambda \in [-1,1]}$ and the family of Mardia copulas $(C_\lambda^{\mathbf{M}})_{\lambda \in [-1,1]}$ are closed under the $*$-product because of

$$C_\lambda^{\mathbf{FGM}} * C_\mu^{\mathbf{FGM}} = C_{\frac{\lambda\mu}{3}}^{\mathbf{FGM}},$$
$$C_\lambda^{\mathbf{M}} * C_\mu^{\mathbf{M}} = C_{\lambda\mu}^{\mathbf{M}}.$$

The $*$-product of copulas has a strong interpretation in the context of Markov processes [Bauer 1981], where the so-called Chapman-Kolmogorov equations play an essential role (in fact, they are a necessary condition for a stochastic process to be a Markov process). To make things precise, consider a stochastic process $(X_t)_{t \in I}$, i.e., a family of random variables where the subset I of $\mathbb{R}$ describes the time and X_t the state of the process at time t, and write briefly

$C_{st} = C_{X_s X_t}$ for $s,t \in I$. Then the corresponding conditional probabilities $P(X_t \le y \mid X_s = x)$ satisfy the Chapman-Kolmogorov equations

$$P(X_t \le y \mid X_s = x) = \int_{-\infty}^{\infty} P(X_t \le y \mid X_u = z) \cdot \frac{\partial P(X_u \le z \mid X_s = x)}{\partial z}\, dz$$

for all $s,t,u \in I$ with $s < u < t$ and almost all $x, y \in \mathbb{R}$ if and only if we have $C_{st} = C_{su} * C_{ut}$ for all $s,t,u \in I$ with $s < u < t$.

As a consequence, e.g., the family of Mardia copulas and suitable discrete subfamilies of the family of Farlie-Gumbel-Morgenstern copulas can be used to construct Markov processes (compare [Darsow *et al.* 1992]).

The impact of strictly monotone transformations of random variables X and Y with continuous distributions on the corresponding copula C_{XY} can be described as follows [Nelsen 1999, Theorems 2.4.3 and 2.4.4] (for the notations $(C_{XY})_{(0,1)}$, $(C_{XY})_{(1,0)}$ and $(C_{XY})\hat{}$ see Example 9.5(iii)):

9.7 THEOREM

Let X and Y be random variables with continuous distribution functions $F_X, F_Y : [-\infty,\infty] \longrightarrow [0,1]$, respectively, and corresponding copula C_{XY}, and assume that $\varphi, \psi : [0,1] \longrightarrow [0,1]$ are strictly monotone functions.

(i) If φ and ψ are both increasing then $C_{\varphi\circ X,\psi\circ Y} = C_{XY}$.
(ii) If φ is increasing and ψ decreasing then $C_{\varphi\circ X,\psi\circ Y} = (C_{XY})_{0,1}$.
(iii) If φ is decreasing and ψ increasing then $C_{\varphi\circ X,\psi\circ Y} = (C_{XY})_{1,0}$.
(iv) If φ and ψ are both decreasing then $C_{\varphi\circ X,\psi\circ Y} = (C_{XY})\hat{}$.

In this sense, the study of rank statistics may be characterized as the study of copulas and copula-invariant properties.

Interestingly enough, each copula itself can be regarded as a joint distribution (if we extend its domain in a natural way to $[-\infty,\infty]^2$) with uniform marginals: for each copula C the function $H_C : [-\infty,\infty]^2 \longrightarrow [0,1]$ given by

$$H_C(x,y) = \begin{cases} 0 & \text{if } \min(x,y) < 0, \\ C(x,y) & \text{if } (x,y) \in [0,1]^2, \\ \min(x,y,1) & \text{otherwise,} \end{cases}$$

is a joint distribution function, and both its marginal distribution functions coincide with the uniform distribution on $[0,1]$.

Sklar's Theorem can be extended to n dimensions with $n > 2$ [Nelsen 1999, Theorem 2.10.11], in which case so-called *n-copulas* [Sklar 1973] are required (for their exact definition see [Schweizer & Sklar 1983, Nelsen 1999]). Similarly as before, an n-copula can be viewed as an n-dimensional joint distribution function with uniform one-dimensional marginals. However, it should be noted that the standard extension of an associative 2-copula does not necessarily lead to an n-copula (e.g., the extension of the Łukasiewicz t-norm $T_{\mathbf{L}}$ given in

Example (1.11) fails to be an n-copula for $n > 2$). A necessary condition for the n-dimensional extension of an associative 2-copula C to be an n-copula for all $n > 2$ is that $T_{\mathbf{P}} \leq C \leq T_{\mathbf{M}}$ (for a necessary and sufficient condition see [Kimberling 1974]).

Although it is not necessarily commutative or associative, each copula has some other interesting properties [Schweizer & Sklar 1983, Lemma 6.1.8 and Lemma 6.3.1]:

9.8 PROPOSITION
For each copula C we have:

(i) $T_{\mathbf{L}} \leq C \leq T_{\mathbf{M}}$.
(ii) C is non-decreasing in each argument.
(iii) C satisfies the Lipschitz property with constant 1, i.e., for all numbers $x_1, x_2, y_1, y_2 \in [0, 1]$ we have

$$|C(x_1, y_1) - C(x_2, y_2)| \leq |x_1 - x_2| + |y_1 - y_2|.$$

An immediate consequence of Proposition 9.8(iii) is that each copula is continuous. Thus, each copula fulfills the axioms of continuous t-norms with the possible exceptions of commutativity (T1) and associativity (T2). If C is an associative copula, however, then $([0, 1], C)$ is an I-semigroup with annihilator 0 and neutral element 1. Therefore, as a consequence of Theorem 2.43 we get:

9.9 COROLLARY
Each associative copula is a continuous t-norm.

Because of Proposition 9.8(i), no t-norm which is weaker than $T_{\mathbf{L}}$ (e.g., no Yager t-norm $T_{\lambda}^{\mathbf{Y}}$ with $\lambda \in]0, 1[$, see Section 4.5) is a copula. The following is a full characterization of all t-norms which are copulas [Moynihan 1978b].

9.10 THEOREM
For a t-norm T the following are equivalent:

(i) T is a copula.
(ii) T satisfies the Lipschitz property with constant 1.

We have already observed that a necessary condition for a t-norm T to be a copula is $T \geq T_{\mathbf{L}}$. Even for continuous Archimedean t-norms this property is not sufficient, as this counterexample shows [Schweizer & Sklar 1983]:

9.11 EXAMPLE
A strict t-norm which is no copula, although it is stronger than $T_{\mathbf{L}}$, is obtained by means of the additive generator $t : [0, 1] \longrightarrow [0, \infty]$ given by

$$t(x) = \begin{cases} \frac{3}{2x} - 1 & \text{if } x \in [0, 0.5], \\ \frac{2}{x} - 2 & \text{otherwise.} \end{cases}$$

A careful analysis of this example shows that the property $T \geq T_{\mathbf{L}}$ is equivalent to the fact that the function $h : [0,\infty] \longrightarrow [0,1]$ given by $h(x) = 1 - t^{(-1)}(x)$, where $t : [0,1] \longrightarrow [0,\infty]$ is an additive generator of T, is subadditive, a property which is not sufficient to guarantee that T satisfies (CO1). In fact, we have the following full characterization of continuous Archimedean t-norms which are copulas (see [Schweizer & Sklar 1963, Moynihan 1978b]) by means of convex generators (compare (5.5)).

9.12 THEOREM
Let T be a continuous Archimedean t-norm and let $t : [0,1] \longrightarrow [0,\infty]$ be an additive generator of T. The following are equivalent:

(i) T is a copula.
(ii) t is a convex function.

Accordingly, a continuous Archimedean t-norm T with a convex additive generator $t : [0,1] \longrightarrow [0,\infty]$ is also called an *Archimedean copula.* If the convex additive generator satisfies $t(0) = \infty$ then one speaks of a *strict copula.*

As an immediate consequence, for each Archimedean copula C property (AP) holds, i.e., for all $x, y \in \,]0,1[$ there exists an $n \in \mathbb{N}$ such that $x_C^{(n)} < y$ (where $x_C^{(n)}$ is defined in analogy to Remark 1.10(i)).

On the other hand, from Proposition 2.15(i) and Corollary 9.9 it readily follows that each associative copula C satisfying $C(x,x) < x$ for all $x \in \,]0,1[$ is necessarily a continuous Archimedean t-norm and, because of Theorem 5.1, also an Archimedean copula.

9.13 EXAMPLE
Theorem 9.12 allows us to identify several well-known t-norms to be copulas.

(i) All Frank t-norms $T_\lambda^{\mathbf{F}}$ (Section 4.4) are copulas [Frank 1979]. They are also interesting from a statistical point of view [Nelsen 1986, Genest 1987].
(ii) The following subfamilies of families of t-norms are copulas:
- $(T_\lambda^{\mathbf{SS}})_{\lambda\in[-\infty,1]}$ (Schweizer-Sklar t-norms, Section 4.2),
- $(T_\lambda^{\mathbf{H}})_{\lambda\in[0,2]}$ (Hamacher t-norms, Section 4.3),
- $(T_\lambda^{\mathbf{Y}})_{\lambda\in[1,\infty]}$ (Yager t-norms, Section 4.5),
- $(T_\lambda^{\mathbf{D}})_{\lambda\in[1,\infty]}$ (Dombi t-norms, Section 4.6),
- $(T_\lambda^{\mathbf{SW}})_{\lambda\in[0,\infty]}$ (Sugeno-Weber t-norms, Section 4.7),
- $(T_\lambda^{\mathbf{AA}})_{\lambda\in[1,\infty]}$ (Aczél-Alsina t-norms, Section 4.8).

(iii) It should be mentioned that, in the context of copulas, $(T_\lambda^{\mathbf{SS}})_{\lambda\in[-\infty,1]}$ is known as the family of *Clayton copulas* [Clayton 1978], and $(T_\lambda^{\mathbf{H}})_{\lambda\in[0,2]}$ is called the family of *Ali-Mikhail-Haq copulas* [Ali *et al.* 1978]. Following [Hutchinson & Lai 1990], $(T_\lambda^{\mathbf{AA}})_{\lambda\in[1,\infty]}$ is referred to as the family of *Gumbel-Hougaard copulas* [Gumbel 1960, Hougaard 1986]. For their role in conjunction with bivariate extreme value distributions see [Genest & Rivest 1989, Nelsen 1999].

(iv) None of the incomparable t-norms $(T^{\mathbf{I}}_{\lambda})_{\lambda\in]0,0.5[}$ considered in Example 3.33 is a copula.

From Theorems 9.10 and 9.12 we get this characterization of continuous Archimedean t-norms which satisfy the Lipschitz property with constant 1:

9.14 COROLLARY
Let T be a continuous Archimedean t-norm and let $t : [0,1] \longrightarrow [0,\infty]$ be an additive generator of T. The following are equivalent:

(i) T satisfies the Lipschitz property with constant 1.
(ii) t is a convex function.

It is not difficult to see that, if T is an ordinal sum of t-norms, i.e., if $T = (\langle a_\alpha, e_\alpha, T_\alpha\rangle)_{\alpha\in A}$, T satisfies the Lipschitz property with constant 1 if anf only if each t-norm T_α does so. Taking into account the representation of continuous t-norm as ordinal sums of continuous Archimedean t-norms in Theorem 5.11, we obtain immediately:

9.15 COROLLARY
Let T be a continuous t-norm which has the uniquely determined ordinal sum representation $T = (\langle a_\alpha, e_\alpha, T_\alpha\rangle)_{\alpha\in A}$. The following are equivalent:

(i) T is a copula.
(ii) For each $\alpha \in A$, the t-norm T_α is a copula.

From Corollary 9.15 it follows in a straightforward manner that, given a continuous t-norm T and a continuous t-conorm S which solve the Frank functional equation (4.2), the t-norm T necessarily is a copula (compare [Frank 1979] or Theorem 5.14).

In analogy to the Frank functional equation (4.2) and the duality between t-norms and t-conorms, one can associate with each copula its dual and its co-copula [Schweizer & Sklar 1983]:

9.16 DEFINITION
Let C be a copula.

(i) The function $\tilde{C} : [0,1]^2 \longrightarrow [0,1]$ given by

$$\tilde{C}(x,y) = x + y - C(x,y) \tag{9.1}$$

is called the *dual* of the copula C.

(ii) The function $C^* : [0,1]^2 \longrightarrow [0,1]$ given by

$$C^*(x,y) = 1 - C(1-x, 1-y)$$

is called the *co-copula* of C.

Obviously, the transition from a copula C to its dual $\tilde{C}$ means exactly the interchanging of the roles of 0 and 1 and the reversing of the inequality sign in axiom (CO1). Therefore, a function $\tilde{C} : [0,1]^2 \longrightarrow [0,1]$ is the dual of some copula C if and only if for all $x, x^*, y, y^* \in [0,1]$ with $x \leq x^*$ and $y \leq y^*$:

(DC1) $\tilde{C}(x,y) + \tilde{C}(x^*,y^*) \leq \tilde{C}(x,y^*) + \tilde{C}(x^*,y)$.
(DC2) $\tilde{C}(x,1) = \tilde{C}(1,x) = 1$.
(DC3) $\tilde{C}(x,0) = \tilde{C}(0,x) = x$.

In general, neither the dual of a copula nor its co-copula is a copula. However, the properties of copulas can be carried over to their duals, if we make the obvious changes where necessary. For instance, for each dual $\tilde{C}$ of a copula C we have $S_{\mathbf{M}} \leq \tilde{C} \leq S_{\mathbf{L}}$, and $\tilde{C}$ satisfies the Lipschitz property with constant 1. Also, the dual $\tilde{C}$ of a copula C is associative if and only if it is a continuous t-conorm which satisfies the Lipschitz property with constant 1, and a continuous Archimedean t-conorm is the dual of a copula if and only if each of its additive generators is a convex function. Moreover, a copula C is associative if and only if its co-copula C^* is associative (in which case C is a t-norm and C^* its dual t-conorm).

Remarkably, if C is an associative copula then its dual $\tilde{C}$ given by (9.1) is not necessarily associative (an example for this situation is the Hamacher product $T_0^{\mathbf{H}}$ as discussed in Section 4.3).

Note that the simultaneous associativity of a copula C and its dual $\tilde{C}$ given by (9.1) means that $(C, \tilde{C})$ is a solution of the Frank functional equation (4.2). In fact, this was the starting point of the work in [Frank 1979].

That means also, as a consequence of Theorem 5.14 and Remark 3.59, that the only pairs of an associative copula and an associative dual thereof, which are linked by (9.1) and which are also dual as t-norms and t-conorms in the sense of (1.9), are given either by $(T_\lambda^{\mathbf{F}}, S_\lambda^{\mathbf{F}})$ for some $\lambda \in [0,\infty]$, or by pairs of ordinal sums $\big(\langle a_\alpha, e_\alpha, T_{\lambda_\alpha}^{\mathbf{F}}\rangle, \langle a_\alpha, e_\alpha, S_{\lambda_\alpha}^{\mathbf{F}}\rangle\big)_{\alpha \in A}$ satisfying that for each $\alpha \in A$ there is a $\beta \in A$ such that $a_\alpha = 1 - e_\beta$, $e_\alpha = 1 - a_\beta$ and $\lambda_\alpha = \lambda_\beta$.

9.2 Triangle functions

Triangular norms are operations on the unit interval turning it into a special semigroup. Triangle functions generalize this concept in the sense that they act on certain distribution functions, and they are crucial for the formulation of the triangle inequality in probabilistic metric spaces (see Section 9.3).

9.17 DEFINITION

A *distance distribution function* is a distribution function whose support is a subset of $[0,\infty]$, i.e., a distribution function $F : [-\infty,\infty] \longrightarrow [0,1]$ with $F(0) = 0$. The class of all distance distribution functions will be denoted Δ^+.

It is clear that $\varepsilon_a \in \Delta^+$ if and only if $a \in [0, \infty]$, and that ε_0 and ε_∞ are the maximal and minimal element, respectively, of the complete lattice $(\Delta^+, \leq)$. This means that the interval $[0, \infty]$ can be canonically embedded into Δ^+ by virtue of the function $\varphi : [0, \infty] \longrightarrow \Delta^+$ given by $\varphi(x) = \varepsilon_x$.

Now it is quite natural (see [Šerstnev 1962, 1964] and [Schweizer & Sklar 1983, Chapter 7]) to consider binary operations τ on Δ^+ such that the triple $(\Delta^+, \tau, \leq)$ forms a commutative, partially ordered semigroup with neutral element ε_0 (compare also Section 2.2).

9.18 Definition

A *triangle function* is a function $\tau : \Delta^+ \times \Delta^+ \longrightarrow \Delta^+$ such that for all distance distribution functions $F, G, H \in \Delta^+$ the following axioms are satisfied:

(TF1)	$\tau(F, G) = \tau(G, F)$.	(*commutativity*)
(TF2)	$\tau(F, \tau(G, H)) = \tau(\tau(F, G), H)$.	(*associativity*)
(TF3)	$\tau(F, G) \leq \tau(F, H)$ whenever $G \leq H$.	(*monotonicity*)
(TF4)	$\tau(F, \varepsilon_0) = F$.	(*neutral element*)

It is readily seen that the minimal unit step ε_∞ is the annihilator of $(\Delta^+, \tau, \leq)$, i.e., we have $\tau(F, \varepsilon_\infty) = \varepsilon_\infty$ for all $F \in \Delta^+$.

Similarly as in the case of t-norms we say that a triangle function τ_1 is *stronger* than the triangle function τ_2 (in symbols $\tau_1 \geq \tau_2$) if for all $F, G \in \Delta^+$ we have $\tau_1(F, G) \geq \tau_2(F, G)$.

In the following we give three examples of important classes of triangle functions which are derived directly from t-norms.

9.19 Example

Let T be a left-continuous t-norm.

(i) The function $\mathbf{T} : \Delta^+ \times \Delta^+ \longrightarrow \Delta^+$ defined by

$$\mathbf{T}(F, G)(x) = T(F(x), G(x)) \tag{9.2}$$

is a triangle function. Obviously, the triangle function $\mathbf{T_M}$ based on the minimum $T_\mathbf{M}$ is the strongest triangle function, i.e., we have $\tau \leq \mathbf{T_M}$ for each triangle function τ. The left-continuity of T cannot be dropped since, e.g., Δ^+ is not closed under $\mathbf{T_D}$.

(ii) Looking at the unit steps in Δ^+, we obtain for all $a, b \in [0, \infty]$

$$\mathbf{T}(\varepsilon_a, \varepsilon_b) = \varepsilon_{\max(a,b)}.$$

Therefore, the function $\varphi : [0, \infty] \longrightarrow \Delta^+$ given by $\varphi(x) = \varepsilon_x$ is an embedding of the semigroup $([0, \infty], \max)$ into $(\Delta^+, \mathbf{T})$.

9.20 Example

Let T be a left-continuous t-norm T and let $L : [0, \infty]^2 \longrightarrow [0, \infty]$ be a surjective function which is continuous on $[0, \infty]^2 \setminus \{(0, \infty), (\infty, 0)\}$, and which satisfies that $L(x_1, y_1) < L(x_2, y_2)$ whenever $x_1 < x_2$ and $y_1 < y_2$, and that $([0, \infty], L)$ is a commutative semigroup with neutral element 0.

(i) The function $\tau_{T,L} : \Delta^+ \times \Delta^+ \longrightarrow \Delta^+$ defined by (recall $\sup \emptyset = 0$)

$$\tau_{T,L}(F,G)(x) = \sup \left\{T(F(u),G(v)) \mid (u,v) \in [0,\infty]^2 \text{ and } L(u,v) = x\right\}$$

is a triangle function [Schweizer & Sklar 1983, Theorem 7.2.4]. In particular, $\tau_{T,\max}$ and $\tau_{T,+}$ are triangle functions for each left-continuous t-norm T, and we have $\tau_{T,L} \leq \tau_{T,\max} = \mathbf{T}$.

(ii) For the unit steps in Δ^+ we obtain for all $a, b \in [0,\infty]$

$$\tau_{T,L}(\varepsilon_a, \varepsilon_b) = \varepsilon_{L(a,b)}, \tag{9.3}$$

i.e., the semigroup $([0,\infty], L)$ can be embedded into $(\Delta^+, \tau_{T,L})$.

(iii) In the case $L = +$ we simply write τ_T instead of $\tau_{T,+}$, i.e.,

$$\tau_T(F,G)(x) = \sup \left\{T(F(u),G(v)) \mid (u,v) \in [0,\infty]^2, u + v = x\right\}. \tag{9.4}$$

In particular, we have $\tau_{T_1} \leq \tau_{T_2}$ if and only if $T_1 \leq T_2$. For an extensive study of the semigroups (Δ^+, τ_T) see, e.g., [Moynihan 1975, 1978]. One of the results is that the cancellation law is not valid in the semigroup (Δ^+, τ_T), but only in the sub-semigroup $(\mathcal{D}^+, \tau_{T_\mathbf{M}}|_{(\mathcal{D}^+)^2})$, where

$$\mathcal{D}^+ = \left\{F \in \Delta^+ \mid \lim_{x \to \infty} F(x) = 1\right\}.$$

(iv) The left-continuity of the t-norm T required in (i) is a sufficient but not a necessary condition: although $T_\mathbf{D}$ is not left-continuous, $\tau_{T_\mathbf{D}}$ is a triangle function given by

$$\begin{aligned}&\tau_{T_\mathbf{D}}(F,G)(x)\\ &\quad = \max\left(F\left(\max\left(x - G^{(-1)}(1), 0\right)\right), G\left(\max\left(x - F^{(-1)}(1), 0\right)\right)\right),\end{aligned}$$

where $F^{(-1)}$ and $G^{(-1)}$ are the pseudo-inverses of F and G (see Definition 3.2), respectively, and where, for the moment, we put $\infty - \infty = \infty$).

9.21 EXAMPLE

Consider a copula C (see Definition 9.4) and a function $L : [0,\infty]^2 \longrightarrow [0,\infty]$ which is surjective and continuous on $[0,\infty]^2 \setminus \{(0,\infty),(\infty,0)\}$, such that, for each $x \in [0,\infty[$, the following set L_x is a bounded subset of $[0,\infty]^2$

$$L_x = \{(u,v) \in [0,\infty]^2 \mid L(u,v) < x\},$$

and such that $([0,\infty], L, \leq)$ is a partially ordered semigroup.

(i) The function $\sigma_{C,L} : \Delta^+ \times \Delta^+ \longrightarrow \Delta^+$ defined by

$$\sigma_{C,L}(F,G)(x) = \begin{cases} 0 & \text{if } x \in [-\infty, 0], \\ \displaystyle\int_{L_x} dC(F(u),G(v)) & \text{if } x \in \,]0,\infty[, \\ 1 & \text{if } x = \infty, \end{cases} \tag{9.5}$$

where the integral is of the Lebesgue-Stieltjes type, is a triangle function if and only if C is an ordinal sum of product t-norms, i.e., if $C = (\langle a_\alpha, e_\alpha, T_\alpha \rangle)_{\alpha \in A}$ with $T_\alpha = T_\mathbf{P}$ for all $\alpha \in A$ [Schweizer & Sklar 1983, Corollary 7.4.4].

(ii) Restricting ourselves to the unit steps in Δ^+, we have for all $a, b \in [0, \infty]$

$$\sigma_{C,L}(\varepsilon_a, \varepsilon_b) = \varepsilon_{L(a,b)} \tag{9.6}$$

for all $a, b \in [0, \infty]$, i.e., the semigroup $([0, \infty], L)$ can be embedded into $(\Delta^+, \sigma_{C,L})$.

(iii) In the special case $C = T_{\mathbf{P}}$ and $L = +$ we just obtain the *convolution of distance distribution functions* $* : \Delta^+ \times \Delta^+ \longrightarrow \Delta^+$, usually defined by (for the equivalence of the formulae (9.5) and (9.7) in this special case see, e.g., [Rényi 1970, Chapter IV, § 10])

$$(F * G)(x) = \begin{cases} 0 & \text{if } x \in [-\infty, 0], \\ \displaystyle\int_{[0,x]} F(x-t)\, dG(t) & \text{if } x \in]0, \infty[, \\ 1 & \text{if } x = \infty. \end{cases} \tag{9.7}$$

Similar as it was done for t-norms in Section 6.3, we can introduce the notion of domination for triangle functions:

9.22 DEFINITION
Let $\tau_1, \tau_2 : \Delta^+ \times \Delta^+ \longrightarrow \Delta^+$ be two triangle functions. Then we say that τ_1 *dominates* τ_2 (in symbols $\tau_1 \gg \tau_2$) if for all $F, G, H, K \in \Delta^+$

$$\tau_1(\tau_2(F, G), \tau_2(H, K)) \geq \tau_2(\tau_1(F, H), \tau_1(G, K)).$$

For the triangle functions $\mathbf{T}$ and τ_T based on some t-norm T (see (9.2) and (9.4)) we have $\mathbf{T_M} \gg \tau$ for each triangle function τ, and $\mathbf{T} \gg \tau_T$ for each continuous t-norm T [Schweizer & Sklar 1983, Lemma 12.7.4]. If T_1 and T_2 are two continuous t-norms, then we have $\tau_{T_1} \gg \tau_{T_2}$ if and only if $T_1 \gg T_2$ [Schweizer & Sklar 1983, Lemma 12.7.3].

9.3 Probabilistic metric spaces

According to [Menger 1942], distance distribution functions as given in Definition 9.1(iii) are a proper tool for measuring distances in statistical metric spaces (which were later called probabilistic metric spaces). The triangle functions (see Definition 9.18) and, implicitly, the t-norms turn out to be indispensable for an appropriate generalization of the classical triangle inequality in this context.

Let us first recall the concept of a classical metric space, as it was introduced in [Fréchet 1906] (see also [Hausdorff 1914, Menger 1928]). Given a non-empty set M, a function $d : M^2 \longrightarrow [0, \infty]$ is called a *metric* on M (and the pair (M, d) a *metric space*) if for all $p, q, r \in M$ we have

(MET1) $\quad d(p, q) = 0 \quad$ if and only if $\quad p = q$.

(MET2) $\quad d(p,q) = d(q,p)$.
(MET3) $\quad d(p,r) \le d(p,q) + d(q,r)$. *(triangle inequality)*

A function $d : M^2 \longrightarrow [0,\infty]$ which satisfies (MET2),(MET3) and

(PMET) $\quad d(p,q) = 0 \quad$ whenever $\quad p = q$,

which is weaker than (MET1), is called a *pseudo-metric* on M.

Based on the original ideas in [Menger 1942], the following definition of probabilistic metric spaces was given in [Šerstnev 1962].

9.23 DEFINITION

(i) Let M be a non-empty set, $\mathfrak{F} : M^2 \longrightarrow \Delta^+$ a function which assigns to each pair $(p,q) \in M^2$ a distance distribution function $F_{pq} \in \Delta^+$, and $\tau : \Delta^+ \times \Delta^+ \longrightarrow \Delta^+$ a triangle function. The triple $(M, \mathfrak{F}, \tau)$ is called a *probabilistic metric space* (briefly *PM space*) if the following properties hold for all $p,q,r \in M$:
   (PM1) $\quad F_{pq} = \varepsilon_0 \quad$ if and only if $\quad p = q$.
   (PM2) $\quad F_{pq} = F_{qp}$.
   (PM3) $\quad F_{pr} \ge \tau(F_{pq}, F_{qr})$. *(probabilistic triangle inequality)*

(ii) A probabilistic metric space $(M, \mathfrak{F}, \tau)$ is said to be *proper* if, for all numbers $a, b \in [0,\infty]$, we have for the corresponding unit steps

$$\tau(\varepsilon_a, \varepsilon_b) \ge \varepsilon_{a+b}.$$

The following are the most important classes of probabilistic metric spaces [Menger 1942, Wald 1943]. For related concepts (and sometimes special cases thereof) see also [Muštari & Šerstnev 1966, Sherwood 1969, Drossos 1977].

9.24 EXAMPLE

Let M be a non-empty set and $\mathfrak{F} : M^2 \longrightarrow \Delta^+$ a function satisfying (PM1) and (PM2).

(i) If, for some t-norm T, $\tau_T : \Delta^+ \times \Delta^+ \longrightarrow \Delta^+$ given by (9.4) is a triangle function then $(M, \mathfrak{F}, \tau_T)$ is a PM space, and it is called a *Menger space*: indeed, for each $x \in [0,\infty]$, for all $p,q,r \in M$ and for all $(u,v) \in [0,\infty]^2$ with $u + v = x$ we obtain $F_{pr}(x) \ge T(F_{pq}(u), F_{qr}(v))$ and, as a direct consequence, $F_{pr}(x) \ge \tau_T(F_{pq}, F_{qr})(x)$.

(ii) Because of (9.3), each Menger space is a proper PM space.

(iii) If $*$ is the convolution of distance distribution functions given in (9.7) then $(M, \mathfrak{F}, *)$ is a probabilistic metric space, and it is called a *Wald space*. Each Wald space is a proper PM space [Schweizer & Sklar 1983, Corollary 8.1.9].

(iv) For each left-continuous t-norm T, $(M, \mathfrak{F}, \mathbf{T})$ is a proper PM space.

Probabilistic metric spaces, in particular Menger and Wald spaces, are generalizations of the classical metric spaces.

9.25 REMARK

If (M,d) is a metric space, if $\mathcal{F}: M^2 \longrightarrow \Delta^+$ is specified by $F_{pq} = \varepsilon_{d(p,q)}$, and if $\tau: \Delta^+ \times \Delta^+ \longrightarrow \Delta^+$ is a triangle function such that $\tau(\varepsilon_a, \varepsilon_b) = \varepsilon_{a+b}$, then $(M, \mathcal{F}, \tau)$ is a proper PM space. Because of (9.3) and (9.6), $(M, \mathcal{F}, \tau)$ is both a Menger and a Wald space. Conversely, if $(M, \mathcal{F}, \tau)$ is a proper PM space and if there is a function $d: M^2 \longrightarrow [0, \infty]$ such that $F_{pq} = \varepsilon_{d(p,q)}$ holds for all $(p,q) \in M^2$ then (M,d) is a metric space.

Now we list a number of relations between special classes of PM spaces, on the one hand, and between t-norms and PM spaces, on the other hand.

9.26 REMARK

Let M be a non-empty set and $\mathcal{F}: M^2 \longrightarrow \Delta^+$ a function satisfying (PM1) and (PM2).

(i) If $\tau_1, \tau_2: \Delta^+ \times \Delta^+ \longrightarrow \Delta^+$ are triangle functions with $\tau_1 \leq \tau_2$ and if $(M, \mathcal{F}, \tau_2)$ is a PM space, then so is $(M, \mathcal{F}, \tau_1)$ [Schweizer & Sklar 1983, Lemma 8.1.7].

(ii) If T is a t-norm, if $L: [0,\infty]^2 \longrightarrow [0,\infty]$ is a function as in Example 9.20 with $L(x,y) \leq x + y$ for all $(x,y) \in [0,\infty]^2$, and if $(M, \mathcal{F}, \tau_{T,L})$ is a PM space, then $(M, \mathcal{F}, \tau_{T,L})$ is a proper probabilistic metric space [Schweizer & Sklar 1983, Theorem 8.1.5].

(iii) If $(M, \mathcal{F}, \tau_{T_{\mathbf{M}}})$ is a Menger space then, for each t-norm T such that τ_T is a triangle function, $(M, \mathcal{F}, \tau_T)$ is a Menger space [Schweizer & Sklar 1983, Theorem 8.1.8(i)].

(iv) The triple $(M, \mathcal{F}, \tau_{T_{\mathbf{P}}})$ is a Menger space if and only if $(M, \mathcal{F}, *)$ is a Wald space [Schweizer & Sklar 1983, Theorem 8.1.8(ii),(iii)]. In [Schweizer & Sklar 1983, Section 8.4] an example of a Menger space $(M, \mathcal{F}, \tau_{T_{\mathbf{M}}})$ is given where $(M, \mathcal{F}, *)$ is not a Wald space.

In order to define the product of PM spaces and to make sure that this product is again a PM space, the domination of triangle functions introduced in Definition 9.22 is needed.

9.27 DEFINITION

Let $(M_1, \mathcal{F}_1, \tau)$ and $(M_2, \mathcal{F}_2, \tau)$ be two PM spaces, and let $\tau_1: \Delta^+ \times \Delta^+ \longrightarrow \Delta^+$ be a triangle function such that $\tau_1 \gg \tau$. Consider the function

$$\mathcal{F}_1 \tau_1 \mathcal{F}_2 : (M_1 \times M_2)^2 \longrightarrow \Delta^+,$$

which assigns to each pair $(\mathbf{p}, \mathbf{q}) \in (M_1 \times M_2)^2$ the distance distribution function $F^{\tau_1}_{\mathbf{pq}} \in \Delta^+$ given by (if $\mathbf{p} = (p_1, p_2)$ and $\mathbf{q} = (q_1, q_2)$)

$$F^{\tau_1}_{\mathbf{pq}} = \tau_1(F_{p_1 q_1}, F_{p_2 q_2}).$$

Then $(M_1 \times M_2, \mathcal{F}_1 \tau_1 \mathcal{F}_2, \tau)$ is a PM space [Schweizer & Sklar 1983, Theorem 12.7.5], and it is called the *product* of $(M_1, \mathcal{F}_1, \tau)$ and $(M_2, \mathcal{F}_2, \tau)$.

In particular, $(M_1 \times M_2, \mathcal{F}_1\tau_1\mathcal{F}_2, \tau)$ is a PM space whenever $\tau_1 = \tau$, or $\tau_1 = \mathbf{T_M}$, or $\tau = \tau_T$ and $\tau_1 = \mathbf{T}$ for some continuous t-norm T.

There are different ways to define natural topologies for probabilistic metric spaces, for example the (ε, λ)-topology [Schweizer & Sklar 1960, Thorp 1960] and the strong topology [Schweizer & Sklar 1983, Chapter 12]).

Of special interest are fixed point theorems in probabilistic metric spaces. The first one is an extension of the Banach contraction mapping theorem [Sehgal & Bharucha-Reid 1972].

9.28 THEOREM
Let $(M, \mathcal{F}, \tau_{T_\mathbf{M}})$ be a Menger space which is complete with respect to the strong topology, and let $f : M \longrightarrow M$ be a probabilistic contraction mapping, i.e., there exists an $\alpha \in]0,1[$ such that for all $p, q \in M$ and for all $x \in [0, \infty[$

$$F_{f(p)f(q)}(x) \geq F_{pq}\left(\tfrac{x}{\alpha}\right). \tag{9.8}$$

Then f has a unique fixed point.

The following fixed point theorem for probabilistic contraction mappings was proved in [Hadžić 1982] (see also [Radu 1985]).

9.29 THEOREM
Let T be a continuous t-norm and $(M, \mathcal{F}, \tau_T)$ be a Menger space. Then the following are equivalent:

(i) Each probabilistic contraction mapping $f : M \longrightarrow M$ has a fixed point.
(ii) For each $a \in]0,1[$ there exists a non-trivial idempotent element b of T with $b \in [a, 1[$.

In [Pap *et al.* 1996] it was shown that a continuous t-norm T satisfies condition (ii) in Theorem 9.29 if and only if $T = (\langle a_\alpha, e_\alpha, T_\alpha \rangle)_{\alpha \in A}$, where each T_α is a continuous Archimedean t-norm, and $\sup\{e_\alpha \mid \alpha \in A\} < 1$ or $\sup\{a_\alpha \mid \alpha \in A\} = 1$ (this class of t-norms was introduced under the name *t-norms of h-type* in [Hadžić 1982], requiring that the sequence $(f_n : [0,1] \longrightarrow [0,1])_{n \in \mathbb{N}}$ defined by $f_n(x) = x_T^{(n)}$ be equicontinuous in $x = 1$). Obviously, no continuous Archimedean t-norm can have this property.

As a consequence of [Radu 1985], each probabilistic contraction mapping has a unique fixed point in a complete Menger space $(M, \mathcal{F}, \tau_{T_\mathbf{M}})$ if and only if, for some t-norm T satisfying condition (ii) in Theorem 9.29, each probabilistic contraction mapping has a unique fixed point in the space $(M, \mathcal{F}, \tau_T)$.

On the other hand, in [Sherwood 1971] it was proved that for each continuous Archimedean t-norm T there exists a complete Menger space $(M, \mathcal{F}, \tau_T)$ and a probabilistic contraction mapping $f : M \longrightarrow M$ which has no fixed point.

An interesting fixed point result where the order of t-norms plays a role was obtained in [Tardiff 1992].

9.30 THEOREM

Let T be a continuous t-norm with $T \geq T_{\mathbf{L}}$ and let $(M, \mathcal{F}, \tau_T)$ be a Menger space which is complete with respect to the strong topology. If for all $p, q \in M$

$$\int_{[1,\infty[} \log x \, dF_{pq}(x) < \infty,$$

then each probabilistic contraction mapping $f : M \longrightarrow M$ has a unique fixed point.

9.31 REMARK

(i) In [Hicks 1983] the notion of an *H-contraction* $f : M \longrightarrow M$ was introduced, replacing in (9.8) the inequality by the implication:

$$F_{f(p)f(q)}(\alpha x) > 1 - \alpha x \quad \text{whenever } F_{pq}(x) > 1 - x.$$

It was shown that each H-contraction $f : M \longrightarrow M$ has a unique fixed point in the complete Menger space $(M, \mathcal{F}, \tau_{T_{\mathbf{L}}})$. This result was generalized in [Radu 1985] to the case of a complete Menger space $(M, \mathcal{F}, \tau_T)$ if the t-norm T is continuous in $(1, 1)$.

(ii) In general, the two notions of probabilistic contraction mappings and H-contractions are independent from each other, a fact which was verified in [Schweizer *et al.* 1988].

Finally, we mention that probabilistic metric spaces also have an interpretation in chaos theory [Schweizer & Smítal 1994].

10. Aggregation operators

> The interpretation of a decision as the intersection of fuzzy sets, computed by applying any t-norm based operator implies that there is no compensation between low and high degrees of membership. If, on the other hand, a decision is defined to be the union of fuzzy sets, represented by some t-conorm, full compensation is assumed. Managerial decisions hardly ever represent any of these extremes.
>
> HANS-JÜRGEN ZIMMERMANN & PAUL ZYSNO

Both t-norms and t-conorms are rather special operations on the unit interval $[0,1]$. However, there are several other useful operations related to and/or generalizing t-norms or t-conorms, either on the unit interval or on an arbitrary closed subinterval $[a,b]$ of the extended real line $[-\infty,\infty]$.

A typical example is an I-semigroup (see Definition 2.40), which generally acts on an interval $[a,b]$ and which is isomorphic to a continuous t-norm or t-conorm because of Theorem 2.43. Another example is the so-called pseudo-addition on $[a,b]$ (in most cases on $[0,\infty]$). A pseudo-addition $\oplus$ on $[a,b]$ is a continuous, associative non-decreasing operation with neutral element a, i.e., $([a,b],\oplus)$ is an I-semigroup with neutral element a and, therefore, can be viewed as a generalization of a continuous t-conorm. The representation given in Corollary 5.12 (in the light of Theorem 2.43) can be carried over in a straightforward way, providing a representation in terms of the operations max and $+$ on $[0,\infty]$.

Because of the natural isomorphism between $[a,b]$ and $[0,1]$, each result for operations on the unit interval $[0,1]$ can be transformed into a result for operators on $[a,b]$, and vice versa. Therefore, the discussion of aggregation operators on $[0,1]$ is sufficiently general, at least from a theoretical point of view, and we shall restrict ourselves to this case.

Focusing on the connection of aggregation operators with t-norms and t-conorms, we discuss, among others, convex-like combinations of t-norms and t-conorms in Section 10.1.

Our general approach to aggregation operators includes also non-associative operators. Prominent examples of associative aggregation operators are the uninorms and the nullnorms presented in Section 10.2, which can be obtained by ordinal sum-like constructions.

10.1 Triangular norm-based aggregation operators

When aggregating data in applications, it is convenient to assign to each tuple of elements in, say, $[0,1]$, a unique number in $[0,1]$. Therefore, an aggregation operator is usually defined as a function from $\bigcup_{n\in\mathbb{N}}[0,1]^n$ to $[0,1]$. Of course, in many special cases, such a general operator can be derived from a binary operation on $[0,1]$ (for instance, from a t-norm or a t-conorm).

10.1 DEFINITION
An *aggregation operator* is a function $\mathbf{A} : \bigcup_{n\in\mathbb{N}} [0,1]^n \longrightarrow [0,1]$ such that:

(i) $\mathbf{A}(x_1,\ldots,x_n) \leq \mathbf{A}(y_1,\ldots,y_n)$ whenever $x_i \leq y_i$ for all $i \in \{1,\ldots,n\}$.
(ii) $\mathbf{A}(x) = x$ for all $x \in [0,1]$.
(iii) $\mathbf{A}(0,\ldots,0) = 0$ and $\mathbf{A}(1,\ldots,1) = 1$.

Each aggregation operator $\mathbf{A}$ can be canonically represented by a family $(A_n)_{n\in\mathbb{N}}$ of n-ary operations, i.e., functions $A_n : [0,1]^n \longrightarrow [0,1]$ given by

$$A_n(x_1,\ldots,x_n) = \mathbf{A}(x_1,\ldots,x_n).$$

Note that $A_1 = \mathrm{id}_{[0,1]}$ and, for $n \geq 2$, each A_n is non-decreasing and satisfies $A_n(0,\ldots,0) = 0$ and $A_n(1,\ldots,1) = 1$. Usually, one identifies an aggregation operator $\mathbf{A}$ and the corresponding family $(A_n)_{n\in\mathbb{N}}$ of n-ary operations.

Most algebraic and analytical properties of an aggregation operator $\mathbf{A}$ are defined via the corresponding properties of the respective n-ary operations A_n.

10.2 DEFINITION
Let $\mathbf{A}$ be an aggregation operator and $(A_n)_{n\in\mathbb{N}}$ the corresponding family of n-ary operations.

(i) The aggregation operator $\mathbf{A}$ is called *commutative*, *idempotent*, *strictly monotone* (on the whole domain) or *continuous* if, for each $n \geq 2$, the n-ary operation A_n is commutative, idempotent, strictly monotone or continuous, respectively.
(ii) The aggregation operator $\mathbf{A}$ is called *associative* if for all $m, n \in \mathbb{N}$ and for all tuples $(x_1, \ldots x_m) \in [0,1]^m$ and $(y_1, \ldots y_n) \in [0,1]^n$

$$\mathbf{A}(x_1,\ldots,x_m,y_1,\ldots,y_n) = \mathbf{A}\left(\mathbf{A}(x_1,\ldots,x_m),\mathbf{A}(y_1,\ldots,y_n)\right).$$

(iii) An element $e \in [0,1]$ is called a *neutral element* and an element $a \in [0,1]$ is called an *annihilator* of the aggregation operator $\mathbf{A}$ if, for each $n \geq 2$, for each $i \in \{1,2,\ldots,n\}$ and for all $x_1,\ldots,x_{i-1},x_{i+1},\ldots,x_n \in [0,1]$, we have, respectively,

$$\mathbf{A}(x_1,\ldots,x_{i-1},e,x_{i+1},\ldots,x_n) = \mathbf{A}(x_1,\ldots,x_{i-1},x_{i+1},\ldots,x_n),$$
$$\mathbf{A}(x_1,\ldots,x_{i-1},a,x_{i+1},\ldots,x_n) = a.$$

Obviously, $\mathbf{A}$ is associative if and only if A_2 is associative and if, for each $n > 2$, A_n equals the canonical n-ary operation derived from A_2 by induction. In such a case, $\mathbf{A}$ is fully and uniquely determined by A_2, and we shall use the same symbol for the aggregation operator $\mathbf{A}$ and for A_2 in this case (as it is done, e.g., for t-norms or t-conorms).

It is interesting to note that, if $\mathbf{A}$ is a commutative, associative aggregation operator, then $a = \mathbf{A}(1,0)$ is an annihilator of $\mathbf{A}$. Also, $\mathbf{A}$ is an idempotent aggregation operator if and only if $T_{\mathbf{M}} \leq \mathbf{A} \leq S_{\mathbf{M}}$.

It is noteworthy that a large class of idempotent aggregation operators can be constructed and represented by means of different types of integrals. For instance, the Lebesgue integral naturally leads to *weighted means*, the Choquet integral [Choquet 1954] to aggregation operators which include the so-called *OWA operators* [Yager 1988, Grabisch 1995], and the Sugeno integral [Sugeno 1974] to aggregation operators including the so-called *generalized OWA operators* [Ralescu & Ralescu 1997]. A recommended recent overview of several types of aggregation operators is [Bouchon-Meunier 1998].

10.3 EXAMPLE

(i) Each t-norm and each t-conorm is a commutative, associative aggregation operator with neutral element 1 and 0, respectively. Moreover, a commutative, associative aggregation operator with neutral element $e \in [0,1]$ is a t-norm if and only if $e = 1$, and a t-conorm if and only if $e = 0$ (and a so-called uninorm, see Definition 10.9, if and only if $e \in]0,1[$).

(ii) The most popular aggregation operators are the arithmetic mean $\mathbf{M}$, the geometric mean $\mathbf{G}$, the harmonic mean $\mathbf{H}$, the quadratic mean $\mathbf{Q}$, and, for $p \in]0,\infty[$, the p-mean $\mathbf{M}_p$ given by, respectively,

$$\mathbf{M}(x_1,\ldots,x_n) = \frac{1}{n}\sum_{i=1}^{n} x_i,$$

$$\mathbf{G}(x_1,\ldots,x_n) = \Big(\prod_{i=1}^{n} x_i\Big)^{\frac{1}{n}},$$

$$\mathbf{H}(x_1,\ldots,x_n) = \frac{n}{\sum\limits_{i=1}^{n}\frac{1}{x_i}},$$

$$\mathbf{Q}(x_1,\ldots,x_n) = \Big(\frac{1}{n}\sum_{i=1}^{n} x_i^2\Big)^{\frac{1}{2}},$$

$$\mathbf{M}_p(x_1,\ldots,x_n) = \Big(\frac{1}{n}\sum_{i=1}^{n} x_i^p\Big)^{\frac{1}{p}}.$$

All these operators are commutative, idempotent, and continuous, none of them is associative, and none of them has a neutral element. The operators $\mathbf{M}$, $\mathbf{Q}$ and $\mathbf{M}_p$ are strictly monotone and have no annihilator, whereas **G** and **H** have annihilator 0, and hence they are not strictly monotone.

A transformation of an aggregation operator by means of a monotone bijection from $[0,1]$ to $[0,1]$ again yields an aggregation operator.

10.4 PROPOSITION

Let **A** *be an aggregation operator and* $\varphi : [0,1] \longrightarrow [0,1]$ *a strictly increasing or a strictly decreasing bijection. Then* $\mathbf{A}_\varphi : \bigcup_{n \in \mathbb{N}} [0,1]^n \longrightarrow [0,1]$ *defined by*

$$\mathbf{A}_\varphi(x_1, \dots, x_n) = \varphi^{-1}\big(\mathbf{A}(\varphi(x_1), \dots, \varphi(x_n))\big)$$

is an aggregation operator.

Trivially, the transformation φ preserves commutativity, idempotency, strict monotonicity, continuity and associativity of aggregation operators. Moreover, if $e \in [0,1]$ is a neutral element and $a \in [0,1]$ is an annihilator of **A** then $\varphi^{-1}(e)$ is a neutral element and $\varphi^{-1}(a)$ is an annihilator of $\mathbf{A}_\varphi$.

10.5 EXAMPLE

(i) If $\varphi : [0,1] \longrightarrow [0,1]$ is given by $\varphi(x) = x^2$ then $\mathbf{M}_\varphi = \mathbf{Q}$ and $\mathbf{G}_\varphi = \mathbf{G}$.

(ii) If $\varphi : [0,1] \longrightarrow [0,1]$ is given by $\varphi(x) = e^{\frac{x-1}{x}}$ then $\mathbf{G}_\varphi = \mathbf{H}$.

(iii) If $\varphi : [0,1] \longrightarrow [0,1]$ is given by $\varphi(x) = 1 - x$ then the aggregation operator $\mathbf{A}_\varphi$ is called the *dual* aggregation operator of **A**, and we shall write **DA** for it, i.e.,

$$\mathbf{DA}(x_1, \dots, x_n) = 1 - \mathbf{A}(1 - x_1, \dots, 1 - x_n).$$

It is trivial that we have always **DDA** = **A**.

(iv) An aggregation operator **A** is called *self-dual* if **DA** = **A**. Self-dual aggregation operators were called *symmetric sums* in [Silvert 1979].

(v) Examples of self-dual aggregation operators are the arithmetic mean **M**, and the operator **C** defined by

$$\mathbf{C}(x_1, \dots, x_n) = \begin{cases} \frac{1}{2} & \text{if } \{0,1\} \subseteq \{x_1, \dots, x_n\} \\ \dfrac{\prod_{i=1}^n x_i}{\prod_{i=1}^n x_i + \prod_{i=1}^n (1 - x_i)} & \text{otherwise.} \end{cases} \tag{10.1}$$

10.6 DEFINITION

If $f : [0,1] \longrightarrow [-\infty, \infty]$ is a continuous, strictly monotone function with $\{f(0), f(1)\} \neq \{-\infty, \infty\}$ then the aggregation operator $\mathbf{M}_f$ given by

$$\mathbf{M}_f(x_1, \dots, x_n) = f^{-1}\Big(\frac{1}{n} \sum_{i=1}^n f(x_i)\Big)$$

is called a *quasi-arithmetic mean*.

It is straightforward that, for all constants $a \in \mathbb{R} \setminus \{0\}$ and $b \in \mathbb{R}$, the functions f and $g = af+b$ lead to the same quasi-arithmetic mean, i.e., we have $\mathbf{M}_f = \mathbf{M}_g$. Consequently, if $f(1)$ is finite (i.e., if 1 is not an annihilator of $\mathbf{M}_f$), then there is always a strictly decreasing continuous function $t : [0,1] \longrightarrow [0,\infty]$ satisfying $t(1) = 0$ (we may take $t = f - f(1)$ if f is decreasing, and $t = f(1) - f$ if f is increasing) such that $\mathbf{M}_f = \mathbf{M}_t$. Since these functions, according to Theorem 5.1, are exactly the additive generators of continuous Archimedean t-norms, this means that there is a one-to-one correspondence between the set of continuous Archimedean t-norms, on the one hand, and the set of quasi-arithmetic means for which 1 is not an annihilator, on the other hand.

In this spirit, $\mathbf{M}$ corresponds to $T_{\mathbf{L}}$ (a generator of which is given by $t(x) = 1 - x$), and $\mathbf{G}$ to $T_{\mathbf{P}}$ (generated by $t(x) = -\log x$). The p-mean $\mathbf{M}_p$ is the counterpart of $T_p^{\mathbf{SS}}$ (generated by $t(x) = 1 - x^p$, see Example 4.3 and Remark 4.4), implying that $\mathbf{Q}$ corresponds to $T_2^{\mathbf{SS}}$, whereas $\mathbf{H}$ is equivalent to $T_0^{\mathbf{H}}$ (generated by $t(x) = \frac{1-x}{x}$, see Example 4.5 and Remark 4.6).

A large class of aggregation operators can be constructed as a combination of t-norms and conorms. From the application point of view, there exist suggestions to use so-called *compensatory operators* as interpretations of the conjunction and the disjunction in certain many-valued logics. To avoid the obvious defects of t-norms and t-conorms with respect to compensation, in [Zimmermann & Zysno 1980] two classes of compensatory operators were suggested.

The first class of compensatory operators investigated was the family of so-called Γ-operators $(\Gamma_\lambda : \bigcup_{n \in \mathbb{N}} [0,1]^n \longrightarrow [0,1])_{\lambda \in [0,1]}$ specified by

$$\Gamma_\lambda(x_1, \ldots, x_n) = \Big(\prod_{i=1}^{n} x_i\Big)^{1-\lambda} \Big(1 - \prod_{i=1}^{n}(1 - x_i)\Big)^{\lambda},$$

which turn out to be a special class of exponential compensatory operators as studied in [Türksen 1992]: for a t-norm T, a t-conorm S and $\lambda \in [0,1]$, the *exponential compensatory operator* $E_{T,S,\lambda} : \bigcup_{n \in \mathbb{N}} [0,1]^n \longrightarrow [0,1]$ is defined by

$$E_{T,S,\lambda}(x_1, \ldots, x_n) = (T(x_1, \ldots, x_n))^{1-\lambda}(S(x_1, \ldots, x_n))^{\lambda}.$$

Obviously, the Γ-operators Γ_λ are based on $T_{\mathbf{P}}$ and $S_{\mathbf{P}}$, i.e., $\Gamma_\lambda = E_{T_{\mathbf{P}}, S_{\mathbf{P}}, \lambda}$. Note that $E_{T,S,\lambda}$ is a logarithmic convex combination of T and S. Also, for all $\lambda_1, \lambda_2 \in [0,1]$ with $\lambda_1 \leq \lambda_2$ we have $E_{T,S,\lambda_1} \leq E_{T,S,\lambda_2}$.

The second class of compensatory operators proposed in [Zimmermann & Zysno 1980] are the so-called convex-linear compensatory operators (compare [Türksen 1992, Alsina *et al.* 1993a]): for each pair (T, S) of a t-norm and a t-conorm and for each index $\lambda \in [0,1]$, the so-called *convex-linear compensatory operator* $L_{T,S,\lambda} : \bigcup_{n \in \mathbb{N}} [0,1]^n \longrightarrow [0,1]$ is given by

$$L_{T,S,\lambda}(x_1, \ldots, x_n) = (1 - \lambda) \cdot T(x_1, \ldots, x_n) + \lambda \cdot S(x_1, \ldots, x_n).$$

As a concrete example, convex-linear compensatory operators of the form $L_{T_{\mathbf{M}},S_{\mathbf{L}},\lambda}$ were used when solving some linear programming problems [Luhandjula 1982]. Again, for $\lambda_1, \lambda_2 \in [0,1]$ with $\lambda_1 \leq \lambda_2$ we get $L_{T,S,\lambda_1} \leq L_{T,S,\lambda_2}$.

Both the exponential compensatory operators $E_{T,S,\lambda}$ and the convex-linear compensatory operators $L_{T,S,\lambda}$ are commutative, and they are idempotent and continuous if and only if T as well as S are both idempotent and continuous, respectively, but they are associative only if $\lambda \in \{0,1\}$ (in which case they coincide with T and S, respectively).

For $\lambda \in [0,1[$, the exponential compensatory operator $E_{T,S,\lambda}$ has annihilator 0, and the only self-dual convex-linear compensatory operators are of the form $L_{T,S,0.5}$, provided that T and S are dual to each other.

A class of associative compensatory operators based on additive generators of strict t-norms and t-conorms was proposed in [Dombi 1982a, Klement *et al.* 1996] which are special uninorms as considered in the following section.

Following [Yager & Filev 1994], we can introduce a non-linear convex combination of a t-norm T and a t-conorm S, where the parameter $\lambda \in [0,1]$ depends on the given inputs. For a given continuous t-norm T^* without zero divisors (which may be different from T), the degree of highness of an input $(x_1,\ldots,x_n)$ is modeled by $T^*(x_1,\ldots,x_n)$ and the degree of lowness is modeled by $T^*(1-x_1,\ldots,1-x_n)$. A *convex-non-linear compensatory operator* $L_{T,S}^{(T^*)}$ is then defined by

$$L_{T,S}^{(T^*)}(x_1,\ldots,x_n) = \begin{cases} a & \text{if } \{0,1\} \subseteq \{x_1,\ldots,x_n\}, \\ (1-\lambda^*)\cdot T(x_1,\ldots,x_n) + \gamma^* \cdot S(x_1,\ldots,x_n) & \\ & \text{otherwise,} \end{cases}$$

where $a \in [0,1]$ and

$$\lambda^* = \frac{T^*(x_1,\ldots,x_n)}{T^*(x_1,\ldots,x_n) + T^*(1-x_1,\ldots,1-x_n)}.$$

If, as a special case, $T = T^*$ and S is the dual t-conorm of T, and if we write $L_{T,S}^{(T^*)} = L_T$, then we have (again with $a \in [0,1]$)

$$L_T(x_1,\ldots,x_n) = \begin{cases} a & \text{if } \{0,1\} \subseteq \{x_1,\ldots,x_n\}, \\ \frac{T(x_1,\ldots,x_n)}{T(x_1,\ldots,x_n)+T(1-x_1,\ldots,1-x_n)} & \text{otherwise.} \end{cases}$$

If $a = 0.5$ then L_T is a self-dual aggregation operator. Note that, for $a = 0.5$, $L_{T_{\mathbf{P}}}$ coincides with the self-dual aggregation operator $\mathbf{C}$ given in (10.1) and, for $a \in \{0,1\}$, $L_{T_{\mathbf{P}}}$ coincides with the uninorm given in (10.4).

An overview of a large number of compensatory and self-dual operators can be found in [Mizumoto 1989b].

Another type of aggregation operators based on t-norms was recently introduced in [Kolesárová & Komorníková 1999] (for a special case thereof see [Moser *et al.* 1999]). The main idea in this approach is to compensate the lack of compensation of a t-norm by increasing the input values.

10.7 DEFINITION

Let T be a continuous t-norm and consider a sequence $\mathcal{F} = ((f_n, g_n))_{n\in\mathbb{N}}$ where, for each $n \in \mathbb{N}$, $f_n, g_n : [0,1] \longrightarrow [0,1]$ are continuous non-decreasing functions such that $\min(f_n, g_n) \geq \mathrm{id}_{[0,1]}$ and $T(f_n(0), g_n(0)) = 0$. An *iterative aggregation operator* $\mathbf{A}_{T,\mathcal{F}} : \bigcup_{n\in\mathbb{N}} [0,1]^n \longrightarrow [0,1]$ is defined recursively by

$$\mathbf{A}_{T,\mathcal{F}}(x) = x,$$
$$\mathbf{A}_{T,\mathcal{F}}(x_1, \ldots, x_{n+1}) = T(f_n(\mathbf{A}_{T,\mathcal{F}}(x_1, \ldots, x_n)), g_n(x_{n+1})).$$

It is evident that each iterative aggregation operator is continuous. The class of iterative aggregation operators contains the class of quasi-arithmetic means for which 1 is not annihilator as a proper subclass. To see this, assume without loss of generality that the quasi-arithmetic mean $\mathbf{M}_t$ is based on the additive generator $t : [0,1] \longrightarrow [0,\infty]$ of some continuous Archimedean t-norm T. Then, to obtain $\mathcal{F}$ it suffices to define the functions $f_n, g_n : [0,1] \longrightarrow [0,1]$ by $f_n(x) = t^{(-1)}\left(\frac{nt(x)}{n+1}\right)$ and $g_n(x) = t^{(-1)}\left(\frac{t(x)}{n+1}\right)$, and we finally get $\mathbf{M}_t = \mathbf{A}_{T,\mathcal{F}}$.

10.8 EXAMPLE

(i) Defining $\mathcal{F}$ by $f_n(x) = \sqrt{x}$ and $g_n(x) = x$ for all $n \in \mathbb{N}$, then the aggregation operator $\mathbf{A}_{T_\mathbf{M},\mathcal{F}}$ is given by

$$\mathbf{A}_{T_\mathbf{M},\mathcal{F}}(x_1, x_2, \ldots, x_n) = \min\left(\sqrt[2^{n-1}]{x_1}, \sqrt[2^{n-2}]{x_2}, \ldots, \sqrt{x_{n-1}}, x_n\right).$$

The aggregation operator $\mathbf{A}_{T_\mathbf{M},\mathcal{F}}$ is continuous and idempotent and has annihilator 0. For each input tuple $(x_1, \ldots, x_n) \in \,]0,1]^n$ and for each value $b \in [0,1[$ we can find $(y_1, \ldots, y_m) \in [0,1]^m$ such that $\mathbf{A}_{T_\mathbf{M},\mathcal{F}}(x_1, \ldots, x_n, y_1, \ldots, y_m) = b$. Moreover, for all $(n+1)$-dimensional tuples $(x_1, \ldots, x_{n+1}) \in [0,1]^{n+1}$ we have

$$\mathbf{A}_{T_\mathbf{M},\mathcal{F}}(x_1, \ldots, x_{n+1}) = \mathbf{A}_{T_\mathbf{M},\mathcal{F}}(\mathbf{A}_{T,\mathcal{F}}(x_1, \ldots, x_n), x_{n+1}).$$

(ii) Define $\mathcal{F}$ by $f_n(x) = \frac{x+1}{2}$ and $g_n(x) = x$ for all $n \in \mathbb{N}$. Then 0 is not an annihilator of the corresponding aggregation operator $\mathbf{A}_{T_\mathbf{P},\mathcal{F}}$, but for all $(x_1, \ldots, x_n) \in [0,1]^n$ we have

$$\mathbf{A}_{T_\mathbf{P},\mathcal{F}}(x_1, \ldots, x_{i-1}, 0, x_{i+1}, \ldots, x_n) = \mathbf{A}_{T_\mathbf{P},\mathcal{F}}(0, x_{i+1}, \ldots, x_n).$$

10.2 Uninorms and nullnorms

Both the neutral element 1 of a t-norm and the neutral element 0 of a t-conorm are boundary points of the unit interval (an analogous statement holds for the neutral element of an I-semigroup $([a,b], *)$ as introduced in Definition 2.40). However, there are many important operations whose neutral element is an interior point of the underlying set, e.g., the multiplication of positive real numbers and the addition of real numbers. The fact that the first three axioms (T1)–(T3) coincide for t-norms and t-conorm, i.e., the only axiomatic difference lies in the location of the neutral element, has led to the introduction of a new class of binary operations closely related to t-norms and t-conorms [Yager & Rybalov 1996].

10.9 DEFINITION

A *uninorm* is a binary operation U on the unit interval $[0,1]$, i.e., a function $U : [0,1]^2 \longrightarrow [0,1]$, which satisfies (T1)–(T3) and

(U4) U has a neutral element $e \in]0,1[$.

Uninorms are the only operations $*$ on the unit interval transforming $([0,1], T_{\mathbf{M}}, *)$ and $([0,1], S_{\mathbf{M}}, *)$ into commutative semirings in the sense of [Kuich & Salomaa 1986] (compare also [Golan 1992]). Also, uninorms play the role of multiplication in some of the integrals considered in 14.3. In some expert systems, special uninorms were used to aggregate local weights into a global one [Hájek *et al.* 1992].

The structure of uninorms was studied extensively in [Fodor *et al.* 1997], where it was proved, among other things, that there is no uninorm which is continuous on the whole unit square. It is evident that, for an arbitrary uninorm U with neutral element $e \in]0,1[$, the operations $T_U, S_U : [0,1]^2 \longrightarrow [0,1]$, which are defined by

$$T_U(x,y) = \frac{1}{e} \cdot U(ex, ey), \tag{10.2}$$

$$S_U(x,y) = \frac{1}{1-e} \cdot \big(U(e + (1-e)x, e + (1-e)y) - e\big), \tag{10.3}$$

are a t-norm and a t-conorm, respectively. This means that we have (recall the ordinal sums of t-norms and t-conorms, and compare Figure 10.1)

$$U|_{[0,e]^2} = (\langle 0, e, T_U \rangle)|_{[0,e]^2},$$
$$U|_{[e,1]^2} = (\langle e, 1, S_U \rangle)|_{[e,1]^2}.$$

On the rest of the unit square U is bounded by the minimum and the maximum, i.e., for all $(x,y) \in [0,1]^2 \setminus \big([0,e]^2 \cup [e,1]^2\big)$ we have

$$\min(x,y) \le U(x,y) \le \max(x,y).$$

The associativity of a uninorm U implies that $U(0,1) \in \{0,1\}$. In the case $U(0,1) = 0$ it is a commutative, associative and monotone extension of the classical Boolean conjunction, and it is called a *conjunctive* uninorm. If $U(0,1) = 1$ (when it extends the Boolean disjunction) we call it a *disjunctive* uninorm.

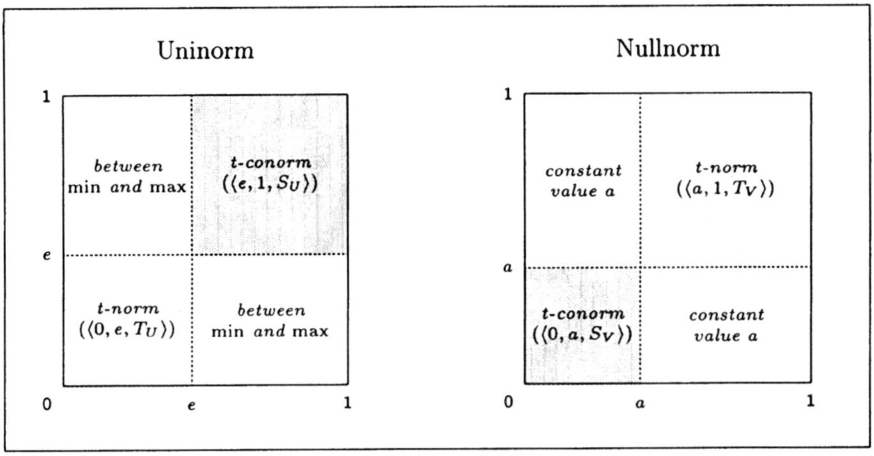

Fig. 10.1. The structure of a uninorm U with neutral element e (left) and of a nullnorm V with annihilator a (see Definition 10.15)

10.10 EXAMPLE

(i) Let T be a t-norm, S be a t-conorm and $e \in]0,1[$. In the spirit of the construction of ordinal sums of fully ordered semigroups presented in [Fuchs 1963, Section 6 of Chapter XI], the following two operations $U^c_{T,S,e} : [0,1]^2 \longrightarrow [0,1]$ and $U^d_{T,S,e} : [0,1]^2 \longrightarrow [0,1]$ make $[0,1]$ into fully ordered semigroups with neutral element e:

$$U^c_{T,S,e}(x,y) = \begin{cases} e \cdot T\left(\frac{x}{e}, \frac{y}{e}\right) & \text{if } (x,y) \in [0,e[^2 \\ e + (1-e) \cdot S\left(\frac{x-e}{1-e}, \frac{y-e}{1-e}\right) & \text{if } (x,y) \in [e,1]^2 \\ \min(x,y) & \text{otherwise,} \end{cases}$$

$$U^d_{T,S,e}(x,y) = \begin{cases} e \cdot T\left(\frac{x}{e}, \frac{y}{e}\right) & \text{if } (x,y) \in [0,e]^2 \\ e + (1-e) \cdot S\left(\frac{x-e}{1-e}, \frac{y-e}{1-e}\right) & \text{if } (x,y) \in]e,1]^2 \\ \max(x,y) & \text{otherwise.} \end{cases}$$

Obviously, $U^c_{T,S,e}$ is a conjunctive, and $U^d_{T,S,e}$ is a disjunctive uninorm. Moreover, for each uninorm U with neutral element $e \in]0,1[$ we have

$$U^c_{T_U,S_U,e} \leq U \leq U^d_{T_U,S_U,e}.$$

(ii) The following conjunctive uninorm $U : [0,1]^2 \longrightarrow [0,1]$ was investigated in [Silvert 1979, Dombi 1982a, Hájek *et al.* 1992, Yager & Filev 1994, Klement *et al.* 1996]:

$$U(x,y) = \begin{cases} \frac{xy}{xy+(1-x)(1-y)} & \text{if } \{x,y\} \neq \{0,1\}, \\ 0 & \text{otherwise.} \end{cases} \tag{10.4}$$

An overview of several classes of uninorms can be found in [De Baets 1998b]. In the following we give representation theorems for two of them. The first result of this type is due to [Klement *et al.* 1996] (compare also [Dombi 1982a]).

10.11 THEOREM
Let $U : [0,1]^2 \longrightarrow [0,1]$ *be a function and* $e \in]0,1[$. *The following are equivalent:*

(i) U *is a uninorm with neutral element* e *which is strictly monotone on* $]0,1[^2$ *and continuous on* $[0,1]^2 \setminus \{(0,1),(1,0)\}$.
(ii) There exists a strictly increasing bijection $u : [0,1] \longrightarrow [-\infty,\infty]$ *with* $u(e) = 0$ *such that for all* $(x,y) \in [0,1]^2$ *we have*

$$U(x,y) = u^{-1}(u(x) + u(y)), \tag{10.5}$$

where, in the case of a conjunctive uninorm U, *we use the convention* $\infty + (-\infty) = -\infty$, *while, in the disjunctive case, we use* $\infty + (-\infty) = \infty$.

If (10.5) holds, then the function u *is uniquely determined by* U *up to a positive multiplicative constant, and it is called an additive generator of the uninorm* U.

If U is a uninorm with additive generator $u : [0,1] \longrightarrow [-\infty,\infty]$ according to Theorem 10.11, then the t-norm T_U and the t-conorm S_U given in (10.2) and (10.3) both are strict, and the functions $t, s : [0,1] \longrightarrow [0,\infty]$ defined by $t(x) = -u(ex)$ and $s(x) = u(e+(1-e)x)$, are additive generators of T_U and S_U, respectively.

Conversely, if T is a strict t-norm and S is a strict t-conorm and if the functions $t, s : [0,1] \longrightarrow [0,\infty]$ are the (uniquely determined) additive generators of T and S, respectively, satisfying $t(0.5) = s(0.5) = 0.5$, then, for each $e \in]0,1[$ and each constant $k \in]0,\infty[$, we can define the strictly increasing bijection $u_{(k)} : [0,1] \longrightarrow [-\infty,\infty]$ by

$$u_{(k)}(x) = \begin{cases} -k \cdot t(\frac{x}{e}) & \text{if } x \in [0,e] \\ s(\frac{x-e}{1-e}) & \text{if } x \in]e,1]. \end{cases} \tag{10.6}$$

Clearly, for each $k \in]0,\infty[$ the construction (10.5) leads to a uninorm $U_{(k)}$ with neutral element e (which, of course, is conjunctive if we use $\infty + (-\infty) = -\infty$, and disjunctive, if we use $\infty + (-\infty) = \infty$). Moreover, we always get $T_{U_{(k)}} = T$ and $S_{U_{(k)}} = S$.

For a fixed $e \in]0,1[$ we get $U_{(k_1)} \geq U_{(k_2)}$ whenever $k_1 \leq k_2$. Using always $\infty + (-\infty) = -\infty$ or always $\infty + (-\infty) = \infty$ we obtain the respective limits

$$\lim_{k \to \infty} U_k = U^c_{T,S,e},$$

$$\lim_{k \searrow 0} U_k = U^d_{T,S,e}.$$

For other limit properties dealing with powers of additive generators of uninorms see [Mesiar & Komorníková 1997].

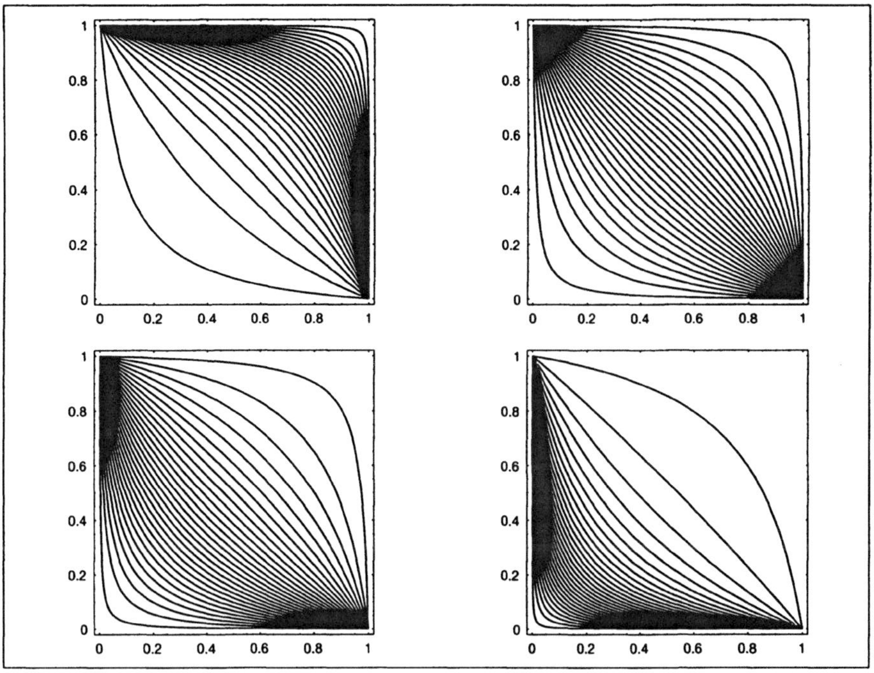

Fig. 10.2. Contour plots of the uninorms $U^{[\frac{1}{9}]}$ (top left), $U^{[1]}$ (top right), $U^{[3]}$ (bottom left), and $U^{[19]}$ (see Example 10.12) with neutral elements 0.9, 0.5, 0.25, and 0.05, respectively

10.12 EXAMPLE
Let, for each $\lambda \in \,]0,\infty[$, $u^{[\lambda]} : [0,1] \longrightarrow [-\infty,\infty]$ be the strictly increasing bijection defined by the formula

$$u^{[\lambda]}(x) = \log\left(\frac{\lambda x}{1-x}\right).$$

Then construction (10.5) leads to the family of uninorms $(U^{[\lambda]})_{\lambda\in]0,\infty[}$ which is given by (in the formula below we use the conventions $\frac{0}{0} = 0$ in the conjunctive case, and $\frac{0}{0} = 1$ in the disjunctive case)

$$U^{[\lambda]}(x,y) = \frac{\lambda xy}{\lambda xy + (1-x)(1-y)},$$

whose neutral element is $\frac{1}{1+\lambda}$. Obviously, in the conjunctive case, $U^{[1]}$ coincides with the conjunctive uninorm given in (10.4), and a simple computation shows that, for each $\lambda \in \,]0,\infty[$, for the t-norm $T_{U^{[\lambda]}}$ we obtain the Hamacher t-norm $T^{\mathbf{H}}_{\frac{\lambda+1}{\lambda}}$, and for the t-conorm $S_{U^{[\lambda]}}$ the Hamacher t-conorm $S^{\mathbf{H}}_{\lambda+1}$ (see Section 4.3). Note that $T^{\mathbf{H}}_{\frac{\lambda+1}{\lambda}}$ and $S^{\mathbf{H}}_{\lambda+1}$ are mutually dual only if $\lambda = 1$. For a visualization of some prototypical members of this family see Figure 10.2.

Starting with an arbitrary continuous Archimedean t-norm T and an arbitrary continuous Archimedean t-conorm S with respective additive generators $t, s : [0,1] \longrightarrow [0,\infty]$ satisfying $t(0.5) = s(0.5) = 1$, and defining $u_{(k)} : [0,1] \longrightarrow [-\infty,\infty]$ as in (10.6), then the construction (10.5) (replacing the inverse $u_{(k)}^{-1}$ by the pseudo-inverse $u_{(k)}^{(-1)}$) leads to the following aggregation operator $U_{(k)} : \bigcup_{n\in\mathbb{N}} [0,1]^n \longrightarrow [0,1]$ specified by

$$U_{(k)}(x_1,\ldots,x_n) = u_{(k)}^{(-1)}\Big(\sum_{i=1}^{n} u_{(k)}(x_i)\Big).$$

However, in general $U_{(k)}$ will not be associative. For instance, take the Łukasiewicz t-norm $T_{\mathbf{L}}$ and its dual t-conorm $S_{\mathbf{L}}$ together with their additive generators specified by $t(x) = 1 - x$ and $s(x) = x$, and fix the neutral element 0.5. Then we obtain $u_{(1)}$ given by $u_{(1)}(x) = 2x - 1$ and, subsequently,

$$U_{(1)}(x_1,\ldots,x_n) = \min\big(1, \max\big(0, \sum_{i=1}^{n} x_i - \tfrac{n-1}{2}\big)\big),$$

i.e., $U_{(1)}$ is a continuous and self-dual aggregation operator which is not associative.

The second representation [De Baets 1999] (compare also [Czogała & Drewniak 1984]) concerns the class of left-continuous idempotent uninorms.

10.13 THEOREM

Let $U : [0,1]^2 \longrightarrow [0,1]$ be a function and $e \in \,]0,1[$. The following are equivalent:

(i) U is a left-continuous idempotent uninorm with neutral element e.

(ii) There exists a left-continuous, uniquely determined non-increasing function $f : [0,1] \longrightarrow [0,1]$ with $f(e) = e$, satisfying $f \circ f(x) \geq x$ for all $x \in [0, f(0)]$ and $f(x) = 0$ for all $x \in \,]f(0), 1]$, such that for all $(x,y) \in [0,1]^2$ we have

$$U(x,y) = \begin{cases} \min(x,y) & \textit{if } x \leq f(0) \textit{ and } y \leq f(x), \\ \max(x,y) & \textit{otherwise.} \end{cases} \tag{10.7}$$

Note that the case of a left-continuous, idempotent conjunctive uninorm corresponds to the requirement $f(0) = 1$ in Theorem 10.13, and that there is a completely analogous result for right-continuous idempotent uninorms.

10.14 EXAMPLE

(i) If we start with an involution $f : [0,1] \longrightarrow [0,1]$ (see Definition 2.37), (10.7) leads to the uninorm U given by

$$U(x,y) = \begin{cases} \min(x,y) & \text{if } y \leq f(x), \\ \max(x,y) & \text{otherwise.} \end{cases}$$

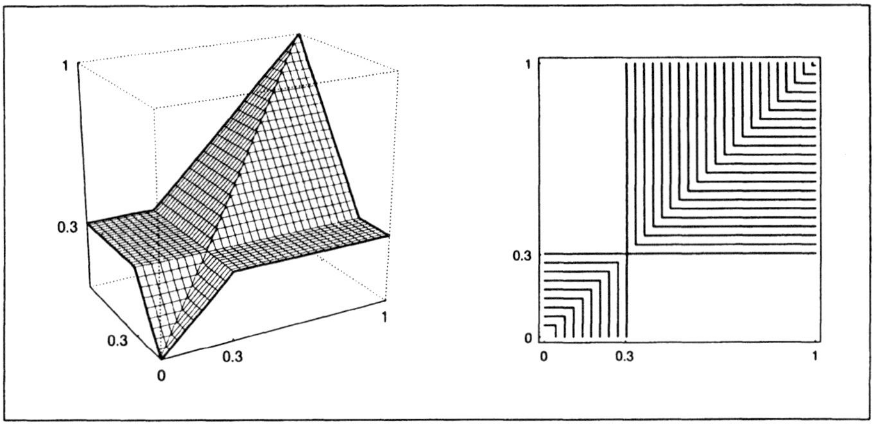

Fig. 10.3. 3D plot (left) and contour plot of the idempotent nullnorm $V_{0.3}$ with annihilator 0.3 as given in (10.8)

(ii) For the disjunctive uninorm $U^d_{T_\mathbf{M},S_\mathbf{M},e}$ with $e \in]0,1[$ introduced in Example 10.10(i), the corresponding function $f : [0,1] \longrightarrow [0,1]$ has the form

$$f(x) = \begin{cases} e & \text{if } x \le e, \\ 0 & \text{otherwise.} \end{cases}$$

Another variation of t-norms and t-conorms (again modifying only the axiom concerning the neutral element) was recently proposed in [Calvo *et al.* 200x].

10.15 Definition

A *nullnorm* is a binary operation V on the unit interval $[0,1]$, i.e., a function $V : [0,1]^2 \longrightarrow [0,1]$, which, for all $x, y, z \in [0,1]$, satisfies (T1)–(T3) and

(V4) there exists $a \in]0,1[$ such that $V(x,0) = x$ for all $x \in [0,a]$ and $V(x,1) = x$ for all $x \in [a,1]$.

It is immediately clear that each nullnorm V satisfies $V(x,a) = a$ for all $a \in [0,1]$, i.e., a is an annihilator of V. Similarly as for uninorms, for a nullnorm V with annihilator $a \in]0,1[$, the operations $T_V, S_V : [0,1]^2 \longrightarrow [0,1]$ defined by

$$T_V(x,y) = \frac{1}{1-a} \cdot \big(V(a+(1-a)x, a+(1-a)y) - a\big),$$
$$S_V(x,y) = \frac{1}{a} \cdot V(ax, ay),$$

are a t-norm and a t-conorm, respectively. This means that (see Figure 10.1)

$$V|_{[0,a]^2} = (\langle 0, a, S_V \rangle)|_{[0,a]^2},$$
$$V|_{[a,1]^2} = (\langle a, 1, T_V \rangle)|_{[a,1]^2},$$

and, for all $(x,y) \in [0,1]^2 \setminus \left([0,a]^2 \cup [a,1]^2\right)$, we get $V(x,y) = a$. Given a t-norm T, a t-conorm S and $a \in \,]0,1[$ then there is a unique nullnorm V with annihilator a such that $T_V = T$ and $S_V = S$. Consequently, for each $a \in \,]0,1[$ there exists a unique idempotent nullnorm V_a with annihilator a which is given by (for a visualization see Figure 10.3)

$$V_a(x,y) = \begin{cases} \max(x,y) & \text{if } (x,y) \in [0,a[^2\,, \\ \min(x,y) & \text{if } (x,y) \in [a,1]^2\,, \\ a & \text{otherwise.} \end{cases} \tag{10.8}$$

Observe that $V_a(x,y)$ is just the standard median of x, y and a (see [Fung & Fu 1975, Czogała & Drewniak 1984]). Note also that $V_{0.5}$ is self-dual (see also [Silvert 1979]).

Recently, in [Mas *et al.* 1999] it was shown that an associative, commutative aggregation operator $F : [0,1]^2 \longrightarrow [0,1]$ such that the horizontal sections $F(.,0), F(.,1) : [0,1] \longrightarrow [0,1]$ are continuous (which means that the graph of the restriction of F to the boundary of the unit square $[0,1]^2$ is a continuous curve) is either a t-norm or a t-conorm or a nullnorm.

The case of associative, non-commutative aggregation operators was discussed, e.g., in [Fodor 1996, Sander 1998]. Again, these operators can be represented in terms of t-norms and t-conorms.

11. Many-valued logics

> The same thing cannot at the same time both belong and not belong to the same object and in the same respect. [...] Of any object, one thing must be either asserted or denied.
>
> ARISTOTLE

> Logic changes from its very foundations if we assume that in addition to truth and falsehood there is also some third logical value or several such values.
>
> JAN ŁUKASIEWICZ

The study of many-valued logic was initiated by Jan Łukasiewicz around 1920. He started with a three-valued logic, introducing in particular an implication for it (see [Łukasiewicz 1920, 1930] and [Łukasiewicz & Tarski 1930], a selection of Łukasiewicz's papers can be found in [Borkowski 1970]). Another important step was the study of intuitionistic logic in [Heyting 1930, Gödel 1932], significant contributions were also made in, e.g., [Post 1921], [Moisil 1940, 1941] and [McNaughton 1951, Chang 1958, Rose & Rosser 1958] (for surveys on many-valued logics we refer to the monographs [Rosser & Turquette 1952, Rescher 1969] and [Gottwald 1989, 200x]).

For several decades, many-valued logic was considered a purely theoretical topic. The introduction of fuzzy sets in [Zadeh 1965] and, subsequently, of fuzzy logics (see, e.g.,[Goguen 6869, Chang & Lee 1975], [Pavelka 1979a, 1979b, 1979c], [Novák 1987], [Hájek 1995b, 1998b]), and [Novák *et al.* 1999, Turunen 1999b, Cignoli *et al.* 2000] produced a new impact to the investigation of many-valued logics, in particular to $[0, 1]$-valued logics based on t-norms, which are often called fuzzy logics.

Among various approaches to fuzzy logics, we describe two of them (studied in [Giles 1976, Hájek *et al.* 1996, Hájek 1998b, Esteva *et al.* 200x], on the one hand, and in [Butnariu *et al.* 1995], on the other hand (for a comparison see [Klement & Navara 1999b]) which are constructed in a similar way. Although starting from different basic logical connectives, they both use interpretations based on continuous t-norms (quite often even based on Frank t-norms, see Section 4.4). Different interpretations of the implication lead to different axiomatizations, but most logics presented here are complete. A comparison of the properties, advantages and disadvantages of the two approaches will be given.

In order to keep this chapter readable it was impossible to describe in detail all important approaches to [0, 1]-valued logics which were presented in the literature. For instance, the concepts introduced and studied in [Pavelka 1979a, 1979b, 1979c], [Höhle 1994, 1995] and [Novák *et al.* 1999] (see also [Höhle & Klement 1995]) have been omitted here.

11.1 Interpretations of connectives in fuzzy logics

A many-valued propositional logic in which the class of truth values is modeled by the unit interval [0, 1], and which forms an extension of the classical Boolean logic, i.e., the two-valued logic with truth values $\{0, 1\}$, is quite often called a fuzzy logic ([Gottwald 1989, 200x] and [Hájek 1995b, 1998b]). In such a logic, the conjunction is usually interpreted by a triangular norm.

11.1 DEFINITION

A *(propositional) fuzzy logic* is described as an ordered pair $\mathcal{P} = (\mathcal{L}, \mathcal{Q})$ of a *language* (*syntax*) $\mathcal{L}$ and a *structure* (*semantics*) $\mathcal{Q}$ specified as follows:

(i) The language of $\mathcal{P}$ is a pair $\mathcal{L} = (A, \mathcal{C})$, where A is a finite or countably infinite set of *atomic symbols* and $\mathcal{C}$ is a tuple of *connectives.*

(ii) The structure of $\mathcal{P}$ is a pair $\mathcal{Q} = ([0, 1], \mathcal{M})$, where $[0, 1]$ is the set of *truth values*, and the tuple $\mathcal{M}$ consists of the *interpretations* (*meanings*) of the connectives in $\mathcal{C}$.

For simplicity, we always consider the same set A of atomic symbols, i.e., all fuzzy logics have the same syntax, they may differ only by their semantics.

11.2 DEFINITION

The class $\mathcal{F}_\mathcal{P}$ of *well-formed formulae* in a fuzzy logic ($\mathcal{P}$*-formulae* for short) is defined inductively as follows:

(i) Each atomic symbol $p \in A$ is a $\mathcal{P}$-formula.

(ii) If $\Diamond$ is an n-ary connective and if $\varphi_1, \varphi_2, \dots, \varphi_n$ are $\mathcal{P}$-formulae, then $\Diamond(\varphi_1, \varphi_2, \dots, \varphi_n)$ is a $\mathcal{P}$-formula.

For each function $\mathbf{t} : A \longrightarrow [0, 1]$, which assigns a truth value to each atomic formula, there exists always a unique natural extension of $\mathbf{t}$ to a *truth assignment* $\overline{\mathbf{t}_\mathcal{P}} : \mathcal{F}_\mathcal{P} \longrightarrow [0, 1]$ which, for each atomic symbol p, for each n-ary connective $\Diamond$, its interpretation $\text{Meaning}_\Diamond$ and for all $\mathcal{P}$-formulae $\varphi_1, \varphi_2, \dots, \varphi_n$, is obtained by induction in the following canonical way:

$$\overline{\mathbf{t}_\mathcal{P}}(p) = \mathbf{t}(p),$$
$$\overline{\mathbf{t}_\mathcal{P}}(\Diamond(\varphi_1, \varphi_2, \dots, \varphi_n)) = \text{Meaning}_\Diamond(\overline{\mathbf{t}_\mathcal{P}}(\varphi_1), \overline{\mathbf{t}_\mathcal{P}}(\varphi_2), \dots, \overline{\mathbf{t}_\mathcal{P}}(\varphi_n)).$$

Throughout this chapter, we will assume that the tuple of connectives contains at least the conjunction $\wedge$ whose interpretation will be given by a t-norm, which, in a canonical way, is a generalization of the interpretation of the conjunction in Boolean logic.

Usually, in a fuzzy logic there are also constants such as the true or the false statement (denoted **1** and **0**, respectively), and other connectives such as the unary connective negation $\neg$, and the binary connectives disjunction $\vee$ and implication $\rightarrow$.

Depending on which of them are considered as basic connectives (and in which way the other ones are derived) one obtains logics with rather different properties. In particular, different interpretations of the implication $\rightarrow$ will lead to different axiomatizations of the respective fuzzy logic.

In the following, we list some possibilities for introducing further connectives and their interpretations which are known from the literature.

Starting with the negation $\neg$, it is clear that its interpretation should map 1 to 0 and 0 to 1, in order to be an extension of the interpretation of the classical, two-valued negation. Another natural property is that the interpretation of the negation $\neg$ be a non-increasing function. To simplify notations, and since there is no confusion possible, an interpretation of the negation $\neg$ will also be called a negation.

11.3 DEFINITION

(i) A non-increasing function $N : [0,1] \longrightarrow [0,1]$ is called a *negation* if

(N1) $N(0) = 1$ and $N(1) = 0$.

(ii) A negation $N : [0,1] \longrightarrow [0,1]$ is called a *strict negation* if, additionally,

(N2) N is continuous.

(N3) N is strictly decreasing.

(iii) A strict negation $N : [0,1] \longrightarrow [0,1]$ is called a *strong negation* if it is an involution (see Definition 2.37), i.e., if

(N4) $N \circ N = \mathrm{id}_{[0,1]}$.

It is obvious that $N : [0,1] \longrightarrow [0,1]$ is a strict negation if and only if it is a strictly decreasing bijection.

11.4 EXAMPLE

(i) The most important and most widely used strong negation is the *standard negation* $N_{\mathbf{s}} : [0,1] \longrightarrow [0,1]$ given by

$$N_{\mathbf{s}}(x) = 1 - x. \tag{11.1}$$

Observe that, as a consequence of Proposition 2.38, each strong negation can be seen as a transformation of the standard negation.

(ii) The negation $N : [0,1] \longrightarrow [0,1]$ given by $N(x) = 1 - x^2$ is strict, but not strong.

(iii) An example of a negation which is not strict and, subsequently, not strong, is the *Gödel negation* $N_{\mathbf{G}} : [0,1] \longrightarrow [0,1]$ given by

$$N_{\mathbf{G}}(x) = \begin{cases} 1 & \text{if } x = 0, \\ 0 & \text{if } x \in]0,1]. \end{cases} \tag{11.2}$$

The standard negation $N_{\mathbf{s}}$ was used, e.g., in [Schweizer & Sklar 1958, 1960] when introducing t-conorms as duals of t-norms, or in [Zadeh 1965] when modeling the complement of a fuzzy set.

An immediate consequence of Proposition 2.34 is that, given a t-norm T and a strict negation N, one obtains a t-conorm $S : [0,1]^2 \longrightarrow [0,1]$, which is N-dual to T in the sense of Definition 2.33, putting

$$S(x,y) = N^{-1}(T(N(x), N(y))). \tag{11.3}$$

Note, however, that if N is a non-strict negation, formula (11.3) cannot be applied.

If N is a strong negation, then, applying the construction in (11.3) to the t-conorm S, we get back the t-norm T we started with.

In a logical sense, a t-conorm is an ideal candidate for the interpretation of the disjunction $\vee$, since it is a canonical extension of the interpretation of the two-valued disjunction. However, there are exceptions from this, i.e., fuzzy logics in which the disjunction $\vee$ is not modeled by a t-conorm (e.g., the Gödel and product R-fuzzy logic in Section 11.3).

A triple (T, S, N), where T is a t-norm, S is a t-conorm and N is a negation is called a *De Morgan triple* if for all $(x,y) \in [0,1]^2$ we have

$$T(x,y) = N^{-1}(S(N(x), N(y))),$$
$$S(x,y) = N^{-1}(T(N(x), N(y))).$$

This means that, given a t-norm T, (T, S, N) is a De Morgan triple if and only if N is a strong negation and S is the N-dual of T.

An interesting result concerning De Morgan triples containing a nilpotent t-conorm is the following: let S be an arbitrary nilpotent t-conorm, define the negation $N : [0,1] \longrightarrow [0,1]$ by

$$N(x) = \inf\{y \in [0,1] \mid S(x,y) = 1\},$$

and let T be the t-norm which is N-dual to S (note that the negation N is necessarily strong). If $s : [0,1] \longrightarrow [0,\infty]$ is the unique additive generator of S with $s(1) = 1$, then for all $(x,y) \in [0,1]^2$ we obtain

$$T(x,y) = s^{-1}(T_{\mathbf{L}}(s(x), s(y))),$$
$$S(x,y) = s^{-1}(S_{\mathbf{L}}(s(x), s(y))),$$
$$N(x) = s^{-1}(N_{\mathbf{s}}(s(x))),$$

which means that the De Morgan triple (T, S, N) is isomorphic to the Łukasiewicz De Morgan triple $(T_{\mathbf{L}}, S_{\mathbf{L}}, N_{\mathbf{s}})$.

Even if (T, S, N) is a De Morgan triple, we do not necessarily have $T(x, N(x)) = 0$ and $S(x, N(x)) = 1$ for all $x \in [0, 1]$, i.e., the law of the excluded middle (which is one of the crucial features of the classical, two-valued Boolean logic) may be violated. For instance, if the t-norm T in the De Morgan triple $(T, S, N_{\mathbf{s}})$ has no zero divisors then the law of the excluded middle never holds. On the other hand, in the De Morgan triple $(T_{\mathbf{L}}, S_{\mathbf{L}}, N_{\mathbf{s}})$ and, a fortiori, in each De Morgan triple $(T, S, N_{\mathbf{s}})$ with $T \leq T_{\mathbf{L}}$, we have a many-valued analogue of the classical law of the excluded middle.

It is noteworthy that, given a De Morgan triple (T, S, N), the tuple $([0, 1], T, S, N, 0, 1)$ can never be a Boolean algebra: in order to satisfy distributivity we must have $T = T_{\mathbf{M}}$ and $S = S_{\mathbf{M}}$ (see Proposition 2.22), in which case it is impossible to have both $T(x, N(x)) = 0$ and $S(x, N(x)) = 1$ for all $x \in [0, 1]$.

Depending on the operations used, one can, however, obtain rather general and useful structures such as, for instance, De Morgan algebras, residuated lattices [Birkhoff 1973], l-monoids (see Section 2.4 and [Birkhoff 1973, Höhle 1995]), Girard algebras (see [Girard 1987], [Weber 1997, 1999] and [Höhle & Weber 1997, 1999]), MV-algebras [Chang 1958], and T-clans or T-tribes [Butnariu & Klement 1993] (see Section 12.4).

Turning to the interpretation of the implication $\rightarrow$ in fuzzy logics, it becomes apparent that there are several logical formulae which, in the Boolean two-valued logic, are equivalent to the implication, but give rise to different interpretations when replacing $\{0, 1\}$ by the unit interval $[0, 1]$ (for early references see, e.g., [Dubois & Prade 1984], [Trillas & Valverde 1985a, 1985b] and [Trillas & Alsina 1993]). As in the case of the negation, we shall, since again no confusion is possible, call an interpretation of the implication $\rightarrow$ also an implication.

We have already discussed, for a left-continuous t-norm T, its T-residuum $\rightarrow_T$ given by (2.17). Taking into account that each such T-residuum generalizes the Boolean implication, we prefer the name R-implication for $\rightarrow_T$ in this logical context.

Given a left-continuous t-norm T, the R-implication $\rightarrow_T$ is left-continuous in its first and right-continuous in its second component, and it is continuous if and only if the underlying t-norm T is nilpotent [Miyakoshi & Shimbo 1985, Smets & Magrez 1987, Bodenhofer 1999].

In particular, we recall the Gödel implication $\rightarrow_{\mathbf{M}}$, the Goguen implication $\rightarrow_{\mathbf{P}}$, and the Łukasiewicz implication $\rightarrow_{\mathbf{L}}$ as presented in Example 2.48.

Several axiomatizations of R-implications can be found in the literature. The first set of axioms was given in [Pedrycz 1982, 1985] (where the R-implication was called Φ-operator) as follows:

$$(\Phi 1) \qquad x \rightarrow_T y \leq x \rightarrow_T z \quad \text{whenever } y \leq z.$$

(Φ2) $T(x, x \rightarrow_T y) \leq y$.
(Φ3) $y \leq x \rightarrow_T T(x, y)$.

The axioms (Φ2) and (Φ3) of the R-implication $\rightarrow_T$ explicitly involve the underlying t-norm T. In [Miyakoshi & Shimbo 1985] the following axioms for R-implications $\rightarrow_T$ with respect to left-continuous t-norms, without any reference to the t-norm, were given:

(I1) $x \rightarrow_T (y \rightarrow_T z) = y \rightarrow_T (x \rightarrow_T z)$.
(I2) $1 \rightarrow_T y = y$.
(I3) $x \rightarrow_T y = 1$ if and only if $x \leq y$.
(I4) $\rightarrow_T$ is right-continuous in its second component.
(I5) $\rightarrow_T$ non-decreasing in its second component.

Although the underlying left-continuous t-norm T does not appear in these axioms, note that it is possible to reconstruct it from $\rightarrow_T$ via (2.16). The underlying t-norm T is necessarily nilpotent if we keep (I1) and (I2), and replace (I3)–(I5) by (I6) and (I7) given by

(I6) $\rightarrow_T$ is non-increasing in its first component.
(I7) $(x \rightarrow_T y) \rightarrow_T y = x$ if and only if $x \geq y$.

If one requires the validity of (I7) only for $y \in]0, 1]$, then the underlying t-norm T must be continuous Archimedean [Mesiar 1994].

Another way to introduce an implication which is an extension of the Boolean implication is to exploit the fact that, in a two-valued logic, the formulae $\varphi \rightarrow \psi$ and $\neg\varphi \vee \psi$ are equivalent.

11.5 Definition

Let S be a t-conorm and N be a negation. Then the corresponding *S-implication* $I_{S,N} : [0,1]^2 \longrightarrow [0,1]$ is defined by

$$I_{S,N}(x, y) = S(N(x), y). \tag{11.4}$$

If N equals the standard negation $N_{\mathbf{s}}$ then we write simply I_S instead of $I_{S,N_{\mathbf{s}}}$.

Note that, if N is a strong negation and if T is the t-norm which is N-dual to S in the sense of Definition 2.33, then we also have for all $(x, y) \in [0,1]^2$

$$I_{S,N}(x, y) = N(T(x, N(y))).$$

Clearly, in this case the law of contraposition holds, i.e., for all $(x, y) \in [0,1]^2$

$$I_{S,N}(x, y) = I_{S,N}(N(y), N(x)).$$

11.6 Example

Fixing the standard negation $N_{\mathbf{s}}$, then for the four basic t-conorms $S_{\mathbf{M}}$, $S_{\mathbf{P}}$, $S_{\mathbf{L}}$ and $S_{\mathbf{D}}$ we obtain the following S-implications (to simplify notations we shall write $I_{\mathbf{M}}$, $I_{\mathbf{P}}$, $I_{\mathbf{L}}$ and $I_{\mathbf{D}}$ rather than the more complicated expressions $I_{S_{\mathbf{M}}}$, $I_{S_{\mathbf{P}}}$, $I_{S_{\mathbf{L}}}$, and $I_{S_{\mathbf{D}}}$, respectively):

$$I_{\mathbf{M}}(x,y) = \max(1-x,y), \qquad \textit{(Kleene-Dienes implication)}$$
$$I_{\mathbf{P}}(x,y) = 1-x+x\cdot y, \qquad \textit{(Reichenbach implication)}$$
$$I_{\mathbf{L}}(x,y) = \min(1-x+y,1), \qquad \textit{(Lukasiewicz implication)}$$
$$I_{\mathbf{D}}(x,y) = \begin{cases} 1-x & \text{if } y=0, \\ y & \text{if } x=1, \\ 1 & \text{otherwise.} \end{cases}$$

Observe that, even if T and S are $N_{\mathbf{s}}$-dual to each other, I_S and $\rightarrow_T$ are different in general, but that, for example, $\rightarrow_{\mathbf{L}}$ and $I_{\mathbf{L}}$ coincide. The same happens if T is the nilpotent minimum $T^{\mathbf{nM}}$ (see Example 3.65) and S its ($N_{\mathbf{s}}$-)dual t-conorm. In general, $\rightarrow_T$ coincides with I_S if and only if the t-conorm S is N-dual to the t-norm T, where the negation N is defined by $N(x) = x \rightarrow_T 0$ (in which case the negation is necessarily a strong one). For more details on the relationship between I_S and $\rightarrow_T$ see, e.g., [Fodor 1995, Jenei 2000].

Finally we only mention that, given a left-continuous t-norm T and the corresponding R-implication $\rightarrow_T$, it is straightforward to define the corresponding biimplication $\leftrightarrow_T : [0,1]^2 \longrightarrow [0,1]$ by

$$x \leftrightarrow_T y = T(x \rightarrow_T y, y \rightarrow_T x). \tag{11.5}$$

It is trivial that for all $(x,y) \in [0,1]^2$

$$x \leftrightarrow_T y = \min(x \rightarrow_T y, y \rightarrow_T x) = \max(x,y) \rightarrow_T \min(x,y).$$

11.7 EXAMPLE

(i) For the left-continuous basic t-norms $T_{\mathbf{M}}$, $T_{\mathbf{P}}$ and $T_{\mathbf{L}}$ we obtain the following biimplications (we shall briefly write $\leftrightarrow_{\mathbf{M}}$, $\leftrightarrow_{\mathbf{P}}$ and $\leftrightarrow_{\mathbf{L}}$ instead of the more complicated expressions $\leftrightarrow_{T_{\mathbf{M}}}$, $\leftrightarrow_{T_{\mathbf{P}}}$ and $\leftrightarrow_{T_{\mathbf{L}}}$, respectively):

$$x \leftrightarrow_{\mathbf{M}} y = \begin{cases} 1 & \text{if } x=y, \\ \min(x,y) & \text{otherwise,} \end{cases}$$
$$x \leftrightarrow_{\mathbf{P}} y = \begin{cases} 1 & \text{if } x=y, \\ \frac{\min(x,y)}{\max(x,y)} & \text{otherwise,} \end{cases}$$
$$x \leftrightarrow_{\mathbf{L}} y = 1-|x-y|.$$

(ii) If T is a continuous Archimedean t-norm and if $t : [0,1] \longrightarrow [0,\infty]$ is an additive generator of T then the biimplication $\leftrightarrow_T$ can be obtained by

$$x \leftrightarrow_T y = t^{(-1)}(|t(y) - t(x)|).$$

11.2 Residuum-based fuzzy logics

We shall start with a continuous t-norm T and use the T-residuum $\rightarrow_T$ as interpretation of the implication $\rightarrow$, which is the reason why we shall speak about residuum-based logics. The following general approach to fuzzy logics with residual implications is due to [Hájek 1998a, 1998b] where the term *basic many-valued logic* is used.

11.8 DEFINITION
Let T be a continuous t-norm. The *residuum-based propositional fuzzy logic* (*R-fuzzy logic* for short) $\mathcal{R}_T = (\mathcal{L}, \mathcal{Q}_T)$ is described as follows:

(i) The language of $\mathcal{R}_T$ is the pair $\mathcal{L} = (A, (\wedge, \rightarrow, \mathbf{0}))$, where A is a finite or countably infinite set of atomic symbols, and the connectives $\wedge$, $\rightarrow$ and $\mathbf{0}$ represent the conjunction, the implication and the false statement (the latter being a constant), respectively.
(ii) The structure of $\mathcal{R}_T$ is the pair $\mathcal{Q}_T = ([0,1], (T, \rightarrow_T, 0))$, where T, $\rightarrow_T$ and 0 (the latter is the zero function) are the interpretations (meanings) of the conjunction $\wedge$, the implication $\rightarrow$ and the false statement $\mathbf{0}$, respectively.

Recall that we assume the set A of atomic symbols to be fixed throughout the whole chapter, so all R-fuzzy logics $\mathcal{R}_T$ will have the same syntax.

In each R-fuzzy logic $\mathcal{R}_T$ with respect to a continuous T, as a consequence of (2.18) and (2.19), it is possible to express the lattice operations $\wedge$ and $\vee$ in the lattice $([0,1], \leq)$ by the t-norm T and the corresponding R-implication $\rightarrow_T$.

The class of well-formed formulae $\mathcal{F}_\mathcal{R}$ in an R-fuzzy logic $\mathcal{R}_T$ ($\mathcal{R}$-formulae for short) is constructed with the help of the binary connectives $\wedge$ and $\rightarrow$ and the constant $\mathbf{0}$. Since $\mathcal{F}_\mathcal{R}$ is independent of the underlying T, no index is needed here.

Using the basic logical connectives $\wedge$, $\rightarrow$ and $\mathbf{0}$, it is possible to define additional logical connectives in an R-fuzzy logic $\mathcal{R}_T$. The negation $\neg$ in $\mathcal{R}_T$ is defined as an implication with consequence $\mathbf{0}$, i.e.,

$$\neg\varphi = \varphi \rightarrow \mathbf{0}.$$

Its interpretation is the negation N_T given by

$$N_T(x) = x \rightarrow_T 0.$$

A disjunction $\vee$ in an R-fuzzy logic $\mathcal{R}_\lambda$ is defined using the De Morgan formula

$$\varphi \vee \psi = \neg(\neg\varphi \wedge \neg\psi).$$

Its interpretation is the operation $D_T : [0,1]^2 \longrightarrow [0,1]$ defined by

$$D_T(x,y) = N_T\big(T(N_T(x), N_T(y))\big).$$

It should be noted that, given an R-fuzzy logic $\mathcal{R}_T$ with a continuous T, then the law of the excluded middle always holds, i.e., for each $\varphi \in \mathcal{F}_{\mathcal{R}}$ we have both $\varphi \wedge \neg\varphi = \mathbf{0}$ and $\varphi \vee \neg\varphi = \mathbf{1}$.

However, we want to stress that, the negation N_T is either strong or non-strict (for instance, if the underlying t-norm T has no zero divisors one always obtains the Gödel negation $N_{\mathbf{G}}$ introduced in (11.2)). As a consequence, N_T is not a strong negation in general, but we always have $N_T \circ N_T \circ N_T = N_T$. Also, if N_T is not a strong negation then D_T is never a t-conorm (which is the case, e.g., in the Gödel logic $\mathcal{R}_{\mathbf{G}}$ and in the product logic $\mathcal{R}_{\mathbf{P}}$ below).

For $\Gamma \subseteq \mathcal{F}_{\mathcal{R}}$ and $K \subseteq [0,1]$, Γ is said to be K-satisfiable in $\mathcal{R}_T$ if there is a truth assignment $\overline{\mathbf{t}_{\mathcal{R}_T}}$ such that $\varphi \in \Gamma$ implies $\overline{\mathbf{t}_{\mathcal{R}_T}}(\varphi) \in K$. The set Γ is called finitely K-satisfiable in $\mathcal{R}_\lambda$ if each finite subset of Γ is K-satisfiable in $\mathcal{R}_\lambda$.

In an R-fuzzy logic $\mathcal{R}_T$, the standard definition of a tautology (called a 1-tautology in [Hájek 1998b]) is used. An $\mathcal{R}$-formula φ is said to be a 1-tautology in $\mathcal{R}_T$ if $\overline{\mathbf{t}_{\mathcal{R}_T}}(\varphi) = 1$ for all $\mathbf{t} \in [0,1]^A$. As some theorems in the classical logic are not 1-tautologies in $\mathcal{R}_T$, it is necessary to adapt the logical axioms in order to obtain a sound logic. The notion of 1-tautology in $\mathcal{R}_T$ depends on the choice of T, hence we shall need different axiomatizations for different R-fuzzy logics. We shall discuss them in some detail when describing the Gödel, Łukasiewicz and product R-fuzzy logic below.

For each continuous t-norm T the R-fuzzy logic $\mathcal{R}_T$ is axiomatizable, and its set of axioms is given as follows (see [Hájek 1998a, 1998b]):

(A1) $(\alpha \to \beta) \to [(\beta \to \gamma) \to (\alpha \to \gamma)]$.

(A2) $(\alpha \wedge \beta) \to \alpha$.

(A3) $(\alpha \wedge \beta) \to (\beta \wedge \alpha)$.

(A4) $[\alpha \wedge (\alpha \to \beta)] \to [\beta \wedge (\beta \to \alpha)]$.

(A5a) $[\alpha \to (\beta \to \gamma)] \to [(\alpha \wedge \beta) \to \gamma]$.

(A5b) $[(\alpha \wedge \beta) \to \gamma] \to [\alpha \to (\beta \to \gamma)]$.

(A6) $[(\alpha \to \beta) \to \gamma] \to [((\beta \to \alpha) \to \gamma) \to \gamma]$.

(A7) $\mathbf{0} \to \alpha$.

Axiom (A1) expresses the fact that the implication is transitive, (A2) that the conjunction implies its first component. (A3) and (A4) say that $\wedge$ and min (which is defined by (A4)) are both commutative, whereas (A5a) and (A5b) state that $\wedge$ and $\to$ are linked by residuation. Finally, (A6) is a variant of the proof by cases, and (A7) states that $\mathbf{0}$ implies everything (*ex falso quodlibet*).

The deduction rule in an R-fuzzy logic $\mathcal{R}_T$ is modus ponens. A set $\Gamma \subseteq \mathcal{F}_{\mathcal{R}}$ is said to be closed under modus ponens if we have $\psi \in \Gamma$ whenever $\varphi \in \Gamma$ and $\varphi \to \psi \in \Gamma$. The closure of a set $\Gamma \subseteq \mathcal{F}_{\mathcal{R}}$ under modus ponens is then the smallest subset of $\mathcal{F}_{\mathcal{R}}$ containing Γ and being closed under modus ponens.

An $\mathcal{R}$-formula φ is called an $\mathcal{R}_T$-theorem if it belongs to the closure of the set of all logical axioms which are closed under modus ponens. A theory $\mathcal{T}$ in an R-fuzzy logic $\mathcal{R}_T$ is a set of $\mathcal{R}$-formulae. An $\mathcal{R}$-formula is said to be $\mathcal{R}_T$-provable in $\mathcal{T}$ (in symbols $\mathcal{T} \vdash_{\mathcal{R}_T} \varphi$) if it belongs to the closure of the union of $\mathcal{T}$ and the set of all axioms under modus ponens.

For all continuous t-norms T, in the R-fuzzy logic $\mathcal{R}_T$ it is possible to prove a weaker form of the classical Deduction Theorem (see [Hájek 1998b, Theorem 2.2.18]) as well as a Completeness Theorem (see [Hájek 1998b, Theorem 2.3.19]). Note that, given an $\mathcal{R}$-formula φ, for all $n \in \mathbb{N}$ the $\mathcal{R}$-formulae φ^n are defined recursively by $\varphi^1 = \varphi$ and $\varphi^{n+1} = \varphi \wedge \varphi^n$.

11.9 THEOREM
Let T be a continuous t-norm, $\mathcal{T}$ be a theory in the R-fuzzy logic $\mathcal{R}_T$, and let φ and ψ be $\mathcal{R}$-formulae. The following are equivalent:

(i) $\mathcal{T} \cup \{\varphi\} \vdash_{\mathcal{R}_T} \psi$.
(ii) There is an $n \in \mathbb{N}$ such that $\mathcal{T} \vdash_{\mathcal{R}_T} \varphi^n \to \psi$.

11.10 THEOREM
For each continuous t-norm T, the R-fuzzy logic $\mathcal{R}_T$ is sound and complete, i.e., the set of $\mathcal{R}_T$-theorems and the set of 1-tautologies in $\mathcal{R}_T$ coincide.

In a fuzzy logic $\mathcal{P}$ the validation set $V_{\mathcal{P}}(\varphi)$ of a $\mathcal{P}$-formula φ is defined by

$$V_{\mathcal{P}}(\varphi) = \{\overline{\mathbf{t}_{\mathcal{P}}}(\varphi) \mid \mathbf{t} \in [0,1]^A\}.$$

In an R-fuzzy logic $\mathcal{R}_T$ this notion, in general, depends on the choice of T. In contrast to the situation of S-fuzzy logics (see Section 11.4), the validation set $V_{\mathcal{R}_T}(\varphi)$ of an $\mathcal{R}$-formula φ in $\mathcal{R}_T$ is not necessarily a subinterval of $[0,1]$, so an analogy of Proposition 11.19 below will hold only for the Łukasiewicz R-fuzzy logic.

An $\mathcal{R}$-formula φ is called an $\mathcal{R}_T$-contradiction if $\neg\varphi$ is an $\mathcal{R}_T$-theorem, and φ is called an $\mathcal{R}_T$-contingency if it is neither an $\mathcal{R}_T$-theorem nor an $\mathcal{R}_T$-contradiction.

The algebraic model of the R-fuzzy logic $\mathcal{R}_T$ is a BL-algebra (see [Hájek 1998a, 1998b] and [Turunen 1999a]), i.e., a tuple $(L, \leq, *, \to, 0, 1)$, where L is a set with $0, 1 \in L$, $\leq$ is a partial order on L, and $*$ and $\to$ are binary operations on L, such that for all $x, y, z \in L$ (compare Definition 2.45)

(i) $(L, \leq)$ is a lattice with universal bounds 0 and 1.
(ii) $(L, \leq, *)$ is an integral commutative, residuated, divisible l-monoid, where $\to$ equals the $*$-residuum $\to_*$.
(iii) $(x \to y) \vee (y \to x) = 1$.

11.3 Residuum-based fuzzy logics: Frank t-norms

We now have a closer look at R-fuzzy logics based on the family of Frank t-norms $(T_\lambda^{\mathbf{F}})_{\lambda\in[0,\infty]}$ introduced in Section 4.4. To simplify notations we shall write briefly $\to_\lambda$ instead of $\to_{T^{\mathbf{F}}_\lambda}$, $\mathcal{R}_\lambda$ rather than $\mathcal{R}_{T^{\mathbf{F}}_\lambda}$, N_λ for $N_{T^{\mathbf{F}}_\lambda}$, and D_λ instead of $D_{T^{\mathbf{F}}_\lambda}$. All other definitions and notations introduced in Section 11.2 will be carried over to this special case, and, obviously, both the variation of the classical Deduction Theorem 11.9 and the Completeness Theorem 11.10 hold for all the R-fuzzy logics $\mathcal{R}_\lambda$.

The logics corresponding to the basic t-norms $T_{\mathbf{M}}, T_{\mathbf{L}}$ and $T_{\mathbf{P}}$, which will play a special role, are the *Gödel R-fuzzy logic* $\mathcal{R}_0 = \mathcal{R}_{\mathbf{M}}$, the *Łukasiewicz R-fuzzy logic* $\mathcal{R}_\infty = \mathcal{R}_{\mathbf{L}}$ and the *product R-fuzzy logic* $\mathcal{R}_1 = \mathcal{R}_{\mathbf{P}}$. In fact, only these three logics are investigated explicitly in [Hájek 1998b].

Looking more carefully at the negations N_λ it is immediate that in the Łukasiewicz R-fuzzy logic $\mathcal{R}_{\mathbf{L}}$, i.e., for $\lambda = \infty$, we obtain the standard negation: $N_\infty = N_{\mathbf{s}}$. Since, with the exception of the Łukasiewicz t-norm $T_{\mathbf{L}}$, no Frank t-norm has zero divisors, we get in all the other cases, i.e., for all $\lambda \in [0, \infty[$, the Gödel negation $N_\lambda = N_{\mathbf{G}}$ which is not strict and, therefore, not strong. As $N_{\mathbf{G}}$ attains only the crisp truth values 0 and 1, this will cause problems in the interpretation of the disjunction D_λ in the case $\lambda \in [0, \infty[$.

For the Łukasiewicz R-fuzzy logic, the disjunction D_λ is just the Łukasiewicz t-conorm, i.e., $D_\infty = S_{\mathbf{L}}$. In all the other cases, i.e., for all $\lambda \in [0, \infty[$,

$$D_\lambda(x, y) = \begin{cases} 0 & \text{if } x = y = 0, \\ 1 & \text{otherwise.} \end{cases}$$

This latter operation attains only the crisp truth values 0 and 1, so it does not satisfy the boundary conditions for a t-conorm. Therefore, it is not an ideal candidate for a reasonable interpretation of the disjunction.

In R-fuzzy logics based on Frank t-norms we have the following general characterization by their respective validation sets:

11.11 THEOREM
Let $\lambda \in [0, \infty]$, $\mathcal{R}_\lambda$ be an R-fuzzy logic, φ an $\mathcal{R}$-formula, and $V_{\mathcal{R}_\lambda}(\varphi)$ its validation set. Then we have:

(i) φ is an $\mathcal{R}_\lambda$-theorem if and only if $V_{\mathcal{R}_\lambda}(\varphi) = \{1\}$.
(ii) φ is an $\mathcal{R}_\lambda$-contradiction if and only if $V_{\mathcal{R}_\lambda}(\varphi) = \{0\}$.
(iii) φ is an $\mathcal{R}_\lambda$-contingency if and only if $V_{\mathcal{R}_\lambda}(\varphi) \neq \{0\}$ and $V_{\mathcal{R}_\lambda}(\varphi) \neq \{1\}$.

We now shall describe the different R-fuzzy logics $\mathcal{R}_\lambda$ in some detail, discussing also the question of axiomatization. In particular, this is done for the Łukasiewicz R-fuzzy logic $\mathcal{R}_{\mathbf{L}}$, the Gödel R-fuzzy logic $\mathcal{R}_{\mathbf{M}}$, and for the product R-fuzzy logic $\mathcal{R}_{\mathbf{P}}$. Some of their main properties are summarized in Table 11.1.

Łukasiewicz logic
Choosing the Łukasiewicz t-norm $T_{\mathbf{L}}$ as the interpretation of the conjunction $\wedge$ in the Łukasiewicz R-fuzzy logic $\mathcal{R}_{\mathbf{L}}$, we obtain the interpretation $\to_{\mathbf{L}}$ (see Example 2.48) of the implication $\to$. In the Łukasiewicz R-fuzzy logic $\mathcal{R}_{\mathbf{L}}$, we have the following Compactness Theorem (see [Hájek 1998b]).

11.12 THEOREM
Let $\Gamma \subseteq \mathfrak{F}_{\mathcal{R}}$, let K be a closed subset of $[0,1]$ and $r \in [0,1]$. Then we have:

(i) The set Γ is K-satisfiable in the Łukasiewicz R-fuzzy logic $\mathcal{R}_{\mathbf{L}}$ if and only if it is finitely K-satisfiable in $\mathcal{R}_{\mathbf{L}}$.
(ii) If Γ is $\{r\}$-satisfiable in $\mathcal{R}_{\mathbf{L}}$, then there exists a maximal $r_0 \in [0,1]$ such that Γ is $\{r_0\}$-satisfiable in $\mathcal{R}_{\mathbf{L}}$.

The Łukasiewicz R-fuzzy logic $\mathcal{R}_{\mathbf{L}}$ is axiomatizable (see [Hájek 1998b]), and its axioms are (A1)–(A7) (as in a general R-fuzzy logic $\mathcal{R}_T$ with respect to a continuous t-norm T) together with

$$\text{(L3)} \qquad [(\alpha \to \beta) \to \beta] \to [(\beta \to \alpha) \to \alpha].$$

The classical Deduction Theorem (like Theorem 11.14 in the case of the Gödel logic) does not hold in the Łukasiewicz R-fuzzy logic, only the weaker Theorem 11.9. The Łukasiewicz R-fuzzy logic $\mathcal{R}_{\mathbf{L}}$ is sound and complete (see [Rose & Rosser 1958, Mundici 1993a] and [Hájek 1995a, 1997]), i.e., the set of $\mathcal{R}_{\mathbf{L}}$-theorems and the set of 1-tautologies in $\mathcal{R}_{\mathbf{L}}$ coincide. The corresponding algebraic model of the Łukasiewicz R-fuzzy logic $\mathcal{R}_{\mathbf{L}}$ is an MV-algebra (see [Mundici 1986, 1991, 1993b]), i.e., a BL-algebra $(L, \leq, *, \to, 0, 1)$ such that, additionally, for each $x \in L$ we have $x = (x \to 0) \to 0$.

We only mention that there is an alternative formulation of the Łukasiewicz fuzzy logic, based only on the implication $\to$ and the false statement $\mathbf{0}$ as basic connectives. The conjunction $\wedge$ is then considered as a derived connective given by $\varphi \wedge \psi = \neg(\varphi \to \neg\psi)$. This conjunction is again interpreted by the Łukasiewicz t-norm $T_{\mathbf{L}}$, so the interpretation remains the same.

In this approach, there is an axiomatization with the four axioms (A1) and (L1)–(L3), where (L1) and (L2) are specified as follows:

$$\text{(L1)} \qquad \alpha \to (\beta \to \alpha).$$
$$\text{(L2)} \qquad (\neg\alpha \to \neg\beta) \to (\beta \to \alpha).$$

Notice that (L1) and (L2) are just the axioms (C1) and (C3) of the classical logic (see also Section 11.4), respectively, and that (A1) is weaker than (C2). The closure of all axioms of the forms (A1) and (L1)–(L3) under modus ponens consists exactly of all the $\mathcal{R}_{\mathbf{L}}$-theorems which do not contain the conjunction $\wedge$.

Gödel logic

Choosing the minimum t-norm $T_{\mathbf{M}}$ as the interpretation of the conjunction $\wedge$ in the Gödel R-fuzzy logic $\mathcal{R}_{\mathbf{G}}$, we obtain the interpretation $\to_{\mathbf{M}}$ (specified in Example 2.48) of the implication $\to$. The R-implication $\to_{\mathbf{M}}$ (also called the Gödel implication) is not continuous in the points (x, x) with $x \in [0, 1[$, and it gives rise to the Gödel negation $N_{\mathbf{G}}$ (recall that this is a non-strict negation). In the Gödel R-fuzzy logic $\mathcal{R}_{\mathbf{G}}$, we have a weaker Compactness Theorem (see [Hájek 1998b]):

11.13 THEOREM
For a set $\Gamma \subseteq \mathfrak{F}_{\mathcal{R}}$ the following are equivalent:

(i) Γ is $\{1\}$-satisfiable in $\mathcal{R}_{\mathbf{G}}$.
(ii) Γ is finitely $\{1\}$-satisfiable in $\mathcal{R}_{\mathbf{G}}$.

The Gödel R-fuzzy logic $\mathcal{R}_{\mathbf{G}}$ is axiomatizable (see [Gödel 1932, Dummett 1959, Hájek 1998b]), and its axioms are (A1)–(A7) (as in a general R-fuzzy logic $\mathcal{R}_T$ with respect to a continuous t-norm T) together with

$$\text{(G)} \qquad \alpha \rightarrow (\alpha \wedge \alpha).$$

The axioms (A2) and (G) imply that the conjunction must be interpreted by an idempotent operation. Since the minimum $T_{\mathbf{M}}$ is the only idempotent t-norm, it is the only t-norm candidate for the interpretation of a logic with these axioms.

The Gödel R-fuzzy logic $\mathcal{R}_{\mathbf{G}}$ is the only R-fuzzy logic in which the classical Deduction Theorem holds, a special case of the general Theorem 11.9:

11.14 THEOREM
Let $\mathcal{T}$ be a theory in the R-fuzzy logic $\mathcal{R}_{\mathbf{G}}$, and let φ and ψ be $\mathcal{R}$-formulae. The following are equivalent:

(i) $\mathcal{T} \cup \{\varphi\} \vdash_{\mathcal{R}_{\mathbf{G}}} \psi$.
(ii) $\mathcal{T} \vdash_{\mathcal{R}_{\mathbf{G}}} \varphi \rightarrow \psi$.

The Gödel R-fuzzy logic $\mathcal{R}_{\mathbf{G}}$ is sound and complete. The corresponding algebraic model is a so-called G-algebra [Hájek 1998b] which is just a BL-algebra $(L, \leq, *, \rightarrow, 0, 1)$ such that for all $x \in L$ we have $x * x = x$, or, equivalently, a Heyting algebra $(L, \leq, *, \rightarrow, 0, 1)$ [Rasiowa & Sikorski 1963] satisfying for all $x, y \in L$ the property $(x \rightarrow y) \vee (y \rightarrow x) = 1$.

Product logic
Choosing the product t-norm $T_{\mathbf{P}}$ as the interpretation of the conjunction $\wedge$ in the product R-fuzzy logic $\mathcal{R}_{\mathbf{P}}$, we obtain the interpretation $\rightarrow_{\mathbf{P}}$ (given in Example 2.48) of the implication $\rightarrow$. The R-implication $\rightarrow_{\mathbf{P}}$ (also called the Goguen implication) is not continuous in the point $(0,0)$. It gives rise to the Gödel negation $N_{\mathbf{G}}$, the same as for the Gödel R-fuzzy logic.

It is an open problem whether a Compactness Theorem analogous to Theorem 11.13 holds for the product R-fuzzy logic.

The product R-fuzzy logic $\mathcal{R}_{\mathbf{P}}$ is axiomatizable [Hájek *et al.* 1996] (compare also [Adillon & Verdú 1998]), and its axioms are (A1)–(A7) (as in a general R-fuzzy logic $\mathcal{R}_T$ with respect to a continuous t-norm T) together with

$$\text{(P1)} \qquad \neg\neg\gamma \rightarrow [[(\alpha \wedge \gamma) \rightarrow (\beta \wedge \gamma)] \rightarrow (\alpha \rightarrow \beta)].$$
$$\text{(P2)} \qquad \neg(\alpha \wedge \alpha) \rightarrow \neg\alpha.$$

The axiom (P1) means that the conjunction satisfies the cancellation law. So only t-norms satisfying the cancellation law (CL) are acceptable candidates for the interpretation of a logic with the axioms (P1) and (P2).

The product R-fuzzy logic does not satisfy the classical Deduction Theorem (Theorem 11.14), only the weaker Theorem 11.9 [Hájek *et al.* 1996, Adillon & Verdú 1998]. The product R-fuzzy logic $\mathcal{R}_{\mathbf{P}}$ is sound and complete. The corresponding algebraic model of the product R-fuzzy logic $\mathcal{R}_{\mathbf{P}}$ is a Π-algebra (product algebra, see [Hájek *et al.* 1996, Adillon & Verdú 1998, Hájek 1998b]), i.e., a BL-algebra $(L, \leq, *, \to, 0, 1)$ satisfying for all $x, y, z \in L$ the additional properties $(x \to 0) \to 0 \leq [(x * z \to y * z) \to (x \to y)]$ and $x \wedge (x \to 0) = 0$.

General R-fuzzy logics $\mathcal{R}_\lambda$ with $\lambda \in \,]0, \infty[$

What was said about the product R-fuzzy logic $\mathcal{R}_{\mathbf{P}} = \mathcal{R}_1$, remains essentially valid also for all R-fuzzy logics $\mathcal{R}_\lambda$ with $\lambda \in \,]0, \infty[$. As a consequence of Corollary 5.4, there is a multiplicative generator (i.e., a strictly increasing bijection) $\theta_\lambda^{\mathbf{F}} : [0,1] \longrightarrow [0,1]$ such that, for all $(x, y) \in [0,1]^2$,

$$\theta_\lambda^{\mathbf{F}}(T_\lambda^{\mathbf{F}}(x, y)) = T_{\mathbf{P}}(\theta_\lambda^{\mathbf{F}}(x), \theta_\lambda^{\mathbf{F}}(y)). \tag{11.6}$$

The multiplicative generator $\theta_\lambda^{\mathbf{F}}$ represents the change of the scale of the unit interval which transforms the t-norm $T_\lambda^{\mathbf{F}}$ into the product t-norm $T_{\mathbf{P}}$. It transforms the corresponding R-implication $\to_\lambda$ into the Goguen implication $\to_{\mathbf{P}}$. However, since the Gödel negation $N_{\mathbf{G}}$ assumes only the values 0 and 1, it is preserved under the strictly increasing bijection $\theta_\lambda^{\mathbf{F}}$. The whole structure (up to the change of scale represented by $\theta_\lambda^{\mathbf{F}}$) remains therefore exactly the same as in the product R-fuzzy logic $\mathcal{R}_{\mathbf{P}}$.

Except for the Łukasiewicz R-fuzzy logic $\mathcal{R}_{\mathbf{L}}$, a connective dual to the conjunction (by the De Morgan formula) does not give a reasonable disjunction. We also have some kind of non-symmetry, because we have a conjunction (interpreted by a t-norm) without a corresponding disjunction interpreted by the dual t-conorm.

In each R-fuzzy logic $\mathcal{R}_\lambda$ with $\lambda \in [0, \infty]$, the formula $\varphi \wedge (\varphi \to \psi)$ has many properties of an (idempotent) conjunction of φ and ψ, and it is interpreted by the minimum t-norm (due to the properties of R-implications). In the Gödel R-fuzzy logic $\mathcal{R}_{\mathbf{G}}$ (and only in it), this formula is semantically equivalent to $\varphi \wedge \psi$. Further, the formula

$$[(\varphi \to \psi) \to \psi] \wedge [[(\varphi \to \psi) \to \psi] \to [(\psi \to \varphi) \to \varphi]]$$

has properties of an (idempotent) disjunction of φ and ψ, and it is interpreted by the maximum t-conorm $S_{\mathbf{M}}$. So we have a disjunction (interpreted by a t-conorm) in each R-fuzzy logic, but it is dual to the (basic) conjunction only in the Gödel R-fuzzy logic $\mathcal{R}_{\mathbf{G}}$. In the Łukasiewicz R-fuzzy logic $\mathcal{R}_{\mathbf{L}}$, we have a disjunction dual to the basic conjunction. So the problems with a disjunction arise only in R-fuzzy logics $\mathcal{R}_\lambda$ with $\lambda \in \,]0, \infty[$ since all these logics are (up to an automorphism of $[0,1]$) equivalent to the product R-fuzzy logic $\mathcal{R}_{\mathbf{P}}$.

R-fuzzy logic	*interpretation of the conjunction*	*corresponding algebraic model*	*axiomatizable sound complete*	*classical Deduction Theorem*
$\mathcal{R}_T$ (T continuous)	T	BL-algebra	yes	no
$\mathcal{R}_\lambda$ ($\lambda \in\,]0,\infty[$)	$T_\lambda^{\mathbf{F}}$	Π-algebra	yes	no
Łukasiewicz ($\mathcal{R}_{\mathbf{L}}$)	$T_{\mathbf{L}}$	MV-algebra	yes	no
Gödel ($\mathcal{R}_{\mathbf{G}}$)	$T_{\mathbf{M}}$	G-algebra	yes	yes
Product ($\mathcal{R}_{\mathbf{P}}$)	$T_{\mathbf{P}}$	Π-algebra	yes	no

Table 11.1. Main examples of R-fuzzy logics

Again the problem remains open whether a Compactness Theorem analogous to Theorem 11.13 holds for the R-fuzzy logics $\mathcal{R}_\lambda$ with $\lambda \in\,]0,\infty[$.

All R-fuzzy logics $\mathcal{R}_\lambda$ with $\lambda \in\,]0,\infty[$ are axiomatizable by the same axioms as the product R-fuzzy logic $\mathcal{R}_{\mathbf{P}}$, i.e., by (A1)–(A7) together with (P1) and (P2). They do not satisfy the classical Deduction Theorem (Theorem 11.17), only Theorem 11.9, but they are sound and complete.

The absence of a disjunction dual to the conjunction in R-fuzzy logics $\mathcal{R}_\lambda$ with $\lambda \in\,]0,\infty[$ recently led to a new concept, an R-fuzzy logic with an involutive negation [Esteva *et al.* 200x]. In this approach, the negation $\neg$ becomes an additional basic connective being interpreted by a strong negation N. This negation can be used in the De Morgan formula defining a dual disjunction $\vee$ which is interpreted by the N-dual t-conorm of the t-norm T interpreting the conjunction $\wedge$. Then $\neg\varphi$ does not necessarily coincide with $\varphi \to \mathbf{0}$. Also this logic is axiomatizable, sound and complete. On the other hand, the system of axioms is more complicated, and also one new deduction rule has to be added.

11.4 S-fuzzy logics

Another way to construct propositional fuzzy logics (using a different interpretation of the implication) was presented in [Butnariu *et al.* 1995], where the full details and proofs of most of the theorems can be found. Since we again restrict ourselves to the case of Frank t-norms $T_\lambda^{\mathbf{F}}$ (and their dual t-conorms $S_\lambda^{\mathbf{F}}$), we shall write briefly I_λ rather than $I_{S_\lambda^{\mathbf{F}}}$.

11.15 Definition

For each $\lambda \in [0,\infty]$, the *S-fuzzy logic* $\mathcal{S}_\lambda = (\mathcal{L}, \mathcal{Q}_\lambda)$ is described as follows:

(i) The language of $\mathcal{S}_\lambda$ is the pair $\mathcal{L} = (A, (\neg, \wedge))$, where A is a finite or countably infinite set of atomic symbols, and the connectives $\neg$ and $\wedge$ represent the negation and the conjunction, respectively.
(ii) The structure of $\mathcal{S}_\lambda$ is the pair $\mathcal{Q}_\lambda = ([0,1], (N_\mathbf{s}, T^\mathbf{F}_\lambda))$, where $N_\mathbf{s}$ and $T^\mathbf{F}_\lambda$ are the interpretations (meanings) of the negation $\neg$ and the conjunction $\wedge$, respectively.

Since we fixed the set A of atomic symbols, all S-fuzzy logics $\mathcal{S}_\lambda$ have the same syntax, and they may differ only by their semantics. Therefore, there is no need to index the language $\mathcal{L}$ by the parameter λ.

For $\lambda = 0$, we obtain the *min-max S-fuzzy logic* $\mathcal{S}_0 = \mathcal{S}_\mathbf{M}$. For $\lambda = \infty$, we obtain the *Łukasiewicz S-fuzzy logic* $\mathcal{S}_\infty = \mathcal{S}_\mathbf{L}$. In these cases, we use the indices $\mathbf{M}$ and $\mathbf{L}$ also for the corresponding structures.

The class of well-formed formulae $\mathfrak{F}_\mathcal{S}$ in an S-fuzzy logic ($\mathcal{S}$-formulae for short) is constructed using the basic connectives $\neg$ and $\wedge$. Since $\mathfrak{F}_\mathcal{S}$ is independent of λ, no index will be used.

Starting with the basic logical connectives $\neg$ and $\wedge$, we can define additional logical connectives in an S-fuzzy logic $\mathcal{S}_\lambda$. The disjunction $\vee$ in $\mathcal{S}_\lambda$ is defined using the De Morgan formula $\varphi \vee \psi = \neg(\neg\varphi \wedge \neg\psi)$. For the interpretation of the disjunction $\vee$, we obtain, subsequently,

$$\overline{\mathbf{t}_{\mathcal{S}_\lambda}}(\varphi \vee \psi) = S^\mathbf{F}_\lambda(\overline{\mathbf{t}_{\mathcal{S}_\lambda}}(\varphi), \overline{\mathbf{t}_{\mathcal{S}_\lambda}}(\psi)),$$

so the disjunction $\vee$ is interpreted by the t-conorm $S^\mathbf{F}_\lambda$ which is dual to $T^\mathbf{F}_\lambda$. The implication $\rightarrow$ in $\mathcal{S}_\lambda$ is defined by $\varphi \rightarrow \psi = \neg(\varphi \wedge \neg\psi)$. For the interpretation of the implication $\rightarrow$, we obtain

$$\overline{\mathbf{t}_{\mathcal{S}_\lambda}}(\varphi \rightarrow \psi) = S^\mathbf{F}_\lambda(N_\mathbf{s}(\overline{\mathbf{t}_{\mathcal{S}_\lambda}}(\varphi)), \overline{\mathbf{t}_{\mathcal{S}_\lambda}}(\psi)).$$

Thus the logical implication $\rightarrow$ in $\mathcal{S}_\lambda$ is interpreted by the corresponding S-implication $I_\lambda : [0,1]^2 \longrightarrow [0,1]$ (see Definition 11.5). This notion is the main reason why this logic is called an S-fuzzy logic. Notice that, for all $\lambda \in [0, \infty[$,

$$I_\lambda(x, y) = 1 \quad \text{if and only if} \quad (x = 0 \text{ or } y = 1). \tag{11.7}$$

Only for the Łukasiewicz S-implication $I_\mathbf{L}$ we have $I_\mathbf{L}(x, y) = 1$ if and only if $x \leq y$ (as for an R-implication with respect to a left-continuous t-norm) because of $I_\mathbf{L} = \rightarrow_\mathbf{L}$.

Since the S-implication $I_\mathbf{L}$ coincides with the R-implication $\rightarrow_\mathbf{L}$, the interpretation of the logical connectives in the Łukasiewicz S-fuzzy logic $\mathcal{S}_\mathbf{L}$ and the Łukasiewicz R-fuzzy logic $\mathcal{R}_\mathbf{L}$ is identical (although not the same connectives are considered as the basic ones).

One difference between the two Łukasiewicz fuzzy logics $\mathcal{R}_\mathbf{L}$ and $\mathcal{S}_\mathbf{L}$ is that the constant $\mathbf{0}$ was not considered a basic $\mathcal{S}$-formula. Nevertheless, it can be introduced as a derived formula putting, e.g., $\mathbf{0} = \neg\varphi \wedge \varphi$ for a fixed $\mathcal{S}$-formula φ, so these formulae are semantically equivalent in the Łukasiewicz S-fuzzy logic

$\mathcal{S}_{\mathbf{L}}$. In all other S-fuzzy logics, i.e., in $\mathcal{S}_\lambda$ for $\lambda \in [0, \infty[$, there is no formula which is evaluated to zero by any truth assignment. This disadvantage can be easily eliminated by adding the constant **0** as a basic formula. No serious problems arise, only the validation sets may become singletons $\{0\}$ or $\{1\}$ and the formulation of Theorems 11.20 and 11.21 has to be generalized to include this case.

Satisfiability (as well as finite satisfiability) in S-fuzzy logics is defined analogously to R-fuzzy logics, i.e., for $\Gamma \subseteq \mathcal{F}_\mathcal{S}$ and $K \subseteq [0,1]$ we say that Γ is K-satisfiable in $\mathcal{S}_\lambda$ if there exists a truth assignment $\overline{\mathbf{t}_{\mathcal{S}_\lambda}}$ such that $\varphi \in \Gamma$ implies $\overline{\mathbf{t}_{\mathcal{S}_\lambda}}(\varphi) \in K$. The set Γ is said to be finitely K-satisfiable in $\mathcal{S}_\lambda$ if each finite subset of Γ is K-satisfiable in $\mathcal{S}_\lambda$.

An important feature of S-fuzzy logics $\mathcal{S}_\lambda$ is that we have the following Compactness Theorem [Butnariu *et al.* 1995]:

11.16 THEOREM

For each $\lambda \in [0, \infty]$, for each $\Gamma \subseteq \mathcal{F}_\mathcal{S}$, for each closed subset K of $[0,1]$, and for each $r \in [0,1]$ we have:

(i) The set Γ is K-satisfiable in the S-fuzzy logic $\mathcal{S}_\lambda$ if and only if it is finitely K-satisfiable in $\mathcal{S}_\lambda$.

(ii) If Γ is $\{r\}$-satisfiable in $\mathcal{S}_\lambda$, then there exists a maximal number $r_0 \in [0,1]$ such that Γ is $\{r_0\}$-satisfiable in $\mathcal{S}_\lambda$.

The following are the axioms of S-fuzzy logics $\mathcal{S}_\lambda$ (observe that they coincide with the axioms of the classical two-valued logic):

(C1) $\alpha \to (\beta \to \alpha)$.

(C2) $[\alpha \to (\beta \to \gamma)] \to [(\alpha \to \beta) \to (\alpha \to \gamma)]$.

(C3) $(\neg\alpha \to \neg\beta) \to (\beta \to \alpha)$.

Note that the implication is considered a derived logical connective in $\mathcal{S}_\lambda$, so it should not appear in the axioms, i.e., we should rewrite the axioms (C1)–(C3) using only the basic connectives $\neg$ and $\wedge$. However, this would lead to much less intuitive expressions. Since the form of axioms makes no difference in the notions depending on the axiomatic system, we use the standard form of axioms known from the classical logic, as it was done in [Butnariu *et al.* 1995].

An $\mathcal{S}$-formula φ is called an $\mathcal{S}$-theorem if it belongs to the closure of the set of all logical axioms under modus ponens. This notion is the same in all S-fuzzy logics $\mathcal{S}_\lambda$, so it is not indexed by λ. The notions of an $\mathcal{S}$-contradiction, of an $\mathcal{S}$-contingency, and of a theory $\mathcal{T}$ in an S-fuzzy logic $\mathcal{S}_\lambda$ are defined in complete analogy to R-fuzzy logics. That a formula φ is $\mathcal{S}$-provable in $\mathcal{T}$ (in symbols $\mathcal{T} \vdash_\mathcal{S} \varphi$) is also defined analogously to R-fuzzy logics. However, the $\mathcal{S}$-provability of a formula φ is independent of the choice of the particular S-fuzzy logic $\mathcal{S}_\lambda$.

For each $\lambda \in [0, \infty]$, the classical Deduction Theorem holds in $\mathcal{S}_\lambda$:

11.17 THEOREM
Let $\lambda \in [0, \infty]$, $\mathcal{T}$ be a theory in the S-fuzzy logic $\mathcal{S}_\lambda$, and let φ, ψ be $\mathcal{S}$-formulae. The following are equivalent:

(i) $\mathcal{T} \cup \{\varphi\} \vdash_{\mathcal{S}} \psi$.
(ii) $\mathcal{T} \vdash_{\mathcal{S}} \varphi \to \psi$.

In any of the S-fuzzy logics $\mathcal{S}_\lambda$, a truth assignment $\overline{\mathbf{t}_{\mathcal{S}_\lambda}}$ evaluates some $\mathcal{S}$-theorems, even the axioms, by values which are less than 1. Therefore, in order to achieve soundness and completeness of an S-fuzzy logic, the notion of tautology has to be adapted accordingly. An $\mathcal{S}$-formula φ is said to be a tautology in $\mathcal{S}_\lambda$ if $\overline{\mathbf{t}_{\mathcal{S}_\lambda}}(\varphi) > 0$ for all $\mathbf{t} \in [0,1]^A$. Notice that this notion depends on the choice of λ.

11.18 THEOREM
For each $\lambda \in [0, \infty[$, the S-fuzzy logic $\mathcal{S}_\lambda$ is sound and complete, i.e., the set of $\mathcal{S}$-theorems and the set of tautologies in $\mathcal{S}_\lambda$ coincide.

An analogue of Theorem 11.18 for the Łukasiewicz S-fuzzy logic $\mathcal{S}_{\mathbf{L}}$ does not hold because, as a consequence of the existence of zero divisors of $T_{\mathbf{L}}$, it is not sound. For instance, the formula $(p \wedge p) \vee (\neg p \wedge \neg p)$, where p is an atomic symbol, is an $\mathcal{S}$-theorem which is not a tautology in $\mathcal{S}_{\mathbf{L}}$. Indeed, if we choose $\mathbf{t}(p) = 0.5$, we obtain $\overline{\mathbf{t}_{\mathcal{S}_{\mathbf{L}}}}((p \wedge p) \vee (\neg p \wedge \neg p)) = 0$.

The non-soundness of the Łukasiewicz S-fuzzy logic $\mathcal{S}_{\mathbf{L}}$ seems to contradict the soundness of the Łukasiewicz R-fuzzy logic $\mathcal{R}_{\mathbf{L}}$ (in particular since, in some sense, these two logics are even "semantically equivalent"[Klement & Navara 1999b]). The reason is that the notions of a theorem are different in the two logics $\mathcal{R}_{\mathbf{L}}$ and $\mathcal{S}_{\mathbf{L}}$ (as a consequence of different axiomatizations), and that the tautologies in $\mathcal{S}_{\mathbf{L}}$ are not necessarily 1-tautologies in $\mathcal{R}_{\mathbf{L}}$.

In S-fuzzy logics, even more can be said about the validation set $V_{\mathcal{S}_\lambda}(\varphi)$ of an $\mathcal{S}$-formula φ in $\mathcal{S}_\lambda$, which is defined in the same manner as in R-fuzzy logics:

11.19 PROPOSITION
For each $\lambda \in [0, \infty]$, each S-fuzzy logic $\mathcal{S}_\lambda$ and for each $\mathcal{S}$-formula φ, the validation set $V_{\mathcal{S}_\lambda}(\varphi)$ is a closed subinterval of $[0,1]$ such that either $0 \in V_{\mathcal{S}_\lambda}(\varphi)$ or $1 \in V_{\mathcal{S}_\lambda}(\varphi)$.

For S-fuzzy logics $\mathcal{S}_\lambda$ which are different from the Łukasiewicz S-fuzzy logic $\mathcal{S}_{\mathbf{L}}$ we can give a more specific characterization of the validation sets:

11.20 THEOREM
Let $\lambda \in [0, \infty[$, $\mathcal{S}_\lambda$ be an S-fuzzy logic, φ an $\mathcal{S}$-formula, and $V_{\mathcal{S}_\lambda}(\varphi)$ its validation set. Then we have:

(i) φ is an $\mathcal{S}$-theorem if and only if, for some $a \in]0,1[$, $V_{\mathcal{S}_\lambda}(\varphi) = [a,1]$.
(ii) φ is an $\mathcal{S}$-contradiction if and only if, for some $b \in]0,1[$, $V_{\mathcal{S}_\lambda}(\varphi) = [0,b]$.
(iii) φ is an $\mathcal{S}$-contingency if and only if $V_{\mathcal{S}_\lambda}(\varphi) = [0,1]$.

For the min-max S-fuzzy logic $\mathcal{S}_{\mathbf{M}}$, we have an even stronger result (see [Butnariu *et al.* 1995]):

11.21 THEOREM
Let φ be an $\mathcal{S}$-formula in the min-max S-fuzzy logic $\mathcal{S}_{\mathbf{M}}$ and $V_{\mathcal{S}_{\mathbf{M}}}(\varphi)$ its validation set. Then we have:

(i) *φ is an $\mathcal{S}$-theorem if and only if $V_{\mathcal{S}_{\mathbf{M}}}(\varphi) = [0.5, 1]$.*
(ii) *φ is an $\mathcal{S}$-contradiction if and only if $V_{\mathcal{S}_{\mathbf{M}}}(\varphi) = [0, 0.5]$.*
(iii) *φ is an $\mathcal{S}$-contingency if and only if $V_{\mathcal{S}_{\mathbf{M}}}(\varphi) = [0, 1]$.*

No analogue of Theorem 11.20 holds for the Łukasiewicz S-fuzzy logic $\mathcal{S}_{\mathbf{L}}$. Also, for $\lambda \in \,]0, \infty[$ there is no chance for a strengthening of Theorem 11.20 since, in contrast to Theorem 11.21(i), we have the following result:

11.22 THEOREM
Let $\lambda \in \,]0, \infty[$, let $\mathcal{S}_\lambda$ be an S-fuzzy logic, and let D be the set of all numbers $a \in [0, 1]$ such that there exists an $\mathcal{S}$-theorem φ_a with $V_{\mathcal{S}_\lambda}(\varphi_a) = [a, 1]$. Then D is a dense subset of $[0, 1]$.

When discussing the R-fuzzy logics $\mathcal{R}_\lambda$ with $\lambda \in \,]0, \infty[$, we argued that they are all equivalent to the product R-fuzzy logic $\mathcal{R}_{\mathbf{P}} = \mathcal{R}_1$ up to a multiplicative generator $\theta_\lambda^{\mathbf{F}} : [0, 1] \longrightarrow [0, 1]$ of $T_\lambda^{\mathbf{F}}$. This argument does not work in S-fuzzy logics. The multiplicative generator $\theta_\lambda^{\mathbf{F}}$ satisfying (11.6) is, of course, the same (up to an exponent). However, it need not preserve the standard negation $N_{\mathbf{s}}$, because the equality $\theta_\lambda^{\mathbf{F}}(N_{\mathbf{s}}(x)) = N_{\mathbf{s}}(\theta_\lambda^{\mathbf{F}}(x))$ does not hold in general. In this case, there is still a t-conorm S satisfying

$$\theta_\lambda^{\mathbf{F}}(S_\lambda^{\mathbf{F}}(x, y)) = S_{\mathbf{P}}(\theta_\lambda^{\mathbf{F}}(x), \theta_\lambda^{\mathbf{F}}(y)),$$

which is not necessarily the ($N_{\mathbf{s}}$-)dual of $T_\lambda^{\mathbf{F}}$, but rather the N-dual of $T_\lambda^{\mathbf{F}}$, where the strong negation N is given by

$$N(x) = (\theta_\lambda^{\mathbf{F}})^{-1}(N_{\mathbf{s}}(\theta_\lambda^{\mathbf{F}}(x))).$$

As a consequence, the S-fuzzy logics $\mathcal{S}_\lambda$ with $\lambda \in \,]0, \infty[$ have basically different semantics.

We only mention here that infinitary S-fuzzy logics were introduced in [Butnariu *et al.* 1995] (see also [Klement & Navara 1999a]). The introduction of infinitary R-fuzzy logics seems to be an open field of research.

12. Fuzzy set theory

> More often than not, the classes of objects encountered in the real physical world do not have precisely defined criteria of membership. [...] Yet, the fact remains that such imprecisely defined "classes" play an important role in human thinking, particularly in the domains of pattern recognition, communication of information, and abstraction.
>
> LOTFI A. ZADEH

The theory of fuzzy sets was introduced in 1965 by Lotfi A. Zadeh as a generalization of the classical Cantorian set theory whose logical basis is the two-valued Boolean logic (compare also [Klaua 1965, 1967]). It was suggested in [Zadeh 1965] to use the minimum $T_{\mathbf{M}}$, the maximum $S_{\mathbf{M}}$, and the standard negation $N_{\mathbf{s}}$ in order to model the intersection, the union, and the complement of fuzzy sets, respectively. However, also the product $T_{\mathbf{P}}$, the probabilistic sum $S_{\mathbf{P}}$ and the Łukasiewicz t-conorm $S_{\mathbf{L}}$ (the latter in a restricted form) were already mentioned as possible candidates for the intersection and the union of fuzzy sets, respectively, in the very first paper on this subject.

Very early traces of (some slight variations of) t-norms and t-conorms in the context of integration of fuzzy sets can be found in [Sugeno 1974], first concepts for a unified theory of fuzzy sets (based on $T_{\mathbf{M}}$ and $S_{\mathbf{M}}$) were presented in [Negoita & Ralescu 1975a] and [Gottwald 1976, 1976a, 1977]. The first papers using general t-norms and t-conorms for operations on fuzzy sets were [Antony & Sherwood 1979, Alsina *et al.* 1980, Dubois 1980] and [Klement 1980, 1981] (see also [Dubois & Prade 1980a]). A full characterization of strong negations as models of the complement of fuzzy sets can be found in [Trillas 1979]. More recent references containing various aspects of fuzzy sets are [Kruse *et al.* 1994a, Dubois & Prade 1999].

In Section 12.1, we start with a short introduction to fuzzy sets in general (compare also the fundamental paper [Höhle & Stout 1991] in this context). In the sequel, because this is a book about t-norms, we concentrate on subjects which strongly depend on the choice of the underlying t-norm. Therefore, we first discuss T-equivalence relations, as introduced in [Valverde 1985, Jacas & Recasens 1992, Klawonn & Kruse 1993] and [Höhle 1992, 1993], which provide one possible approach to fuzzy sets. Next we have a brief look at generalized orderings related to t-norms and T-equivalences as studied in [Bodenhofer 1999], on the basis of earlier work presented in [Zadeh 1971, Ovchinnikov 1991, Fodor & Roubens 1994].

Section 12.3 deals with different concepts of disjointness of fuzzy sets and generalized partitions, going back to suggestions and results in [Ruspini 1969, Bezdek & Harris 1978, Butnariu 1983a, Dubois *et al.* 1992, Kruse *et al.* 1994a, De Baets & Mesiar 1998] and [Höhle 1990, 1993, 1998]. Such partitions play an important role also in applied fields like fuzzy control (see Section 13.3).

Finally, as examples of algebraic structures on classes of fuzzy sets we present T-clans and T-tribes which were introduced and studied in [Klement 1980b, 1982a] and [Butnariu 1983a]. Many important tribes (e.g., those with respect to Frank t-norms with the obvious exception of the minimum $T_{\mathbf{M}}$) turn out to be MV-algebras. However, T-tribes are equipped with an additional semigroup operation (namely, the underlying t-norm) which, in some sense, will be respected by the measures on T-tribes (compare Section 14.1 and the monograph [Butnariu & Klement 1993]).

Many other mathematical branches of fuzzy set theory had to be omitted here for the sake of brevity. A prominent example with an extensive bibliography is fuzzy topology, where usually the minimum $T_{\mathbf{M}}$ is used for the intersection of fuzzy sets (a presentation of generalized topologies involving general continuous t-norms can be found in [Höhle & Šostak 1999]). The reader particularly interested in this topic will find early concepts and results in, e.g., [Chang 1968, Hutton 1975, Lowen 1976, Höhle 1978c, Rodabaugh 1980]. More recent overviews on fuzzy topology are provided by several authors in [Höhle & Rodabaugh 1999]. Also the wide field of fuzzy algebra, in particular the discussion of fuzzy groups (see [Rosenfeld 1971], [Antony & Sherwood 1979, 1982] and [Sessa 1984]), could not be included into this overview.

12.1 Fuzzy subsets of a universe

A well-known fact about Cantor sets (for which we shall sometimes use the term *crisp set* to distinguish them from the fuzzy sets to be defined below) is that, given a universe of discourse X, each subset A of X can be identified with its *characteristic function* $\mathbf{1}_A : X \longrightarrow \{0, 1\}$, assigning the value 1 to all elements of X which belong to A, and the value 0 to all remaining elements of X.

It was exactly this concept of a characteristic function which was generalized in [Zadeh 1965], by simply replacing the two-element set $\{0, 1\}$ by the unit interval $[0, 1]$, and therefore considering a continuum of truth values rather than the two classical values *true* and *false* only (compare the discussion of fuzzy logics and their connectives in Section 11.1).

12.1 Definition

(i) Given a (crisp) universe of discourse X, a *fuzzy subset* A of X (or, briefly, a *fuzzy set* A) is characterized by its *membership function*

$$\mu_A : X \longrightarrow [0,1],$$

where for $x \in X$ the number $\mu_A(x)$ is interpreted as the *degree of membership* of x in the fuzzy set A. The class of all fuzzy subsets of X will be denoted $\mathcal{F}(X)$.

(ii) Given two fuzzy subsets A and B of X, we say that A is *equal* to B (in symbols $A = B$) if $\mu_A = \mu_B$, and that A is a *subset* of B (in symbols $A \subseteq B$) if $\mu_A \leq \mu_B$.

In a logical setting, the degree of membership $\mu_A(x)$ can also be seen as the *truth value* of the statement '*x is element of A*'.

Since the membership function of a fuzzy subset of X generalizes the characteristic function of a crisp subset of X, each crisp subset of X is just a special case of a fuzzy subset of X, and its characteristic function is a special membership function. In particular, the membership functions of the universe X and the empty set $\emptyset$ are the constant functions $\mathbf{1}_X$ and $\mathbf{1}_\emptyset$, mapping each $x \in X$ to 1 and 0, respectively.

Fuzzy sets and membership functions are closely related to the inherent imprecision (fuzziness) of linguistic expressions in natural languages: using Cantorian set theory requires a (usually artificial) threshold to determine whether any given person belongs to the "set of tall persons" or not, in fuzzy logic this is a matter of degree rather than a simple yes/no decision. Nor does probability theory provide a way out: it usually deals with crisp events, and the uncertainty is whether this crisp event will occur or not.

Given a fuzzy subset A of X we can associate with it several types of crisp subsets of X which represent some of the properties of A:

(i) The *kernel* $\ker(A)$ of A is given by

$$\ker(A) = \{x \in X \mid \mu_A(x) = 1\}. \tag{12.1}$$

(ii) The *support* $\operatorname{supp}(A)$ of A is defined by

$$\operatorname{supp}(A) = \{x \in X \mid \mu_A(x) > 0\}. \tag{12.2}$$

(iii) For each $\alpha \in [0,1]$ the *α-cut* (or *α-level set*) $[A]_\alpha$ of A is given by

$$[A]_\alpha = \{x \in X \mid \mu_A(x) \geq \alpha\}. \tag{12.3}$$

A fuzzy subset A of X is called *normal* if its kernel $\ker(A)$ is non-empty, i.e., if there is an element $x_0 \in X$ with

$$\mu_A(x_0) = 1. \tag{12.4}$$

Obviously, A is a crisp subset of X if and only if $\ker(A) = \operatorname{supp}(A) = A$ or, equivalently, if and only if we have $[A]_\alpha = [A]_\beta$ for all $\alpha, \beta \in \,]0,1]$. Also, for each fuzzy subset A of X we have $\ker(A) = [A]_1$ and

$$\operatorname{supp}(A) = \bigcup_{\alpha \in]0,1]} [A]_\alpha.$$

It is immediately clear that the family $([A]_\alpha)_{\alpha \in [0,1]}$ of α-cuts of a fuzzy subset A of X satisfies $[A]_0 = X$ and

$$[A]_\alpha \subseteq [A]_\beta \quad \text{whenever } \alpha \geq \beta, \tag{12.5}$$

$$[A]_\alpha = \bigcap_{\beta \in [0,\alpha[} [A]_\beta. \tag{12.6}$$

The family $([A]_\alpha)_{\alpha \in [0,1]}$ of α-cuts of a fuzzy subset A contains all the information concerning A, since it is possible to reconstruct the membership function of A from the α-cuts [Negoita & Ralescu 1975a, 1975b]: for each $x \in X$ we have

$$\mu_A(x) = \sup\{\min(\alpha, \mathbf{1}_{[A]_\alpha}(x)) \mid \alpha \in \,]0,1]\}.$$

This means that each fuzzy subset A of X can be identified with its family $([A]_\alpha)_{\alpha \in [0,1]}$ of α-cuts (the family $([A]_\alpha)_{\alpha \in [0,1]}$ of its α-cuts was called a vertical representation of the fuzzy subset A in [Kruse *et al.* 1994a], as opposed to the so-called horizontal representation by the graph of its membership function $\{(x, \mu_A(x)) \mid x \in X\}$). Moreover, a family $(M_\alpha)_{\alpha \in]0,1]}$ of crisp subsets of X constitutes the family of α-cuts of some fuzzy subset of X if and only if it satisfies (12.5) and (12.6) [Negoita & Ralescu 1975a, 1975b].

In order to generalize the Boolean set-theoretical operations like intersection, union and complement, it is quite natural to use t-norms, t-conorms, and negations, respectively [Alsina *et al.* 1983, Fodor 1993b].

Given a t-norm T, a t-conorm S and a negation N (see Definition 11.3(i)), it is possible to introduce the operations *intersection* $\cap_T$, *union* $\cup_S$ and *complement* $\complement_N$ on $\mathcal{F}(X)$, where, for A and B in $\mathcal{F}(X)$, the membership functions of the fuzzy subsets $A \cap_T B$, $A \cup_S B$ and $\complement_N A$ of X are given by, respectively,

$$\begin{aligned}
\mu_{A \cap_T B}(x) &= T(\mu_A(x), \mu_B(x)), \\
\mu_{A \cup_S B}(x) &= S(\mu_A(x), \mu_B(x)), \\
\mu_{\complement_N A}(x) &= N(\mu_A(x)).
\end{aligned}$$

In the case of the standard negation, i.e., if $N = N_\mathbf{s}$ we shall briefly write $\complement A$ instead of $\complement_{N_\mathbf{s}} A$. For a justification of this pointwise approach from a categorical point of view see [Lowen 1978, Klement 1982c].

The values $\mu_{A \cap_T B}(x)$, $\mu_{A \cup_S B}(x)$ and $\mu_{\complement_N A}(x)$ describe the truth values of the statements '*x is element of A* AND *x is element of B*', '*x is element of A* OR *x is element of B*', and '*x is* NOT *element of A*', respectively.

The properties of the operations $\cap_T$, $\cap_T$ and $\complement_N$ on $\mathcal{F}(X)$ can be directly derived from the corresponding properties of the underlying t-norm T, t-conorm S, and negation N.

Even if (T, S, N) is a De Morgan triple, we do not necessarily have for all fuzzy subsets A of X the equalities $A \cap_T \complement_N A = \emptyset$ and $A \cup_S \complement_N A = X$. If the t-norm T in the De Morgan triple $(T, S, N_{\mathbf{s}})$ has no zero divisors then these properties never hold unless A is a crisp subset of X. On the other hand, if we work with the De Morgan triple $(T_{\mathbf{L}}, S_{\mathbf{L}}, N_{\mathbf{s}})$ or with a De Morgan triple $(T, S, N_{\mathbf{s}})$ satisfying $T \leq T_{\mathbf{L}}$, then always $A \cap_T \complement A = \emptyset$ and $A \cup_S \complement A = X$.

Algebraically speaking, for an arbitrary De Morgan triple (T, S, N) the tuple $(\mathcal{F}(X), \cap_T, \cup_T, \complement_N, 0, 1)$ is never a Boolean algebra because of the lack of distributivity if $T \neq T_{\mathbf{M}}$ or $S \neq S_{\mathbf{M}}$ and arguments similar to those in the previous paragraph. In fact, the only subclasses of $\mathcal{F}(X)$ which are Boolean algebras are the Boolean subalgebras of the power set $\mathcal{P}(X)$.

However, depending on the underlying operations, it is possible to turn $\mathcal{F}(X)$, or appropriate subclasses thereof, into, for instance, a De Morgan algebra, a residuated lattice [Birkhoff 1973], an l-monoid (see Section 2.4 and [Birkhoff 1973, Höhle 1995]), a Girard algebra [Höhle & Weber 1997, Weber 1997, Weber 1999] and [Weber 1997, 1999], an MV-algebra [Chang 1958], or a T-clan or a T-tribe (see Section 12.4 and [Butnariu & Klement 1993]).

Given a t-norm T and fuzzy subsets A and B of X and Y, respectively, the (Cartesian) *product* $A \times_T B$ is the fuzzy subset of $X \times Y$ given by

$$\mu_{A \times_T B}(x, y) = T\big(\mu_A(x), \mu_B(y)\big).$$

In accordance with the other interpretations of membership functions, the value $\mu_{A \times_T B}(x, y)$ stands for the truth value of the statement '*x is element of A* AND *y is element of B*'.

An interesting and natural question is now whether the α-cuts of the intersection $A \cap_T B$, the union $A \cup_S B$ and the Cartesian product $A \times_T B$ of two fuzzy sets A and B coincide with the intersection, union and Cartesian product, respectively, of the corresponding α-cuts $[A]_\alpha$ and $[B]_\alpha$.

In general, this is not the case, as a consequence of the following result: given a t-norm T, a t-conorm S and $\alpha \in [0, 1]$, then the equalities

$$[A \cap_T B]_\alpha = [A]_\alpha \cap [B]_\alpha, \tag{12.7}$$

$$[A \cup_S B]_\alpha = [A]_\alpha \cup [B]_\alpha, \tag{12.8}$$

$$[A \times_T B]_\alpha = [A]_\alpha \times [B]_\alpha, \tag{12.9}$$

hold for all fuzzy sets A and B if and only if α is an idempotent element of T and S, respectively. In particular, this means that equations (12.7)–(12.9) hold for all $\alpha \in [0, 1]$ and for all fuzzy sets A and B if and only if we have $T = T_{\mathbf{M}}$ and $S = S_{\mathbf{M}}$.

12.2 T-equivalences and T-E-orderings

An important and very intuitive theoretical basis for fuzzy subsets is given by the concept of fuzzy equivalence relations, which, in some sense, measure the degree to which two points of the universe are indistinguishable, and which generalize and relax the concept of classical equivalence relations. A strong motivation for this notion came from the so-called *Poincaré Paradox* (see [Poincaré 1905] and [Höhle 1990, 1999b]): if for three (real world) objects A, B and C we know that A is indistinguishable from B and B is indistinguishable from C, we cannot necessarily conclude that A is indistinguishable from C too.

Fuzzy equivalence relations have been introduced under the name *similarity relation* in [Zadeh 1971] (with respect to the minimum $T_{\mathbf{M}}$, the generalization to t-norms was considered in [Trillas & Valverde 1983]). Other names have been used for this concept in the literature (sometimes in connection with a specific t-norm), such as *likeness relation* [Bezdek & Harris 1978, De Baets & Mesiar 1997], *indistinguishability relation* (see [Trillas & Valverde 1983, 1985a] and [Valverde 1985, Jacas & Recasens 1992, Höhle 1993]), *fuzzy equality* (see [Coulon *et al.* 1992] and [Höhle 1992, 1993]), *equality relation* [Klawonn 1993, Klawonn & Kruse 1993, Kruse *et al.* 1994a], *proximity relation* [Dubois & Prade 1994], and *M-valued equality* [Höhle 1998]. We shall prefer in the sequel the terms T-equivalence relation or T-equivalence which, in our opinion, reflect in the best way the mathematical motivation in the axioms below.

In Cantorian set theory, a relation on a universe X can be identified with a subset of X^2. By analogy, a fuzzy relation on X is then a fuzzy subset of X^2. For early traces of properties of fuzzy relations (in connection with equivalence and order) see [Zadeh 1971] and [Ovchinnikov 1983, 1989, 1991], more recent treatments include [Fodor & Roubens 1994, Bodenhofer 1999].

12.2 Definition

Let T be a t-norm. A *T-equivalence relation* (*T-equivalence* for short) on X is a fuzzy subset E of X^2 such that the following properties are satisfied:

(EQU1) $\mu_E(x,x) = 1.$ (*reflexivity*)

(EQU2) $\mu_E(x,y) = \mu_E(y,x).$ (*symmetry*)

(EQU3) $T\big(\mu_E(x,y), \mu_E(y,z)\big) \leq \mu_E(x,z).$ (*T-transitivity*)

If, in addition, we have

(EQU4) $\mu_E(x,y) = 1$ if and only if $x = y$, (*separation property*)

then the T-equivalence relation is called *separated* or, simply, a *T-equality*.

The value $\mu_E(x,y)$ can be interpreted as the *degree of equality* or the *degree of indistinguishability* of x and y or, equivalently, as the truth value of the statement '*x is equal to y*'. The T-transitivity (EQU3) is a many-valued model of the proposition 'IF *x is equal to y* AND *y is equal to z* THEN *x is equal to z*'.

Note that if E is a T^*-equivalence relation, then it is a T-equivalence relation with respect to any t-norm T which is weaker than T^*. Obviously, for each t-norm T, each crisp equivalence relation is a T-equivalence, and the crisp equality is a T-equality. Moreover, for each T-equivalence E, its kernel $\ker(E)$ is a crisp equivalence relation. Finally, for a $T_{\mathbf{M}}$-equivalence E, for each $\alpha \in [0,1]$ the α-cut $[E]_\alpha$ is a crisp equivalence relation.

An extremely important example of a T-equivalence (in fact, it is even a T-equality) on the unit interval $[0,1]$ is the biimplication $\leftrightarrow_T$, as introduced in (11.5), with respect to a left-continuous T.

Another non-trivial example of a T-equivalence (again it is a T-equality) is the following fuzzy relation measuring the similarity of fuzzy subsets of a universe X which has been considered, at least for the special case $T = T_{\mathbf{L}}$, in [Pappis & Karacapilidis 1993, Kruse *et al.* 1994a, Wang *et al.* 1995]: if T is a left-continuous t-norm then the *T-similarity* SIM_T on $\mathcal{F}(X)$ is given by (compare [Gottwald 1993, De Baets & Mesiar 1997, Bodenhofer 1999])

$$\mu_{\mathrm{SIM}_T}(A,B) = \inf\{\mu_A(x) \leftrightarrow_T \mu_B(x) \mid x \in X\}. \tag{12.10}$$

Given a fuzzy relation R on X which is reflexive and T-transitive, it is always possible to derive a T-equivalence from it with the help of a t-norm which dominates T (see Definition 6.16, for a proof of the following result compare [Valverde 1985, Gottwald 1993, Bodenhofer 1999]).

12.3 PROPOSITION
Let T and T^* be t-norms such that T^* dominates T. If R is a fuzzy relation on X which satisfies (EQU1) and (EQU3) then the following fuzzy relation E_R on X is a T-equivalence, where $\mu_{E_R} : X^2 \longrightarrow [0,1]$ is given by

$$\mu_{E_R}(x,y) = T^*\big(\mu_R(x,y), \mu_R(y,x)\big).$$

Equivalence relations with respect to a continuous Archimedean t-norm are closely related to pseudo-metrics (for a proof of the following result see [Ovchinnikov 1992, Moser 1995, De Baets & Mesiar 1997, Bodenhofer 1999]).

12.4 THEOREM
Let X be a universe of discourse and T a continuous Archimedean t-norm with continuous additive generator $t : [0,1] \longrightarrow [0,\infty]$. Then we have

(i) If $d : X^2 \longrightarrow [0,\infty]$ is a pseudo-metric on X, then the fuzzy relation E_d on X specified by $\mu_{E_d} = t^{(-1)} \circ d$ is a T-equivalence. The T-equivalence E_d is a T-equality if and only if d is a metric.

(ii) If E is a T-equivalence on X, then $d_E : X^2 \longrightarrow [0,\infty]$ given by $d_E = t \circ \mu_E$ is a pseudo-metric on X. The pseudo-metric d_E is a metric if and only if E is a T-equality.

In applications such as fuzzy control (see Section 13.2) one uses quite often linguistic expressions like '*approximately five*' in the formulation of rules in a rule base. These expressions can be viewed, on the one hand, as fuzzy numbers (see Section 13.3) or, on the other hand, as fuzzy points, i.e., as extensions of the concept of a singleton or one-point set to the fuzzy case.

12.5 DEFINITION
Let T be a t-norm and E be a T-equivalence on X.

(i) A fuzzy subset A of X is called *extensional* (with respect to T and E) if for all $x, y \in X$ we have

$$T\big(\mu_A(x), \mu_E(x,y)\big) \leq \mu_A(y). \tag{12.11}$$

(ii) A normal extensional fuzzy subset P of X is called a *fuzzy point* (with respect to T and E) if for all $x, y \in X$

$$T\big(\mu_P(x), \mu_P(y)\big) \leq \mu_E(x,y). \tag{12.12}$$

The extensionality of a fuzzy set A guarantees a certain degree of compatibility between the underlying t-norm T, the T-equivalence E, and the fuzzy set A. It is a many-valued formulation of the statement 'IF *x is element of A* AND *x is equal to y* THEN *y is element of A*'. The inequality (12.12), on the other hand, is the proper extension of the statement 'IF *x is element of P* AND *y is element of P* THEN *x is equal to y*', indicating that P contains at most one element, which, together with the normality (12.4) means that P contains exactly one element of X.

The following result shows that fuzzy points can always be expressed in terms of T-equivalence relations, illustrating the close relationship between these two concepts (see [Kruse *et al.* 1994a]).

12.6 THEOREM
Let T be a t-norm, E be a T-equivalence on X, and P be an extensional fuzzy subset P of X. The following are equivalent:

(i) P is a fuzzy point with respect to T and E.
(ii) There exists an $x_0 \in X$ such that $P = P_{x_0}$ with

$$\mu_{P_{x_0}}(x) = \mu_E(x_0, x). \tag{12.13}$$

Moreover, P_{x_0} is the extensional hull of the crisp singleton $\{x_0\}$, i.e., the smallest extensional fuzzy subset A of X such that $\mu_A(x_0) = 1$.

For special pseudo-metrics and t-norms we obtain very simple and familiar shapes for the membership functions of fuzzy points:

12.7 EXAMPLE

If d is a pseudo-metric on X, then for each $x_0 \in X$ the membership function of the fuzzy point P_{x_0} (with respect to $T_{\mathbf{L}}$, the $T_{\mathbf{L}}$-equivalence E_d, and an additive generator $t : [0,1] \longrightarrow [0,\infty]$ of $T_{\mathbf{L}}$) is given by

$$\mu_{P_{x_0}}(x) = t^{(-1)}(d(x, x_0)) = \max\left(1 - \tfrac{d(x,x_0)}{t(0)}, 0\right).$$

If, in particular, $X = \mathbb{R}$ and d is the usual, i.e., the Euclidean metric on $\mathbb{R}$, then the membership function of a fuzzy point P_{x_0} (with respect to $T_{\mathbf{L}}$ and the $T_{\mathbf{L}}$-equivalence E_d) is visualized by the isosceles triangle whose base is the interval $[x_0 - t(0), x_0 + t(0)]$ and which has height 1.

In real control applications, we usually have an input space of dimension higher than one. It is therefore necessary to consider equivalence relations on Cartesian products. Starting from T-equivalences on X and Y and from a t-norm T^* which dominates T, it is possible to construct a T-equivalence on $X \times Y$ (see [Valverde 1985, Gottwald 1993, De Baets & Mesiar 1997, Bodenhofer 1999]).

12.8 PROPOSITION

Let X and Y be two (crisp) universes of discourse, and let T and T^* be t-norms such that T^* dominates T. If E and F are T-equivalences on X and Y, respectively, then the fuzzy relation $E \times_{T^*} F$ on $X \times Y$ defined by

$$\mu_{E \times_{T^*} F}\big((x_1, y_1), (x_2, y_2)\big) = T^*\big(\mu_E(x_1, x_2), \mu_F(y_1, y_2)\big)$$

is a T-equivalence on $X \times Y$. The T-equivalence $E \times_{T^*} F$ on $X \times Y$ is a T-equality if and only if both E and F are T-equalities.

In the Euclidean case the fuzzy points with respect to the Łukasiewicz t-norm again have membership functions with a simple geometric form:

12.9 EXAMPLE

(i) If d_1, d_2 are pseudo-metrics on X and Y, respectively, and E_{d_1} and E_{d_2} are the corresponding $T_{\mathbf{L}}$-equivalences on X and Y (based on the same additive generator of $T_{\mathbf{L}}$), then the product $T_{\mathbf{L}}$-equivalence $E_{d_1} \times_{T_{\mathbf{L}}} E_{d_2}$ is induced by the pseudo-metric

$$d\big((x_1, x_2), (y_1, y_2)\big) = d_1(x_1, y_1) + d_2(x_2, y_2).$$

(ii) If $(X, d_1) = (Y, d_2) = (\mathbb{R}, d)$, where d is the Euclidean metric on $\mathbb{R}$, this means that the membership function of a fuzzy point $P_{\mathbf{x}_0}$ (with respect to $T_{\mathbf{L}}$ and the $T_{\mathbf{L}}$-equivalence $E_d \times_{T_{\mathbf{L}}} E_d$, using an arbitrary generator $t : [0,1] \longrightarrow [0,\infty]$ of $T_{\mathbf{L}}$) is a regular pyramid of height 1, whose base is the square with center $\mathbf{x}_0 \in \mathbb{R}^2$ and the side length $\sqrt{2} \cdot t(0)$.

From a family of fuzzy subsets of X and a left-continuous t-norm T a T-equivalence can be derived with the help of the biimplication $\leftrightarrow_T$ introduced in (11.5), which turns out to be compatible in some sense with the given family of fuzzy subsets (for a proof compare [Valverde 1985, Klawonn & Castro 1995, Bodenhofer 1999]):

12.10 THEOREM
Let T be a left-continuous t-norm and $(A_i)_{i\in I}$ be a family of fuzzy subsets of X. Then the fuzzy relation E on X defined by

$$\mu_E(x,y) = \inf\{\mu_{A_i}(x) \leftrightarrow_T \mu_{A_i}(y) \mid i \in I\}$$

is a T-equivalence. Moreover, E is the greatest T-equivalence on X such that all fuzzy sets A_i are extensional.

Fuzzy order relations have been considered in the literature in [Zadeh 1971, Ovchinnikov 1991, Fodor & Roubens 1994]. The following definition was given in [Bodenhofer 1999]. Unlike its predecessors, its antisymmetry is not only based on a t-norm T, but also on a T-equivalence.

12.11 DEFINITION
Let T be a t-norm and E be a T-equivalence on a universe X. A T-transitive fuzzy relation O on X is called a *fuzzy ordering* with respect to T and E (a *T-E-ordering* for short) if for all $(x,y) \in X^2$

(ORD1) $\mu_O(x,y) \geq \mu_E(x,y)$, (*$E$-reflexivity*)
(ORD2) $T(\mu_O(x,y), \mu_O(y,x)) \leq \mu_E(x,y)$. (*$T$-$E$-antisymmetry*)

Starting from a fuzzy relation on X which is symmetric and T-transitive with respect to some t-norm T, i.e., which satisfies axioms (EQU1) and (EQU3), the following relationship to T-E-orderings was shown in [Bodenhofer 1999]:

12.12 THEOREM
Let T and T^* be t-norms such that T^* dominates T, let E be a T-equivalence on a universe X, and let O be a fuzzy relation on X satisfying (EQU1) and (EQU3) with respect to T. Then we have:

(i) O is a T-E_O-ordering on X, where the T-equivalence E_O on X is given by

$$\mu_{E_O}(x,y) = T^*\big(\mu_O(x,y), \mu_O(y,x)\big).$$

(ii) O is a T-E-ordering if and only if, for all $(x,y) \in X^2$,

$$T\big(\mu_O(x,y), \mu_O(y,x)\big) \leq \mu_E(x,y) \leq \min\big(\mu_O(x,y), \mu_O(y,x)\big).$$

A simple example is that, for each left-continuous t-norm T, the residuum $\vec{\to}_T$ is a T-$\leftrightarrow_T$-ordering on the unit interval $[0,1]$ (where $\leftrightarrow_T$ is the biimplication introduced in (11.5)).

Another example of a fuzzy ordering is the T-inclusion of fuzzy subsets of a universe X which is a special case of the inclusion considered in [Bandler & Kohout 1980] (compare also [Gottwald 1993] for the case $T = T_{\mathbf{L}}$): if T is a left-continuous t-norm then the *T-inclusion* INCL_T is given by

$$\mu_{\mathrm{INCL}_T}(A,B) = \inf\{\mu_A(x) \vec{\to}_T \mu_B(x) \mid x \in X\}.$$

As was shown in [Bodenhofer 1999], INCL_T is indeed a T-SIM_T-ordering on $\mathcal{F}(X)$ (the T-similarity SIM_T is given in (12.10)).

12.3 T-partitions

A well-known fact in Cantorian set theory is the one-to-one correspondence between equivalence relations on a universe X on the one hand, and partitions on the other hand. In the context of fuzzy sets, several concepts of fuzzy partitions have been suggested. We discuss here one of them which has a close relationship to T-equivalences, and only mention some of the other suggestions.

12.13 DEFINITION
Let T be a t-norm. A subclass $\{A_i \mid i \in I\}$ of $\mathcal{F}(X)$ is called a *T-partition* of X if each A_i is normal, $\bigcup_{i\in I} \ker(A_i) = X$ and if, for all $i,j \in I$, all $x \in \ker(A_i)$ and all $y \in X$ we have the inequality

$$T(\mu_{A_j}(x), \mu_{A_j}(y)) \leq \mu_{A_i}(y). \tag{12.14}$$

Observe that, for a T-partition $\{A_i \mid i \in I\}$ of X, (12.14) implies that the kernels of the A_i's are pairwise disjoint and, taking into account the other requirements, form a (crisp) partition of X. Note also that, for all $i,j \in I$ we have for all $x \in \ker(A_i)$ and for all $y \in \ker(A_j)$ that $\mu_{A_j}(x) = \mu_{A_i}(y)$ [Thiele & Schmechel 1995, De Baets & Mesiar 1998].

The following characterization can be found in [Thiele & Schmechel 1995, Höhle 1996, De Baets & Mesiar 1998]:

12.14 THEOREM
Let T be a t-norm and $\{A_i \mid i \in I\}$ a subclass of $\mathcal{F}(X)$. The following are equivalent:

(i) $\{A_i \mid i \in I\}$ is a T-partition of X.
(ii) There exists a T-equality E on I such that for all $(i,j) \in I^2$ and for all $x \in \ker(A_j)$ we have $\mu_E(i,j) = \mu_{A_i}(x)$.
(iii) There exists a T-equivalence F on X such that $\{A_i \mid i \in I\}$ coincides with the set of all fuzzy points $\{P_{x_0} \mid x_0 \in X\}$ with respect to T and F.

This means that the one-to-one correspondence between equivalence relations and partitions holds also in this case: given a T-partition $\{A_i \mid i \in I\}$ of X, the corresponding uniquely determined T-equivalence F on X is given by

$$\mu_F(x, y) = \sup \left\{T_{\mathbf{D}}(\mu_{A_i}(x), \mu_{A_i}(y)) \mid i \in I\right\}.$$

Conversely, given a T-equivalence F on X, the class of all corresponding fuzzy points $\{P_x \mid x \in X\}$ forms a T-partition of X. It should be mentioned that an equivalent approach to T-partitions for left-continuous t-norms T was discussed in [Höhle 1993, 1998].

12.15 PROPOSITION

Let T be a left-continuous t-norm and $\{A_i \mid i \in I\}$ a subclass of $\mathcal{F}(X)$. The following are equivalent:

(i) $\{A_i \mid i \in I\}$ is a T-partition of X.
(ii) Each A_i is normal, $\bigcup_{i \in I} \ker(A_i) = X$ and for all $i, j \in I$ we have

$$\sup\{\mu_{A_i \cap_T A_j}(x) \mid x \in X\} = \mu_{\mathrm{SIM}_T}(A_i, A_j). \tag{12.15}$$

Roughly speaking, (12.15) states that the degree of overlapping of A_i and A_j (described on the left-hand side) equals their similarity (compare [Höhle 1993, Klawonn & Kruse 1993, Kruse *et al.* 1994a]).

If T is a continuous Archimedean t-norm then, as a consequence of Theorem 12.4, we have the following correspondence between pseudo-metrics and T-partitions:

12.16 COROLLARY

Let T be a continuous Archimedean t-norm and $t : [0, 1] \longrightarrow [0, \infty]$ an additive generator of T. Then we have:

(i) If $d : X^2 \longrightarrow [0, \infty]$ is a pseudo-metric on X then $\{P_{x_0} \mid x_0 \in X\}$ is a T-partition of X, where the fuzzy subsets P_{x_0} of X are given by

$$\mu_{P_{x_0}}(x) = t^{(-1)}(d(x, x_0)).$$

(ii) If $\{A_i \mid i \in I\}$ is a T-partition of X, then $d : X^2 \longrightarrow [0, \infty]$ given by

$$d(x, y) = t\big(\sup \left\{T_{\mathbf{D}}\left(\mu_{A_i}(x), \mu_{A_i}(y)\right) \mid i \in I\right\}\big).$$

is a pseudo-metric on X.

12.17 EXAMPLE

Consider a finite (crisp) partition $\{C_1, \ldots, C_n\} \subseteq \mathcal{P}(X)$ of the universe X and the function $k : X \longrightarrow \{1, \ldots, n\}$ defined by $k(x) = i$ if $x \in C_i$. For each $i \in \{1, 2 \ldots, n\}$ we construct the fuzzy subset A_i of X as follows:

$$\mu_{A_i}(x) = 2^{-|i - k(x)|}.$$

Then $\mathcal{A} = \{A_1, A_2, \ldots, A_n\}$ is a $T_{\mathbf{P}}$-partition of X. The corresponding $T_{\mathbf{P}}$-equivalence relation E and the corresponding pseudo-metric d are given by

$$\mu_E(x,y) = 2^{-|k(x)-k(y)|},$$
$$d(x,y) = |k(x) - k(y)|.$$

Concerning refinements and products of T-partitions we have the following result [De Baets & Mesiar 1998] (for the concept of domination see Section 6.3):

12.18 PROPOSITION
Let T be a t-norm and T^* a t-norm which dominates T. Then we have:

(i) If $\{A_i \mid i \in I\}$ and $\{B_j \mid j \in J\}$ are two T-partitions of X then

$$\{T^*(A_i, B_j) \mid (i,j) \in I \times J \text{ and } \ker(A_i) \cap \ker(A_i) \neq \emptyset\}$$

is a T-partition of X, where for each $(i,j) \in I \times J$ the membership function $\mu_{T^*(A_i,B_j)} : X \longrightarrow [0,1]$ of the subset $T^*(A_i, B_j)$ of X is given by

$$\mu_{T^*(A_i,B_j)}(x) = T^*\big(\mu_{A_i}(x), \mu_{B_j}(x)\big).$$

(ii) If $\{A_i \mid i \in I\}$ is a T-partition of X and $\{B_j \mid j \in J\}$ is a T-partition of Y then $\{T^*(A_i, B_j) \mid (i,j) \in I \times J\}$ is a T-partition of $X \times Y$, where for each $(i,j) \in I \times J$ the membership function $\mu_{T^*(A_i,B_j)} : X \times Y \longrightarrow [0,1]$ of the subset $T^*(A_i, B_j)$ of $X \times Y$ is given by

$$\mu_{T^*(A_i,B_j)}(x,y) = T^*\big(\mu_{A_i}(x), \mu_{B_j}(y)\big).$$

Several other concepts for the definition of a fuzzy partition of a universe X have been proposed in the literature. A purely algebraic approach was given in [Ruspini 1969]: an n-tuple $(A_1, A_2, \ldots, A_n)$ of fuzzy subsets of X is called a *fuzzy partition* if for all $x \in X$ we have

$$\sum_{i=1}^{n} \mu_{A_i}(x) = 1.$$

Another way of looking at fuzzy partitions is to use the set operations (i.e., t-norms and t-conorms) and to replace the classical axioms for a partition (the elements of the partition must cover X and be pairwise disjoint) by appropriate many-valued formulae (see [Butnariu 1983a, Dubois *et al.* 1992, Butnariu & Klement 1993, Mesiar & Rybárik 1995] and [Dumitrescu 1992, 1993]).

12.4 T-clans and T-tribes

We now study certain subclasses of $\mathcal{F}(X)$, the class of all fuzzy subsets of a (crisp) universe of discourse X, which are closed under intersections and unions based on t-norms and t-conorms, respectively, and which can serve as domains of the measures as discussed in Section 14.1.

12.19 DEFINITION
Let T be a t-norm. A subclass $\mathcal{C}$ of $\mathcal{F}(X)$ is called a *T-clan* on X if the following properties are satisfied:

(TC1) $\emptyset \in \mathcal{C}$.
(TC2) For all $A \in \mathcal{C}$ we have $\complement A \in \mathcal{C}$.
(TC3) For all $A, B \in \mathcal{C}$ we have $A \cap_T B \in \mathcal{C}$.

12.20 EXAMPLE
(i) Each algebra of crisp subsets of X [Halmos 1950, Bauer 1981] is a T-clan with respect to an arbitrary t-norm T.
(ii) Each $T_{\mathbf{M}}$-clan is a De Morgan algebra [Höhle & Weber 1997], and each $T_{\mathbf{L}}$-clan is a semi-simple MV-algebra [Chang 1958] (compare also [Belluce 1986, 1992] and [Di Nola 1991, Höhle & Weber 1997]).
(iii) Define for each $n \in \mathbb{N}$ and $k \in \{1, 2, \ldots, n-1\}$ the fuzzy subset $A_{\frac{k}{n}}$ of X by $\mu_{A_{\frac{k}{n}}}(x) = \frac{k}{n}$. Then the class

$$\mathcal{T}_n = \{\emptyset, A_{\frac{1}{n}}, A_{\frac{2}{n}}, \ldots, A_{\frac{n-1}{n}}, X\}$$

is both a $T_{\mathbf{M}}$-clan and a $T_{\mathbf{L}}$-clan, but it is never a T-clan when T is a strict t-norm.

In many situations it is also desirable to have, in addition to the properties of a T-clan, a kind of σ-completeness. The following concept was presented in [Butnariu & Klement 1993], it basically appeared in [Klement 1982b] for arbitrary t-norms T, in [Klement 1980b] for $T = T_{\mathbf{M}}$ and in [Butnariu 1983a] for $T = T_{\mathbf{L}}$ (see also [de Glas 1984]).

12.21 DEFINITION
Let T be a t-norm. A T-clan $\mathcal{T} \subseteq \mathcal{F}(X)$ is called a *T-tribe* if

(TT) for all sequences $(A_n)_{n\in\mathbb{N}}$ in $\mathcal{T}$ we have $\bigcap_{T\, n=1}^{\infty} A_n \in \mathcal{T}$.

12.22 EXAMPLE
(i) Obviously, not each T-clan is a T-tribe (a counterexample for the opposite direction is the $T_{\mathbf{L}}$-clan $\{A \mid \text{for all } x \in X : \mu_A(x) \in \mathbb{Q}\}$ which is not a $T_{\mathbf{L}}$-tribe).
(ii) Each $T_{\mathbf{M}}$-tribe is a σ-complete De Morgan algebra, and each $T_{\mathbf{L}}$-tribe is both a σ-complete semi-simple MV-algebra and a clan in the sense of [Wyler 1966].
(iii) A subclass $\mathcal{A}$ of $\mathcal{P}(X)$, i.e., a class of crisp subsets of X, is a σ-algebra if and only if it is a T-tribe with respect to an arbitrary t-norm T.

(iv) Given a Borel measurable t-norm T and a σ-algebra $\mathcal{A}$ of crisp subsets of X, then the class $\mathcal{A}^\wedge$ of all fuzzy subsets of X with $\mathcal{A}$-measurable membership functions is a T-tribe. Note that $\mathcal{A}^\wedge$ was called a *generated (T-)tribe* in [Klement 1980b, 1982b] and [Butnariu & Klement 1993], and that it was used in [Zadeh 1968] as the class of fuzzy events.

(v) Given a T-tribe $\mathcal{T}$, the set $\mathcal{T}^\vee$ of all crisp sets in $\mathcal{T}$ is a σ-algebra, and hence a T-tribe.

(vi) The class $\mathcal{T}$ consisting of all fuzzy subsets of X, whose membership functions are either constant or have all their values in the interval $\left[\frac{1}{3}, \frac{2}{3}\right]$, is a T-tribe for $T = T_{\mathbf{D}}$ and for $T = T_{\mathbf{M}}$, but no T-tribe whenever T is a continuous Archimedean t-norm [Klement 1982b].

(vii) If in (vi) we additionally require that all the membership functions of the elements of $\mathcal{T}$ be continuous, then $\mathcal{T}$ is a $T_{\mathbf{D}}$-tribe, but no longer a $T_{\mathbf{M}}$-tribe [Klement 1982b].

(viii) Consider a crisp subset Y of X. Given a σ-algebra $\mathcal{A}$ of crisp subsets of X, the class $\mathcal{A}_Y^\wedge$ of all fuzzy subsets of X, whose membership functions are $\mathcal{A}$-measurable and assume on Y only the values 0 and 1, is a T-tribe with respect to an arbitrary Borel measurable t-norm T. Note however that, up to the case $Y = X$, $\mathcal{A}_Y^\wedge$ does not contain any fuzzy subset of X with a non-trivial constant membership function (compare Theorem 12.23(iii)). The tribe $\mathcal{A}_Y^\wedge$ was called a *semigenerated (T-)tribe* in [Mesiar 1993a].

The following result for tribes based on Frank t-norms (see Propositions 2.7 and 3.3 and Theorem 3.2 in [Butnariu & Klement 1993]) demonstrates the special role of the Frank t-norms in this context. The crucial argument is, of course, that the Frank t-norms and their dual t-conorms are solutions of the Frank functional equation (4.2).

12.23 THEOREM
For all $\lambda \in \,]0, \infty]$ the following assertions hold:

(i) Each $T_\lambda^{\mathbf{F}}$-tribe is a $T_{\mathbf{M}}$-tribe (and, hence, a σ-complete De Morgan algebra) as well as a $T_{\mathbf{L}}$-tribe (and, hence, a σ-complete semi-simple MV-algebra).

(ii) For each $T_\lambda^{\mathbf{F}}$-tribe $\mathcal{T}$ we have $\mathcal{T} \subseteq (\mathcal{T}^\vee)^\wedge$, i.e., each element in $\mathcal{T}$ has a $\mathcal{T}^\vee$-measurable membership function.

(iii) A $T_\lambda^{\mathbf{F}}$-tribe $\mathcal{T}$ is generated if and only if it contains all fuzzy subsets of X which have a constant membership function.

The interesting problem to characterize all subclasses of $\mathcal{F}(X)$ which are T-tribes with respect to an arbitrary Borel measurable t-norm T was solved in [Navara 1993] (compare also [Mesiar 1993a, 1993b] and [Mesiar & Navara 1996]). The result turns out to be a generalization of Example 12.22(iii) and Theorem 12.23(i).

12.24 THEOREM
Let X be a non-empty set and $\mathcal{T}$ a subclass of $\mathcal{F}(X)$. Then the following are equivalent:

(i) $\mathcal{T}$ is a T-tribe for all Borel measurable t-norms T.
(ii) There is a $\lambda \in]0,\infty[$ such that $\mathcal{T}$ is a $T_\lambda^{\mathbf{F}}$-tribe.
(iii) $\mathcal{T}$ is a weakly generated tribe, i.e., there is a σ-ideal $\mathcal{D}$ of $\mathcal{T}^\vee$ such that

$$\mathcal{T} = \{A \in (\mathcal{T}^\wedge)^\vee \mid \operatorname{supp}(A) \setminus \ker(A) \in \mathcal{D}\}.$$

This means that a $T_\lambda^{\mathbf{F}}$-tribe for $\lambda \in]0,\infty[$ on X is not only a σ-complete lattice (in particular, it is closed under $T_{\mathbf{M}}$) and a σ-complete semi-simple MV-algebra (in particular, it is closed under $T_{\mathbf{L}}$), but that it is also closed under each Borel measurable t-norm. Observe that this is not true for $T_{\mathbf{L}}$-tribes: for example, the $T_{\mathbf{L}}$-clan mentioned in Example 12.20(iii) is a $T_{\mathbf{L}}$-tribe, but no T-tribe if T is a strict t-norm (see also Example 12.26 below).

We already mentioned that each $T_{\mathbf{L}}$-tribe is a σ-complete semi-simple MV-algebra. In fact, each semi-simple MV-algebra is isomorphic to some $T_{\mathbf{L}}$-clan (see [Belluce 1986, Belluce 1992], a complete characterization of all MV-algebras, involving non-standard real numbers, can be found in [Di Nola 1991]). Another characterization of $T_{\mathbf{L}}$-tribes was given in [Klement & Navara 1997]. Both results are summarized below.

12.25 THEOREM
Let $\mathcal{T}$ be a subclass of $\mathcal{F}(X)$. Then the following are equivalent:

(i) $\mathcal{T}$ is a $T_{\mathbf{L}}$-tribe.
(ii) $\mathcal{T}$ is a σ-complete semi-simple MV-algebra.
(iii) There is a σ-algebra $\mathcal{A}$ of crisp subsets of X and a sequence $(\mathcal{E}_n)_{n\in\mathbb{N}}$ of σ-filters in $\mathcal{A}$, with $\mathcal{E}_m \subseteq \mathcal{E}_n$ whenever n is a divisor of m, such that

$$\mathcal{T} = \bigl\{A \in \mathcal{A}^\wedge \bigm| \{x \in X \mid n \cdot \mu_A(x) \in \mathbb{N}\} \in \mathcal{E}_n \text{ for all } n \in \mathbb{N}\bigr\}.$$

Combining all these results, it is possible to give, as a simple example, a rather complete picture of all possible tribes with respect to Frank t-norms, to arbitrary strict t-norms, and to the weakest t-norm $T_{\mathbf{D}}$, respectively.

12.26 EXAMPLE
Let $\{x_0\}$ be a singleton. Then the only generated tribe on $\{x_0\}$, let us call it $\mathcal{T}_\infty$, is just isomorphic to the unit interval $[0,1]$. There are two semigenerated tribes, namely, $\mathcal{T}_\infty$ and $\mathcal{T}_1$ (which is isomorphic to $\{0,1\}$). Since $\{x_0\}$ is finite, these are also the only weakly generated tribes. Due to Theorem 12.24 each $T_\lambda^{\mathbf{F}}$-tribe with $\lambda \in]0,\infty[$ equals either $\mathcal{T}_\infty$ or $\mathcal{T}_1$, the same being true for each strict t-norm. The only $T_{\mathbf{L}}$-tribes on $\{x_0\}$ are $\mathcal{T}_\infty$ and $\mathcal{T}_n$ with $n \in \mathbb{N}$ (as mentioned in Example 12.20(iii)). Each $T_{\mathbf{M}}$-tribe on $\{x_0\}$ is isomorphic to some closed subset of $[0,1]$ which contains 0 and which is symmetric with respect to the midpoint 0.5, and each $T_{\mathbf{D}}$-tribe on $\{x_0\}$ is isomorphic to some arbitrary subset of $[0,1]$ which contains 0 and which is symmetric with respect to 0.5.

13. Applications of fuzzy logic and fuzzy sets

> All traditional logic habitually assumes that precise symbols are being employed. It is therefore not applicable to this terrestrial life, but only to an imagined celestial existence.
>
> BERTRAND RUSSELL

> Clarity is an intellectual value in itself; exactness and precision, however, are not. Absolute precision is unattainable; and there is no point in trying to be more precise than our problem demands.
>
> KARL POPPER

We start this overview with a discussion of the computational rule of inference [Zadeh 1973] which plays a crucial role in fuzzy control (see below) but also in approximate reasoning [Fullér & Zimmermann 1992]. It is a way to associate with a fuzzy input a fuzzy output by means of a fuzzy relation between the input and the output space. We also give a geometric and a logical interpretation of the computational rule of inference, the latter showing a close relationship to the R-implications studied in Section 11.1. If input and output are known, we present criteria for the solvability of the corresponding fuzzy relational equation (see [Sanchez 1974, 1976, 1977, 1984] and [Gottwald 1986, Di Nola *et al.* 1989]).

Fuzzy numbers [Kaufmann & Gupta 1991, Mareš 1994, Dubois *et al.* 1999a] are used as models in situations where the numbers in question are only approximately known. We present the most important types considered in the literature: fuzzy numbers (whose membership functions are convex [Dubois & Prade 1981, 1987]) and fuzzy reals (whose membership functions are distribution functions [Hutton 1975, Schweizer 1975, Höhle 1978b, Schweizer & Sklar 1983]). Using the Zadeh extension principle [Zadeh 1975a, 1975b, 1975c] it is possible to extend operations on the real line to fuzzy numbers and fuzzy reals.

Fuzzy controllers are essentially rule-based systems assigning (with the help of the fuzzification, a suitable inference and the so-called defuzzification) to each input value the corresponding output value, therefore producing an input-output function. The first laboratory experiment, the control of a steam engine in England, is described in [Mamdani & Assilian 1975], the first industrial application, the control of a cement kiln in Denmark, in [Holmblad & Østergaard 1982]. Here we shall restrict ourselves to the two most important and most widely used fuzzy controllers, the Mamdani [Mamdani & Assilian 1975, Mamdani & Gaines 1981] and the Takagi-Sugeno controller [Sugeno 1985b, Takagi & Sugeno 1985]. For more information about fuzzy controllers see [Sugeno 1985a, Driankov *et al.* 1993, Kruse *et al.* 1994a, Nguyen *et al.* 1995, Bauer *et al.* 1995, Höhle 1998, Nguyen & Sugeno 1998], a number of solutions of real world problems is described in [von Altrock *et al.* 1992].

A considerable number of applications of fuzzy logic and fuzzy sets were not included here because the role of t-norms is not so apparent there (usually only the minimum is used, compare, e.g., [Kruse *et al.* 1994b, Dubois & Prade 1999]), and also to keep the length of this chapter reasonably short. We only mention here some other fields of application together with selected key references: approximate reasoning (see [Zadeh 1975d] and [Dubois & Prade 1991a, 1991b]), data analysis [Bandemer & Näther 1992], pattern recognition [Zadeh 1965, Bellman *et al.* 1966, Bezdek 1981, Bezdek & Pal 1992], image processing (see [Rosenfeld 1984a, 1984b, 1985] and [Rosenfeld & Haber 1985, Bauer *et al.* 1996, Bezdek *et al.* 1999]), decision making [Bellman & Zadeh 1970, Kickert 1978, Zimmermann 1987, Kacprzyk & Fedrizzi 1990, Fodor & Roubens 1994], (linear) optimization [Zimmermann 1978, 1985], economics and game theory (see [Aubin 1974a, 1974b, 1981, 1982], [Aumann & Shapley 1974, Ponsard 1988, Billot 1992], [Butnariu 1985, 1987b, 1989], and [Butnariu & Klement 1993, 1996]), as well as the connection to neural networks [Kosko 1991] and genetic algorithms [Herrera & Verdegay 1996].

13.1 Fuzzy relations, compositional rule of inference

Given two crisp sets X and Y, a crisp relation between elements of X and elements of Y always can be described by a crisp subset of the Cartesian product $X \times Y$. In complete analogy, a fuzzy relation R on $X \times Y$ is defined to be a fuzzy subset of the Cartesian product $X \times Y$, described by its membership function $\mu_R : X \times Y \longrightarrow [0,1]$ (observe that in the case of finite sets X and Y this membership function can be conveniently written as a matrix). The value $\mu_R(x,y)$ can be interpreted as the truth value of the statement '*x is in relation R with y*'.

A fuzzy relation R on $X \times Y$ can be used to associate with each fuzzy subset A of X a certain fuzzy subset of Y, using the so-called *compositional rule of inference* going back to [Zadeh 1973] (see also [Fullér & Zimmermann 1992]):

13.1 DEFINITION

Let X and Y be crisp sets, T be a t-norm, A a fuzzy subset of X, and R be a fuzzy relation on $X \times Y$. The membership function of the fuzzy subset $A \circ_T R$ of Y is given by the formula

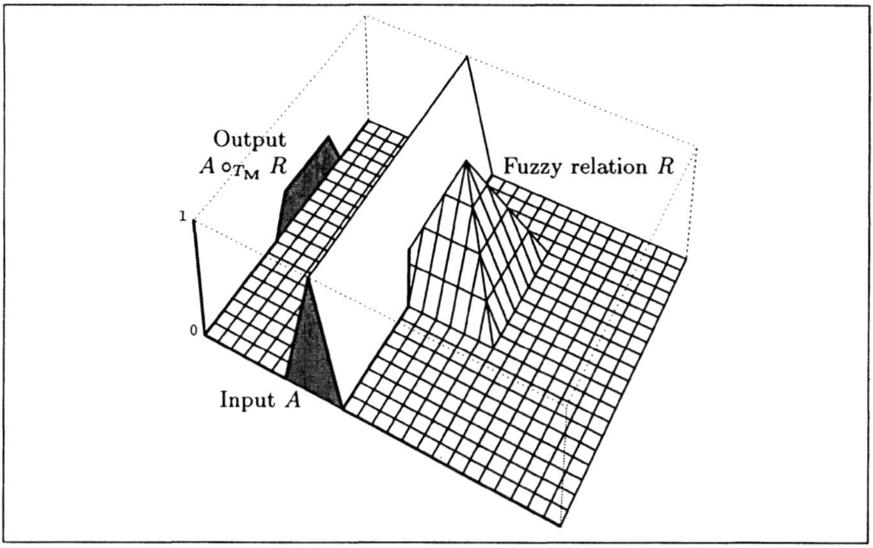

Fig. 13.1. Compositional rule of inference: we start with a triangular input A, build the intersection (based on $T_{\mathbf{M}}$) of its cylindrical extension with the pyramid (representing the fuzzy relation R), and project this intersection to the output space, yielding the trapezoidal output $A \circ_{T_{\mathbf{M}}} R$

$$\mu_{A \circ_T R}(y) = \sup\{T\big(\mu_A(x), \mu_R(x,y)\big) \bigm| x \in X\}.$$

It is immediately clear that, in the special situation if R equals the graph of a function $f : X \longrightarrow Y$ and A is a crisp subset of X, then $A \circ_T R$ equals the image $f(A)$ of the set A under the function f. If the sets X and Y are finite, i.e., if A can be described by a vector and R by a matrix, then $A \circ_T R$ is obtained as a generalized matrix product of A and R, replacing the multiplication by the t-norm T, and the product by the maximum.

Given a fuzzy subset A of X, a fuzzy subset B of Y and a t-norm T, it is quite natural to ask whether there exists a fuzzy relation on $X \times Y$ such that

$$A \circ_T R = B. \tag{13.1}$$

Note that (13.1) is usually called a *fuzzy relational equation.* Such equations were first investigated in [Sanchez 1974, 1976, 1977], for a broader survey see, e.g., [Di Nola *et al.* 1989]. Maximal solutions of (13.1) are of particular interest in the context of fuzzy control (see Section 13.3), minimal solutions were studied in [Sanchez 1977, Sessa 1984, Di Nola *et al.* 1985, Di Nola *et al.* 1989].

The compositional rule of inference has two nice interpretations in terms of geometry and logic. From a geometric point of view, it essentially consists of three following elementary geometric operations.

(i) *Cylindric extension* of A to $A \times Y$, i.e.,

$$\mu_{A\times Y}(x,y) = T\big(\mu_A(x), \mu_Y(y)\big) = T\big(\mu_A(x), 1\big) = \mu_A(x).$$

(ii) *Intersection* of $A \times Y$ with R, i.e.,

$$\begin{aligned}\mu_{(A\times Y)\cap_T R}(x,y) &= T\big(\mu_{A\times Y}(x,y), \mu_R(x,y)\big)\\ &= T\big(\mu_A(x), \mu_R(x,y)\big).\end{aligned}$$

(iii) *Projection* of $(A \times Y) \cap_T R$ onto Y, i.e.,

$$\begin{aligned}\mu_{A\circ_T R}(y) &= \sup\,\big\{\mu_{(A\times Y)\cap_T R}(x,y) \mid x \in X\big\}\\ &= \sup\,\big\{T\big(\mu_A(x), \mu_R(x,y)\big) \mid x \in X\big\}.\end{aligned}$$

This geometric interpretation of the compositional rule of inference is visualized for a simple situation (using $T = T_{\mathbf{M}}$) in Figure 13.1.

The logical interpretation of the compositional rule of inference quite naturally generalizes the classical modus ponens, which is an immediate consequence of the following result [Sanchez 1984, Gottwald 1986]. Recall in this context the definition of an R-implication $\to_T$ (given in (2.17)) based on a left-continuous t-norm T.

13.2 PROPOSITION
Let A, B be fuzzy subsets of the crisp sets X and Y, respectively, and let T be a left-continuous t-norm. The following are equivalent:

(i) There exists a fuzzy relation R on $X \times Y$ which solves the relational equation (13.1).
(ii) The following fuzzy relation $R_T(A,B)$ on $X \times Y$ solves (13.1):

$$\mu_{R_T(A,B)}(x,y) = \mu_A(x) \to_T \mu_B(y). \tag{13.2}$$

Moreover, if (13.1) is solvable, then $R_T(A,B)$ is the largest solution of (13.1).

In the case of systems of fuzzy relational equations the situation is much more complicated. Except for the special case when $T = T_{\mathbf{M}}$ (see [Gottwald 1993, Corollary 3.10]), only lower bounds for the solutions are known [Gottwald 1984b, 1986a] (compare also [Gottwald 1993, Theorem 3.9]). A possible way out is also to consider *solvability degrees* or to work with approximate solutions of such systems of fuzzy relational equations [Gottwald 1993, Section 3.5].

13.2 Fuzzy numbers

Fuzzy numbers intend to model linguistic expressions such as "*approximately five*", on the one hand, and to fuzzify the notion of an interval, on the other hand. Consequently, a fuzzy number is a special fuzzy subset of the real line $\mathbb{R}$. For practical and mathematical reasons we require that for a fuzzy number A and for each $\alpha \in [0,1]$ the corresponding α-cut $[A]_\alpha$ defined in (12.3) is a (non-empty) interval and that its kernel $\ker(A) = [A]_1$ (see (12.1)) is bounded.

Fuzzy quantities were studied in [Mareš 1977a, 1977b, 1989] and [Höhle 1981], and their calculus based on the minimum $T_{\mathbf{M}}$ is extensively discussed in [Mareš 1994].

13.3 DEFINITION

(i) Each fuzzy subset A of $\mathbb{R}$ is called a *fuzzy quantity.* The set of all fuzzy quantities will be denoted $\mathcal{F}(\mathbb{R})$.

(ii) A fuzzy quantity A is called *convex* if for each $\alpha \in [0,1]$ the α-cut $[A]_\alpha$ is a convex subset of $\mathbb{R}$ (i.e., a subinterval of $\mathbb{R}$).

(iii) A fuzzy subset A of $\mathbb{R}$ which is normal, convex and has a bounded kernel, is called a *fuzzy number.* The set of all fuzzy numbers will be denoted $\mathcal{F}_N(\mathbb{R})$.

Note that a fuzzy quantity A is convex [Zadeh 1965, Lowen 1980] if and only if for all $x, y, z \in \mathbb{R}$ with $z \in [x, y]$ we have

$$\mu_A(z) \geq \min\big(\mu_A(x), \mu_A(y)\big).$$

Obviously, if for a fuzzy number A we have $\mu_A(x_0) = 1$ for $x_0 \in \mathbb{R}$, then the restriction of the membership function μ_A to $]-\infty, x_0]$ is a non-decreasing, and the restriction of μ_A to $[x_0, \infty[$ is a non-increasing function.

The definition given above contains several types of fuzzy numbers as special cases (see, e.g., [Dubois & Prade 1981, 1987] and [Fullér & Keresztfalvi 1992, Mareš 1994, Mesiar 1997b]):

13.4 EXAMPLE

(i) An *upper semicontinuous fuzzy number* A has an upper semicontinuous membership function, which means that for all $\alpha \in]0,1]$ the α-cut $[A]_\alpha$ is a closed subinterval of $\mathbb{R}$, and it can be completely described by a quadruple (l, r, F, G) (and we shall briefly write $A = (l, r, F, G)$, see [Mesiar 1997b]), where $l, r \in \mathbb{R}$ with $l \leq r$ and $F, G :]0, \infty[\longrightarrow [0,1[$ are non-increasing left-continuous functions such that for all $x \in \mathbb{R}$

$$\mu_A(x) = \begin{cases} F(l-x) & \text{if } x \in]-\infty, l[\,, \\ 1 & \text{if } x \in [l, r]\,, \\ G(x-r) & \text{if } x \in]r, \infty[\,. \end{cases}$$

Observe that the interval $[l, r]$ then is just the kernel of A, and the functions F and G describe the left and right shape of A, respectively. The class of all upper semicontinuous fuzzy numbers will be denoted $\mathcal{F}_U(\mathbb{R})$. Putting $\mathbf{0}$ for the function which maps everything to 0, for a closed subinterval $[a, b]$ of $\mathbb{R}$ we may write $[a, b] = (a, b, \mathbf{0}, \mathbf{0})$ and, subsequently, for

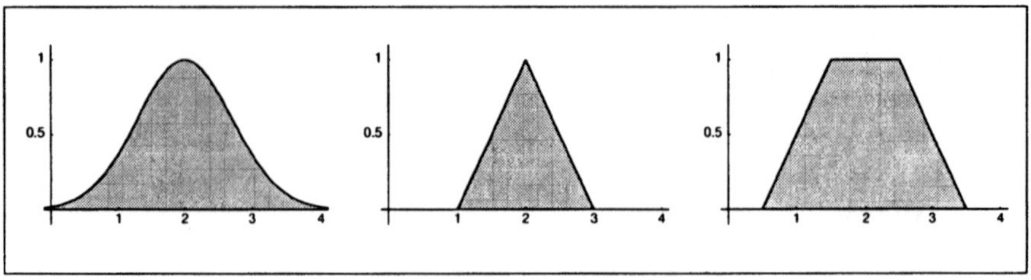

Fig. 13.2. Gaussian (left), triangular (center) and trapezoidal (right) fuzzy numbers

a number $x_0 \in \mathbb{R}$ we may write $(x_0, x_0, \mathbf{0}, \mathbf{0})$ (the corresponding membership functions are just the characteristic functions $\mathbf{1}_{[a,b]}$ and $\mathbf{1}_{\{x_0\}}$, respectively). This means that the class $\mathcal{F}_U(\mathbb{R})$ contains both the class of closed subintervals of $\mathbb{R}$ and $\mathbb{R}$ itself as subclasses.
Semantically speaking, for an upper semicontinuous fuzzy number A the value $\mu_A(x)$ is just the truth value of the (fuzzy) statement 'x *is* A'.
Concrete examples of upper semicontinuous fuzzy numbers are the *Gaussian fuzzy numbers* $A = (x_0, x_0, E_\alpha, E_\alpha)$, where $x_0 \in \mathbb{R}$, $\alpha \in]0, \infty[$ and for all $x \in]0, \infty[$ the shape function $E_\alpha :]0, \infty[\longrightarrow [0, 1[$ is given by

$$E_\alpha(x) = e^{-\frac{x^2}{\alpha}}.$$

Note that in this case for all $x \in \mathbb{R}$ we have (see Figure 13.2)

$$\mu_A(x) = e^{-\frac{(x-x_0)^2}{\alpha}}.$$

(ii) If $l, r \in \mathbb{R}$ with $l \leq r$, $\alpha, \beta \in [0, \infty[$ and if $L, R :]0, 1] \longrightarrow [0, 1[$ are non-increasing left-continuous functions with $L(x) > 0$ and $R(x) > 0$ for all $x \in]0, 1[$, then a fuzzy number A whose membership function $\mu_A : \mathbb{R} \longrightarrow [0, 1]$ is defined by

$$\mu_A(x) = \begin{cases} L\left(\frac{l-x}{\alpha}\right) & \text{if } x \in]l - \alpha, l[\,, \\ 1 & \text{if } x \in [l, r]\,, \\ R\left(\frac{x-r}{\beta}\right) & \text{if } x \in]r, r + \beta[\,, \\ 0 & \text{otherwise,} \end{cases}$$

is called an $(L\text{-}R)$ *fuzzy interval* (see [Dubois & Prade 1981, 1987], [Lowen 1986], [Fullér & Keresztfalvi 1991, 1992], and [Kolesárová 1995, Mesiar 1996a, De Baets & Marková-Stupňanová 1997, Marková 1997]), briefly denoted $A = (l, r, \alpha, \beta)_{LR}$. The class of $(L\text{-}R)$ fuzzy intervals will be denoted $\mathcal{F}_I(\mathbb{R})$. Observe that $\mathcal{F}_I(\mathbb{R})$ just contains the elements of $\mathcal{F}_U(\mathbb{R})$ which have bounded support. Note that again $[l, r]$ is the kernel of of A, and the functions L and R are called the left and right shape

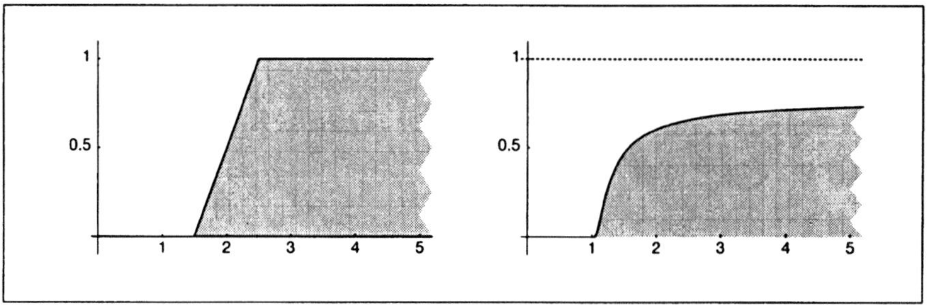

Fig. 13.3. A finite (left) and an infinite (right) fuzzy real

of A, respectively. The values α and β are called the left and the right spread of A, respectively.
The class of fuzzy intervals obviously extends the class of closed subintervals $[a, b]$ of $\mathbb{R}$ (whose membership functions are just the characteristic functions $\mathbf{1}_{[a,b]}$, in our notation $(a, b, 0, 0)_{LR}$). In particular, $\mathcal{F}_I(\mathbb{R})$ contains $\mathbb{R}$ as a subset since for $x_0 \in \mathbb{R}$ we can write $(x_0, x_0, 0, 0)_{LR}$.
Particularly important examples of fuzzy intervals are the *piecewise linear fuzzy numbers* where $L(x) = R(x) = 1 - x$ for all $x \in]0, 1]$. In this case the subscript LR will be omitted in the notation, i.e., we simply write $A = (l, r, \alpha, \beta)$. A piecewise linear fuzzy number (l, r, α, β) is called *triangular* if $l = r$, and *trapezoidal* if $l < r$ (see Figure 13.2). A linguistic interpretation of a triangular fuzzy number $(x_0, x_0, \alpha, \beta)$ is usually given by '*approximately* x_0'.

Another type of fuzzy quantities was studied in, e.g., [Hutton 1975, Schweizer 1975], [Höhle 1978, 1981], [Rodabaugh 1982b, 1982c, 1983b, 1999], [Klement 1983, 1985], and [Lowen 1983, 1984b].

13.5 EXAMPLE
A *fuzzy real* A is a fuzzy quantity with a non-decreasing left-continuous membership function. The class of fuzzy reals will be denoted $\mathcal{F}_R(\mathbb{R})$, and it can be identified with the class of all distribution functions on $[-\infty, \infty]$ (see Definition 9.1). If $\inf\{\mu_A(x) \mid x \in \mathbb{R}\} = 0$ and $\sup\{\mu_A(x) \mid x \in \mathbb{R}\} = 1$ then the fuzzy real A is called a *finite fuzzy real*, in which case its membership function is just the same mathematical object as a distribution function of some random variable. The set $\mathbb{R}$ can be canonically embedded into $\mathcal{F}_I(\mathbb{R})$, identifying each $x_0 \in \mathbb{R}$ with the (crisp) fuzzy real C_{x_0} given by $\mu_{C_{x_0}} = \mathbf{1}_{]x_0,\infty[}$. The set $\mathcal{F}_R(\mathbb{R})$ can be equipped with a fuzzy topology which is a canonical extension of the usual topology on $\mathbb{R}$ (compare [Hutton 1975, Höhle 1981], [Rodabaugh 1982a, 1982c, 1983, 1985, 1999] and [Lowen 1983, 1984a, 1984b, 1984c, 1985, 1986]).
Examples of finite fuzzy reals are the *piecewise linear fuzzy reals*, simply deno-

ted $A = (x_0, x_0 + \varepsilon)$ with $x_0 \in \mathbb{R}$ and $\varepsilon \in [0, \infty[$ (see Figure 13.3), given by

$$\mu_A(x) = \begin{cases} 0 & \text{if } x \in]-\infty, x_0]\,, \\ \frac{x-x_0}{\varepsilon} & \text{if } x \in]x_0, x_0 + \varepsilon]\,, \\ 1 & \text{if } x \in]x_0 + \varepsilon, \infty[\,. \end{cases}$$

The semantic meaning of $\mu_A(x)$ in this case is the truth value of the (fuzzy) statement '*x is larger than x_0*'.
An example of an infinite fuzzy real A is given by (see Figure 13.3)

$$\mu_A(x) = \begin{cases} 0 & \text{if } x \in]-\infty, 1]\,, \\ e^{-\frac{x}{4x-4}} & \text{if } x \in]1, \infty[\,. \end{cases}$$

The *Zadeh extension principle* [Zadeh 1975a, 1975b, 1975c] provides a general tool for extending operations on $\mathbb{R}$ to operations on fuzzy quantities.

13.6 DEFINITION
Let T be a t-norm and $* : \mathbb{R}^2 \longrightarrow \mathbb{R}$ an operation on the real line. Then the operation $\circledast_T : \mathcal{F}(\mathbb{R})^2 \longrightarrow \mathcal{F}(\mathbb{R})$ on the class of fuzzy quantities is defined by (recall that $\sup \emptyset = 0$ in this case)

$$\mu_{A \circledast_T B}(z) = \sup\{T\left(\mu_A(x), \mu_B(y)\right) \mid x, y \in \mathbb{R}, x * y = z\}. \tag{13.3}$$

Clearly, $\circledast_T$ is an extension of $*$ since for any two crisp subsets A, B of $\mathbb{R}$ we obtain (taking into account that the membership function of the set A is just the characteristic function $\mathbf{1}_A$)

$$A \circledast_T B = \{a * b \mid a \in A, b \in B\} = A * B,$$

and, as a special case thereof, $\{a\} \circledast_T \{b\} = \{a * b\}$ for any two numbers $a, b \in \mathbb{R}$. If T_1, T_2 are two t-norms with $T_1 \leq T_2$, then for any two fuzzy quantities A and B and for each operation $* : \mathbb{R}^2 \longrightarrow \mathbb{R}$ we have $\mu_{A \circledast_{T_1} B} \leq \mu_{A \circledast_{T_2} B}$. Consequently, for an arbitrary t-norm T we obtain

$$\mu_{A \circledast_{T_{\mathbf{D}}} B} \leq \mu_{A \circledast_T B} \leq \mu_{A \circledast_{T_{\mathbf{M}}} B}.$$

This explains why the extensions of real operations to fuzzy quantities based on the two limit t-norms $T_{\mathbf{D}}$ and $T_{\mathbf{M}}$ were studied with special attention (see [Dubois & Prade 1981, 1987] and [Mareš 1994, Mesiar 1996a]).

In the special case of the strongest t-norm $T_{\mathbf{M}}$ and a continuous binary operation $*$ on $\mathbb{R}$, for all $A, B \in \mathcal{F}(\mathbb{R})$ the following equality holds for all α-cuts with $\alpha \in [0, 1]$ [Nguyen 1978, Dubois & Prade 1987, Nguyen & Walker 1997]:

$$[A \circledast_{T_{\mathbf{M}}} B]_\alpha = [A]_\alpha * [B]_\alpha. \tag{13.4}$$

Although, for arbitrary operations $* : \mathbb{R}^2 \longrightarrow \mathbb{R}$ and arbitrary t-norms T, the class $\mathcal{F}(\mathbb{R})$ is closed under $\circledast_T$, this is not necessarily the case for subclasses of fuzzy quantities:

13.7 EXAMPLE

(i) Consider the (non-continuous) operation $* : \mathbb{R}^2 \longrightarrow \mathbb{R}$ defined by

$$x * y = \begin{cases} 0 & \text{if } x \cdot y = 0, \\ 1 & \text{otherwise,} \end{cases}$$

then for an arbitrary t-norm T we obtain $[0,1] \circledast_T [0,1] = \{0,1\}$, showing that both the class of closed subintervals of $\mathbb{R}$ and the class of fuzzy numbers $\mathcal{F}_N(\mathbb{R})$ are not closed under $\circledast_T$.

(ii) Take the (non-continuous) nilpotent minimum $T^{\mathbf{nM}}$ introduced in Remark 1.21, the triangular fuzzy number $A = (0,0,1,1)$, and the sum $+$ on $\mathbb{R}$, then the membership function of $A \oplus_{T^{\mathbf{nM}}} A$ given by

$$\mu_{A \oplus_{T^{\mathbf{nM}}} A}(x) = \left(1 - \tfrac{|x|}{2}\right) \cdot \mathbf{1}_{]-1,1[}(x)$$

is not upper semicontinuous, showing that both the class of fuzzy intervals $\mathcal{F}_I(\mathbb{R})$ and the class of upper semicontinuous fuzzy numbers $\mathcal{F}_U(\mathbb{R})$ are not closed under $\oplus_{T^{\mathbf{nM}}}$.

(iii) Consider the (non-continuous) t-norm T defined by

$$T(x,y) = \begin{cases} 0 & \text{if } x + y < 1, \\ \min(x,y) & \text{otherwise,} \end{cases}$$

and the piecewise linear fuzzy real $A = (0,1)$, then the membership function of $A \oplus_T A$ given by

$$\mu_{A \oplus_T A}(x) = \min(\tfrac{x}{2}, 1) \cdot \mathbf{1}_{[1,\infty[}(x)$$

is not left-continuous, thus the class of fuzzy reals $\mathcal{F}_R(\mathbb{R})$ is not closed under $\oplus_T$.

However, we have the following sufficient conditions for our special classes of fuzzy quantities to be closed under the extension of suitable operations on the real line based on a t-norm T (see [Nguyen 1978, Dubois & Prade 1987, Fullér & Keresztfalvi 1991, Nguyen & Walker 1997]):

13.8 REMARK

(i) If the operation $* : \mathbb{R}^2 \longrightarrow \mathbb{R}$ is continuous, then for each t-norm T the class of fuzzy numbers $\mathcal{F}_N(\mathbb{R})$ is closed under $\circledast_T$.

(ii) If the operation $* : \mathbb{R}^2 \longrightarrow \mathbb{R}$ is continuous and if the t-norm T is continuous on the open square $]0,1[^2$, then both the class of upper semicontinuous fuzzy numbers $\mathcal{F}_U(\mathbb{R})$ and the class of fuzzy intervals $\mathcal{F}_I(\mathbb{R})$ are closed under $\circledast_T$.

(iii) If the t-norm T is continuous and if the operation $* : \mathbb{R}^2 \longrightarrow \mathbb{R}$ is continuous, non-decreasing in each component and satisfies $\lim_{x\to\infty} x * x = \infty$, then the class of fuzzy reals $\mathcal{F}_R(\mathbb{R})$ is closed under $\circledast_T$.

Formula (13.3) allows us to introduce, e.g., the addition and multiplication of fuzzy numbers, as well as their maximum and minimum. Note that for arbitrary operations $*$ and t-norms T an exact formula for $A \circledast_T B$ can be rather complicated or even inaccessible. However, in some special cases such formulae exist, some of which will be given below. For a deeper discussion, we refer to [Dubois & Prade 1987, Kaufmann & Gupta 1991, Mareš 1994, Mesiar 1997b, Dubois *et al.* 1999a]. In this overview, however, we shall restrict ourselves to extensions $\oplus_T$ of the addition $+$ on $\mathbb{R}$.

13.9 EXAMPLE
For the two extremal t-norms $T_{\mathbf{M}}$ (using formula (13.4)) and $T_{\mathbf{D}}$ we obtain the following extensions of the sum [Mesiar 1997b]:

(i) If $A = (l_1, r_1, F_1, G_1)$ and $B = (l_2, r_2, F_2, G_2)$ are upper semicontinuous fuzzy numbers, then

$$\begin{aligned} A \oplus_{T_{\mathbf{M}}} B =& (l_1 + l_2, r_1 + r_2, ((-F_1)^{(-1)} + (-F_2)^{(-1)})^{(-1)}, \\ & ((-G_1)^{(-1)} + (-G_2)^{(-1)})^{(-1)}), \\ A \oplus_{T_{\mathbf{D}}} B =& (l_1 + l_2, r_1 + r_2, \max(F_1, F_2), \max(G_1, G_2)), \end{aligned}$$

where $f^{(-1)}$ denotes the pseudo-inverse of the monotone function f as introduced in Definition 3.2 (to obtain closed intervals, define the values $F_1(0) = F_2(0) = G_1(0) = G_2(0) = 1$, $F_i(\infty) = \lim_{n\to\infty} F_i(n)$ and $G_i(\infty) = \lim_{n\to\infty} G_i(n)$ for $i \in \{1, 2\}$, and restrict the result to $]0, \infty[$).

(ii) If $F, G : \,]0, \infty[\longrightarrow [0, 1[$ are non-increasing left-continuous functions, define for $\alpha, \beta \in \,]0, \infty[$ the functions $F_\alpha, G_\beta : \,]0, \infty[\longrightarrow [0, 1[$ by

$$F_\alpha(x) = F(\tfrac{x}{\alpha}) \qquad \text{and} \qquad G_\beta(x) = G(\tfrac{x}{\beta}),$$

then for the upper semicontinuous fuzzy numbers $A = (l_1, r_1, F_{\alpha_1}, G_{\beta_1})$ and $B = (l_2, r_2, F_{\alpha_2}, G_{\beta_2})$ we get

$$\begin{aligned} A \oplus_{T_{\mathbf{M}}} B &= (l_1 + l_2, r_1 + r_2, F_{\alpha_1+\alpha_2}, G_{\beta_1+\beta_2}), \\ A \oplus_{T_{\mathbf{D}}} B &= (l_1 + l_2, r_1 + r_2, F_{\max(\alpha_1,\alpha_2)}, G_{\max(\beta_1,\beta_2)}). \end{aligned}$$

(iii) In the special case of two L-R fuzzy intervals $A = (l_1, r_1, \alpha_1, \beta_1)_{LR}$ and $B = (l_2, r_2, \alpha_2, \beta_2)_{LR}$ with the same shapes L and R, the following well-known formulae hold [Dubois & Prade 1981, Mesiar 1997b] (for a visualization see Figure 13.4:

$$\begin{aligned} A \oplus_{T_{\mathbf{M}}} B &= (l_1 + l_2, r_1 + r_2, \alpha_1 + \alpha_2, \beta_1 + \beta_2)_{LR}, \\ A \oplus_{T_{\mathbf{D}}} B &= (l_1 + l_2, r_1 + r_2, \max(\alpha_1, \alpha_2), \max(\beta_1, \beta_2))_{LR}. \end{aligned}$$

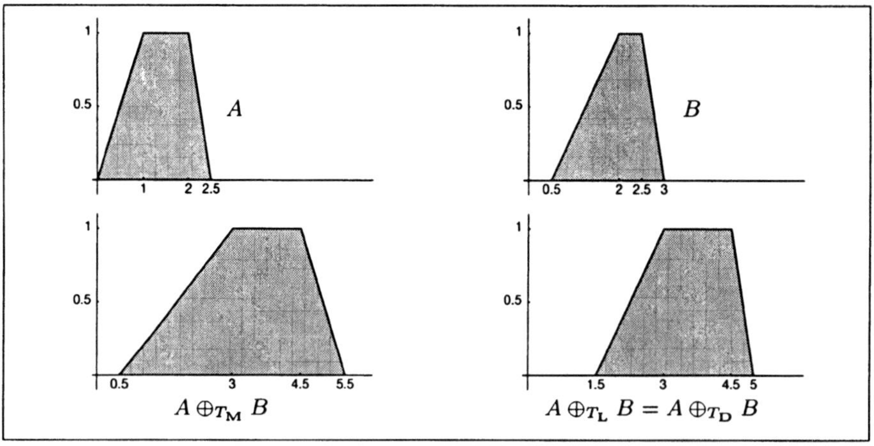

Fig. 13.4. Two *L-R* fuzzy intervals (top) and their sums with respect to $T_{\mathbf{M}}$, $T_{\mathbf{L}}$ and $T_{\mathbf{D}}$ (bottom), as given in Examples 13.9(iii) and 13.11(i) (observe that the two latter sums coincide)

(iv) If A and B are two fuzzy reals then we obtain (compare [Sherwood & Taylor 1974], [Klement 1985, 1987] and [Schweizer & Sklar 1983])

$$\mu_{A\oplus_{T_{\mathbf{M}}} B} = \left({\mu_A}^{(-1)} + {\mu_B}^{(-1)}\right)^{(-1)},$$
$$\mu_{A\oplus_{T_{\mathbf{D}}} B}(x) = \max(\mu_A(x-a), \mu_B(x-b)),$$

where (in order to have closed intervals for the pseudo-inverse) we define the values $\mu_A(-\infty) = \mu_B(-\infty) = 0$, $\mu_A(\infty) = \sup\{\mu_A(x) \mid x \in \mathbb{R}\}$ and $\mu_B(\infty) = \sup\{\mu_B(x) \mid x \in \mathbb{R}\}$, and, for $A \oplus_{T_{\mathbf{D}}} B$, we define also the values $a = \inf\{x \in \mathbb{R} \mid \mu_A(x) = 1\}$ and $b = \inf\{x \in \mathbb{R} \mid \mu_B(x) = 1\}$.

This shows that for the two extremal t-norms $T_{\mathbf{M}}$ and $T_{\mathbf{D}}$, the additions $\oplus_{T_{\mathbf{M}}}$ and $\oplus_{T_{\mathbf{D}}}$ in some sense preserve the shape of the summands. Moreover, for some special shapes there exist also other t-norm-based additions preserving these shapes (see [Kolesárová 1995] and [Mesiar 1997a, 1997b]).

13.10 Proposition

Let T be a continuous Archimedean t-norm, $t : [0,1] \longrightarrow [0,\infty]$ an additive generator of T, and define $F :]0,\infty[\longrightarrow [0,1[$ by $F(x) = t^{(-1)}(x)$. Then for arbitrary upper semicontinuous fuzzy numbers (for the notation F_α see Example 13.9(ii)) $A = (l_1, r_1, F_{\alpha_1}, F_{\beta_1})$ and $B = (l_2, r_2, F_{\alpha_2}, F_{\beta_2})$ we have

$$A \oplus_T B = (l_1 + l_2, r_1 + r_2, F_{\max(\alpha_1,\alpha_2)}, F_{\max(\beta_1,\beta_2)}).$$

Note that for a continuous Archimedean t-norm T and for upper semicontinuous fuzzy numbers A, B satisfying the conditions of Proposition 13.10

$$A \oplus_T B = A \oplus_{T_{\mathbf{D}}} B,$$

which means that we will obtain the same sum for these fuzzy numbers A and B when using an addition based on an arbitrary t-norm T^* which is weaker than T.

13.11 EXAMPLE

(i) An immediate consequence of Proposition 13.10 can be derived for the addition of piecewise linear fuzzy numbers with respect to the Łukasiewicz t-norm $T_{\mathbf{L}}$ (a generator of which is given by $t(x) = 1 - x$): for two piecewise linear fuzzy numbers $A = (l_1, r_1, \alpha_1, \beta_1)$ and $B = (l_2, r_2, \alpha_2, \beta_2)$

$$A \oplus_{T_{\mathbf{L}}} B = (l_1 + l_2, r_1 + r_2, \max(\alpha_1, \alpha_2), \max(\beta_1, \beta_2)),$$

and for two piecewise linear fuzzy reals $A = (x_1, x_1 + \varepsilon_1)$, $B = (x_2, x_2 + \varepsilon_2)$

$$A \oplus_{T_{\mathbf{L}}} B = (x_1 + x_2 + \min(\varepsilon_1, \varepsilon_2), x_1 + x_2 + \varepsilon_1 + \varepsilon_2).$$

(ii) Since for all Yager t-norms $T_\lambda^{\mathbf{Y}}$ with $\lambda \in \,]0,1]$ we have $T_\lambda^{\mathbf{Y}} \leq T_{\mathbf{L}}$, for two piecewise linear fuzzy numbers $A = (l_1, r_1, \alpha_1, \beta_1)$ and $B = (l_2, r_2, \alpha_2, \beta_2)$

$$A \oplus_{T_\lambda^{\mathbf{Y}}} B = (l_1 + l_2, r_1 + r_2, \max(\alpha_1, \alpha_2), \max(\beta_1, \beta_2)).$$

The following generalizes several results concerning the addition of upper semicontinuous fuzzy numbers based on continuous Archimedean t-norms (see [Fullér & Keresztfalvi 1992, Marková 1997] and [Mesiar 1996a, 1997b]).

13.12 PROPOSITION

Let T be a continuous Archimedean t-norm, $t : [0,1] \longrightarrow [0,\infty]$ an additive generator of T, and let $K : [0,\infty[\longrightarrow [0,\infty[$ be a function with $K(0) = 0$ which is continuous and convex. For $\gamma \in \,]0,\infty[$, define $K_\gamma : \,]0,\infty[\longrightarrow [0,1]$ by

$$K_\gamma(x) = t^{(-1)}(\gamma \cdot K(\tfrac{x}{\gamma})).$$

Then for all upper semicontinuous fuzzy numbers $A = (l_1, r_1, K_{\alpha_1}, K_{\beta_1})$ and $B = (l_2, r_2, K_{\alpha_2}, K_{\beta_2})$ we have

$$A \oplus_T B = (l_1 + l_2, r_1 + r_2, K_{\alpha_1 + \alpha_2}, K_{\beta_1 + \beta_2}).$$

The following example gives two immediate consequences of Proposition 13.12.

13.13 EXAMPLE

(i) The sum (based on the product $T_{\mathbf{P}}$) of Gaussian fuzzy numbers is again a Gaussian fuzzy number: take the additive generator $t : [0,1] \longrightarrow [0,\infty]$ of $T_{\mathbf{P}}$ given by $t(x) = -\log x$ and define the function $K : [0,\infty[\longrightarrow [0,\infty[$ by $K(x) = x^2$, then $K_\gamma : \,]0,\infty[\longrightarrow [0,1]$ is given by

$$K_\gamma(x) = e^{-\frac{x^2}{\gamma}},$$

i.e., K_γ coincides with E_γ (see Example 13.4(i)). Therefore, for Gaussian fuzzy numbers $A = (x_1, x_1, E_\alpha, E_\alpha)$ and $B = (x_2, x_2, E_\beta, E_\beta)$ their sum $A \oplus_{T_\mathbf{P}} B$ is again a Gaussian fuzzy number:

$$A \oplus_{T_\mathbf{P}} B = (x_1 + x_2, x_1 + x_2, E_{\alpha+\beta}, E_{\alpha+\beta}).$$

(ii) Similarly, the sum of Gaussian fuzzy numbers based on an Aczél-Alsina t-norm $T_\lambda^{\mathbf{AA}}$ with $\lambda \in [0, \infty]$ (see Section 4.8) is again a Gaussian fuzzy number.

(iii) The addition of fuzzy numbers based on a Yager t-norm $T_\lambda^{\mathbf{Y}}$ (see Section 4.5) preserves the piecewise linearity of fuzzy numbers (see [Kereszt-falvi & Kovács 1992] and [Kolesárová 1995, 1997b, 1998]). For $\lambda \in]0, 1]$ this is just the statement of Example 13.11(ii). For $\lambda \in]1, \infty[$ define the function $K : [0, \infty[\longrightarrow [0, \infty[$ by $K(x) = x^\lambda$ and take the additive generator $t_\lambda^{\mathbf{Y}}$ of the Yager t-norm $T_\lambda^{\mathbf{Y}}$ given by $t_\lambda^{\mathbf{Y}}(x) = (1 - x)^\lambda$. Then

$$K_\gamma(x) = \max\left(0, 1 - \gamma^{\frac{1-\lambda}{\lambda}} x\right)$$

for each $\gamma \in]0, \infty[$, which means that each piecewise linear fuzzy number (l, r, α, β) can be written as

$$(l, r, \alpha, \beta) = \left(l, r, K_{\alpha^{\frac{\lambda}{\lambda-1}}}, K_{\beta^{\frac{\lambda}{\lambda-1}}}\right).$$

Therefore, the sum of $A \oplus_{T_\lambda^{\mathbf{Y}}} B$ of two piecewise linear fuzzy numbers $A = (l_1, r_1, \alpha_1, \beta_1)$ and $B = (l_2, r_2, \alpha_2, \beta_2)$ is again a linear fuzzy number:

$$A \oplus_{T_\lambda^{\mathbf{Y}}} B = \left(l_1 + l_2, r_1 + r_2, \left(\alpha_1^{\frac{\lambda}{\lambda-1}} + \alpha_2^{\frac{\lambda}{\lambda-1}}\right)^{\frac{\lambda-1}{\lambda}}, \left(\beta_1^{\frac{\lambda}{\lambda-1}} + \beta_2^{\frac{\lambda}{\lambda-1}}\right)^{\frac{\lambda-1}{\lambda}}\right).$$

An exhaustive overview of fuzzy numbers and their processing can be found in [Kaufmann & Gupta 1991, Mareš 1994, Dubois *et al.* 1999a].

13.3 Fuzzy control

In Proposition 13.2 we saw that, given a left-continuous t-norm T and fuzzy subsets A and B of X and Y, the fuzzy relation $R_T(A, B)$ (which is closely related to the R-implication $\rightarrow_T$) is a prototypical solution of the relational equation $A \circ_T R = B$. Indeed, the value $\mu_{R_T(A,B)}(x, y)$ can be interpreted as the truth value of the statement

IF x is A THEN y is B,

which is often called a (linguistic) IF... THEN-rule. It therefore makes sense to use a solution of (13.1) to describe the relationship between A and B (compare Proposition 13.18 below).

In [Mamdani & Assilian 1975] a first type of a fuzzy controller was defined, which uses fuzzy sets both in the input and in the output space and, therefore, at least in a technical context, usually needs a defuzzification in order to produce an input-output function.

We shall consider arbitrary (usually higher dimensional) input spaces, and for the output space we shall restrict ourselves to the most usual case $Y = \mathbb{R}$ (it turns out that the case of higher dimensional output spaces can be handled by considering several fuzzy controllers with one-dimensional outputs in parallel, a simplification of the model which unfortunately is not possible in the case of higher dimensional input spaces).

13.14 DEFINITION

Let X be an arbitrary input space, let $A_1, A_2, \ldots, A_n$ and $B_1, B_2, \ldots, B_n$ be normal fuzzy subsets of X and $\mathbb{R}$ with Borel measurable membership functions, respectively, let T be a Borel measurable t-norm, and consider the rulebase

$$\begin{array}{l}\text{IF } x \text{ is } A_1 \text{ THEN } y \text{ is } B_1, \\ \text{IF } x \text{ is } A_2 \text{ THEN } y \text{ is } B_2, \\ \qquad\vdots \\ \text{IF } x \text{ is } A_n \text{ THEN } y \text{ is } B_n.\end{array}$$

The input-output function $F_M : X \longrightarrow \mathbb{R}$ of the *Mamdani controller* is given by

$$F_M(x) = \frac{\int\limits_{\mathbb{R}} \mu_R(x,y) \cdot y \, dy}{\int\limits_{\mathbb{R}} \mu_R(x,y) \, dy}, \tag{13.5}$$

provided that $\int_{\mathbb{R}} \mu_R(x,y)\, dy > 0$ for all $x \in X$, where the membership function $\mu_R : X \times \mathbb{R} \longrightarrow [0,1]$ of the fuzzy relation R on $X \times \mathbb{R}$ is given by

$$\mu_R(x,y) = \max\left[T\left(\mu_{A_1}(x), \mu_{B_1}(y)\right), \ldots, T\left(\mu_{A_n}(x), \mu_{B_n}(y)\right)\right].$$

In a strict mathematical sense, the measurability requirements are necessary for (13.5) being well-defined. In practical situations these hypotheses, however, are usually satisfied.

In Definition 13.14 we have implicitly chosen a special defuzzification method, namely, the so-called *center of gravity*, which is basically contained in equation (13.5). We only mention that there are also other methods of defuzzification, e.g., the *mean of maxima* (see, e.g., [Kruse *et al.* 1994a]).

The t-norms used for the Mamdani controller are most frequently either $T_{\mathbf{M}}$ or $T_{\mathbf{P}}$; in the first case this is also referred to as *max-min-inference*, in the latter case as *max-prod-inference* or *max-dot-inference*. An illustration of a simple Mamdani controller (using the max-min-inference) is given in Figure 13.5.

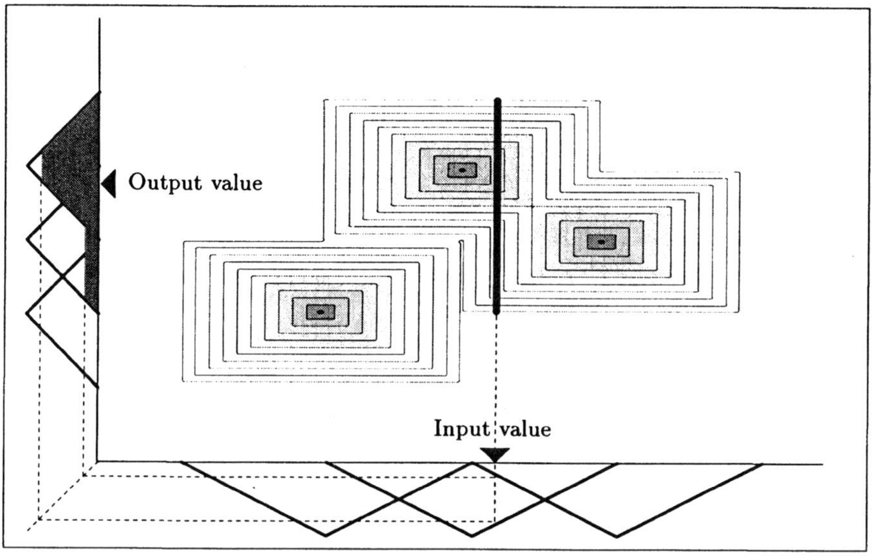

Fig. 13.5. Mamdani controller: the fuzzy subsets of the input and output space are shown at the bottom and left, respectively, the fuzzy relation induced by the Mamdani controller is represented by the contour plot; the output value is the center of area with respect to the grey region on the left

The second important type of fuzzy controllers is the so-called Takagi-Sugeno controller [Sugeno 1985b, Takagi & Sugeno 1985] which uses crisp values in the output space. In a way this means that the inference has a built-in defuzzification.

13.15 DEFINITION
Let X be an input space, let $A_1, A_2, \ldots, A_n$ be normal fuzzy subsets of X with $\sum_{i=1}^{n} \mu_{A_i}(x) > 0$ for all $x \in X$, and $f_1, f_2, \ldots, f_n$ be functions from X to $\mathbb{R}$, and consider the rulebase

$$\begin{aligned} &\text{IF } x \text{ is } A_1 \text{ THEN } y = f_1(x), \\ &\text{IF } x \text{ is } A_2 \text{ THEN } y = f_2(x), \\ &\qquad \vdots \\ &\text{IF } x \text{ is } A_n \text{ THEN } y = f_n(x). \end{aligned}$$

Then the *Takagi-Sugeno controller* defines the following input-output function $F_{TS} : X \longrightarrow \mathbb{R}$ which is given by

$$F_{TS}(x) = \frac{\sum_{i=1}^{n} \mu_{A_i}(x) \cdot f_i(x)}{\sum_{i=1}^{n} \mu_{A_i}(x)}.$$

In the special situation, when for $i = 1,2,\ldots,n$ the functions f_i are constant, i.e., $f_i(x) = u_i$, the Takagi-Sugeno controller can be considered as a special case of the Mamdani controller:

13.16 PROPOSITION

Let X be an input space, let $A_1, A_2, \ldots, A_n$ be normal fuzzy subsets of X with Borel measurable membership functions and with $\sum_{i=1}^{n} \mu_{A_i}(x) > 0$ for all $x \in X$, let $y_1, y_2, \ldots, y_n$ be different elements of $\mathbb{R}$, consider the rulebase

$$\begin{aligned}
&\text{IF } x \text{ is } A_1 \text{ THEN } y = y_1,\\
&\text{IF } x \text{ is } A_2 \text{ THEN } y = y_2,\\
&\qquad\vdots\\
&\text{IF } x \text{ is } A_n \text{ THEN } y = y_n,
\end{aligned}$$

and let F_{TS} be the input-output function of the corresponding Takagi-Sugeno controller. Then there exists a Mamdani controller with input space X such that its corresponding input-output function F_M coincides with F_{TS}.

In $\mathbb{R}$ we can consider crisp points x_0, i.e., one-point sets $\{x_0\}$, as limits of closed intervals $[x_0 - \varepsilon, x_0 + \varepsilon]$ as ε goes to zero. Therefore Proposition 13.16 also says that Takagi-Sugeno controllers (with constant functions f_i) are limits of suitable Mamdani controllers (a result which holds also if other defuzzification methods are used).

If all the fuzzy subsets of both the input space X and the output space $\mathbb{R}$ involved in the rulebase are fuzzy points, then the fuzzy relation R on $X \times \mathbb{R}$ induced by the Mamdani controller can be viewed as the extensional hull of a partial function from X into $\mathbb{R}$:

13.17 PROPOSITION

Let $x_1, x_2, \ldots, x_n$ be different elements of an arbitrary input space X, let $y_1, y_2, \ldots, y_n \in \mathbb{R}$, let $A_1, A_2, \ldots, A_n$ and $B_1, B_2, \ldots, B_n$ be normal fuzzy subsets of X and $\mathbb{R}$, respectively, let T be a t-norm, and consider the rulebase

$$\begin{aligned}
&\text{IF } x \text{ is } A_1 \text{ THEN } y \text{ is } B_1,\\
&\text{IF } x \text{ is } A_2 \text{ THEN } y \text{ is } B_2,\\
&\qquad\vdots\\
&\text{IF } x \text{ is } A_n \text{ THEN } y \text{ is } B_n.
\end{aligned}$$

If the two families $(A_1, A_2, \ldots, A_n)$ and $(B_1, B_2, \ldots, B_n)$ both satisfy (12.15) and if $\mu_{A_i}(x_i) = 1$ and $\mu_{B_i}(y_i) = 1$ for all $i = 1,2,\ldots,n$, then the fuzzy relation R on $X \times \mathbb{R}$ induced by the Mamdani controller is the extensional hull of the crisp set $\{(x_1,y_1),(x_2,y_2),\ldots,(x_n,y_n)\}$ with respect to the equivalence relation $E \times F$, where E and F are equivalence relations on X and $\mathbb{R}$ existing because of Proposition 12.8.

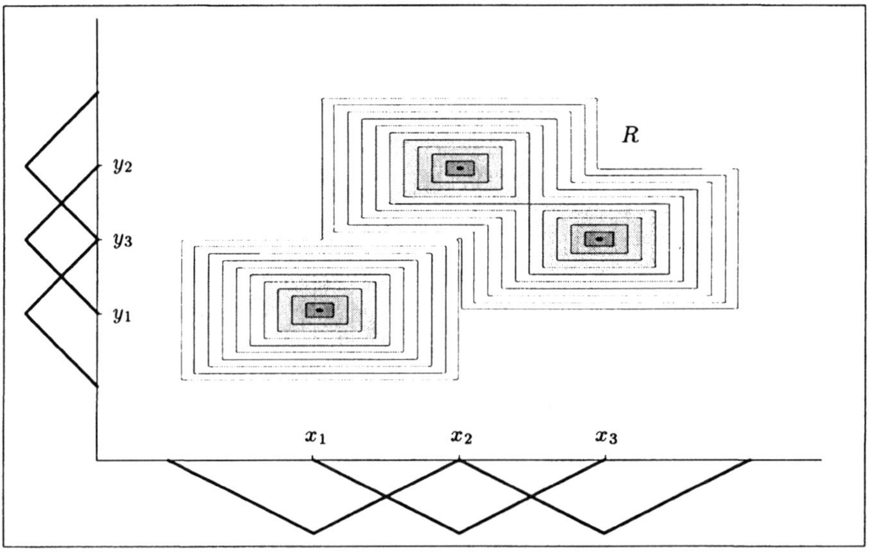

Fig. 13.6. The fuzzy relation R, represented by the contour plot, is the extensional hull of the partial function which assigns y_1 to x_1, y_2 to x_2, and y_3 to x_3 (the fuzzy sets in the bottom and on the left are the corresponding fuzzy points)

Indeed, the crisp set $\{(x_1, y_1), (x_2, y_2), \ldots, (x_n, y_n)\}$ can be regarded as a partial function from X into $\mathbb{R}$ assigning only to the elements x_i (rather than to the whole domain X) a corresponding value, namely, $y_i \in \mathbb{R}$. The fuzzy relation R of the corresponding Mamdani controller then can be viewed as a *fuzzy graph* with the property that the points of the graph of the partial function have degree of membership 1. In this spirit, the rulebase of the Mamdani controller in Proposition 13.17 could be interpreted as

$$
\begin{gathered}
\text{IF } x \text{ is } \textit{approximately } x_1 \text{ THEN } y \text{ is } \textit{approximately } y_1, \\
\text{IF } x \text{ is } \textit{approximately } x_2 \text{ THEN } y \text{ is } \textit{approximately } y_2, \\
\vdots \\
\text{IF } x \text{ is } \textit{approximately } x_n \text{ THEN } y \text{ is } \textit{approximately } y_n.
\end{gathered}
$$

Figure 13.6 shows the contour plot of a rather simple fuzzy graph. The conversion of such a fuzzy graph into a usual, i.e., non-partial function from X into $\mathbb{R}$ is done as follows: start with a crisp input in X, apply the compositional rule of inference (see Section 13.1) and defuzzify.

There is also a strong connection between the Mamdani controller and fuzzy relational equations (see [Czogała & Pedrycz 1981a, 1981b] and [Gottwald 1984a, 1984b]). Starting with one IF... THEN-rule, the fuzzy relation R of the corresponding Mamdani controller is a solution of the relational equation (13.1):

13.18 PROPOSITION

Let A be a normal fuzzy subset of the input space X and B a normal fuzzy subset of $\mathbb{R}$, let T be a left-continuous t-norm, and consider the rule

$$\text{IF } x \text{ is } A \text{ THEN } y \text{ is } B,$$

Then the fuzzy relation R of the corresponding Mamdani controller is a solution of the relational equation (13.1).

In the case of more than one rule we obtain a system of relational equations which, in general, is not solvable.

14. Generalized measures and integrals

> Now, additivity is the most important property of ordinary measures. It is, however, doubtful that an individual uses a "measure" with additivity when he subjectively measures fuzziness. Though a reasonable man is imagined in the theory of subjective probabilities, it would be more realistic to assume that an actual man has no additive measure, because his behaviors are often contradictory to the assumption that he uses an additive measure in evaluating things.
>
> MICHIO SUGENO

From the many generalizations of the concept of a classical (σ-additive) measure which can be found in the literature we shall present two concepts where t-norms and t-conorms are used explicitly.

On the one hand, the domain of the measure has been extended from crisp subsets of some universe to fuzzy subsets thereof in [Klement 1980a, 1982a, 1983], [Klement *et al.* 1981] and [Butnariu 1983a], following some initial ideas of [Zadeh 1968]. The main results in this context are representation theorems for monotone measures based on Frank t-norms (Section 4.4) in terms of Lebesgue integrals with respect to their restriction to the σ-algebra of crisp subsets of their domain, but also Hahn and Jordan Decompositions as well as a Liapounoff Theorem.

A survey of how these concepts fit into a theory of generalized measures can be found in [Klement & Weber 1999]. One application of, e.g., $T_{\mathbf{L}}$-measures is within the theory of games with fuzzy coalitions, allowing the classical Aumann-Shapley value introduced by [Shapley 1953] and [Aumann & Shapley 1974] to be extended to large classes of such games (see [Butnariu 1985, 1987b] and [Butnariu & Klement 1993]).

On the other hand, the usual addition in the range of the measure (and the multiplication when introducing the corresponding integral) have been generalized. An important, very general approach in this context is due to [Sion 1973] who studied integrals based on measures with values in abstract uniform semigroups which, when restricting to closed subintervals of the real line, are just I-semigroups. In this general setting, the distributivity of the multiplication over the addition and some restriction on the measures (called s-boundedness) are necessary to define the integral.

Our presentation starts with a pair (S,T) of a t-conorm and a t-norm, where the t-norm T is only conditionally distributive over S, and gives a survey on the concepts of S-measures and (S,T)-integrals, following the ideas of, e.g., [Shilkret 1971, Sugeno 1974, Dubois & Prade 1982a, Weber 1984, Sugeno & Murofushi 1987, Pap 1990a].

Based on this, generalized convolutions and Laplace transforms [Pap & Teofanov 1998, Pap & Ralević 1998, Pap 1999] are introduced which proved to be useful in the context of some non-linear equations (see [Maslov & Samborskij 1992], [Pap 1992, 1995, 1999] and [Pap & Ralević 1998]), in decision theory [Baccelli *et al.* 1992, Akian *et al.* 1994], and in optimization problems [Pap & Teofanov 1998, Pap & Ralević 1998].

Finally, we look at information measures [Wiener 1948, Shannon & Weaver 1949], in particular at the relationship between composition laws and t-conorms, and we sketch a generalization suggested in [Schweizer & Sklar 1969].

14.1 Measures on T-tribes

Throughout this chapter, the underlying universe (a fixed non-empty set) will be denoted by X, and $\mathcal{A}$ usually denotes a σ-algebra of subsets of X, whereas the σ-algebra of Borel subsets of the real line $\mathbb{R}$ will be denoted $\mathcal{B}$.

When defining measures on T-tribes (as discussed in Section 12.4), a natural starting point is given by the axioms of a classical (σ-additive) measure which are recalled here for convenience.

14.1 DEFINITION
Let $\mathcal{A}$ be a σ-algebra of subsets of X. A function $m : \mathcal{A} \longrightarrow [0,\infty]$ is called a *measure* if the following properties are satisfied:

(M1) $m(\emptyset) = 0$.

(M2) m is continuous from below, i.e., for each non-decreasing sequence $(A_n)_{n\in\mathbb{N}}$ in $\mathcal{A}$ we have

$$\lim_{n\to\infty} m(A_n) = m\Big(\bigcup_{n=1}^{\infty} A_n\Big).$$

(M3) m is additive, i.e., for all $A, B \in \mathcal{A}$ with $A \cap B = \emptyset$ we have

$$m(A \cup B) = m(A) + m(B).$$

If, in addition, we have $m(X) = 1$ then m is called a *probability measure.*

An obvious consequence of the additivity (M3) is that that each measure is monotone in the sense that for all $A, B \in \mathcal{A}$ we have

(M4) $m(A) \leq m(B)$ whenever $A \subseteq B$.

It is now straightforward how to generalize (M1) and (M2) when introducing measures on T-clans and T-tribes (see Section 12.4, Definitions 12.19 and 12.21). Only the generalization of the additivity (M3) is not so clear, since outside of De Morgan algebras, MV-algebras [Chang 1958] and Girard algebras [Girard 1987, Höhle 1995], there is no natural notion of a complement and, subsequently, of disjoint elements (compare, however, the T-disjointness in [Butnariu & Klement 1993]). Therefore it is more reasonable to generalize the classical valuation property, which says that for all elements A, B of the domain of a measure m we have $m(A \cap B) + m(A \cup B) = m(A) + m(B)$. The following definition of measures on T-tribes is essentially due to [Klement 1982b, Butnariu & Klement 1993], for an overview see [Butnariu & Klement 2000].

14.2 DEFINITION
Let T be a t-norm, S the dual t-conorm to T, and let $\mathcal{T}$ be a T-tribe on some non-empty set X.

(i) A function $\mathbf{m} : \mathcal{T} \longrightarrow [-\infty, \infty]$, which assumes at most one of the values $-\infty$ and ∞, is called a *T-measure* if we have

(TM1) $\mathbf{m}(\emptyset) = 0$.

(TM2) For all $A, B \in \mathcal{T}$ we have

$$\mathbf{m}(A \cap_T B) + \mathbf{m}(A \cup_S B) = \mathbf{m}(A) + \mathbf{m}(B).$$

(TM3) For each non-decreasing sequence $(A_n)_{n\in\mathbb{N}}$ in $\mathcal{T}$ and for each $A \in \mathcal{T}$ with $(A_n)_{n\in\mathbb{N}} \nearrow A$ we have

$$\lim_{n\to\infty} \mathbf{m}(A_n) = \mathbf{m}(A).$$

(ii) A T-measure $\mathbf{m}$ is called *monotone* if

(TM4) For all $A, B \in \mathcal{T}$ with $A \subseteq B$ we have $\mathbf{m}(A) \leq \mathbf{m}(B)$.

We shall be working mainly with the family of Frank t-norms $(T_\lambda^{\mathbf{F}})_{\lambda\in[0,\infty]}$ (for details see Section 4.4). Recall in particular that $T_0^{\mathbf{F}} = T_{\mathbf{M}}$ and $T_\infty^{\mathbf{F}} = T_{\mathbf{L}}$.

14.3 REMARK

(i) If $\mathcal{A}$ is a σ-algebra of crisp subsets of X then a function $m : \mathcal{A} \longrightarrow [0, \infty]$ is a measure if and only if it is a T-measure for an arbitrary t-norm T.

(ii) Obviously, if $\mathbf{m} : \mathcal{T} \longrightarrow [-\infty, \infty]$ is a monotone T-measure, then it can assume non-negative values only, i.e., $\mathrm{Ran}(\mathbf{m}) \subseteq [0, \infty]$.

(iii) If $\mathcal{T}$ is both a T-tribe and a $T_{\mathbf{M}}$-tribe, then each T-measure on $\mathcal{T}$ is a $T_{\mathbf{M}}$-measure [Klement 1982a] (i.e., a σ-smooth valuation in the terminology of [Klement & Weber 1999]).

(iv) Each $T_{\mathbf{M}}$-measure on a $T_\lambda^{\mathbf{F}}$-tribe $\mathcal{T}$ with $\lambda \in [0, \infty]$ is a valuation on $\mathcal{T}$ (in the lattice theoretical sense, see [Birkhoff 1973, Höhle & Weber 1997, Klement & Weber 1999]).

(v) Each $T_{\mathbf{L}}$-measure on a $T_\lambda^{\mathbf{F}}$-tribe $\mathcal{T}$ with $\lambda \in]0,\infty]$ is a σ-smooth additive uncertainty measure on (the MV-algebra) $\mathcal{T}$ [Klement & Weber 1999].

(vi) A finite $T_{\mathbf{L}}$-measure $\mathbf{m}$ on a $T_{\mathbf{L}}$-tribe $\mathcal{T}$ is monotone if and only if $\operatorname{Ran}(\mathbf{m}) \subseteq [0,\infty[$ (see [Butnariu 1986]).

The following results (Theorems 14.4–14.6) show that we have essentially three types of monotone $T_\lambda^{\mathbf{F}}$-measures $\mathbf{m}$ on (generated) tribes $\mathcal{T}$ of fuzzy sets: in descending order of generality they are monotone valuations (for arbitrary $\lambda \in [0,\infty]$), monotone $T_\lambda^{\mathbf{F}}$-measures with $\lambda \in]0,\infty]$, and additive uncertainty measures (in the case $\lambda = \infty$, where additivity is understood here as in the context of MV-algebras [Klement & Weber 1999]). According to these three integral representation theorems, $\mathbf{m}(A)$ can be written as a Lebesgue integral with respect to the restriction of $\mathbf{m}$ to the crisp sets in $\mathcal{T}$, and where the integrand depends on A in a way which corresponds to the respective level of generality.

The most general representation theorem uses the powerful concept of a *Markov kernel* [Bauer 1981]. In our situation, a $\mathcal{T}^\vee$-Markov kernel ($\mathcal{T}^\vee$ is the σ-algebra of crisp subsets associated with $\mathcal{T}$, see Example 12.22(v)) is a two-place function $K : X \times (\mathcal{B} \cap [0,1[) \longrightarrow [0,1]$ such that its first component $K(.,A) : X \longrightarrow [0,1]$ is $\mathcal{T}^\vee$-measurable for each $A \in \mathcal{B} \cap [0,1[$, and its second component $K(x,.) : \mathcal{B} \cap [0,1[\longrightarrow [0,1]$ is a probability measure for each $x \in X$. The proof of the following result which goes back to [Klement 1980a] (see also [Butnariu & Klement 1993, Theorem 5.8]) is based on the Radon-Nikodym Theorem [Bauer 1981].

14.4 THEOREM

Consider a generated tribe $\mathcal{T}$ and a function $\mathbf{m} : \mathcal{T} \longrightarrow [0,\infty[$. Then the following assertions are equivalent:

(i) $\mathbf{m}$ is a finite monotone $T_{\mathbf{M}}$-measure on $\mathcal{T}$.

(ii) There exist a finite σ-additive measure m on the σ-algebra $\mathcal{T}^\vee$ and a $\mathcal{T}^\vee$-Markov kernel K such that for all $A \in \mathcal{T}$

$$\mathbf{m}(A) = \int_X K(x, [0, \mu_A(x)[)\, dm(x).$$

Note that the measure m on $\mathcal{T}^\vee$ in Theorem 14.4(ii) is just the restriction of $\mathbf{m}$ to $\mathcal{T}^\vee$ (and, therefore, uniquely determined), and that the $\mathcal{T}^\vee$-Markov kernel K is m-almost everywhere uniquely determined by $\mathbf{m}$.

Recall that each $T_\lambda^{\mathbf{F}}$-measure on a generated tribe (compare Example 12.22(iv)) is also a $T_{\mathbf{M}}$-measure. Therefore, the Markov kernel representation given in Theorem 14.4 applies to all finite monotone $T_\lambda^{\mathbf{F}}$-measures on generated tribes.

In the other extremal case we have $T_{\mathbf{L}}$-measures (i.e., σ-additive uncertainty measures) on a $T_{\mathbf{L}}$-tribe (i.e, on an MV-algebra) $\mathcal{T}$ which turn out to have the simplest possible integral representation (coinciding with the original construction suggested in [Zadeh 1968]), even in the case of a non-generated $T_{\mathbf{L}}$-tribe $\mathcal{T}$. The following result is essentially due to [Butnariu 1987b], an alternative proof is given in [Butnariu & Klement 1993, Theorem 6.2] (compare also [Klement *et al.* 1981, Klement 1982a] and [Butnariu 1983a, 1983b, 1985, 1987a]).

14.5 THEOREM
Let $\mathcal{T}$ be a $T_{\mathbf{L}}$-tribe and $\mathbf{m} : \mathcal{T} \longrightarrow [0, \infty[$ be a function. Then the following assertions are equivalent:

(i) $\mathbf{m}$ is a finite $T_{\mathbf{L}}$-measure on $\mathcal{T}$.
(ii) There exists a finite measure m on $\mathcal{T}^{\vee}$ such that for all $A \in \mathcal{T}$

$$\mathbf{m}(A) = \int_X \mu_A \, dm.$$

Again the measure m on $\mathcal{T}^{\vee}$ in Theorem 14.5(ii) is just the restriction of $\mathbf{m}$ to $\mathcal{T}^{\vee}$ and, therefore, uniquely determined by $\mathbf{m}$.

Comparing Theorems 14.4 and 14.5 it is clear that, in the case of $T_{\mathbf{L}}$-measures on generated tribes, the representing Markov kernel has the simplest form $K(x, [\alpha, \beta[) = \beta - \alpha$, i.e., for each $x \in X$ the probability measure $K(x, .)$ coincides with the Lebesgue measure on $[0, 1[$.

A topological approach to $T_{\mathbf{L}}$-measures on a more general domain makes it possible to study simultaneously $T_{\mathbf{L}}$-measures, group-valued measures on Boolean rings, and linear operators on Riesz spaces, and to prove some extension and decomposition theorems [Barbieri & Weber 1998]. In particular, in [Barbieri *et al.* 200x] a Hahn Decomposition Theorem for $T_{\mathbf{L}}$-measures is given, providing a natural isomorphism between the space of all $T_{\mathbf{L}}$-measures and the space of all measures on a suitable complete Boolean algebra.

The intermediate case, i.e., $\lambda \in]0, \infty[$, leads to a representation of $T^{\mathbf{F}}_{\lambda}$-measures on tribes where the corresponding Markov kernel has a rather special form (which is independent of the index λ). This was proved first for the case of a generated tribe in [Klement 1982a] (see also [Butnariu & Klement 1993, Theorem 7.1]) showing that the integrand, as a consequence of the fact that Frank t-norms and t-conorms satisfy (4.2) in Remark 4.8(viii), can be obtained solving a Gaussian functional equation [Aczél 1966] (for the support of a fuzzy set A see (12.2)). The generalization to arbitrary $T^{\mathbf{F}}_{\lambda}$-tribes, which is based on Theorem 12.24, is due to [Mesiar & Navara 1996], compare also [Mesiar 1993a].

14.6 THEOREM
Let $\mathcal{T}$ be a $T^{\mathbf{F}}_{\lambda_0}$-tribe for some $\lambda_0 \in]0, \infty[$ and $\mathbf{m} : \mathcal{T} \longrightarrow [0, \infty[$ be a function. Then the following assertions are equivalent:

(i) $\mathbf{m}$ is a finite monotone $T^{\mathbf{F}}_{\lambda_0}$-measure on $\mathcal{T}$.
(ii) $\mathbf{m}$ is a finite monotone $T^{\mathbf{F}}_{\lambda}$-measure on $\mathcal{T}$ for each $\lambda \in]0, \infty[$.

(iii) There exist a finite measure m on the σ-algebra $\mathfrak{T}^{\vee}$ and a $\mathfrak{T}^{\vee}$-measurable function $f : X \longrightarrow [0,1]$ such that for all $A \in \mathfrak{T}$

$$\mathbf{m}(A) = \int\limits_{\mathrm{supp}(A)} [f + (1-f) \cdot \mu_A] \, dm.$$

Of course, the measure m on $\mathfrak{T}^{\vee}$ in Theorem 14.6(ii) is uniquely determined (again we have $m = \mathbf{m}|_{\mathfrak{T}^{\vee}}$), and the function f is m-almost everywhere uniquely determined by $\mathbf{m}$. The representing Markov kernel for $T_{\lambda}^{\mathbf{F}}$-measures on generated tribes with $\lambda \in \,]0,\infty[$, which exists because of Theorem 14.4, has the form (which is independent of λ)

$$K(x, [\alpha, \beta[) = \begin{cases} (1 - f(x)) \cdot (\beta - \alpha) & \text{if } \alpha > 0, \\ f(x) + (1 - f(x)) \cdot \beta & \text{if } \alpha = 0, \end{cases}$$

i.e., the probability measure $K(x,.)$ has mass $f(x)$ in the point 0, and the remaining mass $1 - f(x)$ is distributed uniformly on $]0,1[$.

If T is not a Frank t-norm only partial results concerning representation of T-measures by integrals are known in the literature so far [Navara 1999].

As a consequence of the special representation given in Theorem 14.5, many deep measure theoretical results can be carried over to the case of $T_{\mathbf{L}}$-measures. The following Jordan Decomposition Theorem for $T_{\mathbf{L}}$-measures can be derived from [Schmidt 1982] (see also [Schmidt 1989]), and it was proved independently and in a different way in [Butnariu 1986] (see [Butnariu & Klement 1993, Theorem 10.2]).

14.7 THEOREM

Let $\mathfrak{T}$ be a $T_{\mathbf{L}}$-tribe. If $\mathbf{m} : \mathfrak{T} \longrightarrow \,]-\infty,\infty[$ is a finite $T_{\mathbf{L}}$-measure then there exist two non-negative finite $T_{\mathbf{L}}$-measures $\mathbf{m}^+, \mathbf{m}^- : \mathfrak{T} \longrightarrow [0,\infty[$ given by

$$\mathbf{m}^+(A) = \sup\{\mathbf{m}(B) \mid B \in \mathfrak{T}, B \subseteq A\},$$
$$\mathbf{m}^-(A) = \sup\{-\mathbf{m}(B) \mid B \in \mathfrak{T}, B \subseteq A\},$$

such that $\mathbf{m} = \mathbf{m}^+ - \mathbf{m}^-$.

An extremely important result in classical measure theory is that the range of a non-atomic measure, defined on a σ-algebra of subsets of a universe, is always a convex and compact set. In this classical context, this problem was approached in [Poprugenko 1928], and it was solved in [Liapounoff 1940]. In [Halmos 1947, 1948] and [Blackwell 1951, Chernoff 1951, Dvoretzky *et al.* 1951, Marcus 1962, Lindenstrauss 1966] the so-called Liapounoff Theorem was reproved and/or generalized.

Below, first an extension of the result in [Dvoretzky *et al.* 1951] for non-atomic n-vector $T_{\mathbf{L}}$-measures is given [Butnariu 1983b, 1985] (see also [Butnariu & Klement 1993, Theorem 14.2] where the idea of the proof was extracted from [Halmos 1948], and the generalization proved in [Avallone & Barbieri 1997]).

14.8 DEFINITION
Let T be a t-norm.

(i) An *n-vector T-measure* $\mathbf{m} = (\mathbf{m}_1, \mathbf{m}_2, \ldots \mathbf{m}_n)$ on a T-tribe $\mathcal{T}$ is a function from $\mathcal{T}$ to $[-\infty, \infty]^n$ such that each $\mathbf{m}_i$ is a T-measure.
(ii) An n-vector T-measure $\mathbf{m}$ on a T-tribe $\mathcal{T}$ is said to be *non-atomic* if for each $A \in \mathcal{T}$ with $\mathbf{m}(A) > 0$ there exists a $B \in \mathcal{T}$ with $B \subseteq A$ such that $\mathbf{m}(B) \notin \{0, \mathbf{m}(A)\}$.

It should be noted that the non-atomicity of an n-vector $T_{\mathbf{L}}$-measure is equivalent [Butnariu & Klement 1993, Theorem 13.3] to the fact that it has the *Darboux property* [Poprugenko 1928].

14.9 THEOREM
Let $\mathcal{T}$ be a $T_{\mathbf{L}}$-tribe. If $\mathbf{m} : \mathcal{T} \longrightarrow [-\infty, \infty]^n$ is a non-atomic n-vector $T_{\mathbf{L}}$-measure on $\mathcal{T}$, then s Ran($\mathbf{m}$) *is convex and compact.*

This Liapounoff Theorem for non-atomic n-vector $T_{\mathbf{L}}$-measures was one of the basic tools when showing the existence of an *Aumann-Shapley value* [Shapley 1953, Aumann & Shapley 1974] on the space *pFNA* of all games with fuzzy coalitions spanned by positive integer powers of monotone $T_{\mathbf{L}}$-measures [Butnariu & Klement 1993, Theorem 18.4] and, implicitly, for the extrapolation of this Aumann-Shapley value on *pFNA* to much larger classes of games [Butnariu & Klement 1993, Chapter V].

Most recently, in [Barbieri & Weber 200x] a Liapounoff Theorem has been proven for general $T_\lambda^{\mathbf{F}}$-measures. In order to state these results, some preparatory steps are necessary.

14.10 PROPOSITION
Let $\mathcal{T}$ be a $T_{\mathbf{M}}$-clan and $\mathbf{m} : \mathcal{T} \longrightarrow \mathbb{R}$ be a function satisfying (TM1) and (TM2) (with respect to $T_{\mathbf{M}}$). Then the function $d_{\mathbf{m}} : \mathcal{T} \longrightarrow [0, \infty]$ defined by

$$d_{\mathbf{m}}(A, B)$$
$$= \sup \left\{ |\mathbf{m}(C) - \mathbf{m}(D)| \;\middle|\; C, D \in \mathcal{T}, A \cap_{T_{\mathbf{M}}} B \subseteq C \subseteq D \subseteq A \cup_{T_{\mathbf{M}}} B \right\}$$

is a pseudo-metric on the $T_{\mathbf{M}}$-clan $\mathcal{T}$.

14.11 DEFINITION
Let $\mathcal{T}$ be a $T_{\mathbf{M}}$-clan and $\mathbf{m} : \mathcal{T} \longrightarrow \mathbb{R}$ be a function satisfying (TM1) and (TM2) (with respect to $T_{\mathbf{M}}$). Then $\mathcal{T}$ is said to be $\mathbf{m}$-*chained* if, for all $A, B \in \mathcal{T}$ and for each $\varepsilon > 0$, there exist $C_0, C_1, \ldots, C_n$ in $\mathcal{T}$ with $A = C_0 \subseteq C_1 \subseteq \cdots \subseteq C_n = B$ and $d_{\mathbf{m}}(C_{i-1}, C_i) < \varepsilon$ for all $i \in \{1, 2, \ldots, n\}$.

We only mention that, in general, whenever $\mathcal{T}$ is $\mathbf{m}$-chained then $\mathbf{m}$ is atomless, but not conversely. However, a $T_{\mathbf{L}}$-measure $\mathbf{m}$ on a $T_\lambda^{\mathbf{F}}$-tribe $\mathcal{T}$ (and, therefore, its restriction m to the σ-algebra of crisp sets $\mathcal{T}^\vee$) is non-atomic if and only if $\mathcal{T}$ is $\mathbf{m}$-chained or, equivalently, if $\mathcal{T}^{\vee}$ is m-chained. Also note that, in the case of a measure m on a σ-algebra $\mathcal{A}$ of crisp subsets of X, the pseudo-metric space $(\mathcal{A}, d_m)$ is always complete.

14.12 THEOREM
Let $\mathcal{T}$ be a $T_\lambda^{\mathbf{F}}$-tribe and $\mathbf{m} : \mathcal{T} \longrightarrow \mathbb{R}^n$ a non-atomic n-vector $T_\lambda^{\mathbf{F}}$-measure on $\mathcal{T}$ for some $\lambda \in \,]0, \infty[$. Then we have:

(i) If $\mathcal{T}$ is $\mathbf{m}$-chained, then $\mathrm{Ran}(\mathbf{m})$ *is convex.*
(ii) If the space $(\mathcal{T}, d_{\mathbf{m}})$ is complete, then $\mathrm{Ran}(\mathbf{m})$ *is compact.*

The results of Theorem 14.12 do no longer hold if we drop the additional assumptions in (i) and (ii), as the following counterexamples show.

14.13 EXAMPLE
(i) Consider the ($T_\lambda^{\mathbf{F}}$-)tribe $\mathcal{F}(\{x_0\})$ of all fuzzy subsets of the singleton $\{x_0\}$. Then the function $\mathbf{m} : \mathcal{F}(\{x_0\}) \longrightarrow [0, 2]$ defined by

$$\mathbf{m}(A) = \begin{cases} 0 & \text{if } \mu_A(x_0) = 0, \\ \mu_A(x_0) + 1 & \text{otherwise,} \end{cases}$$

is an atomless $T_\lambda^{\mathbf{F}}$-measure for each $\lambda \in \,]0, \infty[$ whose range, however, is not convex.

(ii) If $\boldsymbol{\lambda} : \mathcal{B} \longrightarrow [0, 1]$ denotes the usual Lebesgue measure on the σ-algebra $\mathcal{B}$ of all Borel subsets of $[0, 1]$, then the function $\mathbf{m} : \mathcal{B}^{\wedge} \longrightarrow [0, 1]$ given by

$$\mathbf{m}(A) = \boldsymbol{\lambda}(\mathrm{supp}(A)) - \int_0^1 \mu_A \, d\boldsymbol{\lambda}$$

is a $T_\lambda^{\mathbf{F}}$-measure for each $\lambda \in \,]0, \infty[$ and $\mathcal{B}^{\wedge}$ is $\mathbf{m}$-chained, but $\mathrm{Ran}(\mathbf{m})$ is not compact.

It is noteworthy that most results of this section can be generalized: the values of the measures in question may be not only numbers in the (extended) real line, but in an arbitrary normed commutative group (see [Weber 1996] and [Barbieri & Weber 1998, 1999]).

It would be interesting to know which of the results presented here can be extended to possibility measures (see Definition 14.21 below), to S-measures (as introduced in Definition 14.14 below), or to measures whose values are fuzzy numbers (as described in Section 13.2) rather than crisp numbers (some approaches and partial results in these areas can be found in [Höhle 1976, Klement 1983, Klement & Weber 1991, Mesiar 1992, Kolesárová 1997a]).

14.2 Decomposable measures

The additivity (M3) of classical measures as given in Definition 14.1 can be interpreted as a functional dependence of $m(A \cup B)$ on $m(A)$ and $m(B)$ for all pairs of disjoint sets $A, B \in \mathcal{A}$, which means that there exists a binary operation $*$ on $[0, m(X)]$ such that

$$m(A \cup B) = m(A) * m(B) \quad \text{whenever } A \cap B = \emptyset.$$

For a measure $m : \mathcal{A} \longrightarrow [0,1]$ we obviously have $* = +$. Since, in the case of a probability measure m, (M3) is a condition only for sets A and B satisfying $m(A) + m(B) \leq 1$, and since $S_{\mathbf{L}}(x,y) = x + y$ for all $x, y \in [0,1]$ with $x + y \leq 1$, we may also put $* = S_{\mathbf{L}}$.

In general, we may consider an arbitrary t-conorm, since (M1) and the properties of the union $\cup$ suggest that the operation $*$ should be commutative, associative, monotone, and have neutral element 0. This was the starting point for the following definition (see [Dubois & Prade 1982a, 1982b] and [Weber 1984, Sugeno & Murofushi 1987, Pap 1990a, Klement & Weber 1991, Wang & Klir 1992, Pap 1995]).

14.14 DEFINITION
Let S be a t-conorm and let $\mathcal{A}$ be a σ-algebra of subsets of X. A function $m : \mathcal{A} \longrightarrow [0,1]$ is called an *S-measure* if it satisfies (M1), (M2) and if

(SM) m is *S-decomposable*, i.e., for all $A, B \in \mathcal{A}$ with $A \cap B = \emptyset$

$$m(A \cup B) = S(m(A), m(B)).$$

Of course, from the S-decomposability (SM) it follows immediately that each S-measure satisfies (M4).

14.15 REMARK

(i) If S is a left-continuous t-conorm, then a set function $m : \mathcal{A} \longrightarrow [0,1]$ satisfying (M1) is an S-measure if and only if for each sequence $(A_n)_{n \in \mathbb{N}}$ of pairwise disjoint elements of $\mathcal{A}$ we have

$$m\Big(\bigcup_{n=1}^{\infty} A_n\Big) = \mathop{\mathsf{S}}_{n=1}^{\infty} m(A_n).$$

(ii) If the set X is finite or countably infinite and if S is a left-continuous t-conorm, then each S-measure $m : \mathcal{P}(X) \longrightarrow [0,1]$ is uniquely determined by the values $m(\{x\})$ with $x \in X$.

(iii) A set function $m : \mathcal{A} \longrightarrow [0,1]$ is $S_{\mathbf{M}}$-decomposable if and only if for all $A, B \in \mathcal{A}$ we have $m(A \cup B) = S_{\mathbf{M}}(m(A), m(B))$.

14.16 EXAMPLE

(i) Obviously, each measure $m : \mathcal{A} \longrightarrow [0, \infty]$ with $\operatorname{Ran}(m) \subseteq [0, 1]$ is an $S_{\mathbf{L}}$-measure.

(ii) The function $m : \mathcal{B} \longrightarrow [0, 1]$ defined by $m(A) = \min(\boldsymbol{\lambda}(A), 1)$, where $\boldsymbol{\lambda} : \mathcal{B} \longrightarrow [0, \infty]$ is the Lebesgue measure, is an $S_{\mathbf{L}}$-measure, but not a measure.

(iii) Fix a number $p \in \mathbb{N}$ and define the set function $m_p : \mathcal{P}(\mathbb{N}) \longrightarrow [0, 1]$ by

$$m_p(A) = \min\left(\tfrac{1}{p} \operatorname{card}(A), 1\right).$$

It is easily seen that m_p is an S-measure if and only if the restriction of S to $\Gamma^2_{(p)}$ is an Archimedean divisible discrete t-conorm (see Section 7.3), i.e., if the restrictions of S and $S_{\mathbf{L}}$ to $\Gamma^2_{(p)}$ coincide. Obviously, m_p is always an $S_{\mathbf{L}}$-measure, an $S_{\mathbf{D}}$-measure in the case $p = 2$, and an S-measure for each t-conorm S in the case $p = 1$, but not a measure.

(iv) The function $m : \mathcal{B} \longrightarrow [0, 1]$ defined by $m(A) = 1 - e^{-\boldsymbol{\lambda}(A)}$ is an $S_{\mathbf{P}}$-measure.

(v) For an arbitrary function $f : X \longrightarrow [0, 1]$, the following set function $\mu_f : \mathcal{P}(X) \longrightarrow [0, 1]$ is an $S_{\mathbf{M}}$-measure (compare also [Zadeh 1978]):

$$\mu_f(A) = \sup\{f(x) \mid x \in A\}.$$

(vi) The following set function $m : \mathcal{P}([0, 1]) \longrightarrow [0, 1]$ is an $S_{\mathbf{M}}$-measure:

$$m(A) = \begin{cases} \sup A & \text{if } A \text{ is finite or countably infinite,} \\ 1 & \text{otherwise.} \end{cases}$$

(vii) If $P : \mathcal{A} \longrightarrow [0, 1]$ is a probability measure then for each $\lambda \in \left]-1, \infty\right[\setminus \{0\}$ the set function $m_\lambda : \mathcal{A} \longrightarrow [0, 1]$ defined by

$$m_\lambda(A) = \frac{1}{\lambda}\left((1 + \lambda)^{P(A)} - 1\right)$$

is an $S^{\mathbf{SW}}_\lambda$-measure (for the definition of the Sugeno-Weber t-conorm $S^{\mathbf{SW}}_\lambda$ see Section 4.7). Observe that we have $m_\lambda = (s^{\mathbf{SW}}_\lambda)^{-1} \circ P$, where $s^{\mathbf{SW}}_\lambda$ is the unique additive generator of $S^{\mathbf{SW}}_\lambda$ which satisfies $s^{\mathbf{SW}}_\lambda(1) = 1$. Note also that for disjoint sets $A, B \in \mathcal{A}$ we have (compare [Sugeno 1974])

$$m_\lambda(A \cup B) = m_\lambda(A) + m_\lambda(B) + \lambda \cdot m_\lambda(A) \cdot m_\lambda(B).$$

In particular, this means that m_λ is subadditive if and only if $\lambda \in \left]-1, 0\right[$, and superadditive if and only if $\lambda \in \left]0, \infty\right[$.

(viii) If $m : \mathcal{A} \longrightarrow [0, 1]$ is an S-measure and $\varphi : [0, 1] \longrightarrow [0, 1]$ an increasing bijection, then $\varphi^{-1} \circ m : \mathcal{A} \longrightarrow [0, 1]$ is an S_φ-measure, where S_φ is defined in (2.9).

The S-measures presented in Example 14.16(i)–(iv),(vii) are special cases of the *S-decomposable measures* investigated in [Dubois & Prade 1982a, 1982b] and [Weber 1984], some of which can be constructed as follows:

14.17 PROPOSITION
Let S be a continuous Archimedean t-conorm, $s : [0,1] \longrightarrow [0,\infty]$ an additive generator of S, $\mathcal{A}$ a σ-algebra of subsets of X and $m : \mathcal{A} \longrightarrow [0,\infty]$ a measure. Then the function $s^{(-1)} \circ m : \mathcal{A} \longrightarrow [0,1]$ is an S-measure.

The proper generalization of the concept of *σ-finite measures* [Bauer 1981] is given as follows.

14.18 DEFINITION
Let S be a t-conorm, $\mathcal{A}$ a σ-algebra of subsets of X and $m : \mathcal{A} \longrightarrow [0,1]$ an S-measure.

(i) A set $A \in \mathcal{A}$ is called *S-m-faithful* if for each $B \in \mathcal{A}$ with $B \subseteq A$ we have $S(u,v) < 1$ whenever $u < m(B)$ and $v < m(A \setminus B)$.
(ii) A partition $\mathcal{C} = \{C_k \mid C_k \in \mathcal{A}, k \in K\}$ of X, where K is a finite or countably infinite index set, is called an *S-m-partition* if, for each $k \in K$, the set C_k is S-m-faithful.
(iii) The S-measure m is said to be *S-faithful* if there exists an S-m-faithful partition $\mathcal{C}$ of X.

14.19 REMARK
Let S be a t-conorm, $\mathcal{A}$ a σ-algebra of subsets of X and $m : \mathcal{A} \longrightarrow [0,1]$ an S-measure.

(i) If $A \in \mathcal{A}$ is S-m-faithful then each $B \in \mathcal{A}$ with $B \subseteq A$ is S-m-faithful.
(ii) Each $A \in \mathcal{A}$ with $m(A) < 1$ is S-m-faithful.
(iii) If the t-conorm S satisfies $S(x,y) = 1$ if and only if $\max(x,y) = 1$, then X is S-m-faithful. Consequently, the trivial partition $\{X\}$ (and each partition $\{C_k \mid C_k \in \mathcal{A}, k \in K\}$ of X with a finite or countably infinite index set K) is an S-m-partition. Therefore, each $S_{\mathbf{M}}$-measure is $S_{\mathbf{M}}$-faithful and, if S is a strict t-conorm, then each S-measure is S-faithful too.
(iv) In the case of a nilpotent t-conorm S, the S-faithfulness of an S-measure m is weaker than the requirement that X be m-achievable in the sense of [Weber 1984].

An interesting relationship between measures and S-measures with respect to some continuous Archimedean t-conorm S is given as follows:

14.20 PROPOSITION
Let S be a continuous Archimedean t-conorm, $s : [0,1] \longrightarrow [0,\infty]$ an additive generator of S, $\mathcal{A}$ a σ-algebra of subsets of X and $m : \mathcal{A} \longrightarrow [0,1]$ an S-measure. The following are equivalent:

(i) $s \circ m : \mathcal{A} \longrightarrow [0, \infty]$ *is a measure.*
(ii) X *is* S*-*m*-faithful.*

The assumption that X be S-m-faithful cannot be dropped because of Example 14.16(ii),(iii). Observe that, in Example 14.16(ii), a set A is $S_{\mathbf{L}}$-m-faithful if and only if $\lambda(A) \leq 1$, and in Example 14.16(iii), if and only if $\mathrm{card}(A) \leq p$.

We only mention that, if S, s, $\mathcal{A}$ and m are as in Proposition 14.20, X is not S-m-faithful if and only if $s \circ m$ is not a measure but a set function satisfying (M1) and (M2) and being pseudo-additive (see case (NSP) in [Weber 1984] and case (P) in [Klement & Weber 1991]), i.e., there exist disjoint elements $A, B \in \mathcal{A}$ such that

$$(s \circ m)(A \cup B) < (s \circ m)(A) + (s \circ m)(B).$$

Turning to an example of a non-Archimedean t-conorm, recall that $S_{\mathbf{M}}$-measures were called (σ-)*maxitive measures* in [Shilkret 1971], which appear in the context of *fractal dimensions* (see [Falconer 1990]). Also the *possibility measures*, which are special cases of *upper probabilities* [Dempster 1967] and of *plausibility measures* [Shafer 1976, Höhle & Klement 1983, Dubois & Prade 1988] and which have a close connection to fuzzy sets (see Section 12.1 and [Zadeh 1978]), are of interest.

Since, for an arbitrary (even uncountable) index set I, we naturally have

$$\mathop{\mathsf{S}_{\mathbf{M}}}_{i \in I} a_i = \sup\{a_i \mid i \in I\}$$

for each family $(a_i)_{i \in I} \in [0,1]^I$, we may even consider the following special case of a maxitive, i.e., $S_{\mathbf{M}}$-measure.

14.21 Definition

An $S_{\mathbf{M}}$-measure $m : \mathcal{P}(X) \longrightarrow [0,1]$ is called a *completely maxitive measure* if for each family $(A_i)_{i \in I}$ of subsets of X we have

$$m\Big(\bigcup_{i \in I} A_i\Big) = \sup\{m(A_i) \mid i \in I\}.$$

A completely maxitive measure $\pi : \mathcal{P}(X) \longrightarrow [0,1]$ with $\pi(X) = 1$ is called a *possibility measure.*

14.22 Example

(i) For each function $f : X \longrightarrow [0,1]$, the $S_{\mathbf{M}}$-measure μ_f given in Example 14.16(v) is a completely maxitive measure. Conversely, for each completely maxitive measure $m : \mathcal{P}(X) \longrightarrow [0,1]$ there exists exactly one function $f : X \to [0,1]$, which is given by $f(x) = m(\{x\})$ (and which can be interpreted as the membership function of a fuzzy subset of X in the sense of [Zadeh 1965]), such that $m = \mu_f$.

(ii) The $S_{\mathbf{M}}$-measure m considered in Example 14.16(vi) is not completely maxitive.

(iii) One example of an $S_{\mathbf{M}}$-measure is related to the Hausdorff dimension defined on $\mathcal{P}(\mathbb{R}^n)$ (see, e.g., [Falconer 1990]). First, for each subset $A \subseteq \mathbb{R}^n$ consider its *diameter* $d(A) = \sup\{\|x - y\| \mid x, y \in A\}$ (where $\|\,.\,\|$ is the usual Euclidean norm on $\mathbb{R}^n$), and define for each $\alpha \in\,]0, \infty[$ the function $H^\alpha : \mathcal{P}(\mathbb{R}^n) \longrightarrow [0, \infty]$ by

$$H^\alpha(A) = \lim_{\delta \searrow 0} \Big(\inf \Big\{ \sum_{k=1}^{\infty} d(A_k)^\alpha \;\Big|\; (A_k)_{k\in\mathbb{N}} \in (\mathcal{P}(\mathbb{R}^n))^{\mathbb{N}}, \\ A \subseteq \bigcup_{k=1}^{\infty} A_k, d(A_k) \leq \delta \text{ for all } k \in \mathbb{N} \Big\} \Big).$$

Then the *Hausdorff dimension* $\dim_H : \mathcal{P}(\mathbb{R}^n) \longrightarrow [0, \infty[$ is given by

$$\begin{aligned} \dim_H(A) &= \inf\{\alpha \in\,]0, \infty[\, \mid H^\alpha(A) = 0\} \\ &(= \sup\{\alpha \in\,]0, \infty[\, \mid H^\alpha(A) = \infty\}), \end{aligned}$$

and $\frac{1}{n} \dim_H : \mathcal{P}(\mathbb{R}^n) \longrightarrow [0, 1]$ is an $S_{\mathbf{M}}$-measure (which is not completely maxitive).

If the t-conorm S is continuous in the point $(0, 0)$ then there exists an interesting relationship between S-measures and submeasures (as considered in [Drewnowski 1972, Orlicz 1968]).

14.23 DEFINITION

Let $\mathcal{A}$ be a σ-algebra of subsets of X. A set function $\eta : \mathcal{A} \longrightarrow [0, \infty]$, is called a *submeasure* if it satisfies (M1), (M4) and for all $A, B \in \mathcal{A}$:

$$\eta(A \cup B) \leq \eta(A) + \eta(B).$$

The following result is due to [Pap 1990b, 1991] (compare also [Pap 1995]). It serves as a tool to prove decomposition theorems of the Lebesgue and Saks type [Pap 1990b, 1991].

14.24 THEOREM

Let $\mathcal{A}$ be a σ-algebra of subsets of X, let S be a t-conorm which is continuous in $(0, 0)$, and let $m : \mathcal{A} \longrightarrow [0, 1]$ be an S-measure. Then there exists a submeasure $\eta : \mathcal{A} \longrightarrow [0, \infty]$ such that, for each sequence $(A_n)_{n\in\mathbb{N}}$ in $\mathcal{A}$, we have

$$\lim_{n\to\infty} m(A_n) = 0 \quad \textit{if and only if} \quad \lim_{n\to\infty} \eta(A_n) = 0.$$

For special classes of S-measures with respect to border continuous t-conorms (see Definition 1.23) it is possible to show that they have the Darboux property and that they satisfy a Liapounoff Theorem [Pap 1991].

The construction of the product of two S-measures involves conditionally distributive t-norms (CD) (compare also [Bertoluzza 1993]).

Let T be a continuous Archimedean t-norm and S a continuous Archimedean t-conorm such that T is conditionally distributive over S, and let $\mathcal{A}_1$ and $\mathcal{A}_2$ be σ-algebras of subsets of X_1 and X_2, respectively. If $m_1 : \mathcal{A}_1 \longrightarrow [0,1]$ and $m_2 : \mathcal{A}_2 \longrightarrow [0,1]$ are two S-faithful S-measures then there exists a unique S-faithful S-measure $m_1 \otimes m_2 : \mathcal{A}_1 \otimes \mathcal{A}_2 \longrightarrow [0,1]$ such that for all $A_1 \in \mathcal{A}_1$, $A_2 \in \mathcal{A}_2$ which are S-m_1- and S-m_2-faithful, respectively,

$$m_1 \otimes m_2(A_1 \times A_2) = T(m_1(A_1), m_2(A_2)).$$

In particular, if $\{C_k^{(1)} \mid C_k^{(1)} \in \mathcal{A}_1, k \in K_1\}$ and $\{C_k^{(2)} \mid C_k^{(2)} \in \mathcal{A}_2, k \in K_2\}$ form an S-m_1-partition of X_1 and an S-m_2-partition of X_2, respectively, then $\{C_i^{(1)} \times C_j^{(2)} \mid (i,j) \in K_1 \times K_2\}$ is an S-$(m_1 \otimes m_2)$-partition of $X_1 \times X_2$.

Moreover, if a set $Y_1 \in \mathcal{A}_1$ is S-m_1-faithful and satisfies $m_1(Y_1) = 1$, then m_2 can be seen as a kind of projection of $m_1 \otimes m_2$ in the sense that for all $A \in \mathcal{A}_2$ we have $m_1 \otimes m_2(Y_1 \times A) = m_2(A)$.

14.25 Example

For $p, q \in \mathbb{N}$ consider the $S_{\mathbf{L}}$-measures $m_p, m_q : \mathcal{P}(\mathbb{N}) \longrightarrow [0,1]$ as in Example 14.16(iii). The product measure $m_p \otimes m_q : \mathcal{P}(\mathbb{N} \times \mathbb{N}) \longrightarrow [0,1]$ is given by

$$m_p \otimes m_q(A) = \min\left(\tfrac{1}{pq}\operatorname{card}(A), 1\right).$$

14.3 Integrals based on t-norms and t-conorms

Throughout this section, we shall assume that T is a continuous t-norm and S a continuous t-conorm such that T is conditionally distributive over S, i.e., they satisfy property (CD) in Definition 5.20.

Denote by $\mathcal{M}$ the set of all $\mathcal{A}$-measurable functions from X to $[0,1]$. As usual, a step function $\psi : X \longrightarrow [0,1]$ is a measurable function which assumes only finitely many values, and the set of all step functions will be denoted $\mathcal{S}$. If $\operatorname{Ran}(\psi) = \{\alpha_1, \alpha_2, \ldots, \alpha_n\}$ with $\alpha_i \neq \alpha_j$ whenever $i \neq j$, and if we write $A_i = \psi^{-1}(\{\alpha_i\})$, then there is a canonical representation of ψ given by

$$\psi = \mathop{\mathbf{S}}_{i=1}^{n} \alpha_i \mathbf{1}_{A_i}. \tag{14.1}$$

Observe that representation (14.1) is independent of the t-conorm S under consideration.

The definition of the (S,T)-integral is, in analogy to the Lebesgue integral, first given for step functions and then, using that each measurable is the least upper bound of step functions, for arbitrary measurable functions.

14.26 Definition

Let $m : \mathcal{A} \longrightarrow [0,1]$ be an S-faithful S-measure.

(i) Given an S-m-faithful partition $\mathcal{C} = \{C_k \mid k \in K\}$, the *$(S,T)$-integral* of a measurable step function $\psi : X \longrightarrow [0,1]$ (given in the canonical representation (14.1)) is defined by

$$\int_X^{(S,T)} \mathop{\mathsf{S}}_{i=1}^{n} \alpha_i \mathbf{1}_{A_i}\, dm = \mathop{\mathsf{S}}_{k \in K} \Big(\mathop{\mathsf{S}}_{i=1}^{n} T(\alpha_i, m(A_i \cap C_k)) \Big).$$

Since T is conditionally distributive over S and since the expression on the right-hand side does not depend on the actual choice of $\mathcal{C}$, this integral is well-defined.

(ii) The (S,T)-integral of a measurable function $f : X \longrightarrow [0,1]$ is defined by

$$\int_X^{(S,T)} f\, dm = \sup\Big\{ \int_X^{(S,T)} \psi\, dm \;\Big|\; \psi \in \mathcal{S}, \psi \le f \Big\}.$$

(iii) The (S,T)-integral of a measurable function $f : X \longrightarrow [0,1]$ on an arbitrary set $A \in \mathcal{A}$ is defined by

$$\int_A^{(S,T)} f\, dm = \int_X^{(S,T)} T(\mathbf{1}_A, f)\, dm.$$

14.27 Theorem

Let $m : \mathcal{A} \longrightarrow [0,1]$ be an S-faithful S-measure and $f, g \in \mathcal{M}$.

(i) The (S,T)-integral is monotone, i.e., if $f \le g$ then

$$\int_X^{(S,T)} f\, dm \le \int_X^{(S,T)} g\, dm. \tag{14.2}$$

(ii) If $A, B \in \mathcal{A}$ with $A \cap B = \emptyset$ then

$$\int_{A \cup B}^{(S,T)} f\, dm = S\Big(\int_A^{(S,T)} f\, dm, \int_B^{(S,T)} f\, dm \Big). \tag{14.3}$$

(iii) If $T(S(f,g),a) = S(T(f,a),T(g,a))$ for all $a \in [0,1]$ then

$$\int_X^{(S,T)} S(f,g)\, dm = S\Big(\int_X^{(S,T)} f\, dm, \int_X^{(S,T)} g\, dm \Big). \tag{14.4}$$

It should be mentioned that the condition in Theorem 14.27(iii) is fulfilled for all $f, g \in \mathcal{M}$ only if S equals $S_{\mathbf{M}}$. However, if T is replaced by some other operation $*$ in the sense of Remark 14.31(i) below, then also for other t-conorms condition (14.4) may be satisfied for all $f, g \in \mathcal{M}$. Moreover, because of the conditional distributivity of T over S, (14.4) also holds (for given T, S, f and g) if $S(f(x), g(x)) < 1$ for all $x \in X$.

14.28 THEOREM
Let $m : \mathcal{A} \longrightarrow [0,1]$ be an S-faithful S-measure.

(i) For each $f \in \mathcal{M}$, the function $m_f : \mathcal{A} \longrightarrow [0,1]$ defined by

$$m_f(A) = \int_A^{(S,T)} f\,dm$$

is an S-faithful S-measure.

(ii) For all $f, g \in \mathcal{M}$ we have

$$\int_X^{(S,T)} f\,dm_g = \int_X^{(S,T)} g\,dm_f = \int_X^{(S,T)} T(f,g)\,dm.$$

This shows that the (S,T)-integral is a proper extension of the underlying S-measure because of $m_{\mathbf{1}_X} = m$. Observe also that, for the simple $S_{\mathbf{M}}$-measure $m : \mathcal{P}(X) \longrightarrow [0,1]$ specified by $m(A) = 1$ for all non-empty subsets A of X and for the $S_{\mathbf{M}}$-measure μ_f given in Example14.16(v), we always have $\mu_f = m_f$.

For a S-faithful S-measure $m : \mathcal{A} \longrightarrow [0,1]$, consider the class $\mathcal{M}_m$ given by

$$\mathcal{M}_m = \{f \in \mathcal{M} \mid X \text{ is } S\text{-}m_f\text{-faithfull}\}.$$

Obviously, for each $S_{\mathbf{M}}$-measure $m : \mathcal{A} \longrightarrow [0,1]$ we have $\mathcal{M}_m = \mathcal{M}$. On the family $\mathcal{M}_m$, the (S,T)-integral is homogeneous.

14.29 THEOREM
Let $m : \mathcal{A} \longrightarrow [0,1]$ be an S-faithful S-measure. For all $f \in \mathcal{M}_m$ and for all $a \in [0,1]$ we obtain the equality

$$\int_X^{(S,T)} T(a,f)\,dm = T\Big(a, \int_X^{(S,T)} f\,dm\Big). \tag{14.5}$$

For the (S,T)-integral some classical convergence theorems such as the Fatou Lemma, the Monotone Convergence Theorem, as well as a Radon-Nikodym Theorem can be proved [Klement *et al.* 200x] (compare also [Sugeno & Murofushi 1987, Klement & Weber 1994, Pap 1995]).

For special t-norms T and t-conorms S, we obtain some integrals which are well-known from the literature.

14.30 EXAMPLE

(i) Let $m : \mathcal{A} \longrightarrow [0,1]$ be an $S_{\mathbf{L}}$-measure which is also a measure. Then the $(S_{\mathbf{L}}, T_{\mathbf{P}})$-integral of a function $f \in \mathcal{M}$ coincides with the classical Lebesgue integral of f.

(ii) For each continuous t-norm T, for each $S_{\mathbf{M}}$-measure $m : \mathcal{A} \longrightarrow [0,1]$ and for each function $f \in \mathcal{M}$ we have

$$\int_X^{(S_{\mathbf{M}},T)} f\,dm = \sup\big\{T(a, m(\{x \in X \mid f(x) \geq a\})) \bigm| a \in [0,1]\big\}.$$

If m is a completely maxitive measure then we have

$$\int_X^{(S_{\mathbf{M}},T)} f\,dm = \sup\{T(f(x), m(\{x\})) \mid x \in X\}.$$

In particular, the $(S_{\mathbf{M}}, T_{\mathbf{M}})$-integral coincides with the Sugeno integral introduced in [Sugeno 1974], and the $(S_{\mathbf{M}}, T_{\mathbf{P}})$-integral coincides with the integral studied in [Shilkret 1971].

14.31 Remark

(i) The procedure for the construction of the (S, T)-integral used in Definition 14.26 works also if we replace the t-norm T by an operation $* : [0,1]^2 \longrightarrow [0,1]$ which is left-continuous, monotone, has annihilator 0 and a left neutral element $e \in\,]0,1]$, and which is conditionally distributive over S. Examples of such generalizations are given in Example 14.32 below. Moreover, (14.2) and (14.3) in Theorem 14.27, as well as Theorem 14.28 still hold. If, additionally, the operation $*$ is also associative (e.g., if it is a uninorm [Fodor *et al.* 1997]), then we also have (14.5).

(ii) The restriction to S-measures and measurable functions whose ranges are subsets of $[0,1]$ causes no loss of generality. Let $[a,b] \subseteq [-\infty, \infty]$ such that $([a,b], \oplus)$ is an I-semigroup with neutral element a (see Definition 2.40). Consequently, because of Theorem 2.43, there is a strictly increasing bijection $\varphi : [a,b] \longrightarrow [0,1]$ such that the continuous t-conorm S defined by

$$S(x,y) = \varphi(\varphi^{-1}(x) \oplus \varphi^{-1}(y))$$

is isomorphic to $\oplus$. Let $*$ be a binary operation on $[0,1]$ as in (i), which is conditionally distributive over S, and define $\odot : [a,b]^2 \longrightarrow [a,b]$ by

$$x \odot y = \varphi^{-1}(\varphi(x) * \varphi(y)).$$

Observe that $\odot$ is a left-continuous monotone operation on $[a,b]$ which has annihilator a and possesses some left neutral element, and which satisfies for all $x, y, z \in [a,b]$ with $y \oplus z < b$

$$x \odot (y \oplus z) = (x \odot y) \oplus (x \odot z). \tag{14.6}$$

Then one can introduce, in complete analogy to our constructions, the concept of $\oplus$-faithful $\oplus$-measures $m : \mathcal{A} \longrightarrow [a,b]$ and, for measurable functions $f : X \longrightarrow [a,b]$, the $(\oplus, \odot)$-integral of f with respect to m. The latter ist just a transformation of our $(S, *)$-integral (on $[0,1]$) because of

$$\int_X^{(\oplus,\odot)} f\,dm = \varphi^{-1}\Big(\int_X^{(S,*)} (\varphi \circ f)\, d(\varphi \circ m)\Big)$$

(compare also [Pap 1990a, Maslov & Samborskij 1992]).

(iii) The triple $([a,b], \oplus, \odot)$ considered in (ii) is more general than an abstract semiring [Golan 1992, Kuich & Salomaa 1986] in the sense that we only need a left neutral element and that the distributivity (14.6) is not required to hold for all $x, y, z \in [a,b]$. Observe, however, that in an abstract semiring the addition need not be continuous, and the multiplication need not be left-continuous (both is required in our setting). A rather general approach introducing appropriate triples $([a,b], \oplus, \odot)$ for the construction of integrals can be found in [Benvenuti & Mesiar 200x] where also several nontrivial examples are included.

14.32 EXAMPLE

(i) For $[a,b] = [0,\infty]$ we obtain the integral introduced in [Sugeno & Murofushi 1987], and for $([a,b], \oplus, \odot) = ([0,\infty], +, \cdot)$ we again come back to the classical Lebesgue integral.

(ii) If the operation $\oplus$ in the semiring $([a,b], \oplus, \odot)$ is not idempotent (i.e., if $\oplus \neq \max$), then the operations $\oplus$ and $\odot$ are generated by some uniquely determined strictly increasing bijection $\varphi : [a,b] \longrightarrow [0,\infty]$ via (see [Mesiar & Rybárik 1995])

$$x \oplus y = \varphi^{-1}(\varphi(x) + \varphi(y)),$$
$$x \odot y = \varphi^{-1}(\varphi(x) \cdot \varphi(y)).$$

The corresponding $(\oplus, \odot)$-integral was studied in [Pap 1990a, 1993], and it has the special form

$$\int_X^{(\oplus,\odot)} f \, dm = \varphi^{-1}\Big(\int_X (\varphi \circ f) \, d(\varphi \circ m) \Big),$$

where the integral on the right-hand side is a Lebesgue integral. Under some additional conditions, a generalized derivative can be defined [Pap 1993] (compare also [Marková 1996, Mesiar 1996b]) which turns out to be useful when solving some differential equations (see [Pap 1992, 1999]).

(iii) If S is a continuous Archimedean t-conorm with an additive generator $s : [0,1] \longrightarrow [0,\infty]$, if we replace the t-norm T by the binary operation $* : [0,1]^2 \longrightarrow [0,1]$ which is given by

$$a * b = s^{-1}(a \cdot s(b)),$$

and if we assume that X is m-achievable in the sense of [Weber 1984] in the case when $s(1) < \infty$, we can repeat the whole construction of Definition 14.26 for the $(S,*)$-integral, and we obtain the integral introduced in [Weber 1984].

Further generalizations of the (S,T)-integral can be found in [Murofushi & Sugeno 1991, Maslov & Samborskij 1992] and [Pap 1995, 1999].

The (S,T)-integral considered so far can be understood as a generalization of the classical Lebesgue integral. However, it is possible to define integrals also with respect to monotone set functions vanishing at $\emptyset$ in the spirit of [Choquet 1954], see also [Denneberg 1994, Grabisch *et al.* 1995, Pap 1995]. More precisely, let $m : \mathcal{A} \longrightarrow [0,1]$ be a set function satisfying (M1) and (M4), let $f : X \longrightarrow [0,1]$ be a measurable function, and define the measurable function $h_{m,f} : [0,1] \longrightarrow [0,1]$ by

$$h_{m,f}(z) = m(\{x \in X \mid f(x) \geq z\}).$$

Then $h_{m,f}$ is a non-increasing function, and the *Choquet integral* of f with respect to m is the Riemann integral

$$(C)\int_X f\,dm = \int_0^1 h_{m,f}(z)\,dz.$$

It is well-known [Choquet 1954] that, if m is a measure, then for each measurable function $f : X \longrightarrow [0,1]$ the Choquet and the Lebesgue integral coincide:

$$(C)\int_X f\,dm = \int_X f\,dm.$$

Several generalizations of the Choquet integral based on t-norms, t-conorms or related operations have been studied in [Sugeno 1974, Murofushi & Sugeno 1991, Grabisch *et al.* 1995, Mesiar 1995a, Benvenuti & Vivona 1996]. For an overview of these results see [Benvenuti & Mesiar 2000].

An interesting method of constructing a generalized Choquet integral was recently introduced in [Imaoka 1995, 1997].

14.33 Definition

Let $m : \mathcal{A} \longrightarrow [0,1]$ be a set function satisfying (M1) and (M4), and let $f : X \longrightarrow [0,1]$ be a measurable function. Then the set

$$H_{m,f} = \{(u,v) \in [0,1]^2 \mid v \leq h_{m,f}(u)\}$$

is a Borel measurable subset of $[0,1]^2$. Let C be a copula (see Section 9.1) and let λ be the Lebesgue measure on the Borel subsets of $[0,1]$. Then we can use C to define a joint probability measure P_C on the Borel subsets of the unit square $[0,1]^2$ whose marginal probability measures equal both λ, i.e., for all points $(u,v) \in [0,1]^2$ we have $P_C([0,u] \times [0,v]) = C(u,v)$. The *C-integral* of f with respect to m is then defined by

$$\int_X^{(C)} f\,dm = P_C(H_{m,f}).$$

14.34 EXAMPLE

(i) Starting with the product $T_{\mathbf{P}}$, we obtain $P_{T_{\mathbf{P}}} = \boldsymbol{\lambda} \otimes \boldsymbol{\lambda}$, and the $T_{\mathbf{P}}$-integral is just the Choquet integral (see [Choquet 1954]), i.e., we have

$$\int_X^{(T_{\mathbf{P}})} f\, dm = (C) \int_X f\, dm.$$

(ii) For the minimum $T_{\mathbf{M}}$ we get $P_{T_{\mathbf{M}}}(A) = \boldsymbol{\lambda}(\{u \in [0,1] \mid (u,u) \in A\})$, and the $T_{\mathbf{M}}$-integral coincides with the integral introduced in [Sugeno 1974] (compare also Example 14.30(ii) and [Pap 1995]), i.e.,

$$\int_X^{(T_{\mathbf{M}})} f\, dm = \sup \big\{ \min(u, h_{m,f}(u)) \mid u \in [0,1] \big\}.$$

(iii) Define the function $m : \mathcal{B} \cap [0,1] \longrightarrow [0,1]$ by $m(A) = \sqrt{\boldsymbol{\lambda}(A)}$, and take $f = \mathrm{id}_{[0,1]}$. The C-integral with respect to m depends on the underlying copula C. For instance, when choosing the copulas $T_{\mathbf{M}}$, $T_{\mathbf{P}}$, $T_{\mathbf{L}}$ and C given by $C(u,v) = uv + uv(1-u)(1-v)$ (observe that the latter is not a t-norm, as pointed out in Example 9.5), then we get

$$\int_X^{(T_{\mathbf{M}})} f\, dm = \tfrac{\sqrt{5}-1}{2}, \qquad \int_X^{(T_{\mathbf{P}})} f\, dm = \tfrac{2}{3},$$

$$\int_X^{(T_{\mathbf{L}})} f\, dm = 1, \qquad \int_X^{(C)} f\, dm = \tfrac{19}{30}.$$

14.4 Generalized convolution and Laplace transform

Throughout this section, we assume that S is a continuous t-conorm and T a left-continuous t-norm which is conditionally distributive over S. Let $\mathcal{A}$ be a σ-algebra of subsets of $[0,\infty[$ containing all Borel subsets of $[0,\infty[$ and such that for each $a \in [0,\infty[$ and for each set A of $\mathcal{A}$ also the set $a + A = \{a + x \mid x \in A\}$ belongs to $\mathcal{A}$. Again, the family of all $\mathcal{A}$-measurable functions from $[0,\infty[$ to $[0,1]$ will be denoted $\mathcal{M}$. Finally, we always suppose that $m : \mathcal{A} \longrightarrow [0,1]$ is an S-faithful S-measure which satisfies $m([0,\infty[) = 1$ and which is translation invariant, i.e., satisfies $m(A) = m(a + A)$ for all $a \in [0,\infty[$ and for all $A \in \mathcal{A}$.

14.35 EXAMPLE

(i) The only translation invariant $S_{\mathbf{M}}$-measure $m : \mathcal{P}([0,\infty[) \longrightarrow [0,1]$ with $m([0,\infty[) = 1$ which is completely maxitive is given by (compare also what was said in Example 14.22(i))

$$m(A) = \begin{cases} 0 & \text{if } A = \emptyset, \\ 1 & \text{otherwise.} \end{cases}$$

(ii) For each translation invariant $S_{\mathbf{M}}$-measure $m : \mathcal{B} \cap [0,\infty[\longrightarrow [0,1]$ with $m([0,\infty[) = 1$ (and which is not necessarily completely maxitive) we obtain $m(A) = m(\{0\})$ for all non-empty, finite or countably infinite subsets A of $[0,\infty[$, and $m(A) = 1$ whenever $\lambda(A) > 0$, where λ denotes the Lebesgue measure. The only completely maxitive, translation invariant $S_{\mathbf{M}}$-measure on $\mathcal{B} \cap [0,\infty[$ corresponds to the case $m(A) = m(\{0\}) = 1$ whenever $\lambda(A) > 0$.

(iii) Each translation invariant $S_{\mathbf{L}}$-faithful $S_{\mathbf{L}}$-measure m on the Borel subsets of $[0,\infty[$ with $m([0,\infty[) = 1$ is specified by $m(A) = \min(c \cdot \lambda(A), 1)$ for some $c \in]0,\infty[$.

In probability theory, the proper tool to compute the density function of the sum of two continuous random variables (in the case of their independence) is the *convolution* [Bauer 1981], which for two measurable functions $f, g : [0,\infty[\longrightarrow [0,1]$ is defined by

$$f \circledast g(x) = \int_{[0,x[} f(x-t)g(t)\, d\lambda(t).$$

Also, when solving partial differential or integral equations, the convolution plays an important role [Hoermander 1985, Taylor 1996, Pap *et al.* 1997]. This concept can be extended to our more general setting.

14.36 DEFINITION

The (S,T)-*convolution* $f \overset{(S,T)}{\circledast} g : [0,\infty[\longrightarrow [0,1]$ of two measurable functions $f, g : [0,\infty[\longrightarrow [0,1]$ with respect to an S-measure m is given by

$$f \overset{(S,T)}{\circledast} g(x) = \int_{[0,x]}^{(S,T)} T(f(x-t), g(t))\, dm(t).$$

As an immediate consequence of Theorem 14.28(ii) we obtain the following simpler formula for the (S,T)-convolution of f and g:

$$f \overset{(S,T)}{\circledast} g(x) = \int_{[0,x]}^{(S,T)} f(x-t)\, dm_g(t).$$

14.37 THEOREM

The (S,T)-convolution $\overset{(S,T)}{\circledast}$ is a commutative and monotone binary operation on $\mathcal{M}$. Moreover, it is an associative operation on $\mathcal{M}_m$.

14.38 EXAMPLE

(i) For the function $f : [0,\infty[\longrightarrow [0,1]$ defined by $f(x) = \min(1, \frac{x}{6})$, we get the following convolutions (with respect to the translation invariant $S_{\mathbf{M}}$- and $S_{\mathbf{L}}$-measures mentioned in Example 14.35):

$$f \overset{(S_{\mathbf{M}},T_{\mathbf{M}})}{\circledast} f(x) = \min\left(\tfrac{x}{12}, 1\right),$$
$$f \overset{(S_{\mathbf{M}},T_{\mathbf{L}})}{\circledast} f(x) = \max\left(\min\left(\tfrac{x-12}{6}, 1\right), 0\right),$$
$$f \overset{(S_{\mathbf{L}},T_{\mathbf{P}})}{\circledast} f(x) = \min(c \cdot h(x), 1),$$

where the function $h : [0, \infty[\longrightarrow [0, \infty[$ is given by

$$h(x) = \begin{cases} \frac{x^3}{216} & \text{if } x \in [0, 6], \\ 2x - 10 - \frac{x^3}{216} & \text{if } x \in]6, 12[, \\ x - 6 & \text{otherwise.} \end{cases}$$

(ii) Given a left-continuous t-norm T, then the operation τ_T on the family Δ^+ of all distance distribution functions (see Example 9.20 in Section 9.2) coincides with the $(S_{\mathbf{M}}, T)$-convolution (with respect to the translation invariant $S_{\mathbf{M}}$-measure mentioned in Example 14.35).

It is not difficult to see that the (S, T)-convolution $\overset{(S,T)}{\circledast}$ has a neutral element only if $S = S_{\mathbf{M}}$ and if the measure m is completely maxitive, in which case it is given by $\mathbf{1}_{\{0\}}$ (i.e., by the characteristic function of the set $\{0\}$).

Similarly as in Remark 14.31, we may replace the triple $([0, 1], S, T)$ by $([a, b], \oplus, \odot)$ and carry over the definition of an (S, T)-convolution to the more general case of an $(\oplus, \odot)$-convolution [Pap & Teofanov 1998, Pap & Ralević 1998, Pap & Štajner 1999, Marková-Stupňanová 1999, Pap 2000]. In particular, in the case $([0, \infty], +, \cdot)$ and if $\mathcal{A}$ is the σ-algebra of Borel subsets of $[0, \infty[$ and m equals the Lebesgue measure, then we obtain just the classical convolution.

14.39 EXAMPLE
The $(\max, +)$-convolution $f \overset{(\max,+)}{\circledast} g$ of two functions $f, g : [0, \infty[\longrightarrow [-\infty, \infty]$ is given by the formula

$$f \overset{(\max,+)}{\circledast} g(x) = \sup\{f(x - t) + g(t) \mid t \in [0, x]\}.$$

In analogy to the case of the classical convolution with respect to the Lebesgue measure, it is also possible to work on the whole line $\mathbb{R}$ rather than on $[0, \infty[$, in which case the definition of the (S, T)-convolution given in Definition 14.36 is changed into

$$f \overset{(S,T)}{\circledast} g(x) = \int_{\mathbb{R}}^{(S,T)} T(f(x - t), g(t))\, dm(t).$$

The extension to $(\oplus, \odot)$-convolutions is then obvious.

We only mention that the addition of fuzzy numbers (see Section 13.2) can be read as an $(S_{\mathbf{M}}, T)$-convolution on $\mathbb{R}$, and that $(\min, +)$-convolutions (derived from (max, +)-convolutions using the duality of min and max) have been used successfully in decision theory [Baccelli *et al.* 1992, Akian *et al.* 1994].

The classical *Laplace transform* (which is closely related to the Fourier transform) of a measurable function $f : [0, \infty[\longrightarrow \mathbb{R}$ with respect to the Lebesgue measure is given by [Widder 1941]

$$\big(\mathcal{L}(f)\big)(z) = \int_0^\infty e^{-zx} f(x)\, dx.$$

It is often used to transform some differential or integral equations into algebraic equations [Doetsch 1956, Taylor 1996, Pap *et al.* 1997]. We can generalize the Laplace transform by means of the (S,T)-integral if we consider special t-norms and t-conorms only.

To be precise, we consider a strictly increasing bijection $\varphi : [0,1] \longrightarrow [0,1]$, and we put $T = (T_{\mathbf{P}})_\varphi$ and either $S = (S_{\mathbf{M}})_\varphi = S_{\mathbf{M}}$ or $S = (S_{\mathbf{L}})_\varphi$ (as defined in (2.8) and (2.9), respectively). Note that this means that the pair (S,T) is isomorphic either to $(S_{\mathbf{M}}, T_{\mathbf{P}})$ or to $(S_{\mathbf{L}}, T_{\mathbf{P}})$, and that φ is both a multiplicative generator of the strict t-norm T and (in the latter case) an additive generator of the nilpotent t-conorm S (as studied in Section 3.2). Observe that in this setting T is always conditionally distributive over S because of Theorem 5.21.

Concerning the underlying measure $m : \mathcal{A} \longrightarrow [0,1]$, we continue to suppose that it is an S-faithful S-measure which is translation invariant and (of course only in the case $S = S_{\mathbf{M}}$) completely maxitive.

14.40 Definition

The (S,T)-*Laplace transform* of a measurable function $f : [0,\infty[\longrightarrow [0,1]$ with respect to the S-measure m is defined by

$$\big(\mathcal{L}^{(S,T)}(f)\big)(z) = \int_{[0,\infty[}^{(S,T)} \varphi^{-1}(e^{-zx})\, dm_f(x).$$

14.41 Example

(i) In the special case $(S,T) = (S_{\mathbf{M}}, T_{\mathbf{P}})$ (i.e., where $(T_{\mathbf{P}})_\varphi = T_{\mathbf{P}}$ which means $\varphi(x) = x^p$ for some $p \in \,]0,\infty[$ because of Corollary 5.4) we obtain

$$\big(\mathcal{L}^{(S_{\mathbf{M}},T_{\mathbf{P}})}(f)\big)(z) = \sup\left\{e^{-\frac{zx}{p}} f(x) \mid x \in [0,\infty[\right\}.$$

(ii) In the special case $(S,T) = (S_{\mathbf{L}}, T_{\mathbf{P}})$ the $S_{\mathbf{L}}$-measure m is given by $m(A) = \min(c \cdot \lambda(A), 1)$ for some $c \in \,]0,\infty[$ because of Example 14.35(ii), and we get (here $\mathcal{L}$ is the classical Laplace transform)

$$\big(\mathcal{L}^{(S_{\mathbf{L}},T_{\mathbf{P}})}(f)\big)(z) = \min(c \cdot \mathcal{L}(f), 1).$$

14.42 Remark

As an immediate consequence of Theorems 14.27 and 14.29, for all $f, g \in \mathcal{M}$ and $a \in [0,1]$ we obtain:

(i) If $T(S(f,g),b) = S(T(f,b),T(g,b))$ for all $b \in [0,1]$ then

$$\mathfrak{L}^{(S,T)}(S(f,g)) = S\big(\mathfrak{L}^{(S,T)}(f), \mathfrak{L}^{(S,T)}(g)\big).$$

(ii) If $f \in \mathcal{M}_m$ then we have

$$\mathfrak{L}^{(S,T)}(T(a,f)) = T\big(a, \mathfrak{L}^{(S,T)}(f)\big).$$

Again in complete analogy to Remark 14.31, we may generalize the triple $([0,1], S, T)$ to an appropriate triple $([a,b], \oplus, \odot)$ and extend the definition of an (S,T)-Laplace transform to the more general case of an $(\oplus,\odot)$-Laplace transform [Fenchel 1949, Pap & Ralević 1998, Pap 1999]. In particular, in the case $([0,\infty], +, \cdot)$ and if $\mathcal{A}$ is the σ-algebra of Borel subsets of $[0,\infty]$ and m is the Lebesgue measure, then we obtain just the classical Laplace transform.

14.43 EXAMPLE
For an arbitrary function $f : [0,\infty[\longrightarrow [-\infty,\infty]$, the (max, +)-Laplace transform $\mathfrak{L}^{(\max,+)}(f)$ is given by

$$\big(\mathfrak{L}^{(\max,+)}(f)\big)(z) = \sup\{-xz + f(x) \mid x \in [0,\infty[\}.$$

In [Pap & Ralević 1998] the following generalized exchange formula which, in some special cases, transforms the $(\oplus,\odot)$-convolution into $\odot$ was proved:

$$\mathfrak{L}^{(\oplus,\odot)}\big(f \overset{(\oplus,\odot)}{\circledast} g\big) = \mathfrak{L}^{(\oplus,\odot)}(f) \odot \mathfrak{L}^{(\oplus,\odot)}(g).$$

Here is an example showing that the (max, +)-Laplace transform can be used to simplify an optimization problem [Pap & Ralević 1998, Pap 1999].

14.44 EXAMPLE
If the functions $f_1, f_2, \ldots, f_n : [0,\infty[\longrightarrow \mathbb{R}$ and a number $z \in \,]0,\infty[$ are given, we want to maximize the expression

$$f_1(x_1) + f_2(x_2) + \cdots + f_n(x_n) \tag{14.7}$$

under the condition $x_1 + x_2 + \cdots + x_n = z$. In mathematical economics and operation research (see, e.g., [Bellman & Dreyfus 1962]) the function (14.7) is usually called a utility function. Defining $f, F : [0,\infty[\longrightarrow [-\infty,\infty]$ by

$$f(x) = \sup\{f_1(x_1) + f_2(x_2) + \cdots + f_n(x_n) \mid x_1 + x_2 + \cdots + x_n = x\},$$
$$F(x) = \big(\mathfrak{L}^{(\max,+)}(f)\big)(x) = \sup\{-yx + f(y) \mid y \in [0,\infty[\},$$

then the generalized exchange formula leads to

$$F = \sum_{i=1}^{n} \mathfrak{L}^{(\max,+)}(f_i).$$

Applying the inverse of the (max, +)-Laplace transform, the original problem can be rewritten as follows: minimize

$$zx + \sum_{i=1}^{n} \left(\mathcal{L}^{(\max,+)}(f_i)\right)(x)$$

under the condition $x \geq 0$. Obviously, this is an optimization problem involving a function in one variable only.

Using the (S,T)-integral it is possible to study (S,T)-linear operators. For example, in [Pap & Teofanov 1998] it was shown that $(S_{\mathbf{M}},T)$-linear operators can be represented in terms of $(S_{\mathbf{M}},T)$-integrals.

The machinery of $\oplus$-measures and $(\oplus,\odot)$-integrals has been used to solve some important non-linear equations, such as the Bellman difference equation [Pap 1995] or the Hamilton-Jacobi partial differential equation with a non-smooth Hamiltonian (see [Maslov & Samborskij 1992, Kolokoltsov & Maslov 1997], [Pap 1992, 1997, 1999] and [Pap & Ralević 1998]).

14.5 Information measures

In information theory [Wiener 1948, Shannon & Weaver 1949, Rényi 1962, Guiaşu 1977] a concept of set functions which is in some sense dual to that of classical probability measures turned out to be useful. To be precise, if $(\Omega,\mathcal{A},P)$ is a probability space, then the *Wiener-Shannon information measure* $J : \mathcal{A} \longrightarrow [0,\infty]$ which is given by

$$J(A) = -\log(P(A)) \tag{14.8}$$

describes, for each event $A \in \mathcal{A}$, the amount of information gained by observing this event A. Given a finite measurable partition, $\mathcal{C} = (C_1, C_2 \ldots, C_n)$ of Ω, i.e., a partition of Ω such that each $C_1, C_2 \ldots, C_n$ is a measurable set, then the average information H which is gained by observing $\mathcal{C}$ (the so-called *Shannon entropy*) is given by

$$H(\mathcal{C}) = \sum_{i=1}^{n} P(C_i)J(C_i) = -\sum_{i=1}^{n} P(C_i)\log(P(C_i)).$$

Note that in information theory the concept of information is considered to be prior to that of probability (just take into account that, e.g., a uniform distribution often simply describes a total lack of information concerning the outcome of a statistical experiment).

The following axiomatic approach to information measures, which only makes use of the measurable space $(\Omega,\mathcal{A})$ but not of any probability measure P on $(\Omega,\mathcal{A})$, is essentially due to [Kampé de Fériet & Forte 1967] (compare also

[Kampé de Fériet & Benvenuti 1969, Kampé de Fériet *et al.* 1969]). It involves the concept of logical independence (compare [Kappos 1960]) which is defined in two steps. Two sub-σ-algebras $\mathcal{C}$ and $\mathcal{D}$ of $\mathcal{A}$ are said to be *logically independent* if for all non-empty sets $C \in \mathcal{C}$ and $D \in \mathcal{D}$ we have $C \cap D \neq \emptyset$. Given two logically independent sub-σ-algebras $\mathcal{C}$ and $\mathcal{D}$ of $\mathcal{A}$, then two sets $A, B \in \mathcal{A}$ are said to be *logically independent* (with respect to $\mathcal{C}$ and $\mathcal{D}$) if either $A \in \mathcal{C}$ and $B \in \mathcal{D}$ or $A \in \mathcal{D}$ and $B \in \mathcal{C}$.

We would like to emphasize that, in contrast to the stochastic independence in a probability space $(\Omega, \mathcal{A}, P)$, where two sets $A, B \in \mathcal{A}$ are said to be independent (with respect to P) if $P(A \cap B) = P(A) \cdot P(B)$, the logical independence of sets does not depend on a probability measure, but on two logically independent sub-σ-algebras of $\mathcal{A}$.

If, in a probability space $(\Omega, \mathcal{A}, P)$, two sets $A, B \in \mathcal{A}$ are stochastically independent then there are always two logically independent sub-σ-algebras $\mathcal{C}$ and $\mathcal{D}$ of $\mathcal{A}$ such that A and B are also logically independent with respect to $\mathcal{C}$ and $\mathcal{D}$. Conversely, if A and B are logically independent with respect to $\mathcal{C}$ and $\mathcal{D}$, then there is always a probability measure P on $(\Omega, \mathcal{A})$ such that A and B are stochastically independent (for more details see [Kappos 1960]).

Consider two measurable spaces $(\Omega_1, \mathcal{A}_1)$ and $(\Omega_2, \mathcal{A}_2)$ and their product space $(\Omega_1 \times \Omega_2, \mathcal{A}_1 \otimes \mathcal{A}_2)$. Then the two sub-$\sigma$-algebras $\mathcal{C} = \{\Omega_1, \emptyset\} \otimes \mathcal{A}_2$ and $\mathcal{D} = \mathcal{A}_1 \otimes \{\Omega_2, \emptyset\}$ of $\mathcal{A}_1 \otimes \mathcal{A}_2$ are logically independent and, consequently, for all $A_1 \in \mathcal{A}_1$ and $A_2 \in \mathcal{A}_2$ the sets $A_1 \times \Omega_2$ and $\Omega_1 \times A_2$ are logically independent with respect to $\mathcal{C}$ and $\mathcal{D}$. However, $A_1 \times \Omega_2$ and $\Omega_1 \times A_2$ are stochastically independent (with respect to some probability measure P) for all $A_1 \in \mathcal{A}_1$ and $A_2 \in \mathcal{A}_2$ only if P is a product of two probability measures, i.e., if there are probability measures P_1 and P_2 on $(\Omega_1, \mathcal{A}_1)$ and $(\Omega_2, \mathcal{A}_2)$, respectively, such that $P = P_1 \otimes P_2$.

14.45 DEFINITION

Let $(\Omega, \mathcal{A})$ be measurable space. A function $J : \mathcal{A} \longrightarrow [0, \infty]$ is called an *information measure* if for all $A, B \in \mathcal{A}$ and for each sequence $(A_n)_{n \in \mathbb{N}}$ in $\mathcal{A}$ we have

(J1) $J(\Omega) = 0$ and $J(\emptyset) = \infty$.

(J2) $J(A) \geq J(B)$ whenever $A \subseteq B$.

(J3) $J(A \cap B) = J(A) + J(B)$ whenever A and B are logically independent.

(J4) There exists a commutative, associative, non-decreasing and continuous operation $K : [0, \infty]^2 \longrightarrow [0, \infty]$ with neutral element ∞ such that, whenever $A \cap B = \emptyset$,

$$J(A \cup B) = K(J(A), J(B)).$$

(J5) $J\Big(\bigcup_{n \in \mathbb{N}} A_n\Big) = \lim_{n \to \infty} J\Big(\bigcup_{i=1}^{n} A_i\Big).$

The operation K in (J4) is usually called a *composition law*, and it is obvious that $([0,\infty], K, \leq)$ is an I-semigroup with neutral element ∞ (see Section 2.3). Therefore, as a consequence of Theorem 2.43, the function $\varphi : [0,1] \longrightarrow [0,\infty]$ given by $\varphi(x) = -\log x$ provides a one-to-one correspondence between K and some continuous t-conorm $S_{(K)}$ in the sense that for all $(x,y) \in [0,1]^2$

$$S_{(K)}(x,y) = \varphi^{-1}(K(\varphi(x), \varphi(y)).$$

Now it is immediately clear that, for each information measure $J : \mathcal{A} \longrightarrow [0,\infty]$ with composition law K, the function $\varphi^{-1} \circ J : \mathcal{A} \longrightarrow [0,1]$ is an $S_{(K)}$-measure. Moreover, axiom (J3) is transformed into (writing briefly m instead of $\varphi^{-1} \circ J$)

$$m(A \cap B) = m(A) \cdot m(B)$$

for all $A, B \in \mathcal{A}$ which are logically independent.

14.46 EXAMPLE

(i) In the case of the Wiener-Shannon information measure given by (14.8), (J4) leads to the functional equation

$$K(-\log x, -\log y) = -\log(x+y)$$

for all $x, y \in [0,1]$ with $x + y \leq 1$, a solution of which is the Wiener-Shannon composition law given by

$$K(x,y) = -\log(\min(e^{-x} + e^{-y}, 1)).$$

The corresponding t-conorm is just the Łukasiewicz t-conorm $S_{\mathbf{L}}$, and the corresponding $S_{\mathbf{L}}$-measure is exactly the probability measure P we started with.

(ii) If Ω is an infinite set and $\mathcal{A} = \mathcal{P}(\Omega)$ then $J : \mathcal{A} \longrightarrow [0,\infty]$ defined by

$$J(A) = \frac{1}{\text{card}(A)}$$

is an information measure. The functional equation derived from (J4) is $K(\frac{1}{n}, \frac{1}{m}) = \frac{1}{m+n}$ for all $m, n \in \mathbb{N} \cup \{\infty\}$, and a solution of it is the so-called hyperbolic composition law given by

$$K(x,y) = \frac{xy}{x+y}.$$

Then the function $s : [0,1] \longrightarrow [0,\infty]$ defined by $s(x) = -\frac{1}{\log x}$ is an additive generator of the strict t-conorm $S_{(K)}$, and the corresponding $S_{(K)}$-measure m is given by $m(A) = s^{-1}(\text{card}(A))$ (compare Proposition 14.17).

Note that, in any case, for two logically independent sets A and B with $A, B \notin \{\emptyset, \Omega\}$ we have

$$J(A) = J(\complement A) = J(B) = J(\complement B) = 0.$$

An interesting situation is obtained for $\Omega = \mathbb{N}$ in which case the information measure J is invariant under translations on $\mathbb{N}$, i.e., we have $J(A + n) = J(A)$ for all $A \subseteq \mathbb{N}$ and $n \in \mathbb{N}$. It is remarkable that a Wiener-Shannon information measure never can have this property since there is no probability measure on $(\mathbb{N}, \mathcal{P}(\mathbb{N}))$ which is invariant under translations.

(iii) Another distinguished composition law is the c-Wiener-Shannon composition law K defined by

$$K(x,y) = -c\log\big(\min\big(e^{-\frac{x}{c}} + e^{-\frac{y}{c}}, 1\big)\big),$$

where $c \in\,]0,\infty[$, yielding the information measure $J : \mathcal{A} \longrightarrow [0,\infty]$ specified by $J(A) = -c\log(P(A))$ [Kampé de Fériet & Forte 1967, Kampé de Fériet & Benvenuti 1969] (where P is an arbitrary probability measure on $\mathcal{A}$). The corresponding t-conorm $S_{(K)}$ is just the Yager t-conorm $S_c^{\mathbf{Y}}$ (see Section 4.5), and the corresponding $S_{(K)}$-measure m is given by $m(A) = P(A)^c$ (compare again Proposition 14.17).

An interesting generalization of information measures was introduced in [Schweizer & Sklar 1969]. The idea is similar to the generalization of classical metric spaces to probabilistic metric spaces [Schweizer & Sklar 1983]: the values of the information measure are proposed not to be just non-negative real numbers but rather distance distribution functions on $[0,\infty]$ (see Definition 9.17).

Let $(\Omega, \mathcal{A}, P)$ be a probability space and $\mathbf{J} : \mathcal{A} \longrightarrow \Delta^+$ a random information (with the interpretation that, for each $A \in \mathcal{A}$ and for each $x \in [0,\infty]$, the value $\mathbf{J}(A)(x)$ describes the probability that the quantity of information provided by $\mathbf{J}(A)$ is smaller than x).

Then the axioms (J1)–(J5) for a classical information measure can be easily extended to such a *generalized information measure* $\mathbf{J}$. In particular, the additivity axiom (J4) then becomes

(J4*) There exists a commutative, associative, non-decreasing and continuous operation $\mathbf{K} : \Delta^+ \times \Delta^+ \longrightarrow \Delta^+$ with neutral element ε_0 such that, whenever $A \cap B = \emptyset$,

$$\mathbf{J}(A \cup B) = \mathbf{K}(\mathbf{J}(A), \mathbf{J}(B)).$$

Again, the function $\mathbf{K}$ is called a *composition law*. Observe that these composition laws can be seen as duals of the triangle functions considered in Definition 9.18. Here are two examples of composition laws which are directly derived from continuous t-conorms in a similar way as the triangle functions given in (9.2) and (9.4).

14.47 EXAMPLE

If S is a continuous t-conorm then the functions $\mathbf{S}, \tau_{\mathbf{S}} : \Delta^+ \times \Delta^+ \longrightarrow \Delta^+$ given by

$$\mathbf{S}(G,H)(x) = S(H(x),G(x)),$$
$$\tau_{\mathbf{S}}(G,H)(x) = \sup\{S(G(u),H(v)) \mid (u,v) \in [0,\infty]^2 \text{ and } u+v=x\},$$

respectively, satisfy (J4*), i.e., they are composition laws.

As it should be expected, generalized information measures $\mathbf{J}$ are proper generalizations of the information measures given in Definition 14.45, a fact which is readily seen from the following example.

14.48 EXAMPLE
Let $(\Omega,\mathcal{A},P)$ be a probability space and $J : \mathcal{A} \longrightarrow [0,\infty]$ a standard information measure (i.e., in the sense of Definition 14.45) with composition law $K : [0,\infty]^2 \longrightarrow [0,\infty]$. Fix an arbitrary distribution function $F \in \Delta^+$ with $\lim_{x\to\infty} F(x) = 1$, and define the function $\mathbf{J}_F : \mathcal{A} \longrightarrow \Delta^+$ by

$$\mathbf{J}_F(A)(x) = F\Big(\frac{x}{J(A)}\Big).$$

Then $\mathbf{J}_F$ is a generalized information measure, and the corresponding composition law $\mathbf{K}_F : \Delta^+ \times \Delta^+ \longrightarrow \Delta^+$ satisfies

$$\mathbf{K}_F(\mathbf{J}_F(A),\mathbf{J}_F(B))(x) = F\Big(\frac{x}{K(J(A),J(B))}\Big).$$

In the special case $F = \varepsilon_1$ we obtain $\mathbf{J}_{\varepsilon_1}(A) = \varepsilon_{J(A)}$ for each $A \in \mathcal{A}$, meaning that $\mathbf{J}_{\varepsilon_1}(A)$ can be identified with the crisp information $J(A)$.

If the first moment of F is finite, i.e., if $E(F) < \infty$, then we obtain $E(\mathbf{J}_F(A)) = E(F) \cdot J(A)$ for each $A \in \mathcal{A}$, which means that $E(\mathbf{J}_F(.))$ is a standard information measure which differs from the starting J only by the multiplicative constant $E(F)$.

Let $(I,\mathfrak{I},P)$ be a probability space, where I is interpreted as a set of observers, and assume that for each $i \in I$ a standard information measure J_i is given. Since the informations J_i and J_j for two different observers $i,j \in I$ need not necessarily be the same, it is reasonable to ask what the average information (over the set I) could be. For technical reasons, assume that the one-place function $J(.,A) : I \longrightarrow [0,\infty]$ is $\mathfrak{I}$-measurable for each $A \in \mathcal{A}$ (observe the analogy to the concept of a Markov kernel [Bauer 1981]). Then the average information is given by the function $\overline{J} : \mathcal{A} \longrightarrow [0,\infty]$ defined by

$$\overline{J}(A) = \int_I J_i(A)\, dP(i)$$

where the integral on the right-hand side is a Lebesgue-Stieltjes integral. It is easy to see that $\overline{J}$ fulfills the modifications of the axioms (J1)–(J3) and (J5), while the additivity (J4) may be violated (even if J_i has the same composition law F for each observer $i \in I$, for more details see [Kampé de Fériet 1969]).

14.49 EXAMPLE

Consider the special case $I = \{1,2\}$, $\mathcal{I} = \mathcal{P}(I)$, $\Omega = \mathbb{N}$ and $\mathcal{A} = \mathcal{P}(\mathbb{N})$, and define the two information measures $J_1, J_2 : \mathcal{A} \longrightarrow [0,\infty]$ by

$$J_1(A) = \frac{1}{\mathrm{card}(A_o)} \quad \text{and} \quad J_2(A) = \frac{1}{\mathrm{card}(A_e)},$$

respectively, where A_o is the set of odd numbers in A and A_e is the set of even numbers in A. If the probability measure P on $(I, \mathcal{I})$ is non-trivial, i.e., if $P(\{1\}) \in]0,1[$, then the average information $\overline{J}$ is given by

$$\overline{J}(A) = \frac{P(\{1\})}{\mathrm{card}(A_o)} + \frac{P(\{2\})}{\mathrm{card}(A_e)}.$$

Then there is no composition function $K : [0,\infty]^2 \longrightarrow [0,\infty]$ such that $\overline{J}$ fulfills the additivity (J4).

If, however, we define the generalized information measure $\mathbf{J} : \mathcal{A} \longrightarrow \Delta^+$ in the spirit of [Schweizer & Sklar 1969] by $\mathbf{J}(A)(x) = P(\{i \in I \mid J_i(A) < x\})$, then this has a simple and nice interpretation: obviously for all $A \in \mathcal{A}$

$$\overline{J}(A) = \int_0^\infty x \, d\mathbf{J}(A)(x),$$

i.e., $\mathbf{J}(A)$ equals the distribution function of $J_.(A)$.

APPENDIX

A. Families of t-norms

The most relevant information related to the families of t-norms considered in Chapter 4 is presented here in a condensed form, together with some visualizations, which should be useful for a deeper understanding of the peculiarities of these families of t-norms.

For quick reference, the reader will find here some important data concerning the families of

A.1	*Aczél-Alsina t-norms* $(T_\lambda^{\mathbf{AA}})_{\lambda\in[0,\infty]}$	(Example 4.15)
A.2	*Dombi t-norms* $(T_\lambda^{\mathbf{D}})_{\lambda\in[0,\infty]}$	(Example 4.11)
A.3	*Frank t-norms* $(T_\lambda^{\mathbf{F}})_{\lambda\in[0,\infty]}$	(Example 4.7)
A.4	*Hamacher t-norms* $(T_\lambda^{\mathbf{H}})_{\lambda\in[0,\infty]}$	(Example 4.5)
A.5	*Mayor-Torrens t-norms* $(T_\lambda^{\mathbf{MT}})_{\lambda\in[0,1]}$	(Example 4.17)
A.6	*Schweizer-Sklar t-norms* $(T_\lambda^{\mathbf{SS}})_{\lambda\in[-\infty,\infty]}$	(Example 4.3)
A.7	*Sugeno-Weber t-norms* $(T_\lambda^{\mathbf{SW}})_{\lambda\in[-1,\infty]}$	(Example 4.13)
A.8	*Yager t-norms* $(T_\lambda^{\mathbf{Y}})_{\lambda\in[0,\infty]}$	(Example 4.9)

In each case we have reserved two pages per family: the formulae of t-norms and corresponding t-conorms are given, and a table shows which members of the respective family are continuous, Archimedean, continuous Archimedean, strict or nilpotent, respectively. We also list additive and multiplicative generators of the continuous Archimedean members. Finally we offer three different visualizations of prototypical examples: plots of the diagonal sections of seven members of each family, and pairs of 3D and contour plots of three representative examples.

We sincerely hope that particularly these illustrations give the reader a clearer image of specific properties of these t-norms which are so often quoted in the literature.

A.1 Aczél-Alsina t-norms $(T_\lambda^{\mathbf{AA}})_{\lambda\in[0,\infty]}$

$$T_\lambda^{\mathbf{AA}}(x,y) = \begin{cases} T_{\mathbf{D}}(x,y) & \text{if } \lambda = 0 \\ T_{\mathbf{M}}(x,y) & \text{if } \lambda = \infty \\ e^{-((-\log x)^\lambda + (-\log y)^\lambda)^{\frac{1}{\lambda}}} & \text{if } \lambda \in \,]0,\infty[\end{cases}$$

Indexes for which the t-norm $T_\lambda^{\mathbf{AA}}$ is ...

continuous	*Archimedean*	*continuous Archimedean*	*strict*	*nilpotent*
$\lambda \in \,]0,\infty]$	$\lambda \in [0,\infty[$	$\lambda \in \,]0,\infty[$	$\lambda \in \,]0,\infty[$	none

Generators of the strict t-norms $(T_\lambda^{\mathbf{AA}})_{\lambda\in]0,\infty[}$

additive	$t_\lambda^{\mathbf{AA}}(x) = (-\log x)^\lambda$
multiplicative	$\theta_\lambda^{\mathbf{AA}}(x) = e^{-(-\log x)^\lambda}$

Aczél-Alsina t-conorms $(S_\lambda^{\mathbf{AA}})_{\lambda\in[0,\infty]}$

$$S_\lambda^{\mathbf{AA}}(x,y) = \begin{cases} S_{\mathbf{D}}(x,y) & \text{if } \lambda = 0 \\ S_{\mathbf{M}}(x,y) & \text{if } \lambda = \infty \\ 1 - e^{-((-\log(1-x))^\lambda + (-\log(1-y))^\lambda)^{\frac{1}{\lambda}}} & \text{if } \lambda \in \,]0,\infty[\end{cases}$$

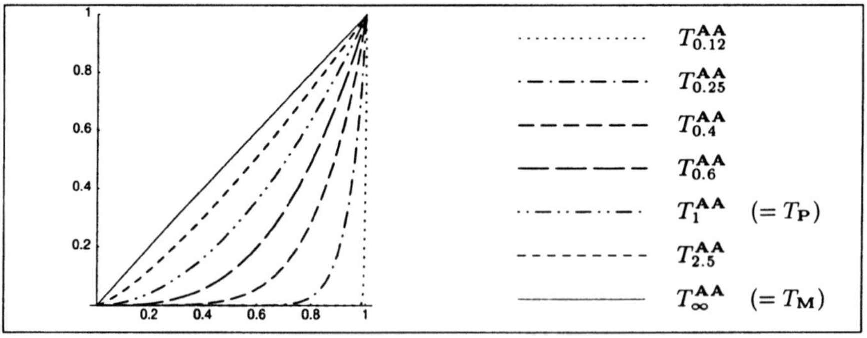

Fig. A.1. Diagonal sections of some Aczél-Alsina t-norms

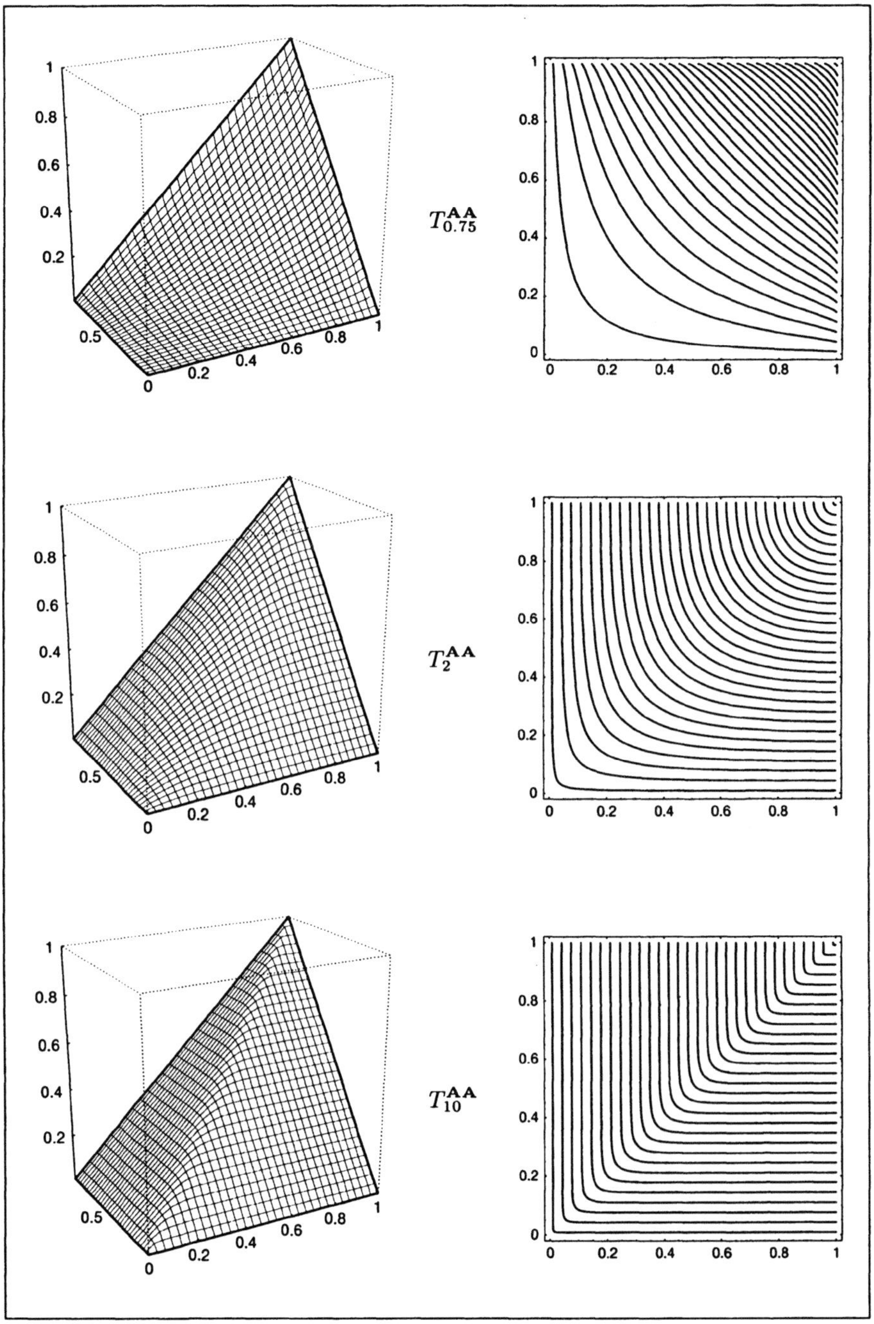

Fig. A.2. 3D and contour plots of the Aczél-Alsina t-norms $T^{\mathbf{AA}}_{0.75}$, $T^{\mathbf{AA}}_{2}$ and $T^{\mathbf{AA}}_{10}$

A.2 Dombi t-norms $(T^{\mathbf{D}}_\lambda)_{\lambda\in[0,\infty]}$

$$T^{\mathbf{D}}_\lambda(x,y) = \begin{cases} T_{\mathbf{D}}(x,y) & \text{if } \lambda = 0 \\ T_{\mathbf{M}}(x,y) & \text{if } \lambda = \infty \\ \frac{1}{1+\left(\left(\frac{1-x}{x}\right)^\lambda + \left(\frac{1-y}{y}\right)^\lambda\right)^{\frac{1}{\lambda}}} & \text{if } \lambda \in \,]0,\infty[\end{cases}$$

Indexes for which the t-norm $T^{\mathbf{D}}_\lambda$ is ...

continuous	*Archimedean*	*continuous Archimedean*	*strict*	*nilpotent*
$\lambda \in \,]0,\infty]$	$\lambda \in [0,\infty[$	$\lambda \in \,]0,\infty[$	$\lambda \in \,]0,\infty[$	none

Generators of the strict t-norms $(T^{\mathbf{D}}_\lambda)_{\lambda\in]0,\infty[}$

additive	$t^{\mathbf{D}}_\lambda(x) = \left(\frac{1-x}{x}\right)^\lambda$
multiplicative	$\theta^{\mathbf{D}}_\lambda(x) = e^{-\left(\frac{1-x}{x}\right)^\lambda}$

Dombi t-conorms $(S^{\mathbf{D}}_\lambda)_{\lambda\in[0,\infty]}$

$$S^{\mathbf{D}}_\lambda(x,y) = \begin{cases} S_{\mathbf{D}}(x,y) & \text{if } \lambda = 0 \\ S_{\mathbf{M}}(x,y) & \text{if } \lambda = \infty \\ 1 - \frac{1}{1+\left(\left(\frac{x}{1-x}\right)^\lambda + \left(\frac{y}{1-y}\right)^\lambda\right)^{\frac{1}{\lambda}}} & \text{if } \lambda \in \,]0,\infty[\end{cases}$$

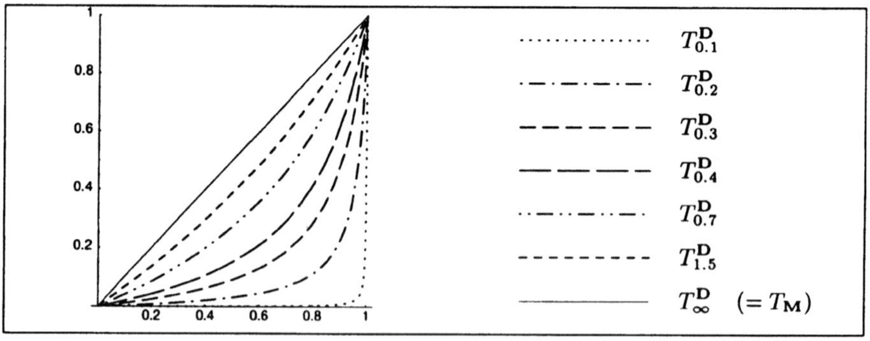

Fig. A.3. Diagonal sections of some Dombi t-norms

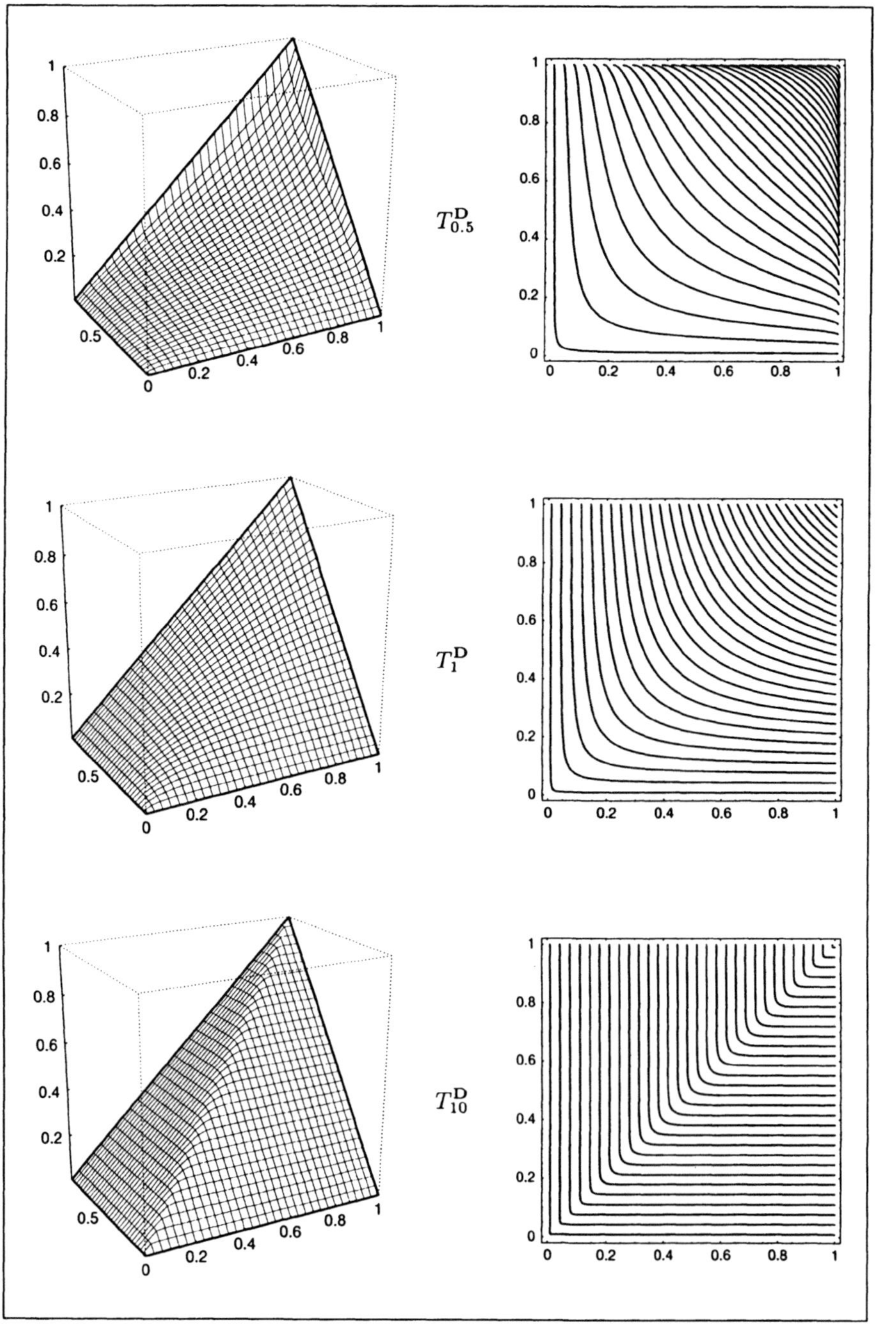

Fig. A.4. 3D and contour plots of the Dombi t-norms $T^{\mathbf{D}}_{0.5}$, $T^{\mathbf{D}}_{1}$ and $T^{\mathbf{D}}_{10}$

A.3 Frank t-norms $(T_\lambda^{\mathbf{F}})_{\lambda\in[0,\infty]}$

$$T_\lambda^{\mathbf{F}}(x,y)=\begin{cases}T_{\mathbf{M}}(x,y) & \text{if } \lambda=0\\ T_{\mathbf{P}}(x,y) & \text{if } \lambda=1\\ T_{\mathbf{L}}(x,y) & \text{if } \lambda=\infty\\ \log_\lambda\left(1+\frac{(\lambda^x-1)(\lambda^y-1)}{\lambda-1}\right) & \text{if } \lambda\in\,]0,1[\,\cup\,]1,\infty[\end{cases}$$

Indexes for which the t-norm $T_\lambda^{\mathbf{F}}$ is ...

continuous	*Archimedean*	*continuous Archimedean*	*strict*	*nilpotent*
$\lambda\in[0,\infty]$	$\lambda\in\,]0,\infty]$	$\lambda\in\,]0,\infty]$	$\lambda\in\,]0,\infty[$	$\lambda=\infty$

Generators of the continuous Archimedean t-norms $(T_\lambda^{\mathbf{F}})_{\lambda\in]0,\infty]}$

additive	$t_\lambda^{\mathbf{F}}(x)=\begin{cases}-\log x & \text{if } \lambda=1\\ 1-x & \text{if } \lambda=\infty\\ \log\frac{\lambda-1}{\lambda^x-1} & \text{if } \lambda\in\,]0,1[\,\cup\,]1,\infty[\end{cases}$
multiplicative	$\theta_\lambda^{\mathbf{F}}(x)=\begin{cases}x & \text{if } \lambda=1\\ e^{x-1} & \text{if } \lambda=\infty\\ \frac{\lambda^x-1}{\lambda-1} & \text{if } \lambda\in\,]0,1[\,\cup\,]1,\infty[\end{cases}$

Frank t-conorms $(S_\lambda^{\mathbf{F}})_{\lambda\in[0,\infty]}$

$$S_\lambda^{\mathbf{F}}(x,y)=\begin{cases}S_{\mathbf{M}}(x,y) & \text{if } \lambda=0\\ S_{\mathbf{P}}(x,y) & \text{if } \lambda=1\\ S_{\mathbf{L}}(x,y) & \text{if } \lambda=\infty\\ 1-\log_\lambda\left(1+\frac{(\lambda^{1-x}-1)(\lambda^{1-y}-1)}{\lambda-1}\right) & \text{if } \lambda\in\,]0,1[\,\cup\,]1,\infty[\end{cases}$$

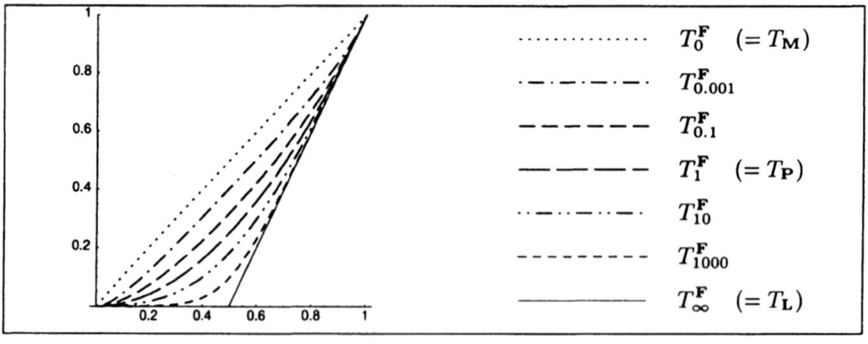

Fig. A.5. Diagonal sections of some Frank t-norms

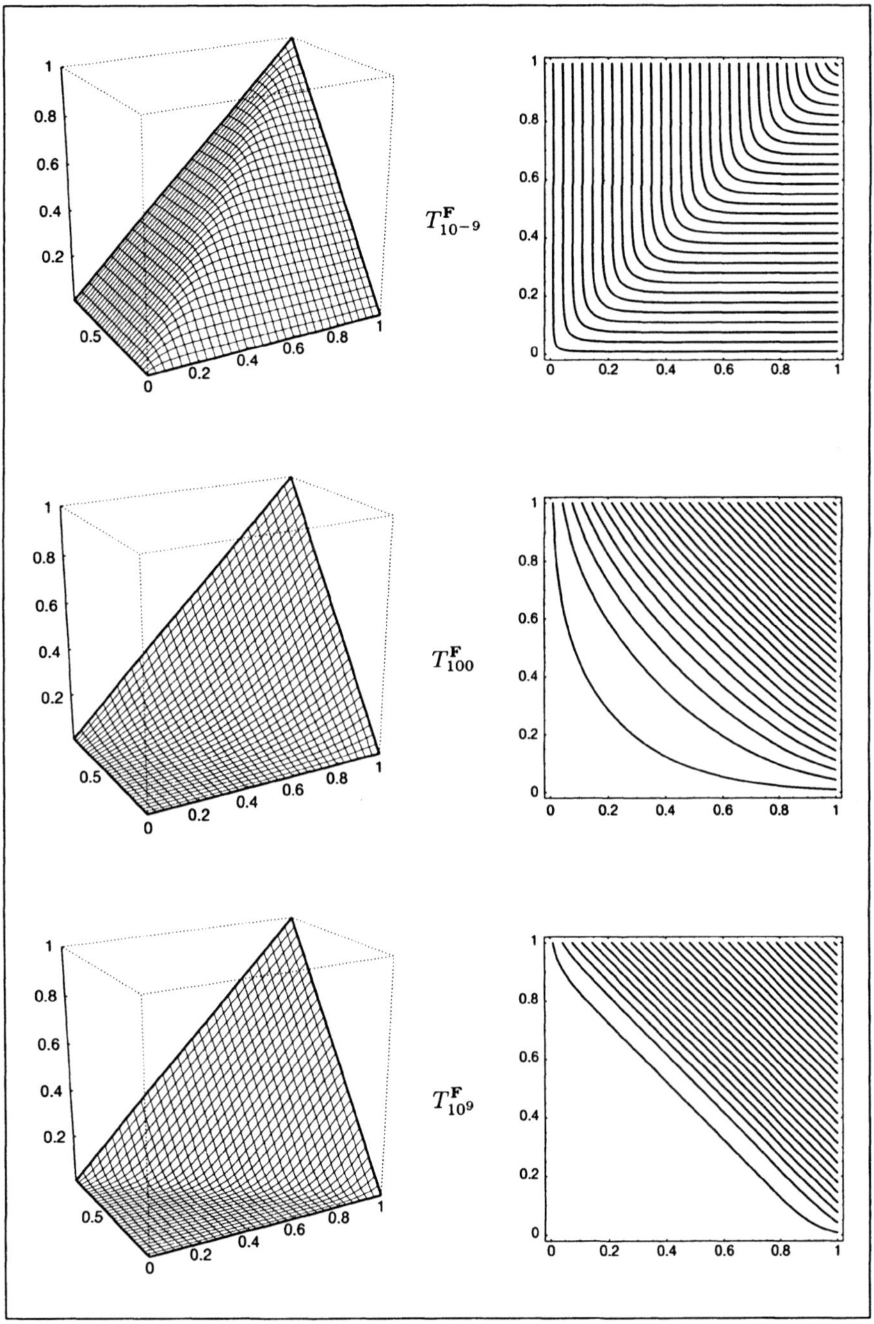

Fig. A.6. 3D and contour plots of the Frank t-norms $T^{\mathbf{F}}_{10^{-9}}$, $T^{\mathbf{F}}_{100}$ and $T^{\mathbf{F}}_{10^{9}}$

A.4 Hamacher t-norms $(T_\lambda^{\mathbf{H}})_{\lambda\in[0,\infty]}$

$$T_\lambda^{\mathbf{H}}(x,y)=\begin{cases}T_{\mathbf{D}}(x,y) & \text{if } \lambda=\infty\\ 0 & \text{if } \lambda=x=y=0\\ \frac{xy}{\lambda+(1-\lambda)(x+y-xy)} & \text{if } \lambda\in[0,\infty[\text{ and } (\lambda,x,y)\neq(0,0,0)\end{cases}$$

Indexes for which the t-norm $T_\lambda^{\mathbf{H}}$ is ...

continuous	*Archimedean*	*continuous Archimedean*	*strict*	*nilpotent*
$\lambda\in[0,\infty[$	$\lambda\in[0,\infty]$	$\lambda\in[0,\infty[$	$\lambda\in[0,\infty[$	none

Generators of the strict t-norms $(T_\lambda^{\mathbf{H}})_{\lambda\in[0,\infty[}$

additive	$t_\lambda^{\mathbf{H}}(x)=\begin{cases}\frac{1-x}{x} & \text{if } \lambda=0\\ \log\left(\frac{\lambda+(1-\lambda)x}{x}\right) & \text{if } \lambda\in\,]0,\infty[\end{cases}$
multiplicative	$\theta_\lambda^{\mathbf{H}}(x)=\begin{cases}e^{\frac{x-1}{x}} & \text{if } \lambda=0\\ \frac{x}{\lambda+(1-\lambda)x} & \text{if } \lambda\in\,]0,\infty[\end{cases}$

Hamacher t-conorms $(S_\lambda^{\mathbf{H}})_{\lambda\in[0,\infty]}$

$$S_\lambda^{\mathbf{H}}(x,y)=\begin{cases}S_{\mathbf{D}}(x,y) & \text{if } \lambda=\infty\\ 1 & \text{if } \lambda=0 \text{ and } x=y=1\\ \frac{x+y+(\lambda-2)xy}{1+(\lambda-1)xy} & \text{if } \lambda\in\,]0,\infty[\text{ and } (\lambda,x,y)\neq(0,1,1)\end{cases}$$

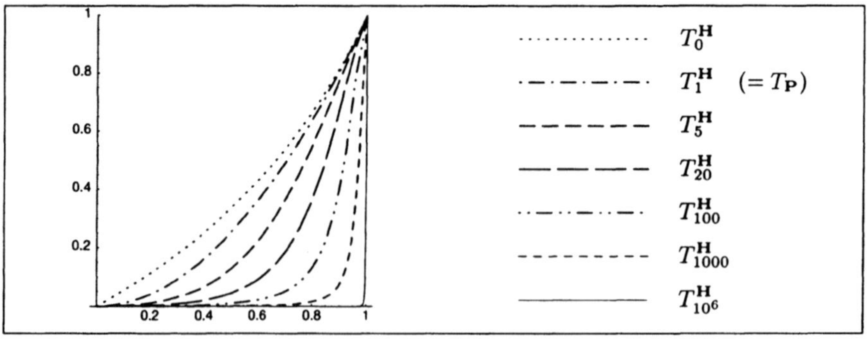

Fig. A.7. Diagonal sections of some Hamacher t-norms

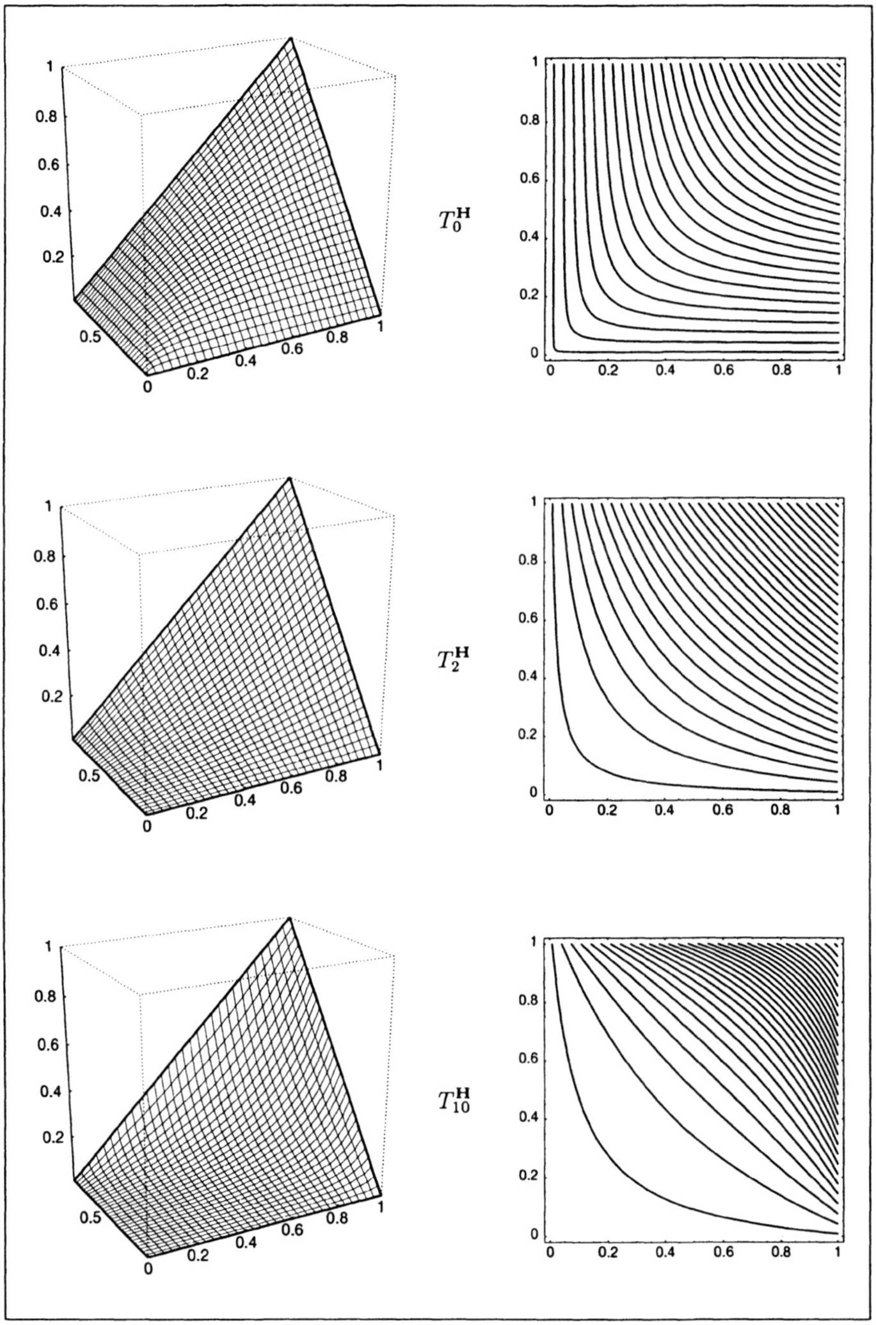

Fig. A.8. 3D and contour plots of the Hamacher t-norms $T_0^{\mathbf{H}}$, $T_2^{\mathbf{H}}$ and $T_{10}^{\mathbf{H}}$

A.5 Mayor-Torrens t-norms $(T_\lambda^{\mathbf{MT}})_{\lambda\in[0,1]}$

$$T_\lambda^{\mathbf{MT}}(x,y)=\begin{cases}\max(x+y-\lambda,0) & \text{if } \lambda\in\,]0,1] \text{ and } (x,y)\in[0,\lambda]^2\\ \min(x,y) & \text{if } \lambda=0 \text{ or } x>\lambda \text{ or } y>\lambda\end{cases}$$

Indexes for which the t-norm $T_\lambda^{\mathbf{MT}}$ is ...

continuous	*Archimedean*	*continuous Archimedean*	*strict*	*nilpotent*
$\lambda\in[0,1]$	$\lambda=1$	$\lambda=1$	none	$\lambda=1$

Generators of the nilpotent t-norm $T_1^{\mathbf{MT}}$

additive	$t_1^{\mathbf{MT}}(x)=1-x$
multiplicative	$\theta_1^{\mathbf{MT}}(x)=e^{x-1}$

Mayor-Torrens t-conorms $(S_\lambda^{\mathbf{MT}})_{\lambda\in[0,1]}$

$$S_\lambda^{\mathbf{MT}}(x,y)=\begin{cases}\min(x+y+\lambda-1,1) & \text{if } \lambda\in\,]0,1] \text{ and } (x,y)\in[1-\lambda,1]^2\\ \max(x,y) & \text{if } \lambda=0 \text{ or } x<1-\lambda \text{ or } y<1-\lambda\end{cases}$$

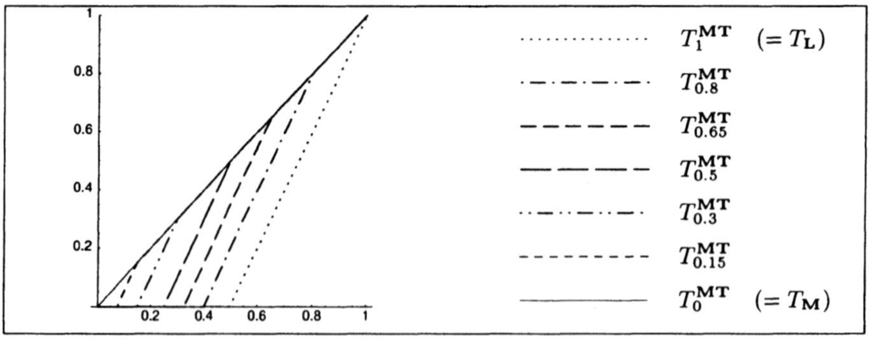

Fig. A.9. Diagonal sections of some Mayor-Torrens t-norms

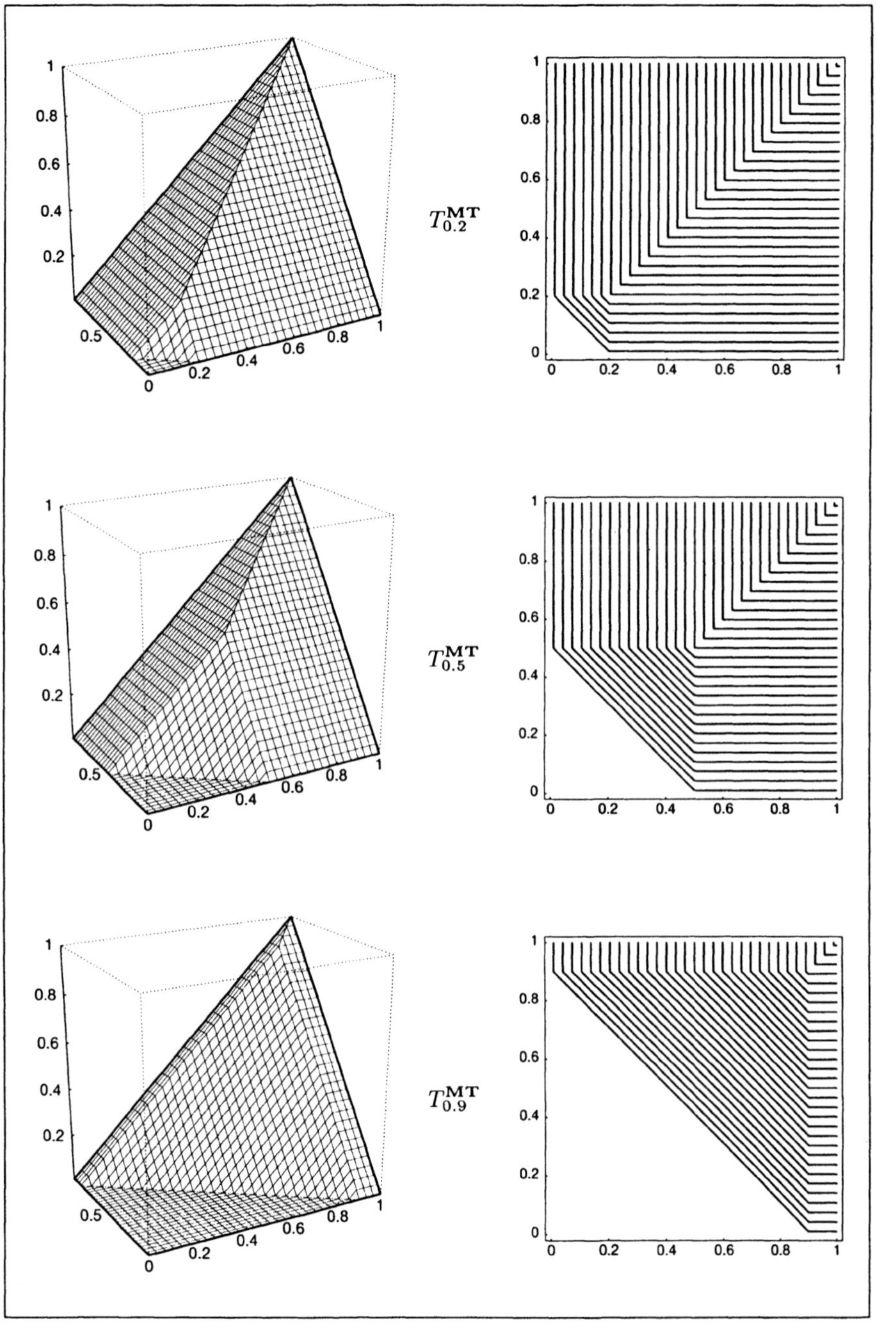

Fig. A.10. 3D and contour plots of the Mayor-Torrens t-norms $T_{0.2}^{\mathbf{MT}}$, $T_{0.5}^{\mathbf{MT}}$ and $T_{0.9}^{\mathbf{MT}}$

A.6 Schweizer-Sklar t-norms $(T_\lambda^{\mathbf{SS}})_{\lambda\in[-\infty,\infty]}$

$$T_\lambda^{\mathbf{SS}}(x,y)=\begin{cases} T_{\mathbf{M}}(x,y) & \text{if } \lambda=-\infty\\ T_{\mathbf{P}}(x,y) & \text{if } \lambda=0\\ T_{\mathbf{D}}(x,y) & \text{if } \lambda=\infty\\ \left(\max\left((x^\lambda+y^\lambda-1),0\right)\right)^{\frac{1}{\lambda}} & \text{if } \lambda\in\,]-\infty,0[\,\cup\,]0,\infty[\end{cases}$$

Indexes for which the t-norm $T_\lambda^{\mathbf{SS}}$ is . . .

continuous	*Archimedean*	*continuous Archimedean*	*strict*	*nilpotent*
$\lambda\in[-\infty,\infty[$	$\lambda\in\,]-\infty,\infty]$	$\lambda\in\,]-\infty,\infty[$	$\lambda\in\,]-\infty,0]$	$\lambda\in\,]0,\infty[$

Generators of the continuous Archimedean t-norms $(T_\lambda^{\mathbf{SS}})_{\lambda\in]-\infty,\infty[}$

additive	$t_\lambda^{\mathbf{SS}}(x)=\begin{cases}-\log x & \text{if } \lambda=0\\ \frac{1-x^\lambda}{\lambda} & \text{if } \lambda\in\,]-\infty,0[\,\cup\,]0,\infty[\end{cases}$
multiplicative	$\theta_\lambda^{\mathbf{SS}}(x)=\begin{cases}x & \text{if } \lambda=0\\ e^{\frac{x^\lambda-1}{\lambda}} & \text{if } \lambda\in\,]-\infty,0[\,\cup\,]0,\infty[\end{cases}$

Schweizer-Sklar t-conorms $(S_\lambda^{\mathbf{SS}})_{\lambda\in[-\infty,\infty]}$

$$S_\lambda^{\mathbf{SS}}(x,y)=\begin{cases} S_{\mathbf{M}}(x,y) & \text{if } \lambda=-\infty\\ S_{\mathbf{P}}(x,y) & \text{if } \lambda=0\\ S_{\mathbf{D}}(x,y) & \text{if } \lambda=\infty\\ 1-\left(\max\left(((1-x)^\lambda+(1-y)^\lambda-1),0\right)\right)^{\frac{1}{\lambda}} & \text{if } \lambda\in\,]-\infty,0[\,\cup\,]0,\infty[\end{cases}$$

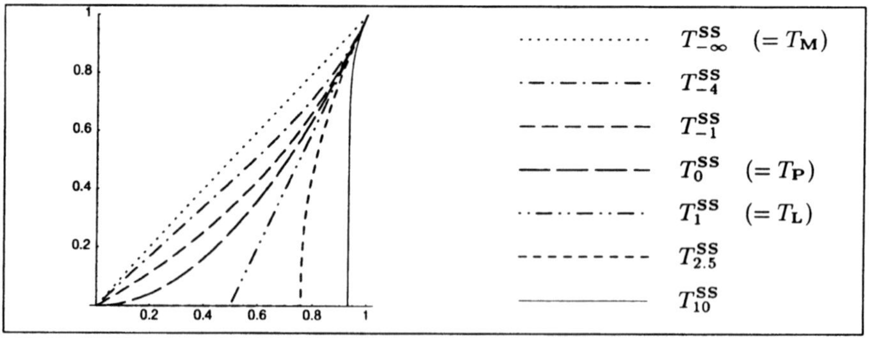

Fig. A.11. Diagonal sections of some Schweizer-Sklar t-norms

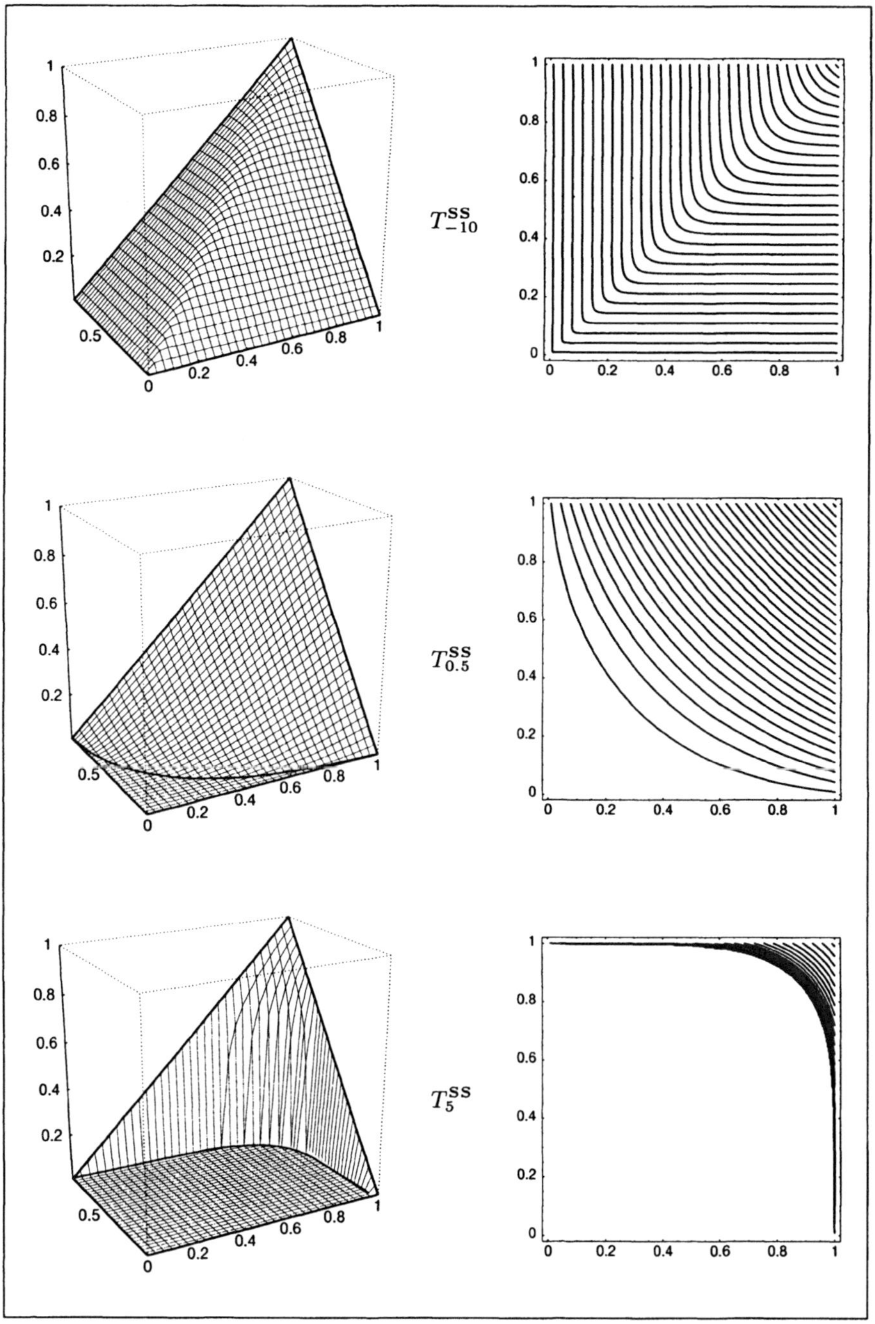

Fig. A.12. 3D and contour plots of the Schweizer-Sklar t-norms $T_{-10}^{\mathbf{SS}}$, $T_{0.5}^{\mathbf{SS}}$ and $T_{5}^{\mathbf{SS}}$

A.7 Sugeno-Weber t-norms $(T_\lambda^{\mathbf{SW}})_{\lambda\in[-1,\infty]}$

$$T_\lambda^{\mathbf{SW}}(x,y) = \begin{cases} T_{\mathbf{D}}(x,y) & \text{if } \lambda = -1 \\ T_{\mathbf{P}}(x,y) & \text{if } \lambda = \infty \\ \max\left(\frac{x+y-1+\lambda xy}{1+\lambda}, 0\right) & \text{if } \lambda \in \,]-1,\infty[\end{cases}$$

Indexes for which the t-norm $T_\lambda^{\mathbf{SW}}$ is . . .

continuous	*Archimedean*	*continuous Archimedean*	*strict*	*nilpotent*
$\lambda \in \,]-1,\infty]$	$\lambda \in [-1,\infty]$	$\lambda \in \,]-1,\infty]$	$\lambda = \infty$	$\lambda \in \,]-1,\infty[$

Generators of the continuous Archimedean t-norms $(T_\lambda^{\mathbf{SW}})_{\lambda\in]-1,\infty]}$

additive	$t_\lambda^{\mathbf{SW}}(x) = \begin{cases} 1-x & \text{if } \lambda = 0 \\ -\log x & \text{if } \lambda = \infty \\ 1 - \frac{\log(1+\lambda x)}{\log(1+\lambda)} & \text{if } \lambda \in \,]-1,0[\,\cup\,]0,\infty[\end{cases}$
multiplicative	$\theta_\lambda^{\mathbf{SW}}(x) = \begin{cases} e^{x-1} & \text{if } \lambda = 0 \\ x & \text{if } \lambda = \infty \\ e^{\frac{\log(1+\lambda x)}{\log(1+\lambda)} - 1} & \text{if } \lambda \in \,]-1,0[\,\cup\,]0,\infty[\end{cases}$

Sugeno-Weber t-conorms $(S_\lambda^{\mathbf{SW}})_{\lambda\in[-1,\infty]}$

$$S_\lambda^{\mathbf{SW}}(x,y) = \begin{cases} S_{\mathbf{P}}(x,y) & \text{if } \lambda = -1 \\ S_{\mathbf{D}}(x,y) & \text{if } \lambda = \infty \\ \min(x+y+\lambda xy, 1) & \text{if } \lambda \in \,]-1,\infty[\end{cases}$$

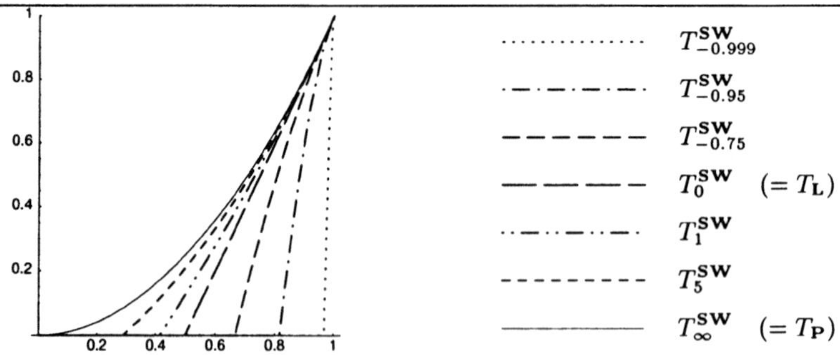

Fig. A.13. Diagonal sections of some Sugeno-Weber t-norms

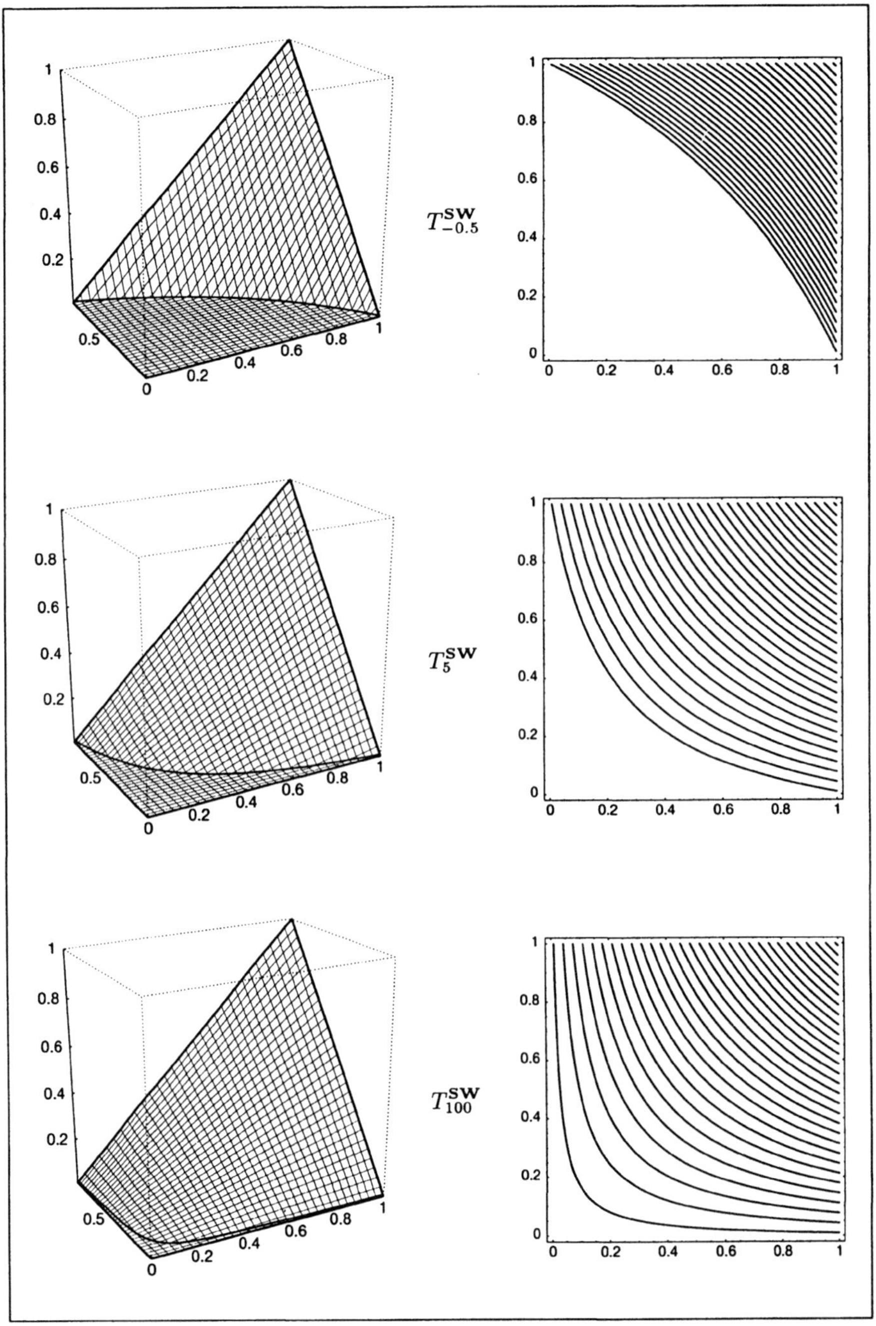

Fig. A.14. 3D and contour plots of the Sugeno-Weber t-norms $T^{\mathbf{SW}}_{-0.5}$, $T^{\mathbf{SW}}_{5}$ and $T^{\mathbf{SW}}_{100}$

A.8 Yager t-norms $(T_\lambda^{\mathbf{Y}})_{\lambda\in[0,\infty]}$

$$T_\lambda^{\mathbf{Y}}(x,y)=\begin{cases}T_{\mathbf{D}}(x,y) & \text{if } \lambda=0\\ T_{\mathbf{M}}(x,y) & \text{if } \lambda=\infty\\ \max\left(1-((1-x)^\lambda+(1-y)^\lambda)^{\frac{1}{\lambda}},0\right) & \text{if } \lambda\in\,]0,\infty[\end{cases}$$

Indexes for which the t-norm $T_\lambda^{\mathbf{Y}}$ is ...

continuous	*Archimedean*	*continuous Archimedean*	*strict*	*nilpotent*
$\lambda\in\,]0,\infty]$	$\lambda\in[0,\infty[$	$\lambda\in\,]0,\infty[$	none	$\lambda\in\,]0,\infty[$

Generators of the nilpotent t-norms $(T_\lambda^{\mathbf{Y}})_{\lambda\in]0,\infty[}$

additive	$t_\lambda^{\mathbf{Y}}(x)=(1-x)^\lambda$
multiplicative	$\theta_\lambda^{\mathbf{Y}}(x)=e^{-(1-x)^\lambda}$

Yager t-conorms $(S_\lambda^{\mathbf{Y}})_{\lambda\in[0,\infty]}$

$$S_\lambda^{\mathbf{Y}}(x,y)=\begin{cases}S_{\mathbf{D}}(x,y) & \text{if } \lambda=0\\ S_{\mathbf{M}}(x,y) & \text{if } \lambda=\infty\\ \min\left((x^\lambda+y^\lambda)^{\frac{1}{\lambda}},1\right) & \text{if } \lambda\in\,]0,\infty[\end{cases}$$

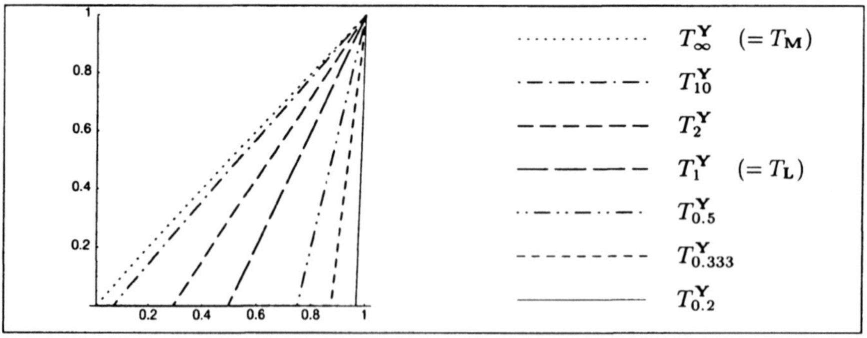

Fig. A.15. Diagonal sections of some Yager t-norms

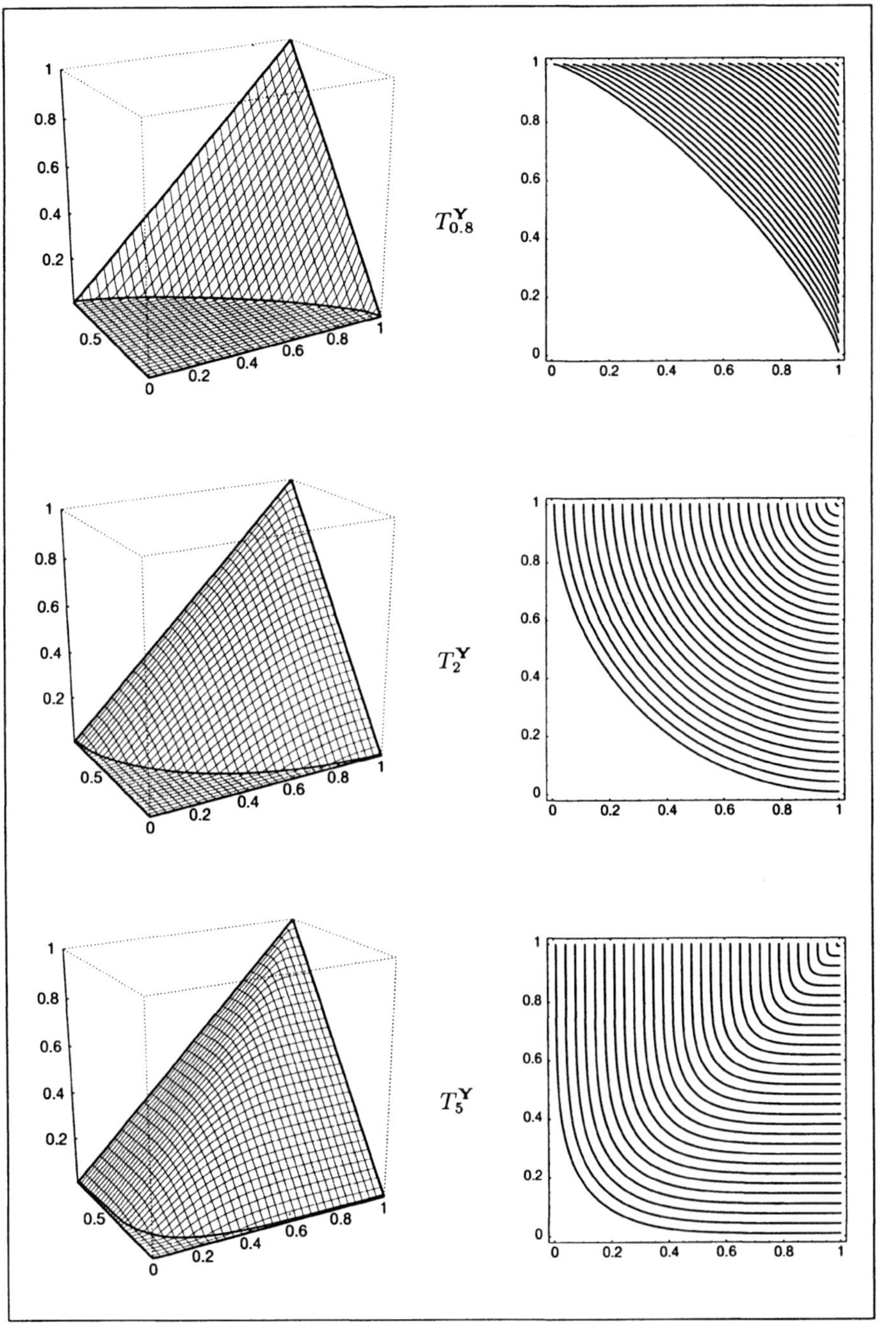

Fig. A.16. 3D and contour plots of the Yager t-norms $T^{\mathbf{Y}}_{0.8}$, $T^{\mathbf{Y}}_{2}$ and $T^{\mathbf{Y}}_{5}$

B. Additional t-norms

The technical details of the Krause t-norm $T^{\mathbf{K}}$ (see Remark 7.20 and [Krause 200x]) which provided a negative answer to Problem 5.8.1 in [Schweizer & Sklar 1983], and of the family of incomparable t-norms $(T_\lambda^{\mathbf{I}})_{\lambda \in [0,0.5[}$ considered in Example 3.33 were too complicated to be included in the main text.

Therefore, we give here a full presentation which contains also the necessary preliminaries concerning the Cantor set and the Farey series, which are both helpful for the construction of the Krause t-norm.

B.1 Krause t-norm

In the following construction of the Krause t-norm we shall make use of the Cantor set and the Farey series, two concepts which we first recall for the convenience of the reader.

The *Cantor set* C is a subset of the unit interval $[0,1]$, which is obtained by removing from $[0,1]$ first its open middle third $]\frac{1}{3}, \frac{2}{3}[$, then from the remaining two intervals $[0, \frac{1}{3}]$ and $[\frac{2}{3}, 1]$ again their open middle thirds $]\frac{1}{9}, \frac{2}{9}[$ and $]\frac{7}{9}, \frac{8}{9}[$, and so on (see Figure B.1).

As the complement of a union of open intervals, C is a closed and compact subset of $[0,1]$. The Cantor set is perfect (i.e., it has no isolated points) and, at the same time, totally disconnected (i.e., each of its components is a singleton). The set C contains uncountably many points, since it can be written as

$$C = \left\{ \sum_{n=1}^{\infty} \frac{c_n}{3^n} \;\middle|\; c_n \in \{0,2\} \right\},$$

but it is a nowhere dense subset of $[0,1]$ (i.e., the interior of its topological closure is empty) and has Lebesgue measure zero (for more details on this remarkable set see [Bourbaki 1966]).

For the construction of the Krause t-norm we need essentially the set C_L of all left endpoints of the deleted intervals. Fix $n \in \mathbb{N}$ and $i \in \{1,2,3,\ldots,2^{n-1}\}$ and observe that the number $i-1$ can be uniquely represented by

$$i - 1 = \sum_{j=0}^{n-2} i_j \cdot 2^j$$

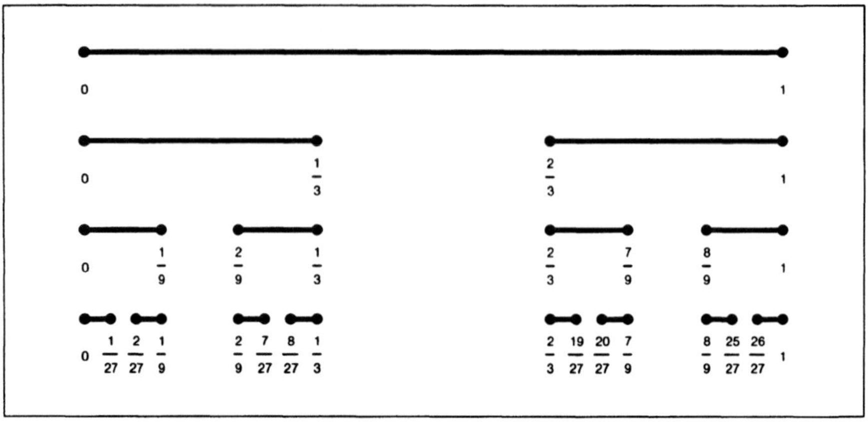

Fig. B.1. Construction of the Cantor set: the first three steps

with $i_j \in \{0,1\}$. Then the left endpoint c_{ni} of the i-th interval which is removed in the n-th step (in which step 2^{n-1} intervals of the length $\frac{1}{3^n}$ are deleted) of the construction of the Cantor set can be written as

$$c_{ni} = \sum_{j=1}^{\infty} \frac{c_j^{(ni)}}{3^j},$$

where the coefficients $c_j^{(ni)}$ are determined by

$$c_j^{(ni)} = \begin{cases} 2 \cdot i_{n-2-j} & \text{if } j < n, \\ 0 & \text{if } j = n, \\ 2 & \text{if } j > n, \end{cases}$$

and we shall write $C_L = \{c_{ni} \mid n \in \mathbb{N}, i \in \{1,2,3,\ldots,2^{n-1}\}\}$.

The (double) Farey series (see, e.g., [Hardy & Wright 1979]) contains all rational numbers of the interval $]0,\infty[$, and it is constructed inductively. We start with the two fractions $\frac{1}{0}$ (in this context also $\infty = \frac{1}{0}$ is considered to be a rational number) and $0 = \frac{0}{1}$ and insert a new fraction a_{11} in between by separately adding the denominators and the numerators, i.e.,

$$a_{11} = \frac{1+0}{0+1} = \frac{1}{1}.$$

Inserting in the same way two rational numbers a_{21} (between $\frac{1}{0}$ and $\frac{1}{1}$) and a_{22} (between $\frac{1}{1}$ and $\frac{0}{1}$), i.e.,

$$a_{21} = \frac{1+1}{0+1} = \frac{2}{1} \qquad \text{and} \qquad a_{22} = \frac{1+0}{1+1} = \frac{1}{2},$$

we get $\frac{1}{0}, \frac{2}{1}, \frac{1}{1}, \frac{1}{2}, \frac{0}{1}$. In the third step we insert the rational numbers $a_{31} = \frac{3}{1}$ (between $\frac{1}{0}$ and $\frac{2}{1}$), $a_{32} = \frac{3}{2}$ (between $\frac{2}{1}$ and $\frac{1}{1}$), $a_{33} = \frac{2}{3}$ (between $\frac{1}{1}$ and $\frac{1}{2}$), and $a_{34} = \frac{1}{3}$ (between $\frac{1}{2}$ and $\frac{0}{1}$), as illustrated in Table B.1.

	$\frac{1}{0}$																$\frac{0}{1}$
a_{11}									$\frac{1}{1}$								
a_{21}, a_{22}					$\frac{2}{1}$								$\frac{1}{2}$				
$a_{31}, a_{32}, a_{33}, a_{34}$			$\frac{3}{1}$				$\frac{3}{2}$				$\frac{2}{3}$				$\frac{1}{3}$		
$a_{41}, a_{42}, \ldots, a_{48}$		$\frac{4}{1}$		$\frac{5}{2}$		$\frac{5}{3}$		$\frac{4}{3}$		$\frac{3}{4}$		$\frac{3}{5}$		$\frac{2}{5}$		$\frac{1}{4}$	

Table B.1. The first four steps of the construction of the Farey series

After the first four steps we obtain the finite ordered sequence (omitting the initial values $\frac{1}{0}$ and $\frac{0}{1}$)

$$\frac{4}{1}, \frac{3}{1}, \frac{5}{2}, \frac{2}{1}, \frac{5}{3}, \frac{3}{2}, \frac{4}{3}, \frac{1}{1}, \frac{3}{4}, \frac{2}{3}, \frac{3}{5}, \frac{1}{2}, \frac{2}{5}, \frac{1}{3}, \frac{1}{4}.$$

The (*double*) *Farey series* F is then the set

$$F = \{a_{ni} \mid n \in \mathbb{N}, i \in \{1, 2, 3, \ldots, 2^{n-1}\}\}.$$

Since F contains all the rational numbers in $]0, \infty[$, it is a dense subset of $[0, \infty]$.

Proceeding now to the construction of the Krause t-norm, recall the set C_L of left endpoints of the intervals removed when constructing the Cantor set, and the (double) Farey series F. Consider first the function $f : C_L \longrightarrow F$ given by $f(c_{ni}) = a_{ni}$, which obviously is strictly decreasing bijection, as a consequence of the construction of the sets C_L and F.

Since F is a dense subset of $[0, \infty]$, there exists a unique non-increasing function $k : [0, 1] \longrightarrow [0, \infty]$ which is an extension of f, i.e., which satisfies $k|_{C_L} = f$. Observe that we have for all $x \in [0, 1]$

$$k(x) = \inf\{f(z) \mid z \in C_L, z \leq x\} = \sup\{f(z) \mid z \in C_L, z \geq x\}$$

and that k is a continuous, non-increasing surjection. In particular, this means that for each $n \in \mathbb{N}$, for each $i \in \{1, 2, 3, \ldots, 2^{n-1}\}$ and for all $x \in \left[c_{ni}, c_{ni} + \frac{1}{3^n}\right]$ we have $k(x) = k(c_{ni})$, i.e., k is constant on the closure of each interval which is deleted during the construction of the Cantor set C. Consequently, the pseudo-inverse $k^{(-1)} : [0, \infty] \longrightarrow [0, 1]$ of k (see Definition 3.2) is strictly decreasing and right-continuous, and we also have $k \circ k^{(-1)} = \mathrm{id}_{[0,\infty]}$ (see Remark 3.4) and $(k^{(-1)} \circ k)|_C = \mathrm{id}_C$. Because of the strict monotonicity of the function $k|_{C_L} = f$, the pseudo-inverse $k^{(-1)}$ is continuous in each point of $[0, \infty] \setminus F$, and we have for each $n \in \mathbb{N}$ and for each $i \in \{1, 2, 3, \ldots, 2^{n-1}\}$

$$k^{(-1)}(a_{ni}{}^{-}) = k^{(-1)}(a_{ni}) + \tfrac{1}{3^n} = c_{ni} + \tfrac{1}{3^n}.$$

Now define the function $K : [0, 1] \longrightarrow [0, 1]$ by

$$K(x) = \begin{cases} 3^n \cdot (x - c_{ni}) & \text{if } x \in \left[c_{ni}, c_{ni} + \frac{1}{3^n}\right], \\ 1 & \text{if } x = 1, \\ 0 & \text{otherwise.} \end{cases}$$

These preliminaries now allow us to formally introduce the Krause t-norm (a representative of a family of t-norms first described in [Krause 200x] where they are referred to as "devil's terraces") as an example of a non-continuous t-norm with a continuous diagonal section, thus providing a negative answer to Problem 5.8.1 in [Schweizer & Sklar 1983]. To keep the formulation of the next result simple, we shall use the convention $k^{(-1)}(0^-) = 1$.

B.1 THEOREM
The function $T^{\mathbf{K}} : [0,1]^2 \longrightarrow [0,1]$ *given by*

$$\begin{aligned} T^{\mathbf{K}}(x,y) &= k^{(-1)}(k(x)+k(y)) \\ &\quad + \min(K(x),K(y)) \cdot \left(k^{(-1)}((k(x)+k(y))^-) - k^{(-1)}(k(x)+k(y))\right) \end{aligned}$$

is a triangular norm (the so-called Krause t-norm) which is neither left- nor right-continuous, but has a continuous diagonal section $\delta_{T^{\mathbf{K}}}$.

Proof: The commutativity (T1) of $T^{\mathbf{K}}$ is obvious. In order to check the boundary condition (T4), note first that

$$T^{\mathbf{K}}(1,1) = k^{(-1)}(0) + (k^{(-1)}(0^-) - k^{(-1)}(0)) = 1.$$

If $x \in \left[c_{ni}, c_{ni} + \frac{1}{3^n}\right]$ for some $n \in \mathbb{N}$ and some $i \in \{1,2,3,\ldots,2^{n-1}\}$, then we have $k^{(-1)}(k(x)) = c_{ni}$ and $k^{(-1)}(k(x)^-) = c_{ni} + \frac{1}{3^n}$ and, subsequently,

$$T^{\mathbf{K}}(x,1) = k^{(-1)}(k(x)) + K(x) \cdot \left(k^{(-1)}(k(x)^-) - k^{(-1)}(k(x))\right) = x.$$

For all remaining points x we have $x \in C$ and $K(x) = 0$ and, therefore,

$$T^{\mathbf{K}}(x,1) = k^{(-1)}(k(x)) = x.$$

Concerning the associativity (T2), note first that for all $(x,y) \in [0,1]^2$ we have $k(T^{\mathbf{K}}(x,y)) = k(x) + k(y)$ and $K(T^{\mathbf{K}}(x,y)) = \min(K(x),K(y))$. Thus we can infer for all $x,y,z \in [0,1]$

$$\begin{aligned} &T^{\mathbf{K}}(T^{\mathbf{K}}(x,y),z) \\ &\qquad = k^{(-1)}(k(x)+k(y)+k(z)) + \min(K(x),K(y),K(z)) \\ &\qquad\quad \cdot \left(k^{(-1)}((k(x)+k(y)+k(z))^-) - k^{(-1)}(k(x)+k(y)+k(z))\right) \\ &\qquad = T^{\mathbf{K}}(x, T^{\mathbf{K}}(y,z)). \end{aligned}$$

Finally, for the monotonicity (T3) fix $x,y,z \in [0,1]$ with $y < z$, i.e., $k(y) \geq k(z)$. If $k(y) > k(z)$ then clearly

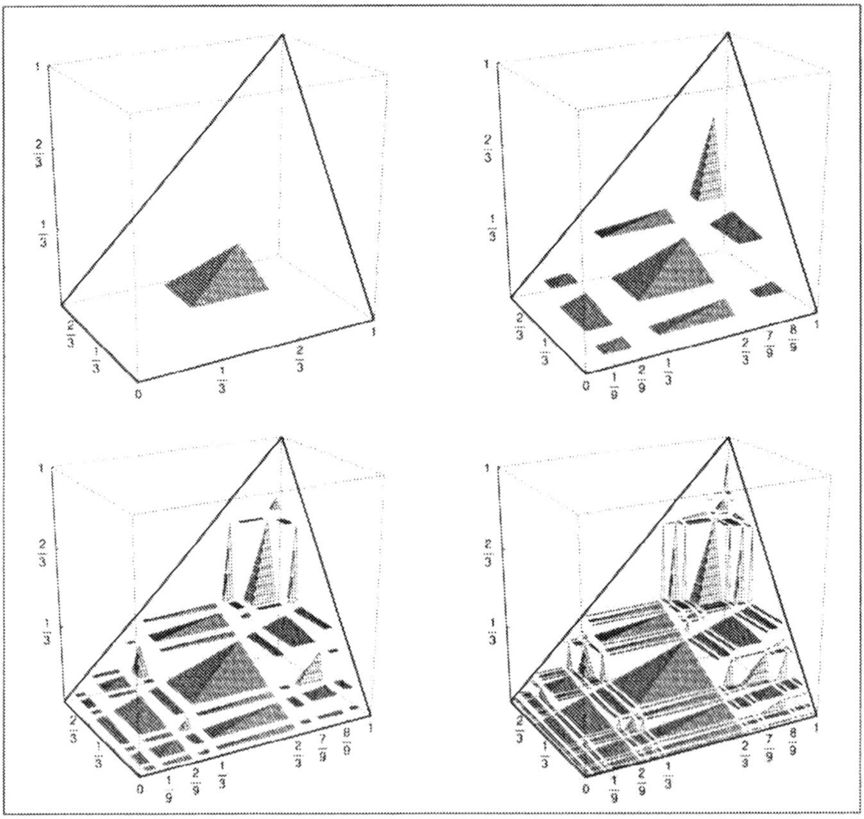

Fig. B.2. Krause t-norm: restrictions of its 3D plot to the subsets of the unit square determined by the middle thirds (intervals of lengths $\frac{1}{3}$, $\frac{1}{9}$, $\frac{1}{27}$, and $\frac{1}{81}$, respectively) which are deleted during the first four steps of the construction of the Cantor set

$$T^{\mathbf{K}}(x,y) \leq k^{(-1)}((k(x)+k(y))^-) \leq T^{\mathbf{K}}(x,z).$$

If $k(y) = k(z)$ then necessarily $c_{ni} \leq y < z \leq c_{ni} + \frac{1}{3^n}$ for some $n \in \mathbb{N}$ and some $i \in \{1,2,3,\ldots,2^{n-1}\}$ and, therefore, $K(y) < K(z)$. Consequently,

$$\begin{aligned} T^{\mathbf{K}}(x,y) &= k^{(-1)}(k(x)+k(y)) \\ &\quad + \min(K(x),K(y)) \cdot \big(k^{(-1)}((k(x)+k(y))^-) - k^{(-1)}(k(x)+k(y))\big) \\ &\leq k^{(-1)}(k(x)+k(z)) \\ &\quad + \min(K(x),K(z)) \cdot \big(k^{(-1)}((k(x)+k(z))^-) - k^{(-1)}(k(x)+k(z))\big) \\ &\leq T^{\mathbf{K}}(x,z), \end{aligned}$$

showing that $T^{\mathbf{K}}$ is indeed a t-norm.

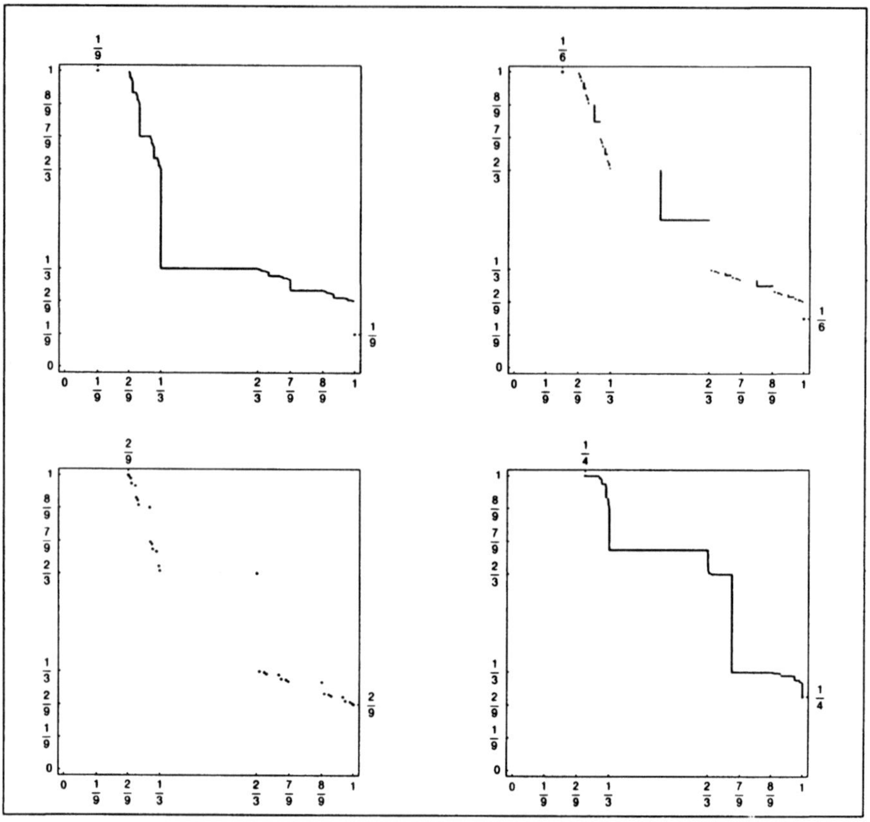

Fig. B.3. Four contour sets of the Krause t-norm $T^{\mathbf{K}}$ at the levels $\frac{1}{9}$ (top left), $\frac{1}{6}$ (top right), $\frac{2}{9}$ (which consists of isolated points only, bottom left), and $\frac{1}{4}$ (which is a connected curve)

The t-norm $T^{\mathbf{K}}$ is not left-continuous in the points of the form $(x, 1)$ or $(1, x)$ with $x \in]0, 1[$ and $K(x) > 0$, which is only possible if $x \in \left]c_{ni}, c_{ni} + \frac{1}{3^n}\right]$ for some $n \in \mathbb{N}$ and some $i \in \{1, 2, 3, \ldots, 2^{n-1}\}$, in which case we obtain

$$T(x, 1^-) = T(1^-, x) = c_{ni} < x.$$

The Krause t-norm is also not right-continuous because of, e.g.,

$$T^{\mathbf{K}}\left(\tfrac{1}{2}, \tfrac{2}{3}\right) = \tfrac{1}{6} < \tfrac{2}{9} = T^{\mathbf{K}}\left(\tfrac{1}{2}, \tfrac{2}{3}^+\right).$$

The set of points in which $T^{\mathbf{K}}$ is not right-continuous consists exactly of the points (x, y) and (y, x), where $x \in [0, 1] \setminus C$ and $y \in \left\{c_{ni} + \frac{1}{3^n} \mid c_{ni} \in C_L\right\}$. However, the diagonal section $\delta_{T^{\mathbf{K}}} : [0, 1] \longrightarrow [0, 1]$ of $T^{\mathbf{K}}$ is continuous. To see this, note first that the continuity of the restrictions of $\delta_{T^{\mathbf{K}}}$ to $\left[c_{ni}, c_{ni} + \frac{1}{3^n}\right]$

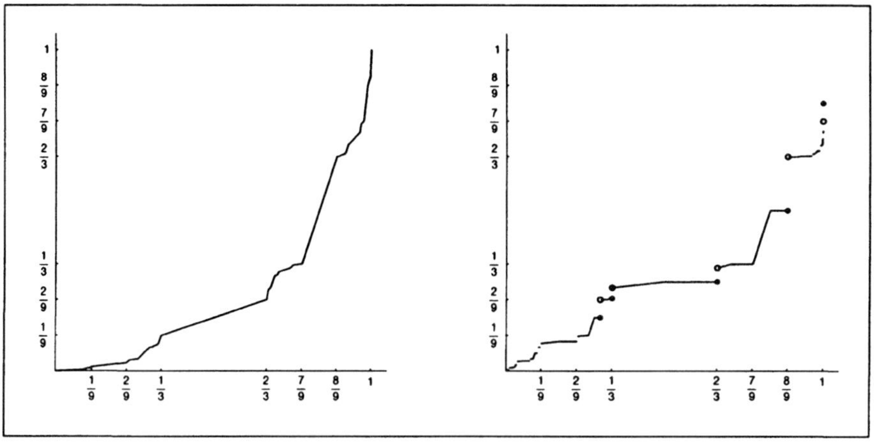

Fig. B.4. Krause t-norm: the continuous diagonal section $\delta_{T^{\mathbf{K}}}$ (left) and the non-continuous horizontal section $T^{\mathbf{K}}(\cdot, \frac{5}{6})$ (selected points of discontinuity are marked)

is implied by the continuity of the corresponding restrictions of K. Moreover, for each $c_{ni} \in C_L$ there are integers $k \in \mathbb{N}$ and $l \in \{1, 2, 3, \ldots, 2^{k-1}\}$ such that

$$\delta_{T^{\mathbf{K}}}(c_{ni}) = k^{(-1)}(2a_{ni}) = f^{-1}(2a_{ni}) = c_{kl} \in C_L,$$
$$\delta_{T^{\mathbf{K}}}\left(c_{ni} + \tfrac{1}{3^n}\right) = k^{(-1)}(2a_{ni}{}^{-}) = c_{kl} + \tfrac{1}{3^k} \in C.$$

Therefore,

$$\{\delta_{T^{\mathbf{K}}}(z) \mid z \in C_L\} = \{f^{-1}(2a_{ni}) \mid a_{ni} \in F\} = \{f^{-1}(a_{ni}) \mid \tfrac{a_{ni}}{2} \in F\} = C_L,$$

and the monotonicity of $\delta_{T^{\mathbf{K}}}$ together with the fact that C_L is a dense subset of the Cantor set C implies the continuity of $\delta_{T^{\mathbf{K}}}$. □

Based on the Krause t-norm it is also possible to give examples of left-continuous (but non-continuous) t-norms with continuous diagonal sections.

B.2 REMARK
Consider the function $K^* : [0,1] \longrightarrow [0,1]$ given by

$$K^*(x) = \begin{cases} K(x) & \text{if } x \in [0,1[, \\ 0 & \text{if } x = 1. \end{cases}$$

Then the function $G : [0,1]^2 \longrightarrow [0,1]$ given by

$$G(x,y) = k^{(-1)}(k(x) + k(y)) \\ \qquad + \min(K^*(x), K^*(y)) \cdot \left(k^{(-1)}((k(x) + k(y))^{-}) - k^{(-1)}(k(x) + k(y))\right)$$

is a left-continuous t-subnorm (see Definition 1.7) coinciding with $T^{\mathbf{K}}$ on the diagonal of the unit square, i.e., $\delta_G = \delta_{T^{\mathbf{K}}}$. Because of Corollary 1.8 and Theorem 3.42, for each $b \in]0,1[$ the function $T : [0,1]^2 \longrightarrow [0,1]$ defined by

$$T(x,y) = \begin{cases} b \cdot G(\frac{x}{b}, \frac{y}{b}) & \text{if } (x,y) \in [0,b]^2, \\ \min(x,y) & \text{otherwise,} \end{cases}$$

is a left-continuous t-norm (which obviously is not continuous). Observe that the semigroup $([0,1],T)$ is an ordinal sum of two semigroups in the sense of Theorem 3.42, but T is not an ordinal sum of t-norms in the sense of Definition 3.44. The t-norm T has a continuous diagonal section δ_T, a fact which follows directly from the continuity of the diagonal section δ_G (together with $\delta_G(0) = 0$ and $\delta_G(1) = 1$, see Theorem B.1) and $\delta_{T_{\mathbf{M}}}$.

B.2 A family of incomparable t-norms

As mentioned in Example 3.33, the family of continuous, strictly decreasing functions $(t^{\mathbf{I}}_\lambda : [0,1] \longrightarrow [0,\infty])_{\lambda\in[0,0.5[}$ defined by (see Figure 3.13)

$$t^{\mathbf{I}}_\lambda(x) = \min\left(1 - \tfrac{x}{2}, \max\left(\tfrac{1-x}{2}, \tfrac{2-3\lambda-(2-2\lambda)x}{2-4\lambda}\right)\right)$$

forms, as a consequence of Theorem 3.23, a family of additive generators of a family of t-norms we denoted $(T^{\mathbf{I}}_\lambda)_{\lambda\in[0,0.5[}$ (see the 3D and contour plots in Figure B.5).

A simple but tedious computation yields the explicit formula for this family of t-norms $(T^{\mathbf{I}}_\lambda)_{\lambda\in[0,0.5[}$, namely,

$$\begin{aligned} T^{\mathbf{I}}_\lambda(x,y) = \max\Big(&1 - 2\Big(\min\left(1-\tfrac{x}{2}, \max\left(\tfrac{1-x}{2}, \tfrac{3\lambda+2x-2\lambda x-2}{4\lambda-2}\right)\right) \\ &\quad + \min\left(1-\tfrac{y}{2}, \max\left(\tfrac{1-y}{2}, \tfrac{3\lambda+2y-2\lambda y-2}{4\lambda-2}\right)\right)\Big), \\ &\min\Big(-2\Big(-1+\min\left(1-\tfrac{x}{2}, \max\left(\tfrac{1-x}{2}, \tfrac{3\lambda+2x-2\lambda x-2}{4\lambda-2}\right)\right) \\ &\quad + \min\left(1-\tfrac{y}{2}, \max\left(\tfrac{1-y}{2}, \tfrac{3\lambda+2y-2\lambda y-2}{4\lambda-2}\right)\right)\Big), \\ &\tfrac{1}{2-2\lambda}\Big(2-3\lambda+2(2\lambda-1)\Big(\min\left(1-\tfrac{x}{2}, \max\left(\tfrac{1-x}{2}, \tfrac{3\lambda+2x-2\lambda x-2}{4\lambda-2}\right)\right) \\ &\quad + \min\left(1-\tfrac{y}{2}, \max\left(\tfrac{1-y}{2}, \tfrac{3\lambda+2y-2\lambda y-2}{4\lambda-2}\right)\right)\Big)\Big)\Big), 0\Big). \end{aligned}$$

It is clear that each $T^{\mathbf{I}}_\lambda$ is a nilpotent t-norm because of Propositions 3.26 and 3.29 and Corollary 3.30.

Obviously, we have $T^{\mathbf{I}}_0 = T_{\mathbf{L}}$ and also $T^{\mathbf{I}}_0 < T^{\mathbf{I}}_\lambda$ for all $\lambda \in]0,0.5[$. However, for $\lambda,\mu \in]0,0.5[$ with $\lambda > \mu$ the corresponding t-norms $T^{\mathbf{I}}_\lambda$ and $T^{\mathbf{I}}_\mu$ turn out to be incomparable because of (for a visualization see the diagonal sections in Figure 3.14)

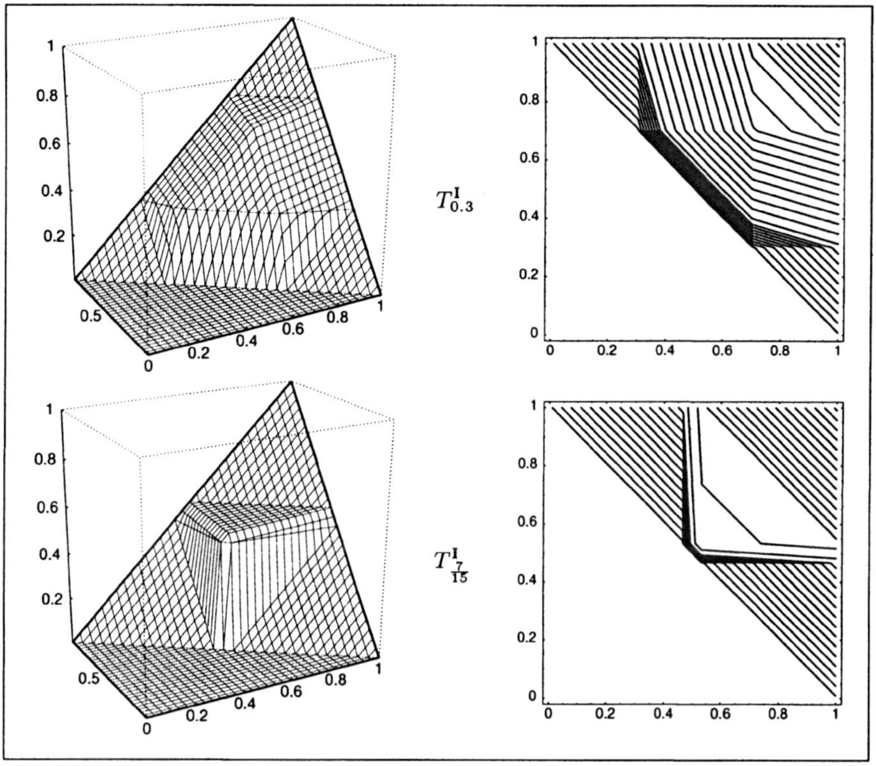

Fig. B.5. 3D and contour plots of the t-norms $T^{\mathbf{I}}_{0.3}$ and $T^{\mathbf{I}}_{\frac{7}{15}}$

$$\begin{aligned}
T^{\mathbf{I}}_\lambda\left(1-\tfrac{\lambda}{2}, 1-\tfrac{\lambda}{2}\right) &= 1-\lambda < 1-\lambda+\tfrac{\lambda-\mu}{2-2\mu} = T^{\mathbf{I}}_\mu\left(1-\tfrac{\lambda}{2}, 1-\tfrac{\lambda}{2}\right),\\
T^{\mathbf{I}}_\lambda\left(\tfrac{1}{2}+\tfrac{\mu(1-2\lambda)}{4-4\lambda}, \tfrac{1}{2}+\tfrac{\mu(1-2\lambda)}{4-4\lambda}\right) &= \mu\\
&> \mu - \tfrac{\mu(\lambda-\mu)}{(1-\lambda)(1-2\mu)}\\
&= T^{\mathbf{I}}_\mu\left(\tfrac{1}{2}+\tfrac{\mu(1-2\lambda)}{4-4\lambda}, \tfrac{1}{2}+\tfrac{\mu(1-2\lambda)}{4-4\lambda}\right).
\end{aligned}$$

Also, the limit $F = \lim\limits_{\lambda \nearrow 0.5} T^{\mathbf{I}}_\lambda$ exists, and it is given by

$$F(x,y) = \begin{cases} 0.5 & \text{if } (x,y) \in [0.5,1]^2 \text{ and } x+y \in \,]1,1.5] \\ T_{\mathbf{L}}(x,y) & \text{otherwise.} \end{cases}$$

Note that F differs only in the point $(0.5,0.5)$ from the t-norm given in Remark 3.40 (see [Viceník 1998a, 1998b]), but that it is not associative and, hence, not a t-norm because of $F(0.5, F(0.6,0.6)) = 0 < 0.5 = F(F(0.5,0.6),0.6)$.

REFERENCE MATERIAL

List of Figures

List of Tables

List of Mathematical Symbols

$\wedge$	conjunction	xvii
$\vee$	disjunction	xvii
$\neg$	negation	xvii
$\rightarrow$	implication	xvii
$\bigwedge, \wedge$	meet in a lattice	xix
$\bigvee, \vee$	join in a lattice	xix
$\subseteq$	subset relation between sets	xvii
$A \cap B$	intersection of sets	xvii
$A \cup B$	union of sets	xvii
$A \setminus B$	difference of sets	xvii
$\complement A$	complement of a set A	xvii
$A \times B$	Cartesian product of sets	xvii
$\emptyset$	empty set	xvii
$\mathcal{P}(X)$	power set of X	xvii
$\mathrm{card}(A)$	cardinality of a set A	xvii
$\mathbf{1}_A$	characteristic function of a crisp set A	xvii
$\mathbb{N}$	set of positive integers $\{1, 2, 3, \dots\}$	xvii
$\mathbb{Z}$	set of integers	xvii
$\mathbb{Q}$	set of rational numbers	xvii
$\mathbb{R}$	set of real numbers (real line)	xvii
$[a, b]$	closed interval	xvii
$]a, b[$	open interval	xvii
$[a, b[\, , \,]a, b]$	half-open intervals	xvii
$[-\infty, \infty]$	extended real line	xvii
$\lfloor x \rfloor$	floor of x	xviii
$\log x$	(natural) logarithm of x, i.e., with respect to the basis e	xviii
$\log_\lambda x$	logarithm of x with respect to the basis λ	xviii
Y^X	set of all functions from X to Y	xvii
$(a_n)_{n \in \mathbb{N}} \quad (\in X^{\mathbb{N}})$	sequence of elements in X	xvii

$(a_i)_{i\in I} \quad (\in X^I)$	family of elements in X with index set I	xvii
$\mathrm{Dom}(f)$	domain of a function f	xvii
$\mathrm{Ran}(f)$	range of a function f	xvii
id_X	identity function on a set X	xviii
$g \circ f$	composition of functions	xviii
$f\vert_A$	restriction of $f: X \longrightarrow Y$ to a subset A of X	xviii
$f(A)$	image of a set A with respect to a function f	xvii
$f^{-1}(A)$	preimage of a set A with respect to a function f	xvii
f^{-1}	inverse function of a bijection f	xviii
$f(x_0,\cdot), f(\cdot,y_0)$	vertical and horizontal section of a function in two variables	xviii
$f(a^+), f(b^-)$	right- and left-hand limit of $f:]a,b[\longrightarrow [-\infty,\infty]$	xviii
$\mathrm{supp}(f)$	support of a function $f: X \longrightarrow [0,\infty]$	xviii
$f^{(-1)}$	pseudo-inverse of a monotone function f	55
$\gg$	domination relation	152
$\mathop{\mathsf{T}}_{i=1}^{n} x_i, \mathop{\mathsf{T}}_{i=1}^{\infty} x_i, \mathop{\mathsf{T}}_{i\in I} x_i$	extensions of a t-norm T	9
$\mathop{\mathsf{S}}_{i=1}^{n} x_i, \mathop{\mathsf{S}}_{i=1}^{\infty} x_i, \mathop{\mathsf{S}}_{i\in I} x_i$	extensions of a t-conorm S	13
$x_T^{(n)}$	n-th power of x with respect to a t-norm T	9
$x_S^{(n)}$	n-th power of x with respect to a t-conorm S	14
$x_T^{(\frac{1}{n})}, x_S^{(\frac{1}{n})}$	n-th root of x with respect to a t-norm T and a t-conorm S	57
δ_T	diagonal section of a t-norm T	164
$T_{\mathbf{M}}$	minimum t-norm	4
$T_{\mathbf{P}}$	product t-norm	4
$T_{\mathbf{L}}$	Łukasiewicz t-norm	4
$T_{\mathbf{D}}$	drastic product	4
$S_{\mathbf{M}}$	maximum t-conorm	11
$S_{\mathbf{P}}$	probabilistic sum	11
$S_{\mathbf{L}}$	Łukasiewicz t-conorm (bounded sum)	11
$S_{\mathbf{D}}$	drastic sum	11
$T^{\mathbf{nM}}$	nilpotent minimum	16
T_φ, S_φ	transformation of a t-norm T (a t-conorm S) with respect to a strictly increasing bijection φ	38
$T^{[n]}$	discretization of a t-norm T	99
$T_\lambda^{\mathbf{AA}}, S_\lambda^{\mathbf{AA}}$	Aczél-Alsina t-norms and t-conorms	116
$T_\lambda^{\mathbf{D}}, S_\lambda^{\mathbf{D}}$	Dombi t-norms and t-conorms	112
$T_\lambda^{\mathbf{DP}}$	Dubois-Prade t-norms	84

$T_\lambda^{\mathbf{F}}, S_\lambda^{\mathbf{F}}$	Frank t-norms and t-conorms	108
$T_\lambda^{\mathbf{H}}, S_\lambda^{\mathbf{H}}$	Hamacher t-norms and t-conorms	106
$T_\lambda^{\mathbf{J}}$	Jenei t-norms	97
$T^{\mathbf{K}}$	Krause t-norm	169
$T_\lambda^{\mathbf{MT}}, S_\lambda^{\mathbf{MT}}$	Mayor-Torrens t-norms and t-conorms	118
$T_\lambda^{\mathbf{SS}}, S_\lambda^{\mathbf{SS}}$	Schweizer-Sklar t-norms and t-conorms	104
$T_\lambda^{\mathbf{SW}}, S_\lambda^{\mathbf{SW}}$	Sugeno-Weber t-norms and t-conorms	114
$T_\lambda^{\mathbf{Y}}, S_\lambda^{\mathbf{Y}}$	Yager t-norms and t-conorms	110
T_{rot}	rotation of a t-norm T	98
$D_{\mathbf{M}}^{(n)}, D_{\mathbf{L}}^{(n)}, D_{\mathbf{D}}^{(n)}$	discretized minimum, Łukasiewicz t-norm, drastic product	171
$(\langle a_\alpha, e_\alpha, T_\alpha\rangle)_{\alpha\in A}$	ordinal sum of t-norms	84
$(\langle a_\alpha, e_\alpha, S_\alpha\rangle)_{\alpha\in A}$	ordinal sum of t-conorms	91
$\to_*$	residuum with respect to a semigroup operation $*$	47
$\to_T$	residuum with respect to a left-continuous t-norm T	49
$\to_{\mathbf{M}}, \to_{\mathbf{P}}, \to_{\mathbf{L}}$	Gödel, Goguen and Łukasiewicz implication	50
$\mathsf{N}, \mathsf{E}, \mathsf{L}$	operators between generators of t-norms and t-conorms	79
Δ	class of distribution functions on $[-\infty, \infty]$	197
Δ^+	class of distance distribution functions	207
ε_a	unit step (Dirac distribution)	197
$F * G$	convolution of distance distribution functions	210
F_X	distribution of a random variable X	198
H_{XY}	joint distribution of a random vector (X, Y)	198
C_{XY}	copula corresponding to a random vector (X, Y)	202
$\hat{C}$	survival copula	199
$\tilde{C}$	dual of a copula C	206
C^*	co-copula of C	206
$C_1 * C_2$	$*$-product of copulas	202
$C_\lambda^{\mathbf{FGM}}$	Farlie-Gumbel-Morgenstern copulas	200
$C_\lambda^{\mathbf{CA}}$	Cuadras-Augé copulas	200
$C_\lambda^{\mathbf{M}}$	Mardia copulas	201
τ	triangle function	208
$\mathbf{T}$	triangle function based on a t-norm T	208
$\tau_{T,L}$	triangle function based on T and L	209
τ_T	triangle function based on T (and $+$)	209
F_{pq}	distance distribution function associated with p and q	211
$\mathbf{A}$	aggregation operator	216
$\mathbf{DA}$	dual aggregation operator	218

Bibliography

ABEL, N. H. (1826). Untersuchungen der Funktionen zweier unabhängigen veränderlichen Größen x und y wie $f(x,y)$, welche die Eigenschaft haben, daß $f(z, f(x,y))$ eine symmetrische Funktion von x, y und z ist. *J. Reine Angew. Math.* 1:11–15.

ACZÉL, J. (1949). Sur les opérations definies pour des nombres réels. *Bull. Soc. Math. France* 76:59–64.

ACZÉL, J. (1961). *Vorlesungen über Funktionalgleichungen und ihre Anwendungen.* Birkhäuser, Basel.

ACZÉL, J. (1966). *Lectures on Functional Equations and their Applications.* Academic Press, New York.

ACZÉL, J. & ALSINA, C. (1984). Characterizations of some classes of quasilinear functions with applications to triangular norms and to synthesizing judgements. *Methods Oper. Res.* 48:3–22.

ADILLON, R. J. & VERDÚ, V. (1998). On product logic. *Soft Computing* 2:141–146.

AKIAN, M., QUADRAT, J.-P., VIOT, M. (1994). Bellman processes. In *Proceedings 11th International Conference on Analysis and Optimization of Sciences: Discrete Event Systems*, volume 199 of *Lecture Notes in Control and Information Sciences*, pages 302–311. Springer, London.

ALI, M. M., MIKHAIL, N. N., HAQ, M. S. (1978). A class of bivariate distributions including the bivariate logistic. *J. Multivariate Anal.* 8:405–412.

ALSINA, C. (1978). On countable products and algebraic convexifications of probabilistic metric spaces. *Pacific J. Math.* 76:91–300.

ALSINA, C. (1983). On convex triangle functions. *Aequationes Math.* 26:91–196.

ALSINA, C. (1985a). On a family of connectives for fuzzy sets. *Fuzzy Sets and Systems* 16:231–235.

ALSINA, C. (1985b). On a functional equation characterizing two binary operations on the space of membership functions. *Fuzzy Sets and Systems* 27:5–9.

ALSINA, C. (1992). On a method of Pi-Calleja for describing additive generators of associative functions. *Aequationes Math.* 43:14–20.

ALSINA, C. & GIMENEZ, J. (1984). Sobre L-órdenes entre t-normas estrictas. *Stochastica* 8:85–89.

ALSINA, C., MAYOR, G., TOMÁS, M. S., TORRENS, J. (1993a). A characterization of a class of aggregation functions. *Fuzzy Sets and Systems* 53:33–38.

ALSINA, C., SCHWEIZER, B., SKLAR, A. (1993b). On the definition of a probabilistic normed space. *Aequationes Math.* 46:91–98.

ALSINA, C. & SKLAR, A. (1987). A characterization of continuous associative operations whose graphs are ruled surfaces. *Aequationes Math.* 33:114–119.

ALSINA, C., TRILLAS, E., VALVERDE, L. (1980). On non-distributive logical connectives for fuzzy sets theory. *BUSEFAL* 3:18–29.

ALSINA, C., TRILLAS, E., VALVERDE, L. (1983). On some logical connectives for fuzzy sets theory. *J. Math. Anal. Appl.* 93:15–26.

ANTONY, J. M. & SHERWOOD, H. (1979). Fuzzy groups redefined. *J. Math. Anal. Appl.* 69:124–130.

ANTONY, J. M. & SHERWOOD, H. (1982). A characterization of fuzzy subgroups. *Fuzzy Sets and Systems* 7:297–305.

APOSTOL, T. M. (1965). *Mathematical Analysis. A Modern Approach to Advanced Calculus.* Addison-Wesley, Reading.

ARNOLD, V. (1957). Concerning the representability of functions of two variables in the form $X[\Phi(x) + \Psi(y)]$. *Uspekhi Mat. Nauk* 12:119–121.

AUBIN, J. P. (1974a). Coeur et équilibres des jeux flous sans paiements latéraux. *C. R. Acad. Sci. Paris Sér. A* 279:963–966.

AUBIN, J. P. (1974b). Coeur et valeur des jeux flous à paiements latéraux. *C. R. Acad. Sci. Paris Sér. A* 279:891–894.

AUBIN, J. P. (1981). Cooperative fuzzy games. *Math. Op. Res.* 6:1–13.

AUBIN, J. P. (1982). *Mathematical Methods of Game and Economic Theory.* North-Holland, Amsterdam.

AUMANN, R. J. & SHAPLEY, L. S. (1974). *Values of Non-Atomic Games.* Princeton University Press, Princeton.

AVALLONE, A. & BARBIERI, G. (1997). Range of finitely additive fuzzy measures. *Fuzzy Sets and Systems* 89:231–241.

BACCELLI, F. L., COHEN, G., OLSDER, G. J., QUADRAT, J.-P. (1992). *Synchronization and Linearity. An Algebra for Discrete Event Systems.* J. Wiley & Sons, Chichester.

BACCHELLI, B. (1986). Representation of continuous associative functions. *Stochastica* 10:13–28.

BANDEMER, H. & NÄTHER, W. (1992). *Fuzzy Data Analysis.* Kluwer Academic Publishers, Dordrecht.

BANDLER, W. & KOHOUT, L. (1980). Fuzzy power sets and fuzzy implication operators. *Fuzzy Sets and Systems* 4:47–58.

BARBIERI, G., LEPELLERE, M. A., WEBER, H. (200x). The Hahn decomposition theorem for fuzzy measures. *Fuzzy Sets and Systems* (in press).

BARBIERI, G. & WEBER, H. (1998). A topological approach to the study of fuzzy measures. In Abramovich, Y., Avgerinos, E., Yannelis, N. C., editors, *Functional Analysis and Economic Theory*, pages 17–46. Springer, Berlin.

BARBIERI, G. & WEBER, H. (200x). A representation theorem and a Lyapunov theorem for T_s-measures: the solution of two problems of Butnariu and Klement. *J. Math. Anal. Appl.* (in press).

BARRO, S., BUGARÍN, A., SOBRINO, A., editors (1998). *Advances of Fuzzy Logics. Selected Papers (with Comments) of Some Spanish Authors.* Universidad de Santiago de Compostela.

BAUER, H. (1981). *Probability Theory and Elements of Measure Theory.* Academic Press, London.

BAUER, P., BODENHOFER, U., KLEMENT, E. P. (1996). A fuzzy algorithm for pixel classification based on the discrepancy norm. In *Proceedings Fifth IEEE International Conference on Fuzzy Systems, FUZZ-IEEE '96, New Orleans*, volume 3, pages 2007–2012. IEEE Press, Piscataway.

BAUER, P., KLEMENT, E. P., LEIKERMOSER, A., MOSER, B. (1995). Modeling of control functions by fuzzy controllers. In [Nguyen *et al.* 1995], chapter 5, pages 91–116.

BELLMAN, R. & GIERTZ, M. (1973). On the analytic formalism of the theory of fuzzy sets. *Inform. Sci.* 5:149–156.

BELLMAN, R. E. & DREYFUS, S. E. (1962). *Applied Dynamical Programming.* Princeton University Press, Princeton.

BELLMAN, R. E., KALABA, R., ZADEH, L. A. (1966). Abstraction and pattern recognition. *J. Math. Anal. Appl.* 13:1–7.

BELLMAN, R. E. & ZADEH, L. A. (1970). Decision-making in a fuzzy environment. *Management Sciences* 17:141–164.

BELLUCE, L. P. (1986). Semi-simple algebras of infinite valued logic and bold fuzzy set theory. *Can. J. Math.* 38:1356–1379.

BELLUCE, L. P. (1992). Semi-simple and complete MV-algebras. *Algebra Universalis* 29:1–9.

BENVENUTI, P. & MESIAR, R. (2000). Integrals with respect to a general fuzzy measure. In [Grabisch *et al.* 2000], pages 205–232.

BENVENUTI, P. & MESIAR, R. (200x). Pseudo-arithmetical operations as a basis for the general measure and integration theory. Submitted for publication.

BENVENUTI, P. & VIVONA, D. (1996). General theory of the fuzzy integral. *Mathware Soft Comput.* 3:199–209.

BERTOLUZZA, C. (1993). On the distributivity between t-norms and t-conorms. In *Proceedings Second International Conference on Fuzzy Systems, FUZZ-IEEE '93, San Francisco*, volume I, pages 140–147. IEEE Press, Piscataway.

BEZDEK, J. C. (1981). *Pattern Recognition with Fuzzy Objective Function Algorithm.* Plenum Press.

BEZDEK, J. C. & HARRIS, J. D. (1978). Fuzzy partitions and relations: An axiomatic basis for clustering. *Fuzzy Sets and Systems* 1:111–127.

BEZDEK, J. C., KELLER, J. M., KRISHNAPURAM, R., PAL, S. K. (1999). *Fuzzy Models and Algorithms for Pattern Recognition and Image Processing.* Kluwer Academic Publishers, Dordrecht.

BEZDEK, J. C. & PAL, S. K., editors (1992). *Fuzzy Models for Pattern Recognition: Methods that Search for Structures in Data.* IEEE Press, New York.

BÉZIVIN, J. P. & TOMÁS, M. S. (1993). On the determination of strict t-norms on some diagonal segments. *Aequationes Math.* 45:239–245.

BILLOT, A. (1992). *Economic Theory of Fuzzy Equilibria: An Axiomatic Analysis*, volume 373 of *Lecture Notes in Economics and Mathematical Systems.* Springer, Berlin.

BIRKHOFF, G. (1973). *Lattice Theory.* American Mathematical Society, Providence.

BLACKWELL, D. (1951). The range of certain vector integrals. *Proc. Amer. Math. Soc.* 2:390–395.

BODENHOFER, U. (1999). *A Similarity-Based Generalization of Fuzzy Orderings*, volume C 26 of *Schriften der Johannes-Kepler-Universität Linz.* Universitätsverlag Rudolf Trauner, Linz.

BORKOWSKI, L., editor (1970). *J. Lukasiewicz: Selected Works.* Studies in Logic and Foundations of Mathematics. North-Holland, Amsterdam.

BOUCHON-MEUNIER, B., editor (1998). *Aggregation and Fusion of Imperfect Information.* Physica-Verlag, Heidelberg.

BOURBAKI, N. (1966). *General Topology.* Addison-Wesley, Reading.

BROUWER, L. E. J. (1909). Die Theorie der endlichen kontinuierlichen Gruppen unabhängig von den Axiomen von Lie. *Math. Ann.* 67:246–267.

BUDINČEVIĆ, M. & KURILIĆ, M. S. (1998). A family of strict and discontinuous triangular norms. *Fuzzy Sets and Systems* 95:381–384.

BURGUÉS, C. (1981). Sobre la sección diagonal y la región cero de una t-norma. *Stochastica* 5:79–87.

BUTNARIU, D. (1983a). Additive fuzzy measures and integrals. I. *J. Math. Anal. Appl.* 93:436–452.

BUTNARIU, D. (1983b). Decomposition and range for additive fuzzy measures. *Fuzzy Sets and Systems* 10:135–155.

BUTNARIU, D. (1985). Non-atomic fuzzy measures and games. *Fuzzy Sets and Systems* 17:39–52.

BUTNARIU, D. (1986). Fuzzy measurability and integrability. *J. Math. Anal. Appl.* 117:385–410.

BUTNARIU, D. (1987a). Additive fuzzy measures and integrals. III. *J. Math. Anal. Appl.* 125:288–303.

BUTNARIU, D. (1987b). Values and cores of fuzzy games with infinitely many players. *Internat. J. Game Theory* 16:43–68.

BUTNARIU, D. (1989). A fuzzy game theoretical approach of the rate problem for services in bulk. In Verdegay, J. L. & Delgado, M., editors, *The Interface between Artificial Intelligence and Operations Research.* Verlag TÜV Rheinland, Basel.

BUTNARIU, D. & KLEMENT, E. P. (1993). *Triangular Norm-Based Measures and Games with Fuzzy Coalitions.* Kluwer Academic Publishers, Dordrecht.

BUTNARIU, D. & KLEMENT, E. P. (1996). Core, value and equilibria for market games: on a problem of Aumann and Shapley. *Internat. J. Game Theory* 25:149–160.

BUTNARIU, D. & KLEMENT, E. P. (2000). Triangular norm-based measures: Properties and integral representations. In [Grabisch *et al.* 2000], pages 233–246.

BUTNARIU, D., KLEMENT, E. P., ZAFRANY, S. (1995). On triangular norm-based propositional fuzzy logics. *Fuzzy Sets and Systems* 69:241–255.

CALVO, T., DE BAETS, B., FODOR, J. C. (200x). The functional equations of Frank and Alsina for uninorms and nullnorms. *Fuzzy Sets and Systems* (in press).

CARRUTH, J. H., HILDEBRANT, J. A., KOCH, R. J. (1983). *The Theory of Topological Semigroups.* Lecture Notes in Mathematics. Marcel Dekker, New York.

CARTAN, É. (1930). *La théorie des groupes finis et continus et l'Analysis Situs*, volume 42 of *Mém. Sci. Math.* Gauthier-Villars, Paris.

CHANG, C. C. (1958). Algebraic analysis of many valued logics. *Trans. Amer. Math. Soc.* 88:467–490.

CHANG, C. L. (1968). Fuzzy topological spaces. *J. Math. Anal. Appl.* 24:182–190.

CHANG, C. L. & LEE, R. C. T. (1975). Some properties of fuzzy logic. *Inform. Control* 19:417–431.

CHERNOFF, H. (1951). An extension of a result of Liapounoff on the range of a vector measure. *Proc. Amer. Math. Soc.* 2:720–726.

CHOQUET, G. (1953–1954). Theory of capacities. *Ann. Inst. Fourier (Grenoble)* 5:131–292.

CIGNOLI, R., D'OTTAVIANO, I. M. L., MUNDICI, D. (2000). *Algebraic Foundations of Many-Valued Reasoning.* Kluwer Academic Publishers, Dordrecht.

CLAYTON, D. G. (1978). A model for association in bivariate life tables and its application in epidemiological studies of familial dependency in chronic disease incidence. *Biometrika* 65:141–151.

CLIFFORD, A. H. (1954). Naturally totally ordered commutative semigroups. *Amer. J. Math.* 76:631–646.

CLIFFORD, A. H. & PRESTON, G. B. (1961). *The algebraic theory of semigroups.* American Mathematical Society, Providence.

CLIMESCU, A. C. (1946). Sur l'équation fonctionelle de l'associativité. *Bull. École Polytechn. Iassy* 1:1–16.

COULON, J., COULON, J.-L., HÖHLE, U. (1992). Classification of extremal subobjects over **SM-SET**. In [Rodabaugh *et al.* 1992], pages 9–31.

CRAIGEN, R. & PÁLES, Z. (1989). The associativity equation revisited. *Aequationes Math.* 37:306–312.

CUADRAS, C. M. & AUGÉ, J. (1981). A continuous general multivariate distribution and its properties. *Comm. Statist. A—Theory Methods* 10:339–353.

CZOGAŁA, E. & DREWNIAK, J. (1984). Associative monotonic operations in fuzzy set theory. *Fuzzy Sets and Systems* 12:249–269.

CZOGAŁA, E. & PEDRYCZ, W. (1981a). On identifications in fuzzy systems and its application in control problems. *Fuzzy Sets and Systems* 6:73–83.

CZOGAŁA, E. & PEDRYCZ, W. (1981b). Some problems concerning the construction of algorithms of decision-making in fuzzy systems. *Intern. J. Man-Machine Stud.* 15:201–211.

DALL'AGLIO, G., KOTZ, S., SALINETTI, G., editors (1991). *Advances in Probability Distributions with Given Marginals.* Kluwer Academic Publishers, Dordrecht.

DARSOW, W. & FRANK, M. (1984). Associative functions and Abel-Schroeder systems. *Publ. Math. Debrecen* 31:253–272.

DARSOW, W., NGUYEN, B., OLSEN, E. T. (1992). Copulas and Markov processes. *Ill. J. Math.* 36:600–642.

DE BAETS, B. (1998a). *Personal communication.* Contained in electronic form in: Database entry A030453, *Sloane's On-Line Encyclopedia of Integer Sequences*, URL `http://www.research.att.com/~njas/sequences/index.html`.

DE BAETS, B. (1998b). Uninorms: the known classes. In Ruan, D., Abderrahim, H. A., D'hondt, P., Kerre, E. E., editors, *Fuzzy Logic and Intelligent Technologies for Nuclear Science and Industry. Proceedings of the Third International FLINS Workshop*, pages 21–28. World Scientific, Singapore.

DE BAETS, B. (1999). Idempotent uninorms. *Europ. J. Oper. Research* 118:631–642.

DE BAETS, B. & FODOR, J. (1999). Residual operators of uninorms. *Soft Computing* 3:89–100.

DE BAETS, B. & MARKOVÁ-STUPŇANOVÁ, A. (1997). Analytical expressions for the addition of fuzzy intervals. *Fuzzy Sets and Systems* 91:203–213.

DE BAETS, B. & MESIAR, R. (1997). Pseudo-metrics and T-equivalences. *J. Fuzzy Math.* 5:471–481.

DE BAETS, B. & MESIAR, R. (1998). $\mathcal{T}$-partitions. *Fuzzy Sets and Systems* 97:211–223.

DE BAETS, B. & MESIAR, R. (1999). Triangular norms on product lattices. *Fuzzy Sets and Systems* 104:61–75.

DE GLAS, M. (1984). Representation of Łukasiewicz' many-valued algebras. The atomic case. *Fuzzy Sets and Systems* 14:175–185.

DEMANT, B. (1994). Deformationen von t-Normen, ihre Symmetrien und Symmetriebrechungen. In *Proceedings Workshop Fuzzy Systeme '94. Klassifikation, Entscheidungssupport & Control, München*. Deutsche Informatik-Akademie, Bonn. 11 pages.

DEMPSTER, A. P. (1967). Upper and lower probabilities induced by a multivalued mapping. *Ann. Math. Stat.* 38:325–339.

DENNEBERG, D. (1994). *Non-additive Measure and Integral*. Kluwer Academic Publishers, Dordrecht.

DI NOLA, A. (1991). Representation and reticulation by quotients of MV algebras. *Ricerche Mat.* 40:291–297.

DI NOLA, A., PEDRYCZ, W., SESSA, S. (1985). On measures of fuzziness of solutions of fuzzy relation equations with generalized connectives. *J. Math. Anal. Appl.* 121:443–453.

DI NOLA, A., SESSA, S., PEDRYCZ, W., SANCHEZ, E. (1989). *Fuzzy Relation Equations and Their Applications to Knowledge Engineering*. Kluwer Academic Publishers, Dordrecht.

DOETSCH, G. (1950–1956). *Handbuch der Laplace-Transformation*, volume I–III. Birkhäuser, Basel.

DOMBI, J. (1982a). Basic concepts for a theory of evaluation: The aggregative operator. *Europ. J. Oper. Research* 10:282–293.

DOMBI, J. (1982b). A general class of fuzzy operators, the De Morgan class of fuzzy operators and fuzziness measures induced by fuzzy operators. *Fuzzy Sets and Systems* 8:149–163.

DOMBI, J. (1986). Properties of the fuzzy connectives in light of the general representations theorem. *Acta Cybernet.* 7:313–321.

DREWNOWSKI, L. (1972). Topological rings of sets, continuous set functions, integration, I, II, III. *Bull. Acad. Polon. Sci. Ser. Math. Astronom. Phys.* 20:269–276, 277–286, 439–445.

DRIANKOV, D., HELLENDOORN, H., REINFRANK, M. (1993). *An Introduction to Fuzzy Control*. Springer, Berlin.

DROSSOS, C. & NAVARA, M. (1996). Generalized t-conorms and closure operators. In *Proceedings EUFIT '96, Aachen*, pages 22–26.

DROSSOS, C. A. (1977). Stochastic Menger spaces and convergence in probability. *Rev. Roumaine Math. Pures Appl.* 22:1069–1076.

DUBOIS, D. (1980). Triangular norms for fuzzy sets. In Klement, E. P., editor, *Proceedings Second International Seminar on Fuzzy Set Theory, Linz*, pages 39–68.

DUBOIS, D., GRABISCH, M., PRADE, H. (1992). Gradual rules and the approximation of functions. In *Proceedings 2nd International Conference on Fuzzy Logic and Neural Networks, Iizuka*, pages 629–632.

DUBOIS, D., KERRE, E., MESIAR, R., PRADE, H. (1999a). Fuzzy interval analysis. In [Dubois & Prade 1999], pages 483–582.

DUBOIS, D., KLEMENT, E. P., PRADE, H., editors (1999b). *Fuzzy Sets, Logics and Reasoning about Knowledge*, volume 15 of *Applied Logic Series*. Kluwer Academic Publishers, Dordrecht.

DUBOIS, D. & PRADE, H. (1980a). *Fuzzy Sets and Systems: Theory and Applications*. Academic Press, New York.

DUBOIS, D. & PRADE, H. (1980b). New results about properties and semantics of fuzzy set-theoretic operators. In Wang, P. & Chang, S., editors, *Fuzzy Sets: Theory and Applications to Policy Analysis and Information Systems*, pages 59–75. Plenum Press, New York.

DUBOIS, D. & PRADE, H. (1981). Additions of interactive fuzzy numbers. *IEEE Trans. Automat. Control* 26:926–936.

DUBOIS, D. & PRADE, H. (1982a). A class of fuzzy measures based on triangular norms — a general framework for the combination of uncertain information. *Internat. J. Gen. Systems* 8:43–61.

DUBOIS, D. & PRADE, H. (1982b). On several representations of uncertain body of evidence. In Gupta, M. M. & Sanchez, E., editors, *Fuzzy Information and Decision Processes*, pages 167–181. North-Holland, Amsterdam.

DUBOIS, D. & PRADE, H. (1984). A theorem on implication functions defined from triangular norms. *Stochastica* 8:267–279.

DUBOIS, D. & PRADE, H. (1985). A review of fuzzy set aggregation connectives. *Inform. Sci.* 36:85–121.

DUBOIS, D. & PRADE, H. (1987). Fuzzy numbers: An overview. In Bezdek, J., editor, *Analysis of Fuzzy Information — Vol. I: Mathematics and Logic*, pages 3–39. CRC Press, Boca Raton.

DUBOIS, D. & PRADE, H. (1988). *Possibility Theory. An Approach to Computerized Processing of Uncertainty.* Plenum Press, New York.

DUBOIS, D. & PRADE, H. (1991a). Fuzzy sets in approximate reasoning — Part 1: Inference with possibility distributions. *Fuzzy Sets and Systems* 40:143–202.

DUBOIS, D. & PRADE, H. (1991b). Fuzzy sets in approximate reasoning — Part 2: Logical approaches. *Fuzzy Sets and Systems* 40:203–244.

DUBOIS, D. & PRADE, H. (1994). Similarity-based approximate reasoning. In Zaruda, J., Marks, R., Robinson, C., editors, *Computational Intelligence Imitating Life*, pages 3–39. IEEE Press, New York.

DUBOIS, D. & PRADE, H., editors (1999). *Fundamentals of Fuzzy Sets.* The Handbook of Fuzzy Sets Series. Kluwer Academic Publishers, Boston.

DUGUNDJI, J. (1966). *Topology.* Allyn & Bacon, Boston.

DUMITRESCU, D. (1992). Fuzzy partitions with the connectives T_∞, S_∞. *Fuzzy Sets and Systems* 47:193–195.

DUMITRESCU, D. (1993). Fuzzy measures and the entropy of fuzzy partitions. *J. Math. Anal. Appl.* 176:359–373.

DUMMETT, M. (1959). A propositional calculus with denumerable matrix. *J. Symb. Logic* 24:97–106.

DVORETZKY, A., WALD, A., WOLFOWITZ, J. (1951). Relations among certain ranges of vector measures. *Pacific J. Math.* 1:59–74.

ESTEVA, F., GODO, L., HÁJEK, P., NAVARA, M. (200x). Residuated fuzzy logics with an involutive negation. *Arch. Math. Logic* (in press).

EYRAUD, H. (1938). Les principes de la mesure des corrélations. *Ann. Univ. Lyon, Sect. A* 1:30–47.

FALCONER, K. (1990). *Fractal Geometry.* J. Wiley & Sons, New York.

FARLIE, D. J. G. (1960). The performance of some correlation coefficients for a general bivariate distribution. *Biometrika* 47:307–323.

FAUCETT, W. M. (1955). Compact semigroups irreducibly connected between two idempotents. *Proc. Amer. Math. Soc.* 6:741–747.

FENCHEL, W. (1949). On the conjugate convex function. *Can. J. Math.* 1:73–77.

FÉRON, R. (1956). Sur les tableaux de corrélation dont les marges sont données. *Publ. Inst. Stat. Univ. Paris* 5:3–12.

FLEGEL, B. (1998). Fuzzy-Spieltheorie. Master's thesis, Technische Universität Braunschweig.

FODOR, J. C. (1991). On fuzzy implication operators. *Fuzzy Sets and Systems* 42:293–300.

FODOR, J. C. (1993a). Fuzzy connectives via matrix logic. *Fuzzy Sets and Systems* 56:67–77.

FODOR, J. C. (1993b). A new look at fuzzy connectives. *Fuzzy Sets and Systems* 57:141–148.

FODOR, J. C. (1995). Contrapositive symmetry of fuzzy implications. *Fuzzy Sets and Systems* 69:141–156.

FODOR, J. C. (1996). An extension of Fung-Fu's theorem. *Internat. J. Uncertain. Fuzziness Knowledge-Based Systems* 4:235–243.

FODOR, J. C. & JENEI, S. (1999). On reversible triangular norms. *Fuzzy Sets and Systems* 104:43–51.

FODOR, J. C. & KERESZTFALVI, T. (1994). A characterization of the Hamacher family of t-norms. *Fuzzy Sets and Systems* 65:51–58.

FODOR, J. C. & ROUBENS, M. (1994). *Fuzzy Preference Modelling and Multicriteria Decision Support.* Kluwer Academic Publishers, Dordrecht.

FODOR, J. C., YAGER, R. R., RYBALOV, A. (1997). Structure of uninorms. *Internat. J. Uncertain. Fuzziness Knowledge-Based Systems* 5:411–427.

FRANK, M. J. (1979). On the simultaneous associativity of $F(x, y)$ and $x + y - F(x, y)$. *Aequationes Math.* 19:194–226.

FRÉCHET, M. (1906). Sur quelques points du calcul fonctionnel. *Rend. Circ. Mat. Palermo* 22:1–74.

FRÉCHET, M. (1951). Sur les tableaux de corrélation dont les marges sont données. *Ann. Univ. Lyon, Sect. A* 9:53–77.

FUCHS, L. (1963). *Partially Ordered Algebraic Systems*. Pergamon Press, Oxford.

FULLÉR, R. & KERESZTFALVI, T. (1991). On generalization of Nguyen's theorem. *Fuzzy Sets and Systems* 41:371–374.

FULLÉR, R. & KERESZTFALVI, T. (1992). t-Norm based addition of fuzzy intervals. *Fuzzy Sets and Systems* 51:155–159.

FULLÉR, R. & ZIMMERMANN, H.-J. (1992). On computation of the compositional rule of inference under triangular norms. *Fuzzy Sets and Systems* 51:267–275.

FUNG, L. W. & FU, K. S. (1975). An axiomatic approach to rational decision making in a fuzzy environment. In Zadeh, L. A., Fu, K. S., Tanaka, K., Shimura, M., editors, *Fuzzy Sets and Their Applications to Cognitive and Decision Processes*, pages 227–256. Academic Press, New York.

GENEST, C. (1987). Frank's family of bivariate distributions. *Biometrika* 74:549–555.

GENEST, C. & RIVEST, L.-P. (1989). A characterization of Gumbel's family of extreme value distributions. *Statist. Probab. Lett.* 8:207–211.

GIERZ, G., HOFMANN, K. H., KEIMEL, K., LAWSON, J. D., MISLOVE, M., SCOTT, D. S. (1980). *A Compendium of Continuous Lattices*. Springer, Berlin.

GILES, R. (1976). Łukasiewicz logic and fuzzy set theory. *Intern. J. Man-Machine Stud.* 8:313–327.

GIRARD, J. Y. (1987). Linear logic. *Theor. Comp. Sci.* 50:1–102.

GÖDEL, K. (1932). Zum intuitionistischen Aussagenkalkül. *Anz. Österr. Akad. Wiss. Math.-Natur. Kl.* 69:65–66.

GODO, L. & SIERRA, C. (1988). A new approach to connective generation in the framework of expert systems using fuzzy logic. In *Proceedings 18th International Symposium on Multiple-Valued Logic, Palma de Mallorca*, pages 157–162. IEEE Computer Society Press.

GOGUEN, J. A. (1968/69). The logic of inexact concepts. *Synthese* 19:325–373.

GOLAN, J. S. (1992). *The Theory of Semirings with Applications in Mathematics and Theoretical Computer Science*. Longman, Essex.

GOTTWALD, S. (1976a). Untersuchungen zur mehrwertigen Mengenlehre. I. *Math. Nachr.* 72:297–303.

GOTTWALD, S. (1976b). Untersuchungen zur mehrwertigen Mengenlehre. II. *Math. Nachr.* 74:329–336.

GOTTWALD, S. (1977). Untersuchungen zur mehrwertigen Mengenlehre. III. *Math. Nachr.* 79:207–217.

GOTTWALD, S. (1984a). Criteria for non-interactivity of fuzzy logic controller rules. In Straszak, H., editor, *Large Scale Systems: Theory and Applications 1983*, volume 10 of *IFAC Proceedings Series*, pages 229–233. Pergamon Press, Oxford.

GOTTWALD, S. (1984b). On the existence of solutions of systems of fuzzy equations. *Fuzzy Sets and Systems* 12:301–302.

GOTTWALD, S. (1986). Characterizations of the solvability of fuzzy equations. *Elektron. Informationsverarb. Kybernet.* 22:67–91.

GOTTWALD, S. (1989). *Mehrwertige Logik. Eine Einführung in Theorie und Anwendungen*. Akademie-Verlag, Berlin.

GOTTWALD, S. (1993). *Fuzzy Sets and Fuzzy Logic. Foundations of Application — from a Mathematical Point of View*. Vieweg, Braunschweig/Wiesbaden.

GOTTWALD, S. (1995). Approximate solutions of fuzzy relational equations and a characterization of t-norms that define metrics for fuzzy sets. *Fuzzy Sets and Systems* 75:189–201.

GOTTWALD, S. (200x). *A Treatise on Many-Valued Logic*. Monograph (in preparation).

GRABISCH, M. (1995). Fuzzy integral in multicriteria decision making. *Fuzzy Sets and Systems* 69:279–298.

GRABISCH, M., MUROFUSHI, T., SUGENO, M., editors (2000). *Fuzzy Measures and Integrals. Theory and Applications*, volume 40 of *Studies in Fuzziness and Soft Computing*. Physica-Verlag, Heidelberg.

GRABISCH, M., NGUYEN, H. T., WALKER, E. A. (1995). *Fundamentals of Uncertainty Calculi with Applications to Fuzzy Inference*. Kluwer Academic Publishers, Dordrecht.

GUIAŞU, S. (1977). *Information Theory with Applications*. McGraw-Hill, New York.

GUMBEL, E. J. (1958). Distributions a plusieurs variables dont les marges sont données. *C. R. Acad. Sci. Paris Sér. A* 246:2717–2719.

GUMBEL, E. J. (1960). Distributions des valeurs extrêmes en plusieurs dimensions. *Publ. Inst. Statist. Univ. Paris* 9:171–173.

HADŽIĆ, O. (1978). On the (ϵ, λ)-topology of probabilistic locally convex spaces. *Glas. Mat.* 13:61–66.

HADŽIĆ, O. (1982). Some theorems on the fixed points in probabilistic metric and random normed spaces. *Boll. Un. Mat. Ital. B(7)* 1:381–391.

HÁJEK, P. (1995a). Fuzzy logic and arithmetical hierarchy. *Fuzzy Sets and Systems* 73:359–363.

HÁJEK, P. (1995b). Fuzzy logic from the logical point of view. In Bartosek, M., Staudek, J., Wiedermann, J., editors, *Theory and Practice of Informatics*, volume 1012 of *Lecture Notes in Computer Science*, pages 31–49. Springer, Berlin.

HÁJEK, P. (1997). Fuzzy logic and arithmetical hierarchy II. *Studia Logica* 58:129–141.

HÁJEK, P. (1998a). Basic fuzzy logic and BL-algebras. *Soft Computing* 2:124–128.

HÁJEK, P. (1998b). *Metamathematics of Fuzzy Logic*. Kluwer Academic Publishers, Dordrecht.

HÁJEK, P., GODO, L., ESTEVA, F. (1996). A complete many-valued logic with product-conjunction. *Arch. Math. Logic* 35:191–208.

HÁJEK, P., HAVRÁNEK, T., JIROUŠEK, R. (1992). *Uncertain Information Processing in Expert Systems*. CRC Press, Boca Raton.

HALMOS, P. R. (1947). On the set of values of a finite measure. *Bull. Amer. Math. Soc.* 53:138–141.

HALMOS, P. R. (1948). The range of a vector measure. *Bull. Amer. Math. Soc.* 54:416–421.

HALMOS, P. R. (1950). *Measure Theory*. Van Nostrand Reinhold, New York.

HAMACHER, H. (1975). Über logische Verknüpfungen unscharfer Aussagen und deren zugehörige Bewertungsfunktionen. In *Progress in Cybernetics and Systems Research*, pages 276–287. Hemisphere Publ. Comp., New York.

HAMACHER, H. (1978). *Über logische Aggregationen nicht-binär explizierter Entscheidungskriterien*. Rita G. Fischer Verlag, Frankfurt.

HARDY, G. H., LITTLEWOOD, J. E., PÓLYA, G. (1952). *Inequalities*. Cambridge University Press, Cambridge.

HARDY, G. H. & WRIGHT, E. M. (1979). *An Introduction to the Theory of Numbers*. Oxford University Press, Oxford.

HAUSDORFF, F. (1914). *Grundzüge der Mengenlehre*. Veit und Comp., Leipzig.

HERRERA, F. & VERDEGAY, J. L., editors (1996). *Genetic Algorithms and Soft Computing*, volume 8 of *Studies in Fuzziness and Soft Computing*. Physica-Verlag, Heidelberg.

HEWITT, E. & STROMBERG, K. (1965). *Real and Abstract Analysis*. Springer, New York.

HEYTING, A. (1930). Die formalen Regeln der intuitionistischen Logik. *Sitzungsberichte Preußische Akademie der Wissenschaften Berlin, Physikal.-Mathemat. Kl.* II:42–56.

HICKS, T. L. (1983). Fixed point theory in probabilistic metric spaces. *Univ. u Novom Sadu Zb. Rad. Prirod.-Mat. Fak. Ser. Mat.* 13:63–72.

HOEFFDING, W. (1940). Maßstabinvariante Korrelationstheorie. *Schriften des Mathematischen Instituts und des Instituts für Angewandte Mathematik der Universität Berlin* 5(3):179–233. (Reprinted as "Scale-invariant correlation theory" in Fisher, N. I. and Sen, P. K., editors, *The Collected Works of Wassily Hoeffding*, pages 57–107. Springer, New York).

HOEFFDING, W. (1941). Maßstabinvariante Korrelationsmaße für diskontinuierliche Verteilungen. *Archiv für mathematische Wirtschafts- und Sozialforschung* 7:49–70. (Reprinted as "Scale-invariant correlation measures for discontinuous distributions" in Fisher, N. I. and Sen, P. K., editors, *The Collected Works of Wassily Hoeffding*, pages 109–133. Springer, New York).

HOERMANDER, L. (1983–1985). *The Analysis of Linear Partial Differential Operators*, volume I–IV. Springer, Berlin.

HÖHLE, U. (1976). Maße auf unscharfen Mengen. *Z. Wahrsch. Verw. Geb.* 36:179–188.

HÖHLE, U. (1978a). Probabilistisch kompakte L-unscharfe mengen. *Manuscripta Math.* 26:345–356.

HÖHLE, U. (1978b). Probabilistische Metriken auf der Menge der nicht negativen Verteilungsfunktionen. *Aequationes Math.* 18:345–356.

HÖHLE, U. (1978c). Probabilistische Topologien. *Manuscripta Math.* 26:223–245.

HÖHLE, U. (1981). Representation theorems for L-fuzzy quantities. *Fuzzy Sets and Systems* 5:83–107.

HÖHLE, U. (1982). Probabilistic metrization of fuzzy topologies. *Fuzzy Sets and Systems* 8:63–69.

HÖHLE, U. (1990). The Poincaré paradox and the cluster problem. In Dress, A. & von Haeseler, A., editors, *Trees and Hierarchical Structures*, volume 84 of *Lecture Notes in Biomathematics*, pages 117–124. Springer, Berlin.

HÖHLE, U. (1992). M-valued sets and sheaves over integral commutative CL-monoids. In [Rodabaugh *et al.* 1992], pages 33–72.

HÖHLE, U. (1993). Fuzzy equalities and indistinguishability. In *Proceedings EUFIT '93, Aachen*, volume 1, pages 358–363.

HÖHLE, U. (1994). Monoidal logic. In [Kruse *et al.* 1994b], pages 233–243.

HÖHLE, U. (1995). Commutative, residuated l-monoids. In [Höhle & Klement 1995], pages 53–106.

HÖHLE, U. (1996). On the fundamentals of fuzzy set theory. *J. Math. Anal. Appl.* 201:786–826.

HÖHLE, U. (1998). Many-valued equalities, singletons and fuzzy partitions. *Soft Computing* 2:134–140.

HÖHLE, U. (1999). The Poincaré paradox and non-classical logics. In [Dubois *et al.* 1999b], pages 7–16.

HÖHLE, U. & KLEMENT, E. P. (1983). Plausibility measures — a general framework for possibility and fuzzy probability measures. In [Skala *et al.* 1983], pages 31–50.

HÖHLE, U. & KLEMENT, E. P., editors (1995). *Non-Classical Logics and Their Applications to Fuzzy Subsets. A Handbook of the Mathematical Foundations of Fuzzy Set Theory.* Kluwer Academic Publishers, Dordrecht.

HÖHLE, U. & RODABAUGH, S. E., editors (1999). *Mathematics of Fuzzy Sets. Logic, Topology, and Measure Theory.* The Handbook of Fuzzy Sets Series. Kluwer Academic Publishers, Boston.

HÖHLE, U. & ŠOSTAK, A. P. (1999). Axiomatic foundations of fixed-basis fuzzy topology. In [Höhle & Rodabaugh 1999], chapter 3, pages 123–272.

HÖHLE, U. & STOUT, L. N. (1991). Foundations of fuzzy sets. *Fuzzy Sets and Systems* 40:257–296.

HÖHLE, U. & WEBER, S. (1997). Uncertainty measures, realizations and entropies. In Goutaias, J., Mahler, R., Nguyen, H., editors, *Random Sets: Theory and Applications*, pages 259–295. Springer, Heidelberg.

HÖHLE, U. & WEBER, S. (1999). On conditioning operators. In [Höhle & Rodabaugh 1999], chapter 12, pages 653–673.

HÖLDER, O. (1901). Die Axiome der Quantität und die Lehre vom Maß. *Leipzig Ber., Math.-Phys. Cl.* 53:1–64.

HOLMBLAD, L. P. & ØSTERGAARD, J. J. (1982). Control of a cement kiln by fuzzy logic. In Gupta, M. & Sanchez, E., editors, *Fuzzy Information and Decision Processes*, pages 389–399. North-Holland, Amsterdam.

HOSSZÚ, M. (1954). Some functional equations related with the associativity law. *Publ. Math. Debrecen* 3:205–214.

HOUGAARD, P. (1986). A class of multivariate failure time distributions. *Biometrika* 73:671–678.

HUTCHINSON, T. P. & LAI, C. D. (1990). *Continuous Bivariate Distributions, Emphasising Applications.* Rumsby Scientific Publishing, Adelaide.

HUTTON, B. (1975). Normality in fuzzy topological spaces. *J. Math. Anal. Appl.* 130:354–364.

IMAOKA, H. (1995). A propsal of Opposite-Sugeno-integral and a uniform expression of fuzzy integrals. In *Proceedings of the International Joint Conference of the Fourth IEEE International Conference on Fuzzy Systems and the Second International Fuzzy Engineering Symposium, Yokohama*, volume II, pages 583–590. IEEE Press.

IMAOKA, H. (1997). On a subjective evaluation model by a generalized fuzzy integral. *Internat. J. Uncertain. Fuzziness Knowledge-Based Systems* 5:517–529.

JACAS, J. & RECASENS, J. (1992). Eigenvectors and generators of fuzzy relations. In *Proceedings IEEE International Conference on Fuzzy Systems 1992, San Diego*, pages 687–694. IEEE Press, Piscataway.

JENEI, S. (1998). On Archimedean triangular norms. *Fuzzy Sets and Systems* 99:179–186.

JENEI, S. (1999). Fibred triangular norms. *Fuzzy Sets and Systems* 103:67–82.

JENEI, S. (2000). New family of triangular norms via contrapositive symmetrization of residuated implications. *Fuzzy Sets and Systems* 110:157–174.

JENEI, S. (200w). Continuity of left-continuous triangular norms with strong induced negations and their boundary conditions. *Fuzzy Sets and Systems* (in press).

JENEI, S. (200x). Structure of Girard monoids on $[0,1]$. In [Rodabaugh & Klement 200x]. (In preparation).

JENEI, S. (200y). Geometry of left-continuous t-norms with strong induced negations. *Belg. J. Oper. Res. Statist. Comput. Sci.* (in press).

JENEI, S. (200z). Structure of left-continuous triangular norms with strong induced negations. (I) Rotation construction. *J. Appl. Non-Classical Logics* (in press).

JENEI, S. & FODOR, J. C. (1998). On continuous triangular norms. *Fuzzy Sets and Systems* 100:273–282.

KACPRZYK, J. & FEDRIZZI, M., editors (1990). *Multiperson Decision Making Models Using Fuzzy Sets and Possibility Theory*. Kluwer Academic Publishers, Dordrecht.

KAMPÉ DE FÉRIET, J. (1969). Mesures de l'information par un ensemble d'observateurs. *C. R. Acad. Sci. Paris Sér. A* 269:1081–1085.

KAMPÉ DE FÉRIET, J. & BENVENUTI, P. (1969). Sur une classe d'informations. *C. R. Acad. Sci. Paris Sér. A* 269:97–101.

KAMPÉ DE FÉRIET, J. & FORTE, B. (1967). Information et probabilité. *C. R. Acad. Sci. Paris Sér. A* 265:110–114, 142–146, 350–353.

KAMPÉ DE FÉRIET, J., FORTE, B., BENVENUTI, P. (1969). Forme génerale de l'opération de composition continue d'une information. *C. R. Acad. Sci. Paris Sér. A* 269:529–534.

KAPPOS, D. A. (1960). *Strukturtheorie der Wahrscheinlichkeits-Felder und Räume*. Springer, Berlin.

KAUFMANN, A. & GUPTA, M. M. (1991). *Introduction to Fuzzy Arithmetic. Theory and Applications*. International Thomson Computer Press, London.

KERESZTFALVI, T. & KOVÁCS, M. (1992). g,p-fuzzification of arithmetic operations. *Tatra Mt. Math. Publ.* 1:65–71.

KICKERT, W. J. M. (1978). *Fuzzy Theories on Decision-Making*, volume 3 of *Frontiers in Systems Research*. Martinus Nijhoff Social Science Division, Leiden.

KIMBERLING, C. (1973). On a class of associative functions. *Publ. Math. Debrecen* 20:21–39.

KIMBERLING, C. H. (1974). A probabilistic interpretation of complete monotonicity. *Aequationes Math.* 10:152–164.

KLAUA, D. (1965). Über einen Ansatz zur mehrwertigen Mengenlehre. *Monatsb. Deutsch. Akad. Wiss. (Berlin)* 7:859–867.

KLAUA, D. (1967). Einbettung der klassischen Mengenlehre in die mehrwertige. *Monatsb. Deutsch. Akad. Wiss. (Berlin)* 9:258–272.

KLAWONN, F. (1993). Mamdani's model in the view of equality relations. In *Proceedings EUFIT '93, Aachen*, volume 1, pages 364–369.

KLAWONN, F. & CASTRO, J. L. (1995). Similarity in fuzzy reasoning. *Mathware Soft Comput.* 3:197–228.

KLAWONN, F. & KRUSE, R. (1993). Equality relations as a basis for fuzzy control. *Fuzzy Sets and Systems* 54:147–156.

KLEMENT, E. P. (1980a). Characterization of finite fuzzy measures using Markoff-kernels. *J. Math. Anal. Appl.* 75:330–339.

KLEMENT, E. P. (1980b). Fuzzy σ-algebras and fuzzy measurable functions. *Fuzzy Sets and Systems* 4:83–93.

KLEMENT, E. P. (1980c). Some remarks on t-norms, fuzzy σ-algebras and fuzzy measures. In Klement, E. P., editor, *Proceedings Second International Seminar on Fuzzy Set Theory, Linz*, pages 125–142.

KLEMENT, E. P. (1981). Operations on fuzzy sets and fuzzy numbers related to triangular norms. In *Proceedings Eleventh International Symposium on Multiple-Valued Logic, Norman*, pages 218–225. IEEE Press, New York.

KLEMENT, E. P. (1982a). Characterization of fuzzy measures constructed by means of triangular norms. *J. Math. Anal. Appl.* 86:345–358.

KLEMENT, E. P. (1982b). Construction of fuzzy σ-algebras using triangular norms. *J. Math. Anal. Appl.* 85:543–565.

KLEMENT, E. P. (1982c). Operations on fuzzy sets — an axiomatic approach. *Inform. Sci.* 27:221–232.

KLEMENT, E. P. (1983). Fuzzy measures assuming their values in the set of fuzzy numbers. *J. Math. Anal. Appl.* 93:312–323.

KLEMENT, E. P. (1985). Integration of fuzzy valued functions. *Rev. Roumaine Math. Pures Appl.* 30:375–384.

KLEMENT, E. P. (1987). Strong law of large numbers for random variables with values in the fuzzy real line. *Communications IFSA Mathematics Chapter* 1:7–11.

KLEMENT, E. P., LOWEN, R., SCHWYHLA, W. (1981). Fuzzy probability measures. *Fuzzy Sets and Systems* 5:21–30.

KLEMENT, E. P., MESIAR, R., PAP, E. (1996). On the relationship of associative compensatory operators to triangular norms and conorms. *Internat. J. Uncertain. Fuzziness Knowledge-Based Systems* 4:129–144.

KLEMENT, E. P., MESIAR, R., PAP, E. (1997). A characterization of the ordering of continuous t-norms. *Fuzzy Sets and Systems* 86:189–195.

KLEMENT, E. P., MESIAR, R., PAP, E. (1999). Quasi- and pseudo-inverses of monotone functions, and the construction of t-norms. *Fuzzy Sets and Systems* 104:3–13.

KLEMENT, E. P., MESIAR, R., PAP, E. (200x). Integration with respect to decomposable measures, based on a conditionally distributive semiring on the unit interval. *Internat. J. Uncertain. Fuzziness Knowledge-Based Systems* (in press).

KLEMENT, E. P. & NAVARA, M. (1997). A characterization of tribes with respect to the Łukasiewicz t-norm. *Czechoslovak Math. J.* 47(122):689–700.

KLEMENT, E. P. & NAVARA, M. (1999a). Propositional fuzzy logics based on Frank t-norms: A comparison. In [Dubois *et al.* 1999b], pages 17–38.

KLEMENT, E. P. & NAVARA, M. (1999b). A survey on different triangular norm-based fuzzy logics. *Fuzzy Sets and Systems* 101:241–251.

KLEMENT, E. P. & WEBER, S. (1991). Generalized measures. *Fuzzy Sets and Systems* 40:375–394.

KLEMENT, E. P. & WEBER, S. (1994). An integral representation for decomposable measures of measurable functions. *Aequationes Math.* 47:255–262.

KLEMENT, E. P. & WEBER, S. (1999). Fundamentals of a generalized measure theory. In [Höhle & Rodabaugh 1999], chapter 11, pages 633–651.

KOLESÁROVÁ, A. (1995). Triangular norm-based addition of linear fuzzy numbers. *Tatra Mt. Math. Publ.* 6:75–81.

KOLESÁROVÁ, A. (1997a). On the integral representation of possibility measures of fuzzy events. *J. Fuzzy Math.* 5:759–766.

KOLESÁROVÁ, A. (1997b). Similarity preserving t-norm-based additions of fuzzy numbers. *Fuzzy Sets and Systems* 91:215–229.

KOLESÁROVÁ, A. (1998). Triangular norm-based addition preserving linearity of *t*-sums of linear intervals. *Mathware Soft Comput.* 5:91–98.

KOLESÁROVÁ, A. (1999). A note on Archimedean triangular norms. *BUSEFAL* 80:57–60.

KOLESÁROVÁ, A. & KOMORNÍKOVÁ, M. (1999). Triangular norm-based iterative compensatory operators. *Fuzzy Sets and Systems* 104:109–120.

KOLOKOLTSOV, V. N. & MASLOV, V. P. (1997). *Idempotent Analysis and Its Applications.* Kluwer Academic Publishers, Dordrecht.

KOSKO, B. (1991). *Neural Networks and Fuzzy Systems.* Prentice Hall, Englewood Cliffs.

KRAUSE, G. M. (1981). *A strengthened form of Ling's theorem on associative functions.* PhD thesis, Illinois Institute of Technology, Chicago.

KRAUSE, G. M. (1983). Interior idempotents and non-representability of groupoids. *Stochastica* 7:5–10.

KRAUSE, G. M. (200x). The devil's terraces: a discontinuous associative function. Submitted for publication.

KRUSE, R., GEBHARDT, J., KLAWONN, F. (1994a). *Foundations of Fuzzy Systems*. J. Wiley & Sons, Chichester.

KRUSE, R., GEBHARDT, J., PALM, R., editors (1994b). *Fuzzy Systems in Computer Science*. Vieweg, Braunschweig.

KUCZMA, M. (1963). *On the Schröder equation*. Rozprawy Math. XXXIV. Panstowowe Wydavnictwo Naukowa, Warsaw.

KUICH, W. & SALOMAA, A. (1986). *Semirings, Automata, Languages*. Springer, Berlin.

LI, H., MIKUSIŃSKI, P., SHERWOOD, H., TAYLOR, M. D. (1997). On approximations of copulas. In Beneš, V. & Štěpán, J., editors, *Distributions with given Marginals and Moment Problems*, pages 107–116. Kluwer Academic Publishers, Dordrecht.

LIAPOUNOFF, A. A. (1940). Sur les fonctions-vecteurs completement additives (in Russian). *Bull. Acad. Sci. URSS Ser. Math.* 4:465–478.

LIDL, R. & PILZ, G. (1998). *Applied Abstract Algebra*. Springer, New York.

LINDENSTRAUSS, J. (1966). A short proof of Liapounoff's convexity theorem. *J. Math. Mech.* 15:971–973.

LING, C. M. (1965). Representation of associative functions. *Publ. Math. Debrecen* 12:189–212.

LOWEN, R. (1976). Fuzzy topological spaces and fuzzy compactness. *J. Math. Anal. Appl.* 56:621–633.

LOWEN, R. (1978). On fuzzy complements. *Inform. Sci.* 14:107–113.

LOWEN, R. (1980). Convex fuzzy sets. *Fuzzy Sets and Systems* 3:291–310.

LOWEN, R. (1983). On $(\mathbb{R}(L), \oplus)$. *Fuzzy Sets and Systems* 10:203–209.

LOWEN, R. (1984a). Compactness properties in the fuzzy real line. *Fuzzy Sets and Systems* 13:193–200.

LOWEN, R. (1984b). Fuzzy integers, fuzzy rationals and other subspaces of the fuzzy real line. *Fuzzy Sets and Systems* 13:193–200.

LOWEN, R. (1984c). On the existence of natural fuzzy topologies on spaces of probability measures. *Math. Nachr.* 115:33–57.

LOWEN, R. (1985). The order aspect of the fuzzy real line. *Manuscripta Math.* 39:293–309.

LOWEN, R. (1986). A complete chartacterization of the level spaces of $\mathbf{R}(I)$ and $I(I)$. *Rocky Mountain J. Math.* 16:669–684.

LOWEN, R. (1996). *Fuzzy Set Theory. Basic Concepts, Techniques and Bibliography*. Kluwer Academic Publishers, Dordrecht.

LUHANDJULA, M. K. (1982). Compensatory operators in fuzzy linear programming with multiple objectives. *Fuzzy Sets and Systems* 8:245–252.

LUKASIEWICZ, J. (1920). O logice trówartosciowej. *Ruch Filozoficzny* 5:170–171. (English translation contained in [Borkowski 1970]).

LUKASIEWICZ, J. (1930). Philosophische Bemerkungen zu mehrwertigen Systemen des Aussagenkalküls. *Comptes Rendus Séances Société des Sciences et Lettres Varsovie cl. III* 23:51–77.

LUKASIEWICZ, J. & TARSKI, A. (1930). Untersuchungen über den Aussagenkalkül. *Comptes Rendus Séances Société des Sciences et Lettres Varsovie cl. III* 23:1–21.

MAMDANI, E. H. & ASSILIAN, S. (1975). An experiment in linguistic synthesis with a fuzzy logic controller. *Intern. J. Man-Machine Stud.* 7:1–13.

MAMDANI, E. H. & GAINES, B. (1981). *Fuzzy Reasoning and Its Applications*. Academic Press, New York.

MARCUS, S. (1962). Atomic measures and their Darboux property. *Rev. Roumaine Math. Pures Appl.* 2:327–333.

MARDIA, K. V. (1970). *Families of Bivariate Distributions*, volume 27 of *Griffin's Statistical Monographs and Courses*. Charles Griffin & Co., London.

MAREŠ, M. (1977a). How to handle fuzzy quantities? *Kybernetika* 13:23–40.

MAREŠ, M. (1977b). On fuzzy quantities with real and integer values. *Kybernetika* 13:41–56.

MAREŠ, M. (1989). Addition of fuzzy quantities: Disjunction-conjunction approach. *Kybernetika* 25:104–116.

MAREŠ, M. (1994). *Computation over Fuzzy Quantities*. CRC Press, Boca Raton.

MARKO, V. & MESIAR, R. (1999). A note on a nilpotent lower bound of nilpotent triangular norms. *Fuzzy Sets and Systems* 104:27–34.

MARKO, V. & MESIAR, R. (200x). Continuous Archimedean t-norms and their bounds. *Fuzzy Sets and Systems* (in press).

MARKOVÁ, A. (1996). A note on g-derivative and g-integral. *Tatra Mt. Math. Publ.* 8:71–76.

MARKOVÁ, A. (1997). T-sum of L-R-fuzzy numbers. *Fuzzy Sets and Systems* 85:379–384.

MARKOVÁ-STUPŇANOVÁ, A. (1999). A note on the idempotent functions with respect to pseudo-convolution. *Fuzzy Sets and Systems* 102:417–421.

MAS, G., MAYOR, G., TORRENS, J. (1999). t-operators. *Internat. J. Uncertain. Fuzziness Knowledge-Based Systems* 7:31–50.

MASLOV, V. P. & SAMBORSKIJ, S. N., editors (1992). *Idempotent Analysis*, volume 13 of *Advances in Soviet Mathematics*. American Mathematical Society, Providence.

MAYOR, G. (1994). Sugeno's negations and t-norms. *Mathware Soft Comput.* 1:93–98.

MAYOR, G. & TORRENS, J. (1991). On a family of t-norms. *Fuzzy Sets and Systems* 41:161–166.

MAYOR, G. & TORRENS, J. (1993). On a class of operators for expert systems. *Int. J. Intell. Syst.* 8:771–778.

MCNAUGHTON, R. (1951). A theorem about infinite-valued sentential logic. *J. Symb. Logic* 16:1–13.

MENGER, K. (1928). Untersuchungen über allgemeine Metrik. *Math. Ann.* 100:75–163.

MENGER, K. (1942). Statistical metrics. *Proc. Nat. Acad. Sci. U.S.A.* 8:535–537.

MENGER, K. (1951a). Ensembles flous et fonctions aleatoires. *C. R. Acad. Sci. Paris Sér. A* 232:2001–2003.

MENGER, K. (1951b). Probabilistic geometry. *Proc. Nat. Acad. Sci. U.S.A.* 37:226–229.

MENGER, K. (1951c). Probabilistic theories of relations. *Proc. Nat. Acad. Sci. U.S.A.* 37:178–180.

MENGER, K. (1979a). Geometry and positivism. A probabilistic microgeometry. In [Menger 1979b], chapter 20, pages 225–234.

MENGER, K. (1979b). *Selected Papers in Logic and Foundations, Didactics, Economics.* Reidel, Dordrecht.

MESIAR, R. (1992). A note to the characterization of possibility measures of fuzzy events using Markov-kernels. *Fuzzy Sets and Systems* 46:301–303.

MESIAR, R. (1993a). Fundamental triangular norm based tribes and measures. *J. Math. Anal. Appl.* 177:633–640.

MESIAR, R. (1993b). On the structure of T_s-tribes. *Tatra Mt. Math. Publ.* 3:167–172.

MESIAR, R. (1994). Fuzzy difference posets and MV-algebras. In *Proceedings IPMU '94, Paris*, pages 208–212.

MESIAR, R. (1995a). Choquet like integrals. *J. Math. Anal. Appl.* 194:477–488.

MESIAR, R. (1995b). On some constructions of new triangular norms. *Mathware Soft Comput.* 2:39–45.

MESIAR, R. (1996a). A note on the T-sum of L-R fuzzy numbers. *Fuzzy Sets and Systems* 79:259–261.

MESIAR, R. (1996b). Pseudo-linear integrals and derivatives based on a generator g. *Tatra Mt. Math. Publ.* 8:67–70.

MESIAR, R. (1997a). Shape preserving additions of fuzzy intervals. *Fuzzy Sets and Systems* 86:73–78.

MESIAR, R. (1997b). Triangular-norm-based addition of fuzzy intervals. *Fuzzy Sets and Systems* 91:231–237.

MESIAR, R. (1997c). Universal triangular conorms. In *Proceedings EUFIT '97, Aachen*, pages 44–47.

MESIAR, R. (1998a). Approximation of continuous t-norms by strict t-norms with smooth generators. *BUSEFAL* 75:72–79.

MESIAR, R. (1998b). On the pointwise convergence of continuous Archimedean t-norms and the convergence of their generators. *BUSEFAL* 75:39–45.

MESIAR, R. & KOMORNÍKOVÁ, M. (1997). Aggregation operators. In Herceg, D. & Surla, K., editors, *Proceedings PRIM '96, XI. Conference on Applied Mathematics*, pages 193–211.

MESIAR, R. & NAVARA, M. (1996). T_s-tribes and T_s-measures. *J. Math. Anal. Appl.* 201:91–102.

MESIAR, R. & NAVARA, M. (1999). Diagonals of continuous triangular norms. *Fuzzy Sets and Systems* 104:35–41.

MESIAR, R. & RYBÁRIK, J. (1995). Pan-operations structure. *Fuzzy Sets and Systems* 74:365–369.

MIYAKOSHI, M. & SHIMBO, M. (1985). Solutions of composite fuzzy relational equations with triangular norms. *Fuzzy Sets and Systems* 16:53–63.

MIZUMOTO, M. (1989a). Pictorial representations of fuzzy connectives, Part 1: Cases of t-norms, t-conorms and averaging operators. *Fuzzy Sets and Systems* 31:217–242.

MIZUMOTO, M. (1989b). Pictorial representations of fuzzy connectives, Part 2: Cases of compensatory operators and self-dual operators. *Fuzzy Sets and Systems* 32:45–79.

MOISIL, G. C. (1940). Recherches sur les logiques non chrysipiennes. *Annales Scientifiques de l'Université de Jassy* 26:431–466.

MOISIL, G. C. (1941). Notes sur les logiques non-chrysipiennes. *Annales Scientifiques de l'Université de Jassy* 27:86–98.

MORGENSTERN, D. (1956). Einfache Beispiele zweidimensionaler Verteilungen. *Mitteilungsblatt für Mathematische Statistik* 8:234–235.

MOSER, B. (1995). *A new approach for representing control surfaces by fuzzy rule bases.* PhD thesis, Johannes Kepler Universität, Linz.

MOSER, B., TSIPORKOVA, E., KLEMENT, E. P. (1999). Convex combinations in terms of triangular norms: A characterization of idempotent, bisymmetrical and self-dual compensatory operators. *Fuzzy Sets and Systems* 104:97–108.

MOSTERT, P. S. & SHIELDS, A. L. (1957). On the structure of semigroups on a compact manifold with boundary. *Ann. of Math.* 65:117–143.

MOYNIHAN, R. (1975). On the class of τ_T semigroups of probability distribution functions. *Aequationes Math.* 12:249–261.

MOYNIHAN, R. (1978a). Infinite τ_T products of probability distribution functions. *J. Austral. Math. Soc. (Ser. A)* 26:227–240.

MOYNIHAN, R. (1978b). On τ_T semigroups of probability distribution functions II. *Aequationes Math.* 17:19–40.

MOYNIHAN, R. & SCHWEIZER, B. (1979). Betweeness relations in PM-spaces. *Pacific J. Math.* 81:175–196.

MUNDICI, D. (1986). Interpretation of AF C^*-algebras in Łukasiewicz sentential calculus. *J. Funct. Anal.* 65:15–63.

MUNDICI, D. (1991). Averaging the truth value in Łukasiewicz logic. *Studia Logica* 5:113–127.

MUNDICI, D. (1993a). An elementary proof of Chang's completeness theorem for the infinite-valued calculus of Łukasiewicz logic, and AF C^*-algebras. *Studia Logica* 58:79–97.

MUNDICI, D. (1993b). Ulam games, Łukasiewicz logic, and AF C^*-algebras. *Fund. Inf.* 18:151–161.

MUROFUSHI, T. & SUGENO, M. (1991). Fuzzy t-conorm integrals with respect to fuzzy measures: generalization of Sugeno integral and Choquet integral. *Fuzzy Sets and Systems* 42:57–71.

MUŠTARI, D. H. & ŠERSTNEV, A. N. (1966). On methods of introducing a topology into random metric spaces. *Izv. Vysh. Mat.* 6(55):99–106.

NAVARA, M. (1993). A characterization of triangular norm based tribes. *Tatra Mt. Math. Publ.* 3:161–166.

NAVARA, M. (1999). Characterization of measures based on strict triangular norms. *J. Math. Anal. Appl.* 236:370–383.

NEGOITA, C. V. & RALESCU, D. A. (1975a). *Applications of Fuzzy Sets to Systems Analysis.* J. Wiley & Sons, New York.

NEGOITA, C. V. & RALESCU, D. A. (1975b). Representation theorems for fuzzy concepts. *Kybernetes* 4:169–174.

NELSEN, R. B. (1986). Properties of a one-parameter family of bivariate distributions with specified marginals. *Comm. Statist.—Theory Methods* 15:3277–3285.

NELSEN, R. B. (1991). Copulas and association. In [Dall'Aglio *et al.* 1991].

NELSEN, R. B. (1995). Copulas, characterization, correlation, and counterexamples. *Math. Mag.* 68:193–198.

NELSEN, R. B. (1999). *An Introduction to Copulas*, volume 139 of *Lecture Notes in Statistics*. Springer, New York.

NGUYEN, H. T. (1978). A note on the extension principle for fuzzy sets. *J. Math. Anal. Appl.* 64:369–380.

NGUYEN, H. T., KREINOVICH, V., WOJCIECHOWSKI, P. (1998). Strict Archimedean t-norms and t-conorms as universal approximators. *Internat. J. Approx. Reason.* 18:239–249.

NGUYEN, H. T. & SUGENO, M., editors (1998). *Fuzzy Systems: Modeling and Control*. The Handbook of Fuzzy Sets Series. Kluwer Academic Publishers, Boston.

NGUYEN, H. T., SUGENO, M., TONG, R., YAGER, R. R., editors (1995). *Theoretical Aspects of Fuzzy Control*. J. Wiley & Sons, New York.

NGUYEN, H. T. & WALKER, E. (1997). *A First Course in Fuzzy Logic*. CRC Press, Boca Raton.

NOVÁK, V. (1987). First-order fuzzy logic. *Studia Logica* 46:87–109.

NOVÁK, V., PERFILIEVA, I., MOČKOŘ, J. (1999). *Mathematical Principles of Fuzzy Logic*. Kluwer Academic Publishers, Norwell.

ORLICZ, W. (1968). On spaces $L^{*\phi}$ based on the notion of a finitely additive integral. *Comment. Math. Prace Mat.* 12:99–113.

OVCHINNIKOV, S. V. (1983). Representations of transitive fuzzy relations. In [Skala *et al.* 1983], pages 105–118.

OVCHINNIKOV, S. V. (1989). Transitive fuzzy orderings of fuzzy numbers. *Fuzzy Sets and Systems* 40:283–295.

OVCHINNIKOV, S. V. (1991). Similarity relations, fuzzy partitions, and fuzzy orderings. *Fuzzy Sets and Systems* 40:107–126.

OVCHINNIKOV, S. V. (1992). Aggregating transitive fuzzy preference relations. In *Proceedings IPMU '92, Cala Blava*, pages 457–460.

PAALMAN-DE MIRANDA, A. B. (1964). *Topological Semigroups*, volume 11 of *Mathematical Centre Tracts*. Matematisch Centrum, Amsterdam.

PAP, E. (1990a). An integral generated by a decomposable measure. *Univ. u Novom Sadu Zb. Rad. Prirod.-Mat. Fak. Ser. Mat.* 20(1):135–144.

PAP, E. (1990b). Lebesgue and Saks decompositions of $\perp$-decomposable measures. *Fuzzy Sets and Systems* 38:345–353.

PAP, E. (1991). On non-additive set functions. *Atti Sem. Mat. Fis. Univ. Modena* 39:345–360.

PAP, E. (1992). Decomposable measures and applications on nonlinear partial differential equations. *Rend. Circ. Mat. Palermo (2) Suppl.* 28:387–403.

PAP, E. (1993). g-calculus. *Univ. u Novom Sadu Zb. Rad. Prirod.-Mat. Fak. Ser. Mat.* 23(1):145–156.

PAP, E. (1995). *Null-Additive Set Functions*. Kluwer Academic Publishers, Dordrecht.

PAP, E. (1997). Decomposable measures and nonlinear equations. *Fuzzy Sets and Systems* 92:205–222.

PAP, E. (1999). Applications of decomposable measures. In [Höhle & Rodabaugh 1999], chapter 13, pages 675–700.

PAP, E. (2000). Pseudo-convolution and its applications. In [Grabisch *et al.* 2000], page ???

PAP, E., HADŽIĆ, O., MESIAR, R. (1996). A fixed point theorem in probabilistic metric spaces and an application. *J. Math. Anal. Appl.* 202:433–449.

PAP, E. & RALEVIĆ, N. (1998). Pseudo-Laplace transform. *Nonlinear Anal.* 33:533–550.

PAP, E. & ŠTAJNER, I. (1999). Generalized pseudo-convolution in the theory of probabilistic metric spaces, information, fuzzy numbers, optimization, system theory. *Fuzzy Sets and Systems* 102:393–415.

PAP, E., TAKAČI, A., TAKAČI, D. (1997). *Partial Differential Equations through Examples and Exercises*. Kluwer Academic Publishers, Dordrecht.

PAP, E. & TEOFANOV, N. (1998). Pseudo-delta functions and sequences in the optimization theory. *Yugosl. J. Oper. Res.* 8:111–128.

PAPPIS, C. & KARACAPILIDIS, N. (1993). A comparative assessment of measures of similarity of fuzzy values. *Fuzzy Sets and Systems* 56:171–174.

PAVELKA, J. (1979a). On fuzzy logic. I. Many-valued rules of inference. *Z. Math. Logik Grundlag. Math.* 25:45–52.

PAVELKA, J. (1979b). On fuzzy logic. II. Enriched residuated lattices and semantics of propositional calculi. *Z. Math. Logik Grundlag. Math.* 25:119–134.

PAVELKA, J. (1979c). On fuzzy logic. III. Semantical completeness of some many-valued propositional calculi. *Z. Math. Logik Grundlag. Math.* 25:447–464.

PEDRYCZ, W. (1982). Fuzzy control and fuzzy systems. Technical Report 82 14, Delft University of Technology, Department of Mathematics.

PEDRYCZ, W. (1983). Fuzzy relational equations with generalized connectives and their applications. *Fuzzy Sets and Systems* 10:185–201.

PEDRYCZ, W. (1985). On generalized fuzzy relational equations and their applications. *J. Math. Anal. Appl.* 107:520–536.

PERNY, P. (1992). *Modélisation, agrégation et exploitation des préférences floues dans une problématique de rangement.* PhD thesis, Université Paris-Dauphine, Paris.

PERNY, P. & ROY, B. (1992). The use of fuzzy outranking relations in preference modelling. *Fuzzy Sets and Systems* 49:33–53.

PI-CALLEJA, P. (1954). Las ecuaciones funcionales de la teoria de magnitudes. In *Proceedings Segundo Symposium de Matemática, Villavicencio, Mendoza, Coni, Buneos Aires*, pages 199–280.

POINCARÉ, H. (1905). *La Valeur de la Science.* Flammarion, Paris.

PONSARD, C. (1988). Fuzzy mathematical models in economics. *Fuzzy Sets and Systems* 28:273–283.

POPRUGENKO, G. (1928). Sur la propriété de Darboux des fonctions continues d'ensemble. *Fundam. Math.* 12:254–268.

POST, E. L. (1921). Introduction to a general theory of elementary propositions. *Amer. J. Math.* 43:163–185.

RADU, V. (1985). On a family of metrics for distribution functions. In *Proceedings Seventh Conference on Probability Theory, Brašov '82*, pages 487–492. VNU Sci. Press, Utrecht.

RALESCU, A. L. & RALESCU, D. A. (1997). Extensions of fuzzy aggregation. *Fuzzy Sets and Systems* 86:321–330.

RASIOWA, H. & SIKORSKI, R. (1963). *The Mathematics of Metamathematics.* Panstwowe Wydawnictwo Naukowe, Warszawa.

RÉNYI, A. (1962). *Wahrscheinlichkeitsrechnung. Mit einem Anhang über Informationstheorie.* Springer, Berlin.

RÉNYI, A. (1970). *Probability Theory.* North-Holland, Amsterdam.

RESCHER, N. (1969). *Many-Valued Logic.* McGraw-Hill, New York.

RODABAUGH, S. E. (1980). The Hausdorff separation axiom for fuzzy topological spaces. *Topology Appl.* 11:319–334.

RODABAUGH, S. E. (1982a). Connectivity and the L-fuzzy unit interval. *Rocky Mountain J. Math.* 12:113–121.

RODABAUGH, S. E. (1982b). Fuzzy addition in the L-fuzzy real line. *Fuzzy Sets and Systems* 8:39–51.

RODABAUGH, S. E. (1982c). The L-fuzzy real line and its subspaces. In [Yager 1982], pages 402–418.

RODABAUGH, S. E. (1983). Separation axioms and the fuzzy real lines. *Fuzzy Sets and Systems* 11:163–183.

RODABAUGH, S. E. (1985). Complete fuzzy topological hyperfields and fuzzy multiplication in the fuzzy real lines. *Fuzzy Sets and Systems* 15:285–310.

RODABAUGH, S. E. (1999). Fuzzy real lines and dual real lines as poslat topological, uniform, and metric ordered semirings with unity. In [Höhle & Rodabaugh 1999], chapter 10, pages 607–631.

RODABAUGH, S. E. & KLEMENT, E. P., editors (200x). *Topological and Algebraic Structures in Fuzzy Sets.* Kluwer Academic Publishers. (In preparation).

RODABAUGH, S. E., KLEMENT, E. P., HÖHLE, U., editors (1992). *Applications of Category Theory to Fuzzy Subsets.* Kluwer Academic Publishers, Dordrecht.

ROSE, A. & ROSSER, J. B. (1958). Fragments of many-valued statement calculi. *Trans. Amer. Math. Soc.* 87:1–53.

ROSENFELD, A. (1971). Fuzzy groups. *J. Math. Anal. Appl.* 35:512–517.

ROSENFELD, A. (1984a). The diameter of a fuzzy set. *Fuzzy Sets and Systems* 13:241–246.

ROSENFELD, A. (1984b). The fuzzy geometry of image subsets. *Pattern Recognition Letters* 2:311–317.

ROSENFELD, A. (1985). Distances between fuzzy sets. *Pattern Recognition Letters* 3:229–233.

ROSENFELD, A. & HABER, S. (1985). The perimeter of a fuzzy set. *Pattern Recognition* 18:125–130.

ROSSER, J. B. (1953). *Logics for Mathematicians*. McGraw-Hill, New York.

ROSSER, J. B. & TURQUETTE, A. R. (1952). *Many-valued Logics*. North-Holland, Amsterdam.

RUSPINI, E. (1969). A new approach to clustering. *Inform. Control* 15:22–32.

SANCHEZ, E. (1974). *Equations de relations floues*. PhD thesis, Université Aix-Marseille II, Faculté de Médecine, Marseille.

SANCHEZ, E. (1976). Resolution of composite fuzzy relation equations. *Inform. Control* 30:38–48.

SANCHEZ, E. (1977). Solutions in composite fuzzy relation equations: application to medical diagnosis in Brouwerian logic. In Gupta, M. M., Saridis, G. N., Gaines, B. R., editors, *Fuzzy Automata and Decision Processes*, pages 221–234. North-Holland, New York.

SANCHEZ, E. (1984). Solution of fuzzy equations with extended operations. *Fuzzy Sets and Systems* 12:237–248.

SANDER, W. (1998). Associativity. In *Leaflets in Mathematics. Proceedings of the Numbers, Functions, Equations '98 International Conference*, pages 139–140. Janus Pannonius University, Pécs.

SCHMIDT, K. D. (1982). A general Jordan decomposition. *Arch. Math.* 38:556–564.

SCHMIDT, K. D. (1989). *Jordan Decompositions of Generalized Vector Measures*. Longman, Essex.

SCHWEIZER, B. (1975). Multiplications on the space of probability distribution functions. *Aequationes Math.* 12:156–183.

SCHWEIZER, B. (1991). Thirty years of copulas. In [Dall'Aglio *et al.* 1991].

SCHWEIZER, B., SHERWOOD, H., TARDIFF, R. M. (1988). Contractions on probabilistic metric spaces: examples and counterexamples. *Stochastica* 12:5–17.

SCHWEIZER, B. & SKLAR, A. (1958). Espaces métriques aléatoires. *C. R. Acad. Sci. Paris Sér. A* 247:2092–2094.

SCHWEIZER, B. & SKLAR, A. (1960). Statistical metric spaces. *Pacific J. Math.* 10:313–334.

SCHWEIZER, B. & SKLAR, A. (1961). Associative functions and statistical triangle inequalities. *Publ. Math. Debrecen* 8:169–186.

SCHWEIZER, B. & SKLAR, A. (1963). Associative functions and abstract semigroups. *Publ. Math. Debrecen* 10:69–81.

SCHWEIZER, B. & SKLAR, A. (1969). Mesures aléatoires de l'information. *C. R. Acad. Sci. Paris Sér. A* 269:721–723.

SCHWEIZER, B. & SKLAR, A. (1983). *Probabilistic Metric Spaces*. North-Holland, New York.

SCHWEIZER, B. & SMÍTAL, J. (1994). Measures of chaos and a spectral decomposition of dynamical systems on the interval. *Trans. Amer. Math. Soc.* 344:737–754.

SEHGAL, V. H. & BHARUCHA-REID, A. T. (1972). Fixed points of contraction mappings in probabilistic metric spaces. *Math. Systems Theory* 6:97–102.

ŠERSTNEV, A. N. (1962). Random normed spaces: problems of completeness. *Kazan. Gos. Univ. Učen. Zap.* 122:3–20.

ŠERSTNEV, A. N. (1964). On a probabilistic generalization of metric spaces. *Kazan. Gos. Univ. Učen. Zap.* 124:3–11.

SESSA, S. (1984). On fuzzy subgroups and fuzzy ideals under triangular norms. *Fuzzy Sets and Systems* 13:95–100.

SHAFER, G. (1976). *A Mathematical Theory of Evidence*. Princeton University Press, Princeton.

SHANNON, C. & WEAVER, W. (1949). *The Mathematical Theory of Communication*. University of Illinois Press, Urbana.

SHAPLEY, L. S. (1953). A value for n-person games. In Kuhn, H. & Tucker, A., editors, *Contributions to the Theory of Games, II*, pages 307–317. Princeton University Press, Princeton.

SHERWOOD, H. (1969). On E-spaces and their relation to other classes of probabilistic metric spaces. *J. London Math. Soc.* 44:441–448.

SHERWOOD, H. (1971). Complete probabilistic metric spaces. *Z. Wahrsch. Verw. Geb.* 20:117–128.

SHERWOOD, H. (1984). Characterizing dominates on a family of triangular norms. *Aequationes Math.* 27:255–273.

SHERWOOD, H. & TAYLOR, M. D. (1974). Some PM structures on the set of distribution functions. *Rev. Roumaine Math. Pures Appl.* 19:1251–1260.

SHILKRET, N. (1971). Maxitive measure and integration. *Indag. Math.* 33:109–116.

SILVERT, W. (1979). Symmetric summation: A class of operations on fuzzy sets. *IEEE Trans. Syst., Man Cybern.* 9:657–659.

SION, M. (1973). *A Theory of Semigroup Valued Measures*, volume 355 of *Lecture Notes in Mathematics*. Springer, Berlin.

SKALA, H. J., TERMINI, S., TRILLAS, E., editors (1983). *Aspects of Vagueness*. Reidel, Dordrecht.

SKLAR, A. (1959). Fonctions de répartition à n dimensions et leurs marges. *Publ. Inst. Statist. Univ. Paris* 8:229–231.

SKLAR, A. (1973). Random variables, joint distribution functions, and copulas. *Kybernetika* 9:449–460.

SKLAR, A. (1996). Random variables, distribution functions, and copulas — a personal look backward and forward. In Rüschendorf, L., Schweizer, B., Taylor, M. D., editors, *Distributions with Fixed Marginals and Related Topics*, pages 1–14. Institute of Mathematical Statistics, Hayward, CA.

SMETS, P. & MAGREZ, P. (1987). Implications in fuzzy logic. *Internat. J. Approx. Reason.* 1:327–347.

SMUTNÁ, D. (1998a). A note on non-continuous t-norms. *BUSEFAL* 76:19–24.

SMUTNÁ, D. (1998b). On a peculiar t-norm. *BUSEFAL* 75:60–67.

SUGENO, M. (1974). *Theory of fuzzy integrals and its applications*. PhD thesis, Tokyo Institute of Technology.

SUGENO, M., editor (1985a). *Industrial Applications of Fuzzy Control*. North-Holland, Amsterdam.

SUGENO, M. (1985b). An introductory survey of fuzzy control. *Inform. Sci.* 36:59–83.

SUGENO, M. & MUROFUSHI, T. (1987). Pseudo-additive measures and integrals. *J. Math. Anal. Appl.* 122:197–222.

TAKACS, R. (1991). *Personal communication*.

TAKAGI, T. & SUGENO, M. (1985). Fuzzy identification of systems and its application to modelling and control. *IEEE Trans. Syst., Man Cybern.* 15:116–132.

TARDIFF, R. M. (1976). Topologies for probabilistic metric spaces. *Pacific J. Math.* 65:233–251.

TARDIFF, R. M. (1980). On a functional inequality arising in the construction of the product of several metric spaces. *Aequationes Math.* 20:51–58.

TARDIFF, R. M. (1984). On a generalized Minkowski inequality and its relation to dominates for t-norms. *Aequationes Math.* 27:308–316.

TARDIFF, R. M. (1992). Contraction maps on probabilistic metric spaces. *J. Math. Anal. Appl.* 165:517–523.

TAYLOR, A. E. (1958). *Introduction to Functional Analysis*. J. Wiley & Sons, New York.

TAYLOR, M. E. (1996). *Partial Differential Equations*, volume 1–3. Springer, Berlin.

THIELE, H. & SCHMECHEL, N. (1995). On the mutual definability of fuzzy equivalence relations and fuzzy partitions. In *Proceedings of the International Joint Conference of the Fourth IEEE International Conference on Fuzzy Systems and the Second International Fuzzy Engineering Symposium, Yokohama*, volume III, pages 1383–1390. IEEE Press.

THORP, E. (1960). The metrization of statistical metric spaces. *Pacific J. Math.* 10:673–675.

TKADLEC, J. (1999). Triangular norms with continuous diagonals. *Tatra Mt. Math. Publ.* 16:187–195.

TRILLAS, E. (1979). Sobre funciones de negación en la teoría de conjuntas difusos. *Stochastica* 3:47–60.

TRILLAS, E. & ALSINA, C. (1993). Logic: Going farther from Tarski? *Fuzzy Sets and Systems* 53:1–13.

TRILLAS, E. & VALVERDE, L. (1983). An inquiry into indistinguishability operators. In [Skala *et al.* 1983], pages 231–256.

TRILLAS, E. & VALVERDE, L. (1985a). On implication and indistinguishability in the setting of fuzzy logic. In Kacprzyk, J. & Yager, R. R., editors, *Management of Decision Support Systems Using Fuzzy Sets and Possibility Theory*, pages 198–212. Verlag TÜV Rheinland, Köln.

TRILLAS, E. & VALVERDE, L. (1985b). On mode and implication in approximate reasoning. In Gupta, M. M., Kandel, A., Bandler, W., editors, *Approximate Reasoning in Expert Systems*, pages 157–166. North-Holland, Amsterdam.

TÜRKSEN, I. B. (1992). Inter-valued fuzzy sets and 'compensatory AND'. *Fuzzy Sets and Systems* 51:295–307.

TURUNEN, E. (1999a). BL-algebras of basic fuzzy logic. *Mathware Soft Comput.* 6:49–61.

TURUNEN, E. (1999b). *Mathematics Behind Fuzzy Logic*, volume 1 of *Advances in Soft Computing*. Physica-Verlag, Heidelberg.

VALVERDE, L. (1985). On the structure of F-indistinguishability operators. *Fuzzy Sets and Systems* 17:313–328.

VICENÍK, P. (1998a). Additive generators and discontinuity. *BUSEFAL* 76:25–28.

VICENÍK, P. (1998b). A note on generators of t-norms. *BUSEFAL* 75:33–38.

VICENÍK, P. (1999a). Additive generators of triangular norms with an infinite set of discontinuity points. In De Baets, B., Fodor, J., Kóczy, L. T., editors, *Proceedings of EUROFUSE — SIC '99, Budapest*, pages 412–416.

VICENÍK, P. (1999b). Generated t-norms and the Archimedean property. In *Proceedings EUFIT '99, Aachen.* (CD-Rom).

VICENÍK, P. (1999c). A note to a construction of t-norms based on pseudo-inverses of monotone functions. *Fuzzy Sets and Systems* 104:15–18.

VON ALTROCK, C., KRAUSE, B., ZIMMERMANN, H.-J. (1992). Advanced fuzzy logic control of a model car in extreme situations. *Fuzzy Sets and Systems* 48:41–52.

WALD, A. (1943). On a statistical generalization of metric spaces. *Proc. Nat. Acad. Sci. U.S.A.* 29:196–197.

WANG, X., DE BAETS, B., KERRE, E. E. (1995). A comparative study of similarity measures. *Fuzzy Sets and Systems* 73:259–268.

WANG, Z. & KLIR, G. J. (1992). *Fuzzy Measure Theory.* Plenum Press, New York.

WEBER, H. (1996). On modular functions. *Funct. et Approx.* 24:35–52.

WEBER, S. (1983). A general concept of fuzzy connectives, negations and implications based on t-norms and t-conorms. *Fuzzy Sets and Systems* 11:115–134.

WEBER, S. (1984). $\perp$-decomposable measures and integrals for Archimedean t-conorms $\perp$. *J. Math. Anal. Appl.* 101:114–138.

WEBER, S. (1997). Conditioning on MV-algebras and additive measures, Part I. *Fuzzy Sets and Systems* 92:241–250.

WEBER, S. (1999). Conditioning on MV-algebras and additive measures — further results. In [Dubois *et al.* 1999b], pages 175–199.

WIDDER, D. V. (1941). *The Laplace Transform.* Princeton University Press, Princeton.

WIENER, N. (1948). *Cybernetics.* Hermann, Paris.

WYLER, O. (1966). Clans. *Compositio Math.* 17:172–189.

YAGER, R. R. (1980). On a general class of fuzzy connectives. *Fuzzy Sets and Systems* 4:235–242.

YAGER, R. R., editor (1982). *Fuzzy Set and Possibility Theory. Recent Developments.* Pergamon Press, Elmsford.

YAGER, R. R. (1988). On ordered weighted averaging aggregation operators in multicriteria decisionmaking. *IEEE Trans. Syst., Man Cybern.* 18:183–190.

YAGER, R. R. & FILEV, D. P. (1994). *Essentials of Fuzzy Modelling and Control.* J. Wiley & Sons, New York.

YAGER, R. R. & RYBALOV, A. (1996). Uninorm aggregation operators. *Fuzzy Sets and Systems* 80:111–120.

ZADEH, L. A. (1965). Fuzzy sets. *Inform. Control* 8:338–353.

ZADEH, L. A. (1968). Probability measures of fuzzy events. *J. Math. Anal. Appl.* 23:421–427.

ZADEH, L. A. (1971). Similarity relations and fuzzy orderings. *Inform. Sci.* 3:177–200.

ZADEH, L. A. (1973). Outline of a new approach to the analysis of complex systems and decision processes. *IEEE Trans. Syst., Man Cybern.* 3:28–44.

ZADEH, L. A. (1975a). The concept of a linguistic variable and its applications to approximate reasoning. Part I. *Inform. Sci.* 8:199–251.

ZADEH, L. A. (1975b). The concept of a linguistic variable and its applications to approximate reasoning. Part II. *Inform. Sci.* 8:301–357.

ZADEH, L. A. (1975c). The concept of a linguistic variable and its applications to approximate reasoning. Part III. *Inform. Sci.* 9:43–80.

ZADEH, L. A. (1975d). Fuzzy logic and approximate reasoning. *Synthese* 30:407–428.

ZADEH, L. A. (1978). Fuzzy sets as a basis for a theory of possibility. *Fuzzy Sets and Systems* 1:3–28.

ZIMMERMANN, H.-J. (1978). Fuzzy programming and LP with several objective functions. *Fuzzy Sets and Systems* 1:45–55.

ZIMMERMANN, H.-J. (1985). Applications of fuzzy set theory to mathematical programming. *Inform. Sci.* 36:29–58.

ZIMMERMANN, H.-J. (1987). *Fuzzy Set, Decision Making and Expert Systems.* Kluwer Academic Publishers, Boston.

ZIMMERMANN, H.-J. & ZYSNO, P. (1980). Latent connectives in human decision making. *Fuzzy Sets and Systems* 4:37–51.

Index

D

T

TRENDS IN LOGIC

1. G. Schurz: *The Is-Ought Problem.* An Investigation in Philosophical Logic. 1997 ISBN 0-7923-4410-3
2. E. Ejerhed and S. Lindström (eds.): *Logic, Action and Cognition.* Essays in Philosophical Logic. 1997 ISBN 0-7923-4560-6
3. H. Wansing: *Displaying Modal Logic.* 1998 ISBN 0-7923-5205-X
4. P. Hájek: *Metamathematics of Fuzzy Logic.* 1998 ISBN 0-7923-5238-6
5. H.J. Ohlbach and U. Reyle (eds.): *Logic, Language and Reasoning.* Essays in Honour of Dov Gabbay. 1999 ISBN 0-7923-5687-X
6. K. Došen: *Cut Elimination in Categories.* 2000 ISBN 0-7923-5720-5
7. R.L.O. Cignoli, I.M.L. D'Ottaviano and D. Mundici (eds.): *Algebraic Foundations of many-valued Reasoning.* 2000 ISBN 0-7923-6009-5
8. E.P. Klement, R. Mesiar and E. Pap: *Triangular Norms.* 2000 ISBN 0-7923-6416-3

KLUWER ACADEMIC PUBLISHERS – DORDRECHT / BOSTON / LONDON

Printed in the United States
78277LV00001B/12

9 780792 364160